INTRODUCTION TO
SUPERSYMMETRY

(Second Edition)

World Scientific Lecture Notes in Physics

World Scientific Lecture Notes in Physics – Vol. 80

INTRODUCTION TO
SUPERSYMMETRY

(Second Edition)

Harald J W Müller-Kirsten
University of Kaiserslautern, Germany

Armin Wiedemann
Baden-Württemberg Cooperative State University Mannheim, Germany

World Scientific

NEW JERSEY · LONDON · SINGAPORE · BEIJING · SHANGHAI · HONG KONG · TAIPEI · CHENNAI

Published by

World Scientific Publishing Co. Pte. Ltd.

5 Toh Tuck Link, Singapore 596224

USA office: 27 Warren Street, Suite 401-402, Hackensack, NJ 07601

UK office: 57 Shelton Street, Covent Garden, London WC2H 9HE

British Library Cataloguing-in-Publication Data
A catalogue record for this book is available from the British Library.

World Scientific Lecture Notes in Physics — Vol. 80
INTRODUCTION TO SUPERSYMMETRY (2nd Edition)

ISBN-13 978-981-4293-41-9
ISBN-10 981-4293-41-5
ISBN-13 978-981-4293-42-6 (pbk)
ISBN-10 981-4293-42-3 (pbk)

Preface to the Second Edition

The positive response to our text published two decades ago and its quick sell-out encouraged us after careful consideration to produce this second and LATEX-typed version. We observed that our text was used or recommended in courses at several universities. Naturally, the most positive response (particularly from the US) came from those for whom the text was written, namely students and others without prior knowledge of supersymmetry. We thank our readers for this feedback and other comments. This success assured us that indeed there was a definite demand for a text like our's with explicit calculational details. In this second edition we have corrected writing errors we became aware of. We have also made minor changes here and there, and we cite more literature.

In the course of the last 20 years since the publication of the First Edition, a vast amount of literature on supersymmetry and related topics has been published. Various non-technical books on supersymmetry, *e.g.* D. Hooper [58], G. Kane [62], L. Randall [95], and the books by B. Greene [52, 53] are now available. There are several introductory texts on supersymmetry, *e.g.* I.J.A. Aitchinson [1, 2], A. Bilal [15], M. Drees [32], N. Polonsky [89], S.P. Misra [73], U. Lindström [69], J.D. Lykken [70], S.P. Martin [71]. Books on superstrings generally include brief introductions to supersymmetry, see *e.g.* the monographs by M.B. Green, J.H. Schwarz and E. Witten [51], M. Kaku [61] and J. Polchinski [88]. Phenomenological implications and astrophysical and cosmological aspects of supersymmetry — subjects which we do not cover in this text — can be found in M. Dine [28], K.A. Olive [80] and S.P. Martin [71]. The Minimal Supersymmetric Standard Model (MSSM) — *i.e.* the minimal extension to the Standard Model of particle physics that realizes supersymmetry with a strong emphasis on phenomenological aspects — is discussed in I.J.A. Aitchinson [1] and in the review article by S.P. Martin [71]. This subject is also treated in the monograph by S. Weinberg [118], Chapter

28.4. Non-perturbative methods of supersymmetry are discussed in A. Wipf [129], J. Terning [111] and M. Bianchi *et al.* [13]. More advanced topics such as the Seiberg–Witten duality — an important development of non-perturbative methods in supersymmetric Yang–Mills Theory — is presented in the pedagogical reviews of L. Alvarez–Gaumé and S.L. Hassan [5], A. Bilal [14], and W. Lerche [68]. The book by D.I. Olive and P.C. West [79] is also devoted to this topic. Recently the books of S. Duplij, W. Siegel, and J. Bagger [33], J. Terning [112], M. Dine [29], and P. Binetruy and K. Hentschel [16] appeared, which cover the developments of the last two decades in supersymmetry.

A few years ago, the discovery of supersymmetry celebrated its 30th anniversary. The exceptional history of supersymmetry is discussed in the book by G. Kane and M. Shifman [63], in particular we recommend the comprehensive contribution of R. Di Stefano [30]. See also the recollections of J. Wess [122] and the overview of D. Olive and P. West in [79].

One of us (A.W.) thanks Mees de Roo, University of Groningen, for numerous suggestions and his encouragement.

Finally we thank the editors of World Scientific for their support and cooperation.

Summer 2009

H.J.W. Müller-Kirsten
University of Kaiserslautern

A. Wiedemann
Baden-Württemberg Cooperative State University Mannheim

Contents

Introduction

Symmetries are of fundamental importance in the description of physical phenomena. In the realm of particle physics symmetries are believed to permit ultimately a classification of all observed particles. A fundamental symmetry of particle physics, which has been firmly established both theoretically and experimentally is that of the Poincaré group, *i.e.* of rotations and translations in four-dimensional Minkowski space. Besides this fundamental symmetry there are other so-called internal symmetries (such as the symmetry of the $SU(3)$ flavor group) which have also been firmly established over the last few decades, although their manifestation in Nature is not exact. As is well known, the consistent search for more fundamental symmetries led to the development of non-Abelian gauge theories and the spectacular experimental confirmation of several predictions of the latter in recent years.

In the course of time several attempts have been made to unify the spacetime symmetry of the Poincaré group with the symmetry of some internal group ([22], [81], [82], [83]). Such attempts have, however, been shown to be futile if the theory, which necessarily has to be a quantum field theory, is expected to satisfy certain basic requirements. In fact, the so-called "no-go"-theorem of S. Coleman and J. Mandula [23] shows that if one makes the plausible assumptions of locality, causality, positivity of energy and finiteness of the number of particles (and one more technical assumption) the invariance group of the theory can at best be the direct product of the Poincaré group and a compact internal group, and this therefore does not offer a genuine unification of one group with the other.

The generators of the Poincaré group satisfy well known commutation relations and Noether's theorem relates these to conserved currents. In their turn these conserved currents are functions of relativistic fields. The commutation relations of the field operators which quantize these fields are therefore directly related to those of the generators. It was realized by J. Wess and B. Zumino [124], [125] that if one allows also anticommutation relations of generators of supersymmetry transformations which transform bosons into fermions and *vice versa*, then the unification of the spacetime symmetries

1

of the Poincaré group with this internal symmetry can be achieved. The formal proof of this discovery, *i.e.* the proof that anticommuting generators which respect the other assumptions of the theorem of S. Coleman and J. Mandula [23] do exist, was established by R. Haag, J.T. Łopuszański and M.F. Sohnius [54].

Supersymmetry thus arises as a symmetry which combines bosons and fermions in the same representation or multiplet of the enlarged group which encompasses both the transformations of the Poincaré group and the appropriate supersymmetry transformations. Thus every bosonic particle must have a fermionic partner and *vice versa*. In view of the fact that such a spectrum of particles is not compatible with observation, supersymmetry must be badly broken at the level of presently available energies. Clearly only experimental observation can decide whether supersymmetry is indeed inherent in Nature. It can be argued that one of the most immediate ways to observe evidence of supersymmetry is to see if there is a missing energy and momentum in the final e^+e^- spectrum of the reaction

$$e^+ + e^- \longrightarrow \gamma \longrightarrow \tilde{e}^+ + \tilde{e}^- \longrightarrow e^+ + e^- + \tilde{\gamma} + \tilde{\gamma}$$

where $\tilde{e}^+, \tilde{e}^-$ and $\tilde{\gamma}$ are the supersymmetric partners of e^+, e^- and γ respectively. If there is such a missing energy and momentum it could be that carried away by the neutral photino $\tilde{\gamma}$. Charged supersymmetry particles at energies presently available would have been detected long ago. Since supersymmetry must be broken, the photinos $\tilde{\gamma}$ would not be massless.

However, supersymmetry does not only open the possibility of a much more complex spectrum of particles than heretofore envisaged; supersymmetry also has some intriguing theoretical consequences which could make it a desirable theory. It is well kown that a realistic quantum field theory in the traditional sense is plagued by the problem of ultraviolet divergences and the consequent necessity of renormalization. Supersymmetry, however, provides a mechanism for the cancellation of such divergences in view of the same number of bosonic and fermionic degrees of freedom in each particle multiplet. Clearly, such a built-in cancellation of divergent terms is a highly desirable feature of a quantum field theory.

In Chapter 1 we begin with a recapitulation of basic aspects of the Lorentz group, including a discussion of Casimir operators and the classification of representations in terms of their eigenvalues. We then consider the group $SL(2, \mathbb{C})$ and its basic representations, *i.e.* the self-representation and the complex conjugate self-representation. The elements of the appropriate representation spaces are the undotted and dotted Weyl spinors. In view of the importance of Weyl spinors throughout the entire text, we consider these here in more detail than is generally done in the literature. We then introduce the concept of Grassmann numbers and perform some basic manipulations involving Weyl spinors, thereby deriving a number of useful formulas. In the subsequent section the connection between the special linear group $SL(2,\mathbb{C})$ and the proper orthochronous Lorentz group is established. It is then natural to discuss four-component Dirac spinors and the Weyl representation. The connection with two-component Weyl spinors is obtained by introducing four-component Majorana spinors. Then again various formulas are derived which are useful in later calculations.

Chapter 2 begins with a discussion of the "no-go" theorems of Coleman and Mandula [23] and Haag, Łopuszański and Sohnius [54]. The latter leads to a consideration of graded Lie algebras which we approach in successive steps by defining first the characteristics of a Lie algebra, then those of a graded algebra and finally those of a graded Lie algebra, *i.e.* the properties of *grading, supersymmetrization* and *generalized Jacobi identities*. As an example we construct the graded Lie algebra of the algebra $su(2,\mathbb{C})$. The final section of Chapter 2 deals with graded matrices and their properties.

The following Chapter 3 deals with the grading, *i.e.* supersymmetrization of the Poincaré algebra. We demonstrate explicitly that for the grading chosen all possible Jacobi identities are satisfied. This turns out to be a crucial point of consistency of a grading. Having established the algebra of the Super–Poincaré group with the fermionic generators in the Dirac four-component form, we then decompose it into the appropriate relations of the two-component Weyl formalism.

In Chapter 4 we use the method of Casimir operators to classify the irreducible representations of the Super–Poincaré algebra, and it is shown that supersymmetry implies an equal number of bosonic and fermionic degrees of freedom.

Chapter 5 deals with the most immediate field theoretical realization of the Super–Poincaré algebra, the *Wess–Zumino model*, which is a field theory involving a scalar field, a pseudoscalar field and one spinor field, all with the same mass. We demonstrate by explicit calculation that the spinor charges of the theory, considered as linear operators in Fock space, satisfy the commutation and anticommutation relations of the Super–Poincaré algebra.

In Chapter 6 we introduce the concepts of superspace and superfields, and define differentiation with respect to Grassmann numbers. Then three different but related operators are constructed which describe three different but equivalent actions of the supersymmetry group on functions in superspace. These operators define three different types of superfields. By considering infinitesimal supersymmetry transformations we obtain the corresponding three differential operator representations of the fermionic generators of the

Super–Poincaré group. Then covariant derivatives are introduced as a prerequisite for the construction of manifestly supersymmetric action integrals. These covariant derivatives also permit the definition of projection operators. The search for irreducible representations of the Super–Poincaré algebra then becomes a search for solutions of constraint equations expressed in terms of these projection operators. The final section of Chapter 6 is devoted to the derivation of the explicit supersymmetry transformations of the component fields of the supermultiplet. In this context it is seen that the highest order component field always transforms into a total Minkowski derivative and thus is a candidate for a supersymmetric Lagrangian density.

In Chapter 7 we begin with an investigation of the constraint equations which define left-handed and right-handed chiral superfields (also known as scalar superfields). Then vector superfields are defined by an appropriate constraint equation, and the supersymmetric generalization of the Abelian gauge transformation is discussed. Finally, left-handed and right-handed spinor superfields are discussed which represent the components of the supersymmetric field strength for an arbitrary vector superfield.

Chapter 8 deals with the construction of supersymmetric action integrals. We begin with the definition of integration over Grassmann numbers. Then Lagrangians are constructed from scalar superfields and from vector superfields (*i.e.* the supersymmetric field strength). The case of the former is shown to contain the Wess–Zumino model as a special case, whereas the case of the latter yields the supersymmetric generalization of the pure Maxwell theory (*i.e.* with no interaction with matter fields) which contains in addition to the massless vector field also the massless spinor field of the photino.

Chapter 9 deals with the spontaneous breaking of supersymmetry. For the convenience of discussions the concept of superpotential is introduced. In view of the necessity of evaluating action integrals over superspace an equivalent and convenient Grassmann projection technique is developed. Some general aspects of spontaneous symmetry breaking are then discussed and, in particular, the Goldstone theorem is established for the general case of the breaking of supersymmetry and some other symmetry. Finally, the O'Raifeartaigh model, which is a specific theory involving three scalar superfields, is considered and the spectrum resulting from the spontaneous breaking of supersymmetry is investigated. In this case supersymmetry breaking results from the nonvanishing vacuum expectation value of some auxiliary field of a superfield.

Finally, in Chapter 10, we consider supersymmetric gauge theories. Introducing first global and local $U(1)$ gauge transformations of scalar superfields and the corresponding supersymmetric version of minimal coupling, we consider super quantum electrodynamics. We then investigate the Fayet–Iliopoulos mechanism of spontaneous breaking of supersymmetry in which the latter results from the nonvanishing vacuum expectation value of the highest order component of a vector superfield. The last section contains a brief introduction to non-Abelian gauge transformations for superfields with the appropriate tensorial transformation properties.

Chapter 1

Lorentz and Poincaré Group, $SL(2, \mathbb{C})$, Dirac and Majorana Spinors

1.1 The Lorentz Group

A point in the *spacetime manifold*[1] is denoted by

$$(x^\mu) = (x^0, x^1, x^2, x^3),$$

where $x^0 = t$, and x^1, x^2, x^3 are the space components of the four-vector x^μ. This space is called *Minkowski space* and denoted by $\mathbb{M}_4$. The laws of physics, formulated in $\mathbb{M}_4$, are invariant under the *Lorentz group*. Transformations of this group are linear transformations acting on four-vectors[2]

$$x'^\mu = \Lambda^\mu{}_\nu x^\nu, \tag{1.1}$$

leaving the quadratic form

$$x^2 = x^\mu x_\mu = \eta_{\mu\nu} x^\mu x^\nu = (x^0)^2 - (\mathbf{x})^2 \tag{1.2}$$

invariant, *i.e.*

$$x'^2 = x'^\mu x'_\mu = \eta_{\mu\nu} x'^\mu x'^\nu = \eta_{\mu\nu} \Lambda^\mu{}_\rho x^\rho \Lambda^\nu{}_\tau x^\tau \overset{!}{=} \eta_{\rho\tau} x^\rho x^\tau.$$

---

[1]Secs. 1.1 and 1.2 serve mainly the purpose of completeness, to define notation and to recollect some formulas which will be needed later in the text. The reader familiar with Secs. 1.1 and 1.2 could start immediately with Sec. 1.3. Readers unfamiliar with the subjects treated here can consult [75], Chap. 12, or [98] for more details.

[2]In this book we adopt the Einstein summation convention.

Hence the Lorentz transformations $\Lambda^{\mu}{}_{\nu}$ have to satisfy the condition:

$$\eta_{\mu\nu}\Lambda^{\mu}{}_{\rho}\Lambda^{\nu}{}_{\tau} = \eta_{\rho\tau}. \tag{1.3}$$

Here

$$(\eta_{\mu\nu}) = \begin{pmatrix} 1 & 0 & 0 & 0 \\ 0 & -1 & 0 & 0 \\ 0 & 0 & -1 & 0 \\ 0 & 0 & 0 & -1 \end{pmatrix} \tag{1.4}$$

is the metric tensor; it lowers indices and its inverse $\eta^{\mu\nu}$ raises indices.

Proposition 1.1:

The constraints

$$\det \Lambda = \pm 1, \qquad |\Lambda^{0}{}_{0}| \geq 1 \tag{1.5}$$

define four disconnected pieces in the parameter space.

Proof: The determinant of a product of matrices is the product of the determinants. Hence, taking the determinant of Eq. (1.3) yields

$$\left(\det \Lambda\right)^2 = 1 \quad \text{or} \quad \det \Lambda = \pm 1.$$

Taking the 00-component of Eq. (1.3) we obtain:

$$\eta_{00} = \eta_{\mu\nu}\Lambda^{\mu}{}_{0}\Lambda^{\nu}{}_{0} = \eta_{00}\Lambda^{0}{}_{0}\Lambda^{0}{}_{0} + \eta_{ii}\Lambda^{i}{}_{0}\Lambda^{i}{}_{0} = \left(\Lambda^{0}{}_{0}\right)^2 - \left(\Lambda^{k}{}_{0}\right)^2 \overset{!}{=} 1,$$

or

$$\left(\Lambda^{0}{}_{0}\right)^2 = 1 + \left(\Lambda^{k}{}_{0}\right)^2.$$

Hence $\left(\Lambda^{0}{}_{0}\right)^2 \geq 1$.

The second of constraints (1.5) distinguishes so-called *orthochronous* Lorentz transformations with $\Lambda^{0}{}_{0} \geq 1$ from *non-orthochronous* Lorentz transformations with $\Lambda^{0}{}_{0} \leq -1$.

Proposition 1.2:

The matrices $(\Lambda^{\mu}{}_{\nu})$ form a non-compact Lie group, the *Lorentz group*[3]

$$L = O(1,3;\mathbb{R}) = \{\Lambda \in GL(4,\mathbb{R}) \mid \Lambda^{\top}\eta\Lambda = \eta\}$$

---

[3]Recall the notation for the groups means the following: The O stands for *orthogonal* which is related to the fact that the transformations that it represents preserve the orthogonality of coordinate axes. The numbers $1,3$ stand for the fact, that the transformations obey Eq. (1.3) with the metric $+1$ for the time component and a -1 for each space component. Later we shall encounter *special orthogonal groups* denoted by $SO(1,3;\mathbb{R})$. Special means determinant equal to $+1$, and the geometric meaning is that orientation reversing transformations such as reflections are excluded.

with Lie algebra

$$o(1,3;\mathbb{R}) = \{\mathfrak{A} \in M_{4\times4}(\mathbb{R}) \mid \mathfrak{A}^\top = -\eta\mathfrak{A}\eta\},$$

where $GL(4;\mathbb{R})$ denotes the set of all invertible 4×4–matrices with real entries,[4] and $M_{4\times4}(\mathbb{R})$ is the set of all 4×4–matrices with real elements.

Proof: From Lie algebra theory we know that each $\Lambda \in O(1,3;\mathbb{R})$ can be written in the form[5]

$$\Lambda(t) = \exp(t\mathfrak{A}),$$

where t is a real parameter and $\mathfrak{A} \in o(1,3;\mathbb{R})$ is an element of the Lie algebra. Matrices of $O(1,3;\mathbb{R})$ are subject to the condition

$$\Lambda^\top(t)\eta\Lambda(t) = \eta.$$

Inserting the above expression we obtain:

$$\big[\exp(t\mathfrak{A})\big]^\top \eta \big[\exp(t\mathfrak{A})\big] = \eta,$$

and considering

$$\frac{d}{dt}\Big\{ \big[\exp(t\mathfrak{A})\big]^\top \eta \big[\exp(t\mathfrak{A})\big] \Big\}\Big|_{t=0} = 0,$$

we obtain the conditions for the Lie algebra elements, since the Lie algebra of any Lie group is isomorphic to the tangent space at the identity of the group. It follows that

$$0 = \frac{d}{dt}\big[\exp(t\mathfrak{A})\big]^\top \eta \exp(t\mathfrak{A})\Big|_{t=0} + \big[\exp(t\mathfrak{A})\big]^\top \eta \frac{d}{dt}\exp(t\mathfrak{A})\Big|_{t=0},$$

which leads to

$$\mathfrak{A}^\top\eta + \eta\mathfrak{A} = 0,$$

or

$$\mathfrak{A}^\top = -\eta\mathfrak{A}\eta, \quad \forall\,\mathfrak{A} \in o(1,3,\mathbb{R}).$$

In summary we have the following classification: Let Λ be any invertible 4×4-matrix with real elements, *i.e.* $\Lambda \in GL(4,\mathbb{R})$, then:

---

[4]The set $GL(4;\mathbb{R})$ forms a group and is also called the *general linear group*.
[5]See for example W. Miller [72], Chap. 5 or V.S. Varadarajan [115], Chap. 2.

(i) The full Lorentz group is

$$L := O(1,3;\mathbb{R}) = \{\Lambda \in GL(4,\mathbb{R}) \mid \Lambda^\top \eta \Lambda = \eta\}.$$

(ii) Proper Lorentz transformations are

$$L_+ := SO(1,3;\mathbb{R}) = \{\Lambda \in O(1,3;\mathbb{R}) \mid \det \Lambda = +1\},$$

L_+ being a subgroup of L.

(iii) Improper Lorentz transformations are

$$L_- := \{\Lambda \in O(1,3;\mathbb{R}) \mid \det \Lambda = -1\}.$$

L_- is not a subgroup of L, since the identity element is not an element of L_-. Note, however, that discrete transformations such as time- or space-reflections are elements of L_-.

(iv) Orthochronous Lorentz transformations are

$$L^\uparrow := \{\Lambda \in O(1,3;\mathbb{R}) \mid \Lambda^0{}_0 \geq +1\}.$$

$L^\uparrow$ is a subgroup of L.

(v) Non-orthochronous Lorentz transformations are

$$L^\downarrow := \{\Lambda \in O(1,3;\mathbb{R}) \mid \Lambda^0{}_0 \leq -1\}.$$

(vi) The restricted Lorentz group is

$$L^\uparrow_+ = L^\uparrow \cap L_+$$
$$= \{\Lambda \in O(1,3;\mathbb{R}) \mid \det \Lambda = +1, \Lambda^0{}_0 \geq +1\}.$$

This subgroup of L is also called the proper orthochronous Lorentz group; it does not contain time- or space-reflections.

Remark: Lorentz transformations which are orthochronous map the forward light-cone onto itself, and $\Lambda \in L^\downarrow$ maps the backward light-cone onto itself.

Generators of the Lorentz group

In the neighborhood of the identity $\mathbb{1}_{SO(1,3;\mathbb{R})}$, a Lorentz transformation $\Lambda \in L^\uparrow_+$ can be written in the form

$$\Lambda = \mathbb{1}_{4\times4} + \omega. \tag{1.6}$$

where $\mathbb{1}_{4\times 4}$ denotes the 4×4-unit matrix and ω is a 4×4-matrix with infinitesimal parameters. The aim is to obtain the constraints which these parameters have to satisfy in order to obey Eq. (1.3). Inserting expression (1.6) into Eq. (1.3) we obtain

$$
\begin{aligned}
\eta_{\mu\nu}\Lambda^\mu{}_\rho\Lambda^\nu{}_\sigma &= \eta_{\mu\nu}\left\{\delta^\mu{}_\rho + \omega^\mu{}_\rho\right\}\left\{\delta^\nu{}_\sigma + \omega^\nu{}_\sigma\right\} \\
&= \eta_{\mu\nu}\delta^\mu{}_\rho\delta^\nu{}_\sigma + \eta_{\mu\nu}\delta^\mu{}_\rho\omega^\nu{}_\sigma + \eta_{\mu\nu}\omega^\mu{}_\rho\delta^\nu{}_\sigma \\
&= \eta_{\rho\sigma} + \omega_{\rho\sigma} + \omega_{\sigma\rho} \overset{!}{=} \eta_{\rho\sigma}.
\end{aligned}
$$

Hence:

$$\omega_{\rho\sigma} = -\omega_{\sigma\rho}, \tag{1.7}$$

i.e. the matrix ω with its infinitesimal parameters is antisymmetric in the indices ρ and σ. Now

$$x'^\mu \overset{(1.1)}{=} \Lambda^\mu{}_\nu x^\nu \overset{(1.6)}{=} (1+\omega)^\mu{}_\nu x^\nu = x^\mu + \omega^\mu{}_\nu x^\nu,$$

and the variation of the four-vector x^μ due to infinitesimal transformations is:

$$\delta x^\mu := x'^\mu - x^\mu = \omega^\mu{}_\nu x^\nu$$

with antisymmetric infinitesimal parameters as demonstrated in Eq. (1.7). On the other hand we may consider the vector representation of the restricted Lorentz group $L_+^\uparrow$, and we write $\Lambda \in L_+^\uparrow$ in the form

$$\Lambda^\mu{}_\nu = \left[\exp(-\frac{i}{2}\omega^{\rho\sigma}M_{\rho\sigma})\right]^\mu{}_\nu \tag{1.8}$$

where the 4×4-matrices $(M_{\rho\sigma})$ constitute a basis of the Lie algebra $o(1,3;\mathbb{R})$, to be verified later. $(M_{\rho\sigma})$ are antisymmetric in ρ and σ and the factor i is chosen in such a way that the $(M_{\rho\sigma})$ are Hermitian. For infinitesimal transformations, *i.e.* $\omega^{\rho\sigma}$ infinitesimal, we consider

$$
\begin{aligned}
\delta x^\mu = x'^\mu - x^\mu = \Lambda^\mu{}_\nu x^\nu - x^\mu &\overset{(1.8)}{=} \left[\exp(-\frac{i}{2}\omega^{\rho\sigma}M_{\rho\sigma})\right]^\mu{}_\nu x^\nu - x^\mu \\
&= \left\{\mathbb{1}_{4\times 4} - \frac{i}{2}\omega^{\rho\sigma}M_{\rho\sigma}\right\}^\mu{}_\nu x^\nu - x^\mu = -\frac{i}{2}\omega^{\rho\sigma}(M_{\rho\sigma})^\mu{}_\nu x^\nu, \tag{1.9}
\end{aligned}
$$

and we conclude

$$\omega^\mu{}_\nu = -\frac{i}{2}\omega^{\rho\sigma}(M_{\rho\sigma})^\mu{}_\nu. \tag{1.10}$$

Therefore the generators $M_{\rho\sigma}$ of the Lorentz group have the matrix form:

$$(M_{\rho\sigma})^\mu{}_\nu = i(\eta_{\sigma\nu}\delta_\rho{}^\mu - \eta_{\rho\nu}\delta_\sigma{}^\mu). \tag{1.11}$$

Checking:

$$-\frac{i}{2}\omega^{\rho\sigma}(M_{\rho\sigma})^{\mu}{}_{\nu} = \frac{1}{2}\omega^{\rho\sigma}(\eta_{\sigma\nu}\delta_{\rho}{}^{\mu} - \eta_{\rho\nu}\delta_{\sigma}{}^{\mu}) = \frac{1}{2}(\omega^{\mu}{}_{\nu} - \omega_{\nu}{}^{\mu}) \overset{(1.7)}{=} \omega^{\mu}{}_{\nu}.$$

We now derive the matrix form of the generators of the Lorentz group, *i.e.* Eq. (1.11), explicitly.

According to the definition of the Lie algebra $o(1,3;\mathbb{R})$, any $\mathfrak{A} \in o(1,3;\mathbb{R})$ satisfies the relation

$$\mathfrak{A}^{\top} = -\eta\mathfrak{A}\eta.$$

Explicitly this condition implies for the elements of such a matrix:

$$\begin{pmatrix} a_{00} & a_{01} & a_{02} & a_{03} \\ a_{10} & a_{11} & a_{12} & a_{13} \\ a_{20} & a_{21} & a_{22} & a_{23} \\ a_{30} & a_{31} & a_{32} & a_{33} \end{pmatrix}^{\top}$$

$$= -\begin{pmatrix} 1 & & & \\ & -1 & & 0 \\ 0 & & -1 & \\ & & & -1 \end{pmatrix}\begin{pmatrix} a_{00} & a_{01} & a_{02} & a_{03} \\ a_{10} & a_{11} & a_{12} & a_{13} \\ a_{20} & a_{21} & a_{22} & a_{23} \\ a_{30} & a_{31} & a_{32} & a_{33} \end{pmatrix}\begin{pmatrix} 1 & & & \\ & -1 & & 0 \\ 0 & & -1 & \\ & & & -1 \end{pmatrix}$$

which leads to the set of equations:

$$a_{00} = a_{11} = a_{22} = a_{33} = 0,$$
$$a_{10} = a_{01}, \quad a_{20} = a_{02}, \quad a_{30} = a_{03},$$
$$a_{21} = -a_{12}, \quad a_{31} = -a_{13}, \quad a_{32} = -a_{23}.$$

Therefore a typical matrix $\mathfrak{A} \in o(1,3;\mathbb{R})$ has the form:

$$\mathfrak{A} = \begin{pmatrix} 0 & a_{01} & a_{02} & a_{03} \\ a_{01} & 0 & -a_{12} & a_{13} \\ a_{02} & a_{12} & 0 & -a_{23} \\ a_{03} & -a_{13} & a_{23} & 0 \end{pmatrix}.$$

We choose a basis of the Lie algebra $o(1,3;\mathbb{R})$ of the following form:

$$M_1 := \begin{pmatrix} 0 & 0 & 0 & 0 \\ 0 & 0 & 0 & 0 \\ 0 & 0 & 0 & -1 \\ 0 & 0 & 1 & 0 \end{pmatrix}, \quad M_2 := \begin{pmatrix} 0 & 0 & 0 & 0 \\ 0 & 0 & 0 & 1 \\ 0 & 0 & 0 & 0 \\ 0 & -1 & 0 & 0 \end{pmatrix},$$

$$M_3 := \begin{pmatrix} 0 & 0 & 0 & 0 \\ 0 & 0 & -1 & 0 \\ 0 & 1 & 0 & 0 \\ 0 & 0 & 0 & 0 \end{pmatrix}, \quad N_1 := \begin{pmatrix} 0 & 1 & 0 & 0 \\ 1 & 0 & 0 & 0 \\ 0 & 0 & 0 & 0 \\ 0 & 0 & 0 & 0 \end{pmatrix},$$

$$N_2 := \begin{pmatrix} 0 & 0 & 1 & 0 \\ 0 & 0 & 0 & 0 \\ 1 & 0 & 0 & 0 \\ 0 & 0 & 0 & 0 \end{pmatrix}, \quad N_3 := \begin{pmatrix} 0 & 0 & 0 & 1 \\ 0 & 0 & 0 & 0 \\ 0 & 0 & 0 & 0 \\ 1 & 0 & 0 & 0 \end{pmatrix}.$$

The three matrices $M_i, i = 1, 2, 3$ generate the rotation group $SO(3; \mathbb{R})$ as a subgroup of $SO(1, 3; \mathbb{R})$, the three matrices $N_i, i = 1, 2, 3$ generate Lorentz boosts. As can be shown by explicit calculation, these generators obey the following commutation relations:

$$[M_i, M_j] = \epsilon_{ijk} M_k,$$
$$[N_i, M_j] = -\epsilon_{ijk} N_k,$$
$$[N_i, N_j] = -\epsilon_{ijk} M_k, \quad i, j, k = 1, 2, 3,$$

where ϵ_{ijk} is totally antisymmetric in i, j, k. The matrices $M_i, i = 1, 2, 3$, are antisymmetric, *i.e.*

$$M_i^\top = -M_i,$$

whereas the Lorentz boost generators N_i are symmetric:

$$N_i^\top = N_i.$$

We now construct Hermitian matrices

$$J_l := i M_l, \quad l = 1, 2, 3,$$

with

$$J_l^\dagger = (i M_l)^\dagger = -i M_l^\top = i M_l = J_l,$$

and anti-Hermitian matrices

$$K_l := i N_l, \quad l = 1, 2, 3,$$

such that

$$K_l^\dagger = (i N_l)^\dagger = -i N_l^\top = -i N_l = -K_l.$$

These matrices obey the following commutation relations:

$$[J_i, J_k] = i \epsilon_{ikl} J_l,$$
$$[J_i, K_l] = i \epsilon_{ilm} K_m,$$
$$[K_i, K_j] = -i \epsilon_{ijk} J_k.$$

The first commutation relation shows that $J_i, i = 1, 2, 3$, generate the rotation subgroup of $L_+^\uparrow$. Usually physicists take the matrices K_i and J_i as generators of Lorentz boosts and rotations respectively.

Starting with matrices K_i and J_i one can construct a covariant formalism, defining an antisymmetric 4×4-matrix $M_{\mu\nu}, \mu, \nu = 0, 1, 2, 3$:

$$\epsilon_{kij} M_{ij} := J_k, \quad (i, j, k = 1, 2, 3 \text{ in cyclic order}),$$
$$M_{0i} := -K_i, \quad (i = 1, 2, 3).$$

Explicitly:

$$(M_{\mu\nu}) := \begin{pmatrix} 0 & -K_1 & -K_2 & -K_3 \\ K_1 & 0 & J_3 & -J_2 \\ K_2 & -J_3 & 0 & J_1 \\ K_3 & J_2 & -J_1 & 0 \end{pmatrix}. \tag{1.12}$$

Proposition 1.3:

In covariant notation the matrix $M_{\mu\nu}$ is given by Eq. (1.11), *i.e.*

$$(M_{\rho\sigma})^\mu{}_\nu = i(\eta_{\sigma\nu}\delta_\rho{}^\mu - \eta_{\rho\nu}\delta_\sigma{}^\mu).$$

Proof: In order to prove this we consider separately:

(i) $\rho = 0, \sigma = 1$. Using Eq. (1.11) we have

$$(M_{01})^\mu{}_\nu = i(\eta_{1\nu}\delta_0{}^\mu - \eta_{0\nu}\delta_1{}^\mu),$$

and the nonvanishing elements of the matrix M_{01} are

$$(M_{01})^0{}_1 = -i, \quad (M_{01})^1{}_0 = -i.$$

On the other hand using Eq. (1.12),

$$(M_{01})^\mu{}_\nu = -(K_1)^\mu{}_\nu = -i(N_1)^\mu{}_\nu,$$

and the explicit form of N_1 gives the only nonvanishing elements

$$(N_1)^0{}_1 = 1 = (N_1)^1{}_0.$$

Hence

$$(M_{01})^0{}_1 = -i = (M_{01})^1{}_0.$$

(ii) $\rho = 2, \sigma = 3$. Using Eq. (1.11) we have

$$(M_{23})^{\mu}{}_{\nu} = i(\eta_{3\nu}\delta_2{}^{\mu} - \eta_{2\nu}\delta_3{}^{\mu}),$$

and the nonvanishing elements of M_{23} are

$$(M_{23})^2{}_3 = -i, \quad (M_{23})^3{}_2 = +i.$$

On the other hand using Eq. (1.12),

$$(M_{23})^{\mu}{}_{\nu} = (J_1)^{\mu}{}_{\nu} = i(M_1)^{\mu}{}_{\nu}$$

and from the explicit form of M_1 we obtain

$$(M_{23})^2{}_3 = i(M_1)^2{}_3 = -i,$$
$$(M_{23})^3{}_2 = i(M_1)^3{}_2 = +i.$$

The other matrices can be checked analogously.

We now define

$$(M_{\rho\sigma})_{\mu\nu} = \eta_{\mu\tau}(M_{\rho\sigma})^{\tau}{}_{\nu} = i\eta_{\mu\tau}(\eta_{\sigma\nu}\delta_\rho{}^{\tau} - \eta_{\rho\nu}\delta_\sigma{}^{\tau}) = i(\eta_{\rho\mu}\eta_{\sigma\nu} - \eta_{\rho\nu}\eta_{\sigma\mu}),$$
$$(1.13)$$

where we used Eq. (1.11).

Proposition 1.4:

The matrices

$$[(M_{\rho\sigma})_{\mu\nu}]$$

defined by Eq. (1.13), are Hermitian matrices, *i.e.*

$$[(M_{\rho\sigma})_{\mu\nu}]^{\dagger} = [(M_{\rho\sigma})_{\mu\nu}].$$

Proof: Consider

$$[(M_{\rho\sigma})_{\mu\nu}]^{\dagger} = \left[i(\eta_{\rho\mu}\eta_{\sigma\nu} - \eta_{\rho\nu}\eta_{\sigma\mu})\right]^{\dagger} = -i(\eta_{\rho\mu}\eta_{\sigma\nu} - \eta_{\rho\nu}\eta_{\sigma\mu})^{\top}.$$

We observe that transposition means interchanging the matrix indices μ and ν, *i.e.*

$$[(M_{\rho\sigma})_{\mu\nu}]^{\dagger} = -i(\eta_{\rho\nu}\eta_{\sigma\mu} - \eta_{\rho\mu}\eta_{\sigma\nu}) = i(\eta_{\rho\mu}\eta_{\sigma\nu} - \eta_{\rho\nu}\eta_{\sigma\mu}) = [(M_{\rho\sigma})_{\mu\nu}]$$

(using Eq. (1.13)).

The Hermiticity of the matrices $[(M_{\rho\sigma})_{\mu\nu}]$ implies for the generators K_i and J_i:

$$\left[(J_i)_{\mu\nu}\right]^\dagger = \left[(J_i)_{\mu\nu}\right], \qquad \left[(K_i)_{\mu\nu}\right]^\dagger = \left[(K_i)_{\mu\nu}\right].$$

Explicititly we have:

$$(J_1)_{\mu\nu} = \eta_{\mu\rho}(J_1)^\rho{}_\nu,$$
$$(J_1)_{23} = \eta_{2\rho}(J_1)^\rho{}_3 = \eta_{22}(J_1)^2{}_3 = -i(M_1)^2{}_3 = i,$$
$$(J_1)_{32} = \eta_{3\rho}(J_1)^\rho{}_2 = \eta_{33}(J_1)^3{}_2 = -i(M_1)^3{}_2 = -i.$$

All other elements of this matrix are zero. Hence

$$\left[(J_1)_{\mu\nu}\right] = \begin{pmatrix} 0 & 0 & 0 & 0 \\ 0 & 0 & 0 & 0 \\ 0 & 0 & 0 & i \\ 0 & 0 & -i & 0 \end{pmatrix}.$$

Similarly one shows that

$$\left[(J_2)_{\mu\nu}\right] = \begin{pmatrix} 0 & 0 & 0 & 0 \\ 0 & 0 & 0 & -i \\ 0 & 0 & 0 & 0 \\ 0 & i & 0 & 0 \end{pmatrix}, \qquad \left[(J_3)_{\mu\nu}\right] = \begin{pmatrix} 0 & 0 & 0 & 0 \\ 0 & 0 & i & 0 \\ 0 & -i & 0 & 0 \\ 0 & 0 & 0 & 0 \end{pmatrix}.$$

For the boost generators we have

$$(K_1)_{\mu\nu} = \eta_{\mu\rho}(K_1)^\rho{}_\nu,$$
$$(K_1)_{01} = \eta_{0\rho}(K_1)^\rho{}_1 = \eta_{00}(K_1)^0{}_1 = i(N_1)^0{}_1 = i,$$
$$(K_1)_{10} = \eta_{1\rho}(K_1)^\rho{}_0 = \eta_{11}(K_1)^1{}_0 = -i(N_1)^1{}_0 = -i,$$

and all other elements of this matrix are zero. Therefore the explicit form of $[(K_1)_{\mu\nu}]$ is:

$$\left[(K_1)_{\mu\nu}\right] = \begin{pmatrix} 0 & i & 0 & 0 \\ -i & 0 & 0 & 0 \\ 0 & 0 & 0 & 0 \\ 0 & 0 & 0 & 0 \end{pmatrix}.$$

The boost generators K_2 and K_3 are given by

$$\left[(K_2)_{\mu\nu}\right] = \begin{pmatrix} 0 & 0 & i & 0 \\ 0 & 0 & 0 & 0 \\ -i & 0 & 0 & 0 \\ 0 & 0 & 0 & 0 \end{pmatrix}, \qquad \left[(K_3)_{\mu\nu}\right] = \begin{pmatrix} 0 & 0 & 0 & i \\ 0 & 0 & 0 & 0 \\ 0 & 0 & 0 & 0 \\ -i & 0 & 0 & 0 \end{pmatrix}.$$

Hence in this form the six matrices

$$\{K_i\}_{i=1,2,3}, \quad \text{and} \quad \{J_i\}_{i=1,2,3},$$

form a set of Hermitian generators of the Lorentz group. The Lie algebra $o(1,3;\mathbb{R})$ describes the Lorentz group locally and is determined by the commutator of the basis elements.

Proposition 1.5:

The generators $M_{\mu\nu}$ of the Lorentz group obey the following commutation relation:

$$[M_{\mu\nu}, M_{\rho\sigma}] = -i(\eta_{\mu\rho}M_{\nu\sigma} - \eta_{\mu\sigma}M_{\nu\rho} - \eta_{\nu\rho}M_{\mu\sigma} + \eta_{\nu\sigma}M_{\mu\rho}). \quad (1.14)$$

Proof: Using Eqs. (1.11) and (1.13) we have:

$$
\begin{aligned}
[M_{\mu\nu}, M_{\rho\sigma}]_{\alpha\beta} &= (M_{\mu\nu})_{\alpha\gamma}(M_{\rho\sigma})^{\gamma}{}_{\beta} - (M_{\rho\sigma})_{\alpha\gamma}(M_{\mu\nu})^{\gamma}{}_{\beta} \\
&= i(\eta_{\mu\alpha}\eta_{\nu\gamma} - \eta_{\mu\gamma}\eta_{\nu\alpha})i(\delta_{\rho}{}^{\gamma}\eta_{\sigma\beta} - \eta_{\rho\beta}\delta_{\sigma}{}^{\gamma}) \\
&\quad -i(\eta_{\rho\alpha}\eta_{\sigma\gamma} - \eta_{\rho\gamma}\eta_{\sigma\alpha})i(\delta_{\mu}{}^{\gamma}\eta_{\nu\beta} - \eta_{\mu\beta}\delta_{\nu}{}^{\gamma}) \\
&= -(\eta_{\mu\alpha}\eta_{\nu\rho}\eta_{\sigma\beta} - \eta_{\mu\alpha}\eta_{\nu\sigma}\eta_{\rho\beta} - \eta_{\mu\rho}\eta_{\nu\alpha}\eta_{\sigma\beta} + \eta_{\mu\sigma}\eta_{\nu\alpha}\eta_{\rho\beta}) \\
&\quad +(\eta_{\rho\alpha}\eta_{\sigma\mu}\eta_{\nu\beta} - \eta_{\rho\alpha}\eta_{\sigma\nu}\eta_{\mu\beta} - \eta_{\rho\mu}\eta_{\sigma\alpha}\eta_{\nu\beta} + \eta_{\rho\nu}\eta_{\sigma\alpha}\eta_{\mu\beta}) \\
&= -\Big[\eta_{\mu\alpha}(\eta_{\nu\rho}\eta_{\sigma\beta} - \eta_{\nu\sigma}\eta_{\rho\beta}) - \eta_{\nu\alpha}(\eta_{\mu\rho}\eta_{\sigma\beta} - \eta_{\mu\sigma}\eta_{\rho\beta}) \\
&\quad -\eta_{\rho\alpha}(\eta_{\sigma\mu}\eta_{\nu\beta} - \eta_{\sigma\nu}\eta_{\mu\beta}) + \eta_{\sigma\alpha}(\eta_{\rho\mu}\eta_{\nu\beta} - \eta_{\rho\nu}\eta_{\mu\beta})\Big] \\
&= -i\eta_{\mu\rho}(i)(\eta_{\nu\alpha}\eta_{\sigma\beta} - \eta_{\nu\beta}\eta_{\sigma\alpha}) + i\eta_{\mu\sigma}(i)(\eta_{\nu\alpha}\eta_{\rho\beta} - \eta_{\rho\alpha}\eta_{\nu\beta}) \\
&\quad +i\eta_{\nu\rho}(i)(\eta_{\mu\alpha}\eta_{\sigma\beta} - \eta_{\sigma\alpha}\eta_{\mu\beta}) - i\eta_{\nu\sigma}(i)(\eta_{\mu\alpha}\eta_{\rho\beta} - \eta_{\rho\alpha}\eta_{\mu\beta}) \\
&= -i\eta_{\mu\rho}(M_{\nu\sigma})_{\alpha\beta} + i\eta_{\mu\sigma}(M_{\nu\rho})_{\alpha\beta} \\
&\quad +i\eta_{\nu\rho}(M_{\mu\sigma})_{\alpha\beta} - i\eta_{\nu\sigma}(M_{\mu\rho})_{\alpha\beta}.
\end{aligned}
$$

Dropping the matrix indices α, β, we obtain Eq. (1.14).

Next we want to rederive (as a consistency check) the commutation relations of the generators K_i and J_j, starting from the commutation relations Eq. (1.14). We had:

$$M_{mn} = \epsilon_{mni}J_i, \quad (1.15)$$
$$M_{0i} = -K_i. \quad (1.16)$$

Equation (1.14) reads for the indices $\mu = \rho = 0, \nu = i, \sigma = j; i,j = 1,2,3$:

$$[M_{0i}, M_{0j}] \overset{(1.16)}{=} [K_i, K_j] = -i(\eta_{00}M_{ij} - \eta_{0j}M_{i0} - \eta_{i0}M_{0j} + \eta_{ij}M_{00})$$
$$\overset{(1.14)}{=} -i\eta_{00}M_{ij} = -i\epsilon_{ijk}J_k.$$

Hence:

$$[K_i, K_j] = -i\epsilon_{ijk}J_k. \tag{1.17}$$

For $\mu = 0, \nu = i, \rho = k, \sigma = l;\ i, k, l = 1, 2, 3$ we have:

$$[M_{0i}, M_{kl}] \overset{(1.15)}{\underset{(1.16)}{=}} -[K_i, \epsilon_{klm}J_m] = -i(\eta_{0k}M_{il} - \eta_{0l}M_{ik} - \eta_{ik}M_{0l} + \eta_{il}M_{0k})$$

$$\overset{(1.14)}{=} -i\eta_{ik}K_l + \eta_{il}K_k = i(\delta_{ik}K_l - \delta_{il}K_k),$$

where in the last step we used $\eta_{ik} = -\delta_{ik}$. Hence:

$$\epsilon_{klm}[K_i, J_m] = -i(\delta_{ik}K_l - \delta_{il}K_k). \tag{1.18}$$

Then using

$$\epsilon_{kln}\epsilon_{klm} = 2\delta_{nm}, \tag{1.19}$$

we obtain (with Eqs. (1.19) and (1.18))

$$\epsilon_{kln}\epsilon_{klm}[K_i, J_m] = 2\delta_{nm}[K_i, J_m] = 2[K_i, J_n]$$

$$= -i\epsilon_{kln}(\delta_{ik}K_l - \delta_{il}K_k) = -i\epsilon_{iln}K_l + i\epsilon_{kin}K_k = 2i\epsilon_{inl}K_l.$$

Hence

$$[K_i, J_k] = i\epsilon_{ikl}K_l, \tag{1.20}$$

and finally we have to evaluate Eq. (1.14) for the case $\mu = i, \nu = j, \rho = k, \sigma = l;\ i, j, k, l = 1, 2, 3$. Using Eq. (1.14) we have

$$[M_{ij}, M_{kl}] = [\epsilon_{ijm}J_m, \epsilon_{kln}J_n] = \epsilon_{ijm}\epsilon_{kln}[J_m, J_n]$$

$$= -i(\eta_{ik}M_{jl} - \eta_{il}M_{jk} - \eta_{jk}M_{il} + \eta_{jl}M_{ik})$$

$$= -i(\eta_{ik}\epsilon_{jla} - \eta_{il}\epsilon_{jka} - \eta_{jk}\epsilon_{ila} + \eta_{jl}\epsilon_{ika})J_a.$$

Then with Eq. (1.19)

$$\frac{1}{4}\epsilon_{ijf}\epsilon_{klg}\epsilon_{ijm}\epsilon_{kln}[J_m, J_n] = \delta_{fm}\delta_{gn}[J_m, J_n] = [J_f, J_g]$$

$$= -\frac{i}{4}\epsilon_{ijf}\epsilon_{klg}(\eta_{ik}\epsilon_{jla} - \eta_{il}\epsilon_{jka} - \eta_{jk}\epsilon_{ila} + \eta_{jl}\epsilon_{ika})J_a$$

$$= -\frac{i}{4}\{-\epsilon_{ijf}\epsilon_{ilg}\epsilon_{jla} + \epsilon_{ijf}\epsilon_{kig}\epsilon_{jka}$$

$$+ \epsilon_{ijf}\epsilon_{jlg}\epsilon_{ila} - \epsilon_{ijf}\epsilon_{kjg}\epsilon_{ika}\}J_a$$

$$= \frac{i}{4}\{(\delta_{jl}\delta_{fg} - \delta_{jg}\delta_{fl})\epsilon_{jla} + (\delta_{jk}\delta_{fg} - \delta_{jg}\delta_{fk})\epsilon_{jka}$$

$$+ (\delta_{il}\delta_{fg} - \delta_{ig}\delta_{lf})\epsilon_{ila} + (\delta_{ik}\delta_{fg} - \delta_{ig}\delta_{kf})\epsilon_{ika}\}J_a$$

$$= -\frac{i}{4}\{\epsilon_{gfa} + \epsilon_{gfa} + \epsilon_{gfa} + \epsilon_{gfa}\}J_a = i\epsilon_{fga}J_a,$$

where we made use of $\eta_{ij} = -\delta_{ij}$ and

$$\epsilon_{ijk}\epsilon_{ilm} = \delta_{jl}\delta_{km} - \delta_{jm}\delta_{kl}.$$

Hence:

$$[J_i, J_j] = i\epsilon_{ijk}J_k. \tag{1.21}$$

Equation (1.21) defines the rotation group $SO(3;\mathbb{R})$ as a subgroup of $L_+^\uparrow$, whereas Eq. (1.20) states that $\mathbf{K}$ is a vector under the Lorentz group. The minus-sign in Eq. (1.17) is significant; it expresses the difference between the non-compact group $SO(1,3;\mathbb{R})$ and its compact form $SO(4;\mathbb{R})$ or between $SL(2,\mathbb{C})$ and $SU(2,\mathbb{C}) \times SU(2,\mathbb{C})$, since locally homomorphic Lie groups[6] have homomorphic Lie algebras.

In order to be able to classify the irreducible, finite-dimensional, non-unitary representations [103] of the restricted Lorentz group $L_+^\uparrow$, we have to change the basis K_i, J_i of the Lie algebra $so(1,3;\mathbb{R})$ by introducing the complex linear combinations

$$S_i := \frac{1}{2}(J_i + iK_i), \tag{1.22}$$

$$T_i := \frac{1}{2}(J_i - iK_i). \tag{1.23}$$

One can verify, using Eqs. (1.17), (1.20) and (1.21), that these non-Hermitian generators decouple the commutation relations of K_i and J_i so that

$$\begin{aligned} [S_i, S_j] &= i\epsilon_{ijk}S_k, \\ [T_i, T_j] &= i\epsilon_{ijk}T_k, \\ [T_i, S_j] &= 0. \end{aligned} \tag{1.24}$$

This means that the generators S_i and T_j obey the commutation relations of the Lie algebra of $SU(2,\mathbb{C})$. In addition, the commutation relations of Eqs. (1.24) show that the Lie algebra $so(1,3;\mathbb{R})$ decomposes into the direct sum of two $su(2,\mathbb{C})$ Lie algebras. However, this decomposition holds only for the complexified Lie algebra $so(1,3;\mathbb{R})^\mathbb{C}$, *i.e.* considering the set of real 4×4-matrices $\mathfrak{A}$ satisfying

$$\mathfrak{A}^\top = -\eta\mathfrak{A}\eta$$

as a complex vector space, this allows complex linear combinations of the form S_i and T_i. Hence the decomposition

$$so(1,3;\mathbb{R})^\mathbb{C} \cong su(2,\mathbb{C}) \times su(2,\mathbb{C})$$

---

[6]The groups $SO(1,3;\mathbb{R})$ and $SL(2,\mathbb{C})$ are locally homomorphic as we shall see explicitly in Sec. 1.3.3, as well as the groups $SO(4,\mathbb{R})$ and $SU(2,\mathbb{C}) \times SU(2,\mathbb{C})$.

is valid only for the complexified Lie algebra of the Lorentz group. However, we can use the classification of irreducible represensions of the complex Lie algebra $so(1,3;\mathbb{R})^{\mathbb{C}}$ to find the irreducible representations of the real Lie algebra $so(1,3,\mathbb{R}) \cong sl(2,\mathbb{C})$, since there is a one-to-one correspondence between representations of a complex Lie algebra and representations of any of its real forms [19, 72].

But the classification of finite-dimensional irreducible representations of the algebra $su(2,\mathbb{C}) \times su(2,\mathbb{C})$ is well known. According to the theorem of Racah,[7] there is one *Casimir operator* for every $su(2,\mathbb{C})$ subalgebra, *i.e.*

$$\sum_{i=1}^{3} S_i S_i \quad \text{and} \quad \sum_{i=1}^{3} T_i T_i$$

are Casimir operators, commuting with any element of the algebra, with eigenvalues $n(n+1)$ and $m(m+1)$ respectively. Here n and m are eigenvalues of S_3 and T_3 respectively with values $n, m = 0, 1/2, 1, 3/2, \ldots$. Therefore we can label representations of $so(1,3;\mathbb{R})$ by the pair (n,m), and since

$$J_3 = S_3 + T_3,$$

we can identify the spin of the representation with $n + m$. The dimension of the representation space is $(2n+1) \cdot (2m+1)$ for an (n,m)-representation of the Lorentz group.

It is important to note that the two $su(2,\mathbb{C})$ subalgebras are not independent since they can be interchanged by the operation of *parity*. Parity acts as follows on rotation and boost generators:

$$J_i \longrightarrow J_i, \qquad K_i \longrightarrow -K_i,$$

and Eqs. (1.22) and (1.23) show that parity transforms S_i into T_i and T_i into S_i. In addition the operation of Hermitian conjugation also interchanges S_i and T_i since, as demonstrated above, J_i and K_i can be chosen to be Hermitian:

$$S_i^{\dagger} = \frac{1}{2}(J_i + iK_i)^{\dagger} = \frac{1}{2}(J_i - iK_i) = T_i,$$

$$T_i^{\dagger} = \frac{1}{2}(J_i - iK_i)^{\dagger} = \frac{1}{2}(J_i + iK_i) = S_i.$$

Hence, the parity operation is equivalent to Hermitian conjugation.

As examples we consider the following representations:

---

[7]See *e.g.* W. Miller [72], p. 395.

(a) $(0,0)$ with total spin zero is the *scalar representation*; the dimension of the representation space is one.

(b) $(1/2,0)$ with total spin $1/2$ is called the *left-handed spinor representation*; the dimension of the representation space is two.

(c) $(0,1/2)$ with total spin $1/2$ is called the *right-handed spinor representation*; the dimension of the representation space is again two.

The handedness is a convention. In subsequent sections we shall discuss these two spinor representations in great detail. Since parity switches S_i to T_i and T_i to S_i, representations of the Lorentz group in general are not parity eigenstates. In particular, the left-handed spinor representation $(1/2,0)$ transforms under parity into the $(0,1/2)$ representation and *vice versa*. Therefore to obtain a representation such that parity acts as a linear transformation, one has to consider the direct sum of the spinor representations (b) and (c), $(1/2,0) \oplus (0,1/2)$, which yields a *Dirac spinor representation*. This representation of the Lorentz group will be considered in detail in Sec. 1.4.

The importance of the representations (b) and (c) is due to the fact, that any other representation of the Lorentz group can be generated from these spinor representations. In Sec. 1.3 we shall consider a few examples explicitly. For instance, the Kronecker product of the representations (b) and (c)

$$(1/2,0) \otimes (0,1/2) = (1/2,1/2)$$

gives a spin 1 representation (see Eq. (1.152a)) with four components, and the Kronecker product of two left-handed spinor representations decomposes into a scalar representation and a spin 1 representation, given by an antisymmetric, selfdual second rank tensor (see Eq. (1.152b)), *i.e.*

$$(1/2,0) \otimes (1/2,0) = (0,0) \oplus (1,0).$$

1.2 The Poincaré Group

As stated above, the Lorentz group leaves the interval $(x-y)^2$ in Minkowski space $\mathbb{M}_4$ invariant. On the other hand, the translations

$$x^\mu \longrightarrow x'^\mu = x^\mu + a^\mu,$$

where a^μ is a constant four-vector, also leaves the length squared $(x-y)^2$ invariant. This leads to the definition of the *Poincaré group P* as the group of all real transformations in Minkowski space of the form:[8]

$$x^\mu \longrightarrow x'^\mu = \Lambda^\mu{}_\nu x^\nu + a^\mu, \tag{1.25}$$

which leaves the length squared $(x - y)^2$ invariant. Definition (1.25) leads to the following composition law for the elements of P: Let

$$
\begin{aligned}
x' &= \Lambda_1 x + a_1, \\
x'' &= \Lambda_2 x' + a_2 \\
&= \Lambda_2 \{\Lambda_1 x + a_1\} + a_2 \\
&= \Lambda_2 \Lambda_1 x + \Lambda_2 a_1 + a_2.
\end{aligned}
$$

Writing (Λ, a) for an element of P, we have for the composition of two Poincaré transformations:

$$(\Lambda_2, a_2) \circ (\Lambda_1, a_1) = (\Lambda_2 \Lambda_1, \Lambda_2 a_1 + a_2). \tag{1.26}$$

This demonstrates that the Poincaré group P is the semi-direct product $L \overset{s}{\otimes} \mathbb{T}_4$ of the Lorentz group L and the translation group in four spacetime dimensions $\mathbb{T}_4$ in Minkowski space. Similarly, as for the Lorentz group, the Poincaré group P decomposes into four pieces identified by $\det \Lambda$ and $\Lambda^0{}_0$; *i.e.*

$$P^\uparrow_+, \quad P^\downarrow_+, \quad P^\uparrow_-, \quad P^\downarrow_-.$$

The identity element of the Poincaré group P is the element $(\mathbb{1}_{4\times4}, 0)$, and the inverse of the transformation $(\Lambda, a) \in P$ is $(\Lambda^{-1}, -\Lambda^{-1}a)$ such that (applying the composition (1.26))

$$
\begin{aligned}
(\Lambda, a) \circ (\Lambda^{-1}, -\Lambda^{-1}a) &= (\Lambda\Lambda^{-1}, -\Lambda\Lambda^{-1}a + a) = (\mathbb{1}_{4\times4}, 0), \\
(\Lambda^{-1}, -\Lambda^{-1}a) \circ (\Lambda, a) &= (\Lambda^{-1}\Lambda, \Lambda^{-1}a - \Lambda^{-1}a) = (\mathbb{1}_{4\times4}, 0).
\end{aligned}
$$

The Lie algebra of $P^\uparrow_+$ is determined by the commutation relation (1.14) of the Lorentz group $L^\uparrow_+$, the trivial commutation relation of the translation group (observe that the translation group $\mathbb{T}_4$ in Minkowski space is Abelian), and the commutator of translations and Lorentz transformations, still to be determined. In order to obtain this commutator, we consider a faithful representation of $P^\uparrow_+$:

$$(\Lambda, a) \longrightarrow g(\Lambda, a) \tag{1.27}$$

---

[8]In the mathematical literature transformations acting on a manifold that preserve the distance between pairs of ponts are called *isometries*. Therefore, the Poincaré group is the *isometry group* of Minkowski space, see *e.g.* [72], p. 20.

in a vector space V such that

$$g(\Lambda_2, a_2)g(\Lambda_1, a_1) = g(\Lambda_2\Lambda_1, \Lambda_2 a_1 + a_2), \qquad (1.28)$$
$$g^{-1}(\Lambda, a) = g(\Lambda^{-1}, -\Lambda^{-1}a). \qquad (1.29)$$

This means, g is a bijective linear map of a vector space V onto V, satisfying Eqs. (1.28) and (1.29). Infinitesimally we can write[9]

$$g(\Lambda, a) = \mathbb{1}_V - \frac{i}{2}\omega_{\rho\sigma}M^{\rho\sigma} + ia_\mu P^\mu, \qquad (1.30)$$

where $\omega_{\rho\sigma} = -\omega_{\sigma\rho}$ are six infinitesimal parameters leading to an infinitesimal Lorentz transformation $\Lambda \in L_+^\uparrow$, and a_μ denotes four infinitesimal parameters, leading to an infinitesimal translation.

Now $M^{\rho\sigma} = -M^{\sigma\rho}$ and P^μ are generators of Lorentz transformations and translations respectively in the corresponding representation. Consider[10]

$$g^{-1}(\Lambda, 0)g(\Lambda', a')g(\Lambda, 0) \stackrel{(1.28)}{=} g^{-1}(\Lambda, 0)g(\Lambda'\Lambda, a')$$
$$\stackrel{(1.29)}{=} g(\Lambda^{-1}, 0)g(\Lambda'\Lambda, a')$$
$$\stackrel{(1.28)}{=} g(\Lambda^{-1}\Lambda'\Lambda, \Lambda^{-1}a'). \qquad (1.31)$$

For infinitesimal $(\Lambda', a') \in P_+^\uparrow$ the left hand side of Eq. (1.31) gives

$$g^{-1}(\Lambda, 0)g(\Lambda', a')g(\Lambda, 0) \stackrel{(1.30)}{=} g^{-1}(\Lambda, 0)\left\{\mathbb{1}_V - \frac{i}{2}\omega'_{\mu\nu}M^{\mu\nu} + ia'_\mu P^\mu\right\}g(\Lambda, 0)$$
$$= \mathbb{1}_V - \frac{i}{2}\omega'_{\mu\nu}g^{-1}(\Lambda, 0)M^{\mu\nu}g(\Lambda, 0)$$
$$+ ia'_\mu g^{-1}(\Lambda, 0)P^\mu g(\Lambda, 0).$$

The right hand side of Eq. (1.31) may be expanded as

$$g(\Lambda^{-1}\Lambda'\Lambda, \Lambda^{-1}a') = \mathbb{1}_V - \frac{i}{2}(\Lambda^{-1}\omega'\Lambda)_{\rho\sigma}M^{\rho\sigma} + i(\Lambda^{-1}a')_\rho P^\rho$$
$$= \mathbb{1}_V - \frac{i}{2}(\Lambda^{-1})_\rho{}^\mu \omega'_{\mu\nu}\Lambda^\nu{}_\sigma M^{\rho\sigma} + i(\Lambda^{-1})_\rho{}^\mu a'_\mu P^\rho$$
$$= \mathbb{1}_V - \frac{i}{2}\omega'_{\mu\nu}\Lambda^\mu{}_\rho \Lambda^\nu{}_\sigma M^{\rho\sigma} + ia'_\mu \Lambda^\mu{}_\rho P^\rho.$$

---

[9]We denote by $\mathbb{1}_V$ the identity map of the vector space V.
[10]We apply the properties (1.28) and (1.29) of the faithful representation g of the Poincaré group.

Hence we obtain:

$$
\begin{aligned}
g^{-1}(\Lambda, 0) M^{\mu\nu} g(\Lambda, 0) &= \Lambda^{\mu}{}_{\rho} \Lambda^{\nu}{}_{\sigma} M^{\rho\sigma}, \\
g^{-1}(\Lambda, 0) P^{\mu} g(\Lambda, 0) &= \Lambda^{\mu}{}_{\rho} P^{\rho}.
\end{aligned}
\tag{1.32}
$$

The first equation states that $M^{\mu\nu}$ is an antisymmetric tensor operator under $L_{+}^{\uparrow}$; the second relation shows that P^{μ} is a vector operator under $L_{+}^{\uparrow}$.

Now consider infinitesimal $\Lambda \in L_{+}^{\uparrow}$; for Eq. (1.32) we obtain:

$$
\begin{aligned}
g^{-1}(\Lambda, 0) M^{\mu\nu} g(\Lambda, 0) &= g(\Lambda^{-1}, 0) M^{\mu\nu} g(\Lambda, 0) \\
&= \left(\mathbb{1}_V + \frac{i}{2} \omega_{\rho\sigma} M^{\rho\sigma} \right) M^{\mu\nu} \left(\mathbb{1}_V - \frac{i}{2} \omega_{\rho\sigma} M^{\rho\sigma} \right) \\
&= M^{\mu\nu} + \frac{i}{2} \omega_{\rho\sigma} \left[M^{\rho\sigma}, M^{\mu\nu} \right].
\end{aligned}
$$

On the other hand, the right hand side of Eq. (1.32) can be expanded as:

$$
\begin{aligned}
\Lambda^{\mu}{}_{\rho} \Lambda^{\nu}{}_{\sigma} M^{\rho\sigma} &= \left(\delta^{\mu}{}_{\rho} + \omega^{\mu}{}_{\rho} \right) \left(\delta^{\nu}{}_{\sigma} + \omega^{\nu}{}_{\sigma} \right) M^{\rho\sigma} \\
&= M^{\mu\nu} + \omega^{\mu}{}_{\rho} \delta^{\nu}{}_{\sigma} M^{\rho\sigma} + \delta^{\mu}{}_{\rho} \omega^{\nu}{}_{\sigma} M^{\rho\sigma} \\
&= M^{\mu\nu} + \omega^{\mu}{}_{\rho} M^{\rho\nu} + \omega^{\nu}{}_{\sigma} M^{\mu\sigma} \\
&= M^{\mu\nu} + \eta^{\mu\sigma} \omega_{\sigma\rho} M^{\rho\nu} + \eta^{\nu\rho} \omega_{\rho\sigma} M^{\mu\sigma} \\
&= M^{\mu\nu} + \frac{1}{2} \{ \eta^{\mu\sigma} \omega_{\sigma\rho} M^{\rho\nu} + \eta^{\mu\rho} \omega_{\rho\sigma} M^{\sigma\nu} \\
&\quad + \eta^{\nu\rho} \omega_{\rho\sigma} M^{\mu\sigma} + \eta^{\nu\sigma} \omega_{\sigma\rho} M^{\mu\rho} \} \\
&= M^{\mu\nu} + \frac{1}{2} \omega_{\rho\sigma} \{ \eta^{\mu\rho} M^{\sigma\nu} - \eta^{\mu\sigma} M^{\rho\nu} \\
&\quad - \eta^{\nu\sigma} M^{\mu\rho} + \eta^{\nu\rho} M^{\mu\sigma} \}.
\end{aligned}
$$

Hence

$$
\left[M^{\rho\sigma}, M^{\mu\nu} \right] = -i \left(\eta^{\mu\rho} M^{\sigma\nu} - \eta^{\mu\sigma} M^{\rho\nu} - \eta^{\nu\sigma} M^{\mu\rho} + \eta^{\nu\rho} M^{\mu\sigma} \right),
$$

or

$$
\left[M^{\mu\nu}, M^{\rho\sigma} \right] = -i \left(\eta^{\mu\rho} M^{\nu\sigma} - \eta^{\mu\sigma} M^{\nu\rho} - \eta^{\nu\rho} M^{\mu\sigma} + \eta^{\nu\sigma} M^{\mu\rho} \right).
$$

In this way we have rederived the commutation relation of the Lie algebra $so(1, 3; \mathbb{R})$, *i.e.* Eq. (1.14).

The second of Eqs. (1.32) leads to

$$
\begin{aligned}
g^{-1}(\Lambda,0)P^\rho g(\Lambda,0) &= \left(\mathbb{1}_V + \frac{i}{2}\omega_{\mu\nu}M^{\mu\nu}\right)P^\rho\left(\mathbb{1}_V - \frac{i}{2}\omega_{\mu\nu}M^{\mu\nu}\right) \\
&= P^\rho + \frac{i}{2}\omega_{\mu\nu}\left[M^{\mu\nu},P^\rho\right] \\
&\overset{!}{=} \Lambda^\rho{}_\nu P^\nu \qquad \text{(using Eq. (1.32))} \\
&= \left(\delta^\rho{}_\nu + \omega^\rho{}_\nu\right)P^\nu \\
&= P^\rho + \omega^\rho{}_\nu p^\nu \\
&= P^\rho + \eta^{\rho\mu}\omega_{\mu\nu}P^\nu \\
&= P^\rho + \frac{1}{2}\left\{\eta^{\rho\mu}\omega_{\mu\nu}P^\nu + \eta^{\rho\nu}\omega_{\nu\mu}P^\mu\right\} \\
&= P^\rho + \frac{1}{2}\omega_{\mu\nu}\left\{\eta^{\rho\mu}P^\nu - \eta^{\rho\nu}P^\mu\right\}.
\end{aligned}
$$

Therefore we obtain for the commutator of the operator $M_{\mu\nu}$ with P_ρ:

$$
\left[M^{\mu\nu},P^\rho\right] = -i\left(\eta^{\mu\rho}P^\nu - \eta^{\nu\rho}P^\mu\right). \tag{1.33}
$$

In summary we see that the Poincaré algebra is given by the following set of commutators:

$$
\begin{aligned}
\left[M_{\mu\nu},M_{\rho\sigma}\right] &= -i\left(\eta_{\mu\rho}M_{\nu\sigma} - \eta_{\mu\sigma}M_{\nu\rho} - \eta_{\nu\rho}M_{\mu\sigma} + \eta_{\nu\sigma}M_{\mu\rho}\right), \\
\left[M_{\mu\nu},P_\rho\right] &= -i\left(\eta_{\mu\rho}P_\nu - \eta_{\nu\rho}P_\mu\right), \\
\left[P_\mu,P_\nu\right] &= 0.
\end{aligned} \tag{1.34}
$$

Proposition 1.6:

The square of the energy-momentum vector $P^2 = P_\mu P^\mu$ is a Casimir operator of the Poincaré algebra (1.34), *i.e.*

$$
\left[M_{\mu\nu},P^2\right] = 0, \tag{1.35}
$$

$$
\left[P_\mu,P^2\right] = 0. \tag{1.36}
$$

Proof: The verification of relation (1.36) is trivial. Hence we consider the relation (1.35). Using Eq. (1.33) we have:

$$
\begin{aligned}
\left[M_{\mu\nu}, P^2\right] &= \left[M_{\mu\nu}, P^\rho P_\rho\right] = \left[M_{\mu\nu}, P^\rho\right] P_\rho + P^\rho \left[M_{\mu\nu}, P_\rho\right] \\
&= \left[M_{\mu\nu}, \eta^{\rho\sigma} P_\sigma\right] P_\rho + P^\rho \left[M_{\mu\nu}, P_\rho\right] \\
&= -\eta^{\rho\sigma} i \left(\eta_{\mu\sigma} P_\nu - \eta_{\nu\sigma} P_\mu\right) P_\rho - P^\rho i \left(\eta_{\mu\rho} P_\nu - \eta_{\nu\rho} P_\mu\right) \\
&= -i\eta^{\rho\sigma} \eta_{\mu\sigma} P_\nu P_\rho + i\eta^{\rho\sigma} \eta_{\nu\sigma} P_\mu P_\rho - iP^\rho \eta_{\mu\rho} P_\nu + iP^\rho \eta_{\nu\rho} P_\mu \\
&= -i\delta^\rho{}_\mu P_\nu P_\rho + i\delta^\rho{}_\nu P_\mu P_\rho - iP_\mu P_\nu + iP_\nu P_\mu \\
&= -i\left[P_\nu, P_\mu\right] - i\left[P_\mu, P_\nu\right] = 0.
\end{aligned}
$$

The second Casimir operator of the Poincaré algebra (1.34) can be constructed from the *Pauli–Ljubanski polarization vector*[11] defined by

$$
W_\mu := \frac{1}{2} \epsilon_{\mu\nu\rho\sigma} P^\nu M^{\rho\sigma}, \tag{1.37}
$$

where $\epsilon_{\mu\nu\rho\sigma}$ is the Levi–Civita tensor with $\epsilon_{0123} = +1$.

Proposition 1.7:

The Pauli–Ljubanski vector is invariant under translations, *i.e.*

$$
\left[P_\mu, W_\nu\right] = 0. \tag{1.38a}
$$

It can be written in the form

$$
W^\mu = \left[I, P^\mu\right], \tag{1.38b}
$$

where

$$
I := \frac{i}{8} \epsilon_{\mu\nu\rho\sigma} M^{\mu\nu} M^{\rho\sigma},
$$

and is a vector under $L_+^\uparrow$:

$$
\left[M_{\mu\nu}, W_\rho\right] = -i\left(\eta_{\mu\rho} W_\nu - \eta_{\nu\rho} W_\mu\right). \tag{1.39}
$$

Proof: We first demonstrate Eq. (1.38a). With the help of Eqs. (1.37) and

---

[11] See the discussion in R. Penrose [87], Chap. 22.12.

(1.33) we have:

$$
\begin{aligned}
[P_\mu, W_\nu] &= \frac{1}{2}\left[P_\mu, \epsilon_{\nu\rho\sigma\tau}P^\rho M^{\sigma\tau}\right] = \frac{1}{2}\epsilon_{\nu\rho\sigma\tau}\left[P_\mu, P^\rho M^{\sigma\tau}\right] \\
&= \frac{1}{2}\epsilon_{\nu\rho\sigma\tau}\eta_{\mu\gamma}\left\{\left[P^\gamma, P^\rho\right]M^{\sigma\tau} + P^\rho\left[P^\gamma, M^{\sigma\tau}\right]\right\} \\
&= \frac{i}{2}\epsilon_{\nu\rho\sigma\tau}\eta_{\mu\gamma}P^\rho\left\{\eta^{\sigma\gamma}P^\tau - \eta^{\tau\gamma}P^\sigma\right\} \\
&= \frac{i}{2}\left[\epsilon_{\nu\rho\mu\tau}P^\rho P^\tau - \epsilon_{\nu\rho\sigma\mu}P^\rho P^\sigma\right] \\
&= \frac{i}{4}\left[\epsilon_{\nu\rho\mu\tau}P^\rho P^\tau + \epsilon_{\nu\tau\mu\rho}P^\tau P^\rho - \epsilon_{\nu\rho\sigma\mu}P^\rho P^\sigma - \epsilon_{\nu\sigma\rho\mu}P^\sigma P^\rho\right] \\
&= 0.
\end{aligned}
$$

Proof of Eq. (1.38b):

$$
\begin{aligned}
[I, P^\mu] &= [\frac{i}{8}\epsilon_{\alpha\beta\gamma\delta}M^{\alpha\beta}M^{\gamma\delta}, P^\mu] = \frac{i}{8}\epsilon_{\alpha\beta\gamma\delta}[M^{\alpha\beta}M^{\gamma\delta}, P^\mu] \\
&= \frac{i}{8}\epsilon_{\alpha\beta\gamma\delta}\left\{M^{\alpha\beta}[M^{\gamma\delta}, P^\mu] + [M^{\alpha\beta}, P^\mu]M^{\gamma\delta}\right\} \\
&= \frac{i}{8}\epsilon_{\alpha\beta\gamma\delta}\left\{M^{\alpha\beta}(-i)(\eta^{\gamma\mu}P^\delta - \eta^{\delta\mu}P^\gamma)\right. \\
&\qquad\left. -i(\eta^{\alpha\mu}P^\beta - \eta^{\beta\mu}P^\alpha)M^{\gamma\delta}\right\} \\
&= \frac{1}{8}\left\{\epsilon_{\alpha\beta\gamma\delta}\eta^{\gamma\mu}M^{\alpha\beta}P^\delta - \epsilon_{\alpha\beta\gamma\delta}\eta^{\delta\mu}M^{\alpha\beta}P^\gamma\right. \\
&\qquad\left. +\epsilon^\mu{}_{\beta\gamma\delta}P^\beta M^{\gamma\delta} - \epsilon_\alpha{}^\mu{}_{\gamma\delta}P^\alpha M^{\gamma\delta}\right\} \\
&= \frac{1}{4}\epsilon^\mu{}_{\alpha\beta\delta}M^{\alpha\beta}P^\delta + \frac{1}{4}\epsilon^\mu{}_{\alpha\gamma\delta}P^\alpha M^{\gamma\delta}.
\end{aligned}
$$

Now owing to the commutator given in Eq. (1.33) we can write:

$$
M^{\alpha\beta}P^\delta = P^\delta M^{\alpha\beta} - i(\eta^{\alpha\delta}P^\beta - \eta^{\beta\delta}P^\alpha),
$$

and contracting with $\epsilon^\mu{}_{\alpha\beta\delta}$ yields

$$
\frac{1}{4}\epsilon^\mu{}_{\alpha\beta\delta}M^{\alpha\beta}P^\delta = \frac{1}{4}\epsilon^\mu{}_{\alpha\beta\delta}P^\delta M^{\alpha\beta} = \frac{1}{4}\epsilon^\mu{}_{\delta\alpha\beta}P^\delta M^{\alpha\beta},
$$

so that

$$
[I, P^\mu] = \frac{1}{2}\epsilon^\mu{}_{\alpha\gamma\delta}P^\alpha M^{\gamma\delta} = W^\mu.
$$

In order to prove Eq. (1.39) we consider the following commutator and use the Jacobi identity:

$$
\begin{aligned}
\left[M^{\mu\nu}, W^\rho\right] &= \left[M^{\mu\nu}, [I, P^\rho]\right] \\
&= -\left[I, [P^\rho, M^{\mu\nu}]\right] - \left[P^\rho, [M^{\mu\nu}, I]\right] \\
&= -\left[I, i(\eta^{\mu\rho}P^\nu - \eta^{\nu\rho}P^\mu)\right] - 0 \\
&= -i\left(\eta^{\mu\rho}[I, P^\mu] - \eta^{\nu\rho}[I, P^\mu]\right) \\
&= -i\left(\eta^{\mu\rho}W^\nu - \eta^{\nu\rho}W^\mu\right),
\end{aligned}
$$

where

$$
\left[M^{\mu\nu}, I\right] = 0,
$$

since I is invariant under Lorentz transformations.

Proposition 1.8:

The square of the Pauli–Ljubanski vector is given by the expression

$$
W^2 = W_\mu W^\mu = -\frac{1}{2}M_{\mu\nu}M^{\mu\nu}P^2 + M^{\rho\sigma}M_{\nu\sigma}P_\rho P^\nu. \tag{1.40}
$$

Proof:

$$
\begin{aligned}
W^2 = W_\mu W^\mu &= \eta^{\mu\nu}W_\mu W_\nu \\
&= \eta^{\mu\nu}\left\{\frac{1}{2}\epsilon_{\mu\alpha\beta\gamma}P^\alpha M^{\beta\gamma}\right\}\left\{\frac{1}{2}\epsilon_{\nu\rho\sigma\tau}P^\rho M^{\sigma\tau}\right\} \\
&= \frac{1}{4}\epsilon^\mu{}_{\alpha\beta\gamma}\epsilon_{\mu\rho\sigma\tau}P^\alpha M^{\beta\gamma}P^\rho M^{\sigma\tau}.
\end{aligned}
$$

Applying the following contraction of two totally antisymmetric tensors in four-dimensional Minkowski space,

$$
\begin{aligned}
\epsilon^\mu{}_{\alpha\beta\gamma}\epsilon_{\mu\rho\sigma\tau} =\ & \eta_{\alpha\rho}\left(\eta_{\beta\sigma}\eta_{\gamma\tau} - \eta_{\beta\tau}\eta_{\gamma\sigma}\right) - \eta_{\alpha\sigma}\left(\eta_{\beta\rho}\eta_{\gamma\tau} - \eta_{\beta\tau}\eta_{\gamma\rho}\right) \\
& + \eta_{\alpha\tau}\left(\eta_{\beta\rho}\eta_{\gamma\sigma} - \eta_{\beta\sigma}\eta_{\gamma\rho}\right),
\end{aligned}
$$

we find

$$
\begin{aligned}
W^2 =\ & -\frac{1}{4}\Big\{\eta_{\alpha\rho}\left(\eta_{\beta\sigma}\eta_{\gamma\tau} - \eta_{\beta\tau}\eta_{\gamma\sigma}\right) - \eta_{\alpha\sigma}\left(\eta_{\beta\rho}\eta_{\gamma\tau} - \eta_{\beta\tau}\eta_{\gamma\rho}\right) \\
& + \eta_{\alpha\tau}\left(\eta_{\beta\rho}\eta_{\gamma\sigma} - \eta_{\beta\sigma}\eta_{\gamma\rho}\right)\Big\}P^\alpha M^{\beta\gamma}P^\rho M^{\sigma\tau} \\
=\ & -\frac{1}{4}\Big\{P_\rho M_{\sigma\tau}P^\rho M^{\sigma\tau} - P_\rho M_{\tau\sigma}P^\rho M^{\sigma\tau} - P_\sigma M_{\rho\tau}P^\rho M^{\sigma\tau} \\
& + P_\sigma M_{\tau\rho}P^\rho M^{\sigma\tau} + P_\tau M_{\rho\sigma}P^\rho M^{\sigma\tau} - P_\tau M_{\sigma\rho}P^\rho M^{\sigma\tau}\Big\}.
\end{aligned}
$$

Now using Eq. (1.34),

$$
\begin{aligned}
P_\rho M_{\sigma\tau} P^\rho M^{\sigma\tau} &= \{M_{\sigma\tau} P_\rho - i(\eta_{\tau\rho} P_\sigma - \eta_{\sigma\rho} P_\tau)\} P^\rho M^{\sigma\tau} \\
&= M_{\sigma\tau} P_\rho P^\rho M^{\sigma\tau} - i(P_\sigma P_\tau - P_\tau P_\sigma) M^{\sigma\tau} \\
&= M_{\sigma\tau} P^2 M^{\sigma\tau} - i[P_\sigma, P_\tau] M^{\sigma\tau} \\
&= M_{\sigma\tau} M^{\sigma\tau} P^2,
\end{aligned}
$$

where we applied Eqs. (1.34) and (1.35). Using the antisymmetry of the generators $M^{\sigma\tau}$ we get:

$$
\begin{aligned}
-P_\sigma M_{\rho\tau} P^\rho M^{\sigma\tau} &+ P_\sigma M_{\tau\rho} P^\rho M^{\sigma\tau} + P_\tau M_{\rho\sigma} P^\rho M^{\sigma\tau} - P_\tau M_{\sigma\rho} P^\rho M^{\sigma\tau} \\
&= -2P_\sigma M_{\rho\tau} P^\rho M^{\sigma\tau} + 2P_\tau M_{\rho\sigma} P^\rho M^{\sigma\tau} \\
&= -4P_\sigma M_{\rho\tau} P^\rho M^{\sigma\tau} \\
&= -4\{M_{\rho\tau} P_\sigma - i(\eta_{\tau\sigma} P_\rho - \eta_{\rho\sigma} P_\tau)\} P^\rho M^{\sigma\tau} \\
&= -4M_{\rho\tau} P_\sigma P^\rho M^{\sigma\tau} + 4iP^2 M_\tau{}^\tau - 4iP_\tau P_\sigma M^{\sigma\tau} \\
&= -4M_{\rho\tau} P_\sigma P^\rho M^{\sigma\tau},
\end{aligned}
$$

since $M_\tau{}^\tau = 0$, and because the antisymmetry of $M^{\sigma\tau}$ implies that $P_\tau P_\sigma M^{\tau\sigma}$ vanishes.

In the same way one can show that

$$
-4M_{\rho\tau} P_\sigma P^\rho M^{\sigma\tau} = -4M_{\rho\tau} M^{\sigma\tau} P_\sigma P^\rho,
$$

and therefore

$$
W^2 = W_\mu W^\mu = -\frac{1}{2} M_{\mu\nu} M^{\mu\nu} P^2 + M^{\mu\rho} M_{\nu\rho} P_\mu P^\nu.
$$

The importance of the Pauli–Ljubanski vector is due to the fact that its square W^2 is the second Casimir operator of the Poincaré algebra.

Proposition 1.9:

The square of the Pauli–Ljubanski vector, W^2, commutes with the generators of the Poincaré group P_μ and $M_{\mu\nu}$, i.e.

$$
[M_{\mu\nu}, W^2] = 0, \quad [P_\mu, W^2] = 0. \tag{1.41}
$$

Proof: Using Eq. (1.39) we have:

$$
\begin{aligned}
[M_{\mu\nu}, W^2] &= [M_{\mu\nu}, W_\rho W^\rho] \\
&= [M_{\mu\nu}, W_\rho] W^\rho + W^\rho [M_{\mu\nu}, W_\rho] \\
&= -i(\eta_{\mu\rho} W_\nu - \eta_{\nu\rho} W_\mu) W^\rho - iW^\rho (\eta_{\mu\rho} W_\nu - \eta_{\nu\rho} W_\mu) \\
&= -i(W_\nu W_\mu - W_\mu W_\nu + W_\mu W_\nu - W_\nu W_\mu) \\
&= 0.
\end{aligned}
$$

With Eq. (1.38a) the second of relations (1.41) is trivial, *i.e.*

$$[P_\mu, W^2] = [P_\mu, W_\rho]W^\rho + W^\rho[P_\mu, W_\rho] = 0.$$

Proposition 1.10:

The Pauli–Ljubanski polarization vector has the additional property

$$W_\mu P^\mu = 0. \qquad (1.42)$$

Proof:

$$W_\mu P^\mu \overset{(1.37)}{=} \frac{1}{2}\epsilon_{\mu\nu\rho\sigma}P^\nu M^{\rho\sigma}P^\mu$$

$$\overset{(1.34)}{=} \frac{1}{2}\epsilon_{\mu\nu\rho\sigma}P^\nu P^\mu M^{\rho\sigma} + \frac{i}{2}\epsilon_{\mu\nu\rho\sigma}P^\nu \eta^{\sigma\mu}P^\rho - \frac{i}{2}\epsilon_{\mu\nu\rho\sigma}P^\nu \eta^{\rho\mu}P^\sigma$$

$$= \frac{1}{4}\epsilon_{\mu\nu\rho\sigma}P^\nu P^\mu M^{\rho\sigma} + \frac{1}{4}\epsilon_{\nu\mu\rho\sigma}P^\mu P^\nu M^{\rho\sigma}$$

$$+ \frac{i}{2}\epsilon_{\mu\nu\rho}{}^\mu P^\nu P^\rho - \frac{i}{2}\epsilon_{\mu\nu}{}^\mu{}_\sigma P^\nu P^\sigma$$

$$= \frac{1}{4}\epsilon_{\mu\nu\rho\sigma}[P^\nu, P^\mu]M^{\rho\sigma} \overset{(1.34)}{=} 0$$

The representation theory of the Poincaré group has been discussed extensively in the literature using the formalism of induced representations and the concept of little groups.[12] We do not go into details here, therefore, and simply quote the following results [103], [106]:

The unitary (infinite-dimensional) representations of the Poincaré group can be split into three main classes. These are:

(a)

$$P^2 = P_\mu P^\mu = m^2 > 0; \quad W^2 = -m^2 s(s+1). \qquad (1.43)$$

The eigenvalue of the second Casimir operator W^2 is $-m^2 s(s+1)$, where s denotes the spin which assumes discrete values

$$s = 0, 1/2, 1, 3/2, \ldots.$$

---

[12]In the mathematical context, little groups are called isotropy groups or stability subgroups. See *e.g.* the books by V.S. Varadarajan [115], Chap. 2.9, T. Bröcker and T. tom Dieck [19], and A.O Barut and R. Rączka [8], Chap. 16.

From Eq. (1.42) one deduces that in the rest frame ($P^\mu = (m, \mathbf{0})$) the zero component of the Pauli–Ljubanski vector must vanish, and the space components in the rest frame are given by

$$W_i = \frac{1}{2}\epsilon_{i0jk}P^0 S^{jk},$$

such that

$$W^2 = -\mathbf{W}^2 = -m^2\mathbf{S}^2, \qquad (1.44)$$

where

$$S^i = \frac{1}{2}\epsilon^{ijk}S_{jk} \qquad (1.45)$$

is the *spin operator*. This representation is specified in terms of the mass m and spin s. Physically a state in a representation (m, s) corresponds to a particle of rest mass m and spin s; moreover, since the spin projection s_3 can take on any value from $-s$ to $+s$, massive particles fall into $(2s + 1)$-dimensional multiplets.

(b)

$$P^2 = 0; \quad W^2 = 0. \qquad (1.46)$$

In this case, W and P are linearly dependent:

$$W_\mu = \lambda P_\mu. \qquad (1.47)$$

The constant of proportionality is called the *helicity* and is equal to $\pm s$ where $s = 0, 1/2, 1, \ldots$ is the spin of the representation. The time component of W^μ is

$$W^0 = \frac{1}{2}\epsilon^{0ijk}P_i M_{jk} \overset{(1.12)}{=} \mathbf{P}\cdot\mathbf{J}, \qquad (1.48)$$

so that Eq. (1.47) implies

$$\lambda = \frac{\mathbf{P}\cdot\mathbf{J}}{P_0}, \qquad (1.49)$$

which is the definition of the helicity of a massless particle. Examples of particles which fall into this category are the photon with spin 1 and helicity states ± 1, and the neutrino with spin 1/2 and helicity states $\pm 1/2$.

(c)

$$P^2 = 0; \quad W^2 = -\rho^2. \qquad (1.50)$$

This type of representation describes a particle of rest mass zero with an infinite number of polarization states labeled by the continuous variable ρ. These representations do not seem to be realized in nature.

Remark: For this case the calculations of case (a) do not work since we cannot make a transformation to the rest frame. However, we can always transform to a system where

$$P_\mu = (P_0, 0, 0, P_0).$$

If ω_μ is the eigenvalue of the Pauli–Ljubanski vector W_μ, Eq. (1.42) implies

$$0 = \omega_\mu P^\mu = \omega^0 P^0 - \omega^3 P^0,$$

i.e.

$$\omega^0 = \omega^3$$

and

$$\omega^2 = \eta^{\mu\nu}\omega_\mu\omega_\nu = \omega_0^2 - \boldsymbol{\omega}^2 = -(\omega_1^2 + \omega_2^2) = -\rho^2.$$

Thus in this case the eigenvalues of the Casimir operator W^2 can assume any value.

1.3   SL(2, ℂ), Dotted and Undotted Indices

1.3.1   Spinor Algebra

We consider the special linear group in two dimensions with complex parameters,

$$SL(2, \mathbb{C}) := \{ M \in GL(2, \mathbb{C}) \mid \det M = +1 \}. \qquad (1.51)$$

A *linear representation* of this group is a map from the group $SL(2, \mathbb{C})$ into the automorphim group of a certain vector space[13] F. This means

$$M \in SL(2, \mathbb{C}) \longrightarrow D(M). \qquad (1.52)$$

The automorphism group being defined as the set of linear bijective maps from F to F, the group multiplication being the composition of maps. As

---

[13]The reader who wants to refresh his memory of definitions of mathematical terms without delving into mathematical texts is advised to consult the article by A.S. Sciarrino and P. Sorba [104] which is a "Junior Dictionary" of group theory concepts commonly met in particle physics. A rigorous mathematical treatment of these subjects can be found in the textbook by S. Lang [66].

usual we demand the representation properties

$$D(\mathbb{1}_{SL(2,\mathbb{C})}) = \mathbb{1}_F, \tag{1.53}$$

$$D(M_1)D(M_2) = D(M_1 \cdot M_2), \quad \forall M_1, M_2 \in SL(2, \mathbb{C}), \tag{1.54}$$

where $\mathbb{1}_{SL(2,\mathbb{C})}$ is the unit element of the group $SL(2, \mathbb{C})$ and $\mathbb{1}_F$ is the identity map in the vector space F.

Let ψ be any element of F and $\{\widehat{e}_i\}_{i=1,\dots,\dim F}$ the canonical basis in F. Then

$$\psi = \sum_{n=1}^{\dim F} \psi_n \widehat{e}_n, \tag{1.55}$$

and representation matrices act on ψ in the following way:

$$D(M)\psi = \sum_{n=1}^{\dim F} \psi'_n \widehat{e}_n, \tag{1.56}$$

where

$$\psi'_n = \sum_{i=1}^{\dim F} D_n{}^i(M)\psi_i. \tag{1.57}$$

The set $(D_n{}^m(M))$ are $\dim F \times \dim F$–dimensional matrices called *representation matrices*; $\dim F$ is called the *dimension of the representation*.

Two representations $D^{(1)}, D^{(2)}$ are called *equivalent*, if an invertible $\dim F \times \dim F$–matrix U can be found, such that

$$D^{(1)}(M) = UD^{(2)}(M)U^{-1}, \quad U \in GL(F, K), \quad K = \mathbb{R} \text{ or } \mathbb{C}. \tag{1.58}$$

The special linear group $SL(2, \mathbb{C})$ admits two inequivalent spinor representations, as we shall see:[14]

(a) The self-representation:
The self-representation is defined by

$$D(M) := M, \quad \forall M \in SL(2, \mathbb{C}). \tag{1.59}$$

The dimension of the self-representation is therefore two. According to Eq. (1.57) elements of the representation space $\psi \in F$ transform under the self-representation as[15]

$$\psi'_A = M_A{}^B \psi_B, \quad A, B = 1, 2. \tag{1.60}$$

---

[14]In the context of supergravity and superstrings one considers supersymmetric theories in more than four spacetime dimensions. For a discussion of properties of spinors and γ-matrices in an arbitrary number of spacetime dimensions see A. Pais [85] or P. van Nieuwenhuizen [114].

[15]We adopt the Einstein summation convention also for spinor indices $A, B, \dots$ which means summation over repeated indices.

In the literature the elements $\psi \in F$ transforming according to Eq. (1.60) under $SL(2,\mathbb{C})$ are called *left-handed Weyl spinors* or *covariant Weyl spinors*, and this representation[16] is denoted by $(1/2, 0)$.

(b) The complex conjugate self-representation:
The so-called complex conjugate self-representation is defined by

$$D(M) := M^*, \quad \forall M \in SL(2,\mathbb{C}), \tag{1.61}$$

where M^* means complex conjugation. As in case (a), the dimension of the representation space is two. We call the corresponding representation space $\dot{F}$. Elements of this representation space are denoted by $\overline{\psi}$ and transform under $SL(2,\mathbb{C})$ according to

$$\overline{\psi}'_{\dot{A}} = (M^*)_{\dot{A}}{}^{\dot{B}}\,\overline{\psi}_{\dot{B}}, \quad \dot{A}, \dot{B} = \dot{1}, \dot{2}, \tag{1.62}$$

where $\overline{\psi}_{\dot{B}} \in \dot{F}$. The elements $\overline{\psi}_{\dot{A}} \in \dot{F}$ transforming according to Eq. (1.62) are called *right-handed Weyl spinors*, and the representation is denoted by $(0, 1/2)$.

Left- and right-handed Weyl spinors are related by complex conjugation, *i.e.* taking $\psi_A \in F$, then (see also Eq. (1.200))

$$(\psi_A)^* =: \overline{\psi}^{\dot{A}} \in \dot{F}. \tag{1.63}$$

It should be observed that this equation does not exhibit the same index structure on both sides. As long as we deal only with Weyl spinors this does not give rise to difficulties. A relation consistent with Eq. (1.63) which does exhibit the same index structure on both sides will be given in Sec. 1.4.4. The self-representation and its complex conjugate representation are inequivalent, *i.e.* it is not possible to find a 2×2-matrix C, such that

$$M = CM^*C^{-1}.$$

Proposition 1.11:

The representation

$$D(M) = M^{-1\top} \tag{1.64}$$

is equivalent to the self-representation given by Eq. (1.59).

---

[16]See *e.g.* the article by H.P. Nilles [77] or the textbook by R.U Sexl and H.K. Urbantke [106], Chap. 8.3.

Proof: We have to show the existence of a matrix ϵ such that

$$\epsilon M \epsilon^{-1} = M^{-1\top}, \qquad \epsilon \in GL(2,\mathbb{C}). \tag{1.65}$$

Let

$$\epsilon = (\epsilon_{AB}) = \begin{pmatrix} \epsilon_{11} & \epsilon_{12} \\ \epsilon_{21} & \epsilon_{22} \end{pmatrix},$$

and

$$\epsilon^{-1} = (\epsilon_{AB})^{-1} = \frac{1}{\det \epsilon} \begin{pmatrix} \epsilon_{22} & -\epsilon_{12} \\ -\epsilon_{21} & \epsilon_{11} \end{pmatrix},$$

and

$$M := \begin{pmatrix} M_{11} & M_{12} \\ M_{21} & M_{22} \end{pmatrix}.$$

Then

$$M^{-1\top} = \frac{1}{\det M} \begin{pmatrix} M_{22} & -M_{12} \\ -M_{21} & M_{11} \end{pmatrix}^{\top} = \begin{pmatrix} M_{22} & -M_{21} \\ -M_{12} & M_{11} \end{pmatrix},$$

since $M \in SL(2,\mathbb{C})$, and so $\det M = +1$. Then we have to show that the matrix equation

$$\frac{1}{\det \epsilon} \begin{pmatrix} \epsilon_{11} & \epsilon_{12} \\ \epsilon_{21} & \epsilon_{22} \end{pmatrix} \begin{pmatrix} M_{11} & M_{12} \\ M_{21} & M_{22} \end{pmatrix} \begin{pmatrix} \epsilon_{22} & -\epsilon_{12} \\ -\epsilon_{21} & \epsilon_{11} \end{pmatrix} \overset{!}{=} \begin{pmatrix} M_{22} & -M_{21} \\ -M_{12} & M_{11} \end{pmatrix}$$

leads to a consistent set of matrix elements ϵ_{AB}. Evaluating the product of the three matrices we have:

$$(\det \epsilon)^{-1} \begin{pmatrix} a & b \\ c & d \end{pmatrix} \overset{!}{=} \begin{pmatrix} M_{22} & -M_{21} \\ -M_{12} & M_{11} \end{pmatrix},$$

with

$$a = M_{11}\epsilon_{11}\epsilon_{22} - M_{12}\epsilon_{11}\epsilon_{21} + M_{21}\epsilon_{12}\epsilon_{22} - M_{22}\epsilon_{12}\epsilon_{21},$$
$$b = -M_{11}\epsilon_{11}\epsilon_{12} + M_{12}\epsilon_{11}\epsilon_{22} - M_{21}\epsilon_{12}^2 + M_{22}\epsilon_{12}\epsilon_{22},$$
$$c = M_{11}\epsilon_{21}\epsilon_{22} - M_{12}\epsilon_{21}^2 + M_{21}\epsilon_{22}^2 - M_{22}\epsilon_{22}\epsilon_{21},$$
$$d = -M_{11}\epsilon_{12}\epsilon_{21} + M_{12}\epsilon_{21}\epsilon_{22} - M_{21}\epsilon_{12}\epsilon_{22} + M_{22}\epsilon_{22}^2.$$

This leads to the following set of equations:

$$M_{11}\epsilon_{11}\epsilon_{22} - M_{12}\epsilon_{11}\epsilon_{21} + M_{21}\epsilon_{12}\epsilon_{22} - M_{22}\epsilon_{12}\epsilon_{21} = M_{22} \det \epsilon,$$
$$M_{11}\epsilon_{21}\epsilon_{22} - M_{12}\epsilon_{21}^2 + M_{21}\epsilon_{22}^2 - M_{22}\epsilon_{22}\epsilon_{21} = -M_{12} \det \epsilon,$$
$$-M_{11}\epsilon_{11}\epsilon_{12} + M_{12}\epsilon_{11}\epsilon_{22} - M_{21}\epsilon_{12}^2 + M_{22}\epsilon_{12}\epsilon_{22} = -M_{21} \det \epsilon,$$
$$-M_{11}\epsilon_{12}\epsilon_{21} + M_{12}\epsilon_{21}\epsilon_{22} - M_{21}\epsilon_{12}\epsilon_{22} + M_{22}\epsilon_{22}^2 = M_{11} \det \epsilon.$$

From the first equation we deduce by equating coefficients of M_{AB} on both sides

$$\epsilon_{11} = \epsilon_{22} = 0 \qquad \text{and} \qquad -\epsilon_{21}\epsilon_{12} = \det \epsilon.$$

From the second:

$$\epsilon_{21}^2 = \det \epsilon,$$

and from the third

$$\epsilon_{12}^2 = \det \epsilon.$$

Then

$$-\epsilon_{12}\epsilon_{21} = \det \epsilon = \epsilon_{12}^2,$$

i.e.

$$-\epsilon_{12} = \epsilon_{21}.$$

We choose $\epsilon_{12} = -1$ so that $\epsilon_{21} = +1$, and we have:

$$(\epsilon_{AB}) = \begin{pmatrix} 0 & -1 \\ 1 & 0 \end{pmatrix} =: (\epsilon^{AB})^{-1}, \tag{1.66}$$

where the right hand side defines a matrix with upper indices. Then the matrix ϵ with upper indices is:

$$(\epsilon^{AB}) = \begin{pmatrix} 0 & 1 \\ -1 & 0 \end{pmatrix} = (\epsilon_{AB})^{\top}, \tag{1.67}$$

such that

$$(\epsilon^{AB})(\epsilon^{AB})^{-1} \equiv \epsilon \cdot \epsilon^{-1} = \mathbb{1}_{2 \times 2}$$

implies

$$\begin{pmatrix} 0 & 1 \\ -1 & 0 \end{pmatrix} \begin{pmatrix} 0 & -1 \\ 1 & 0 \end{pmatrix} = \begin{pmatrix} 1 & 0 \\ 0 & 1 \end{pmatrix},$$

or in mixed form:

$$\epsilon^{AB}\epsilon_{BC} = \delta^A{}_C, \qquad \epsilon_{AB}\epsilon^{BC} = \delta_A{}^C, \qquad \epsilon^{\top}_{AB}\epsilon^{BC} = -\delta_A{}^C. \tag{1.68}$$

With this index convention, *i.e.* writing ϵ with upper and ϵ^{-1} with lower indices, so that ϵ plays the role of a metric, we can write Eq. (1.65)

$$\epsilon^{AB}M_B{}^C\epsilon_{CD} = (M^{-1\top})^A{}_D. \tag{1.69}$$

Thus we have constructed a 2×2-matrix ϵ, such that

$$\epsilon M \epsilon^{-1} = M^{-1\top}.$$

Hence, the two representations M and $M^{-1\top}$ are equivalent.

Multiplying Eq. (1.69) by ϵ_{EA} from the left and ϵ^{DF} from the right, we obtain:

$$\epsilon_{EA}\epsilon^{AB}M_B{}^C\epsilon_{CD}\epsilon^{DF} = \epsilon_{EA}(M^{-1\top})^A{}_D\epsilon^{DF},$$

i.e.

$$\delta_E{}^B M_B{}^C \delta_C{}^F = \epsilon_{EA}(M^{-1\top})^A{}_D\epsilon^{DF},$$

i.e.

$$M_E{}^F = \epsilon_{EA}(M^{-1\top})^A{}_D\epsilon^{DF}, \tag{1.70}$$

where we used Eq. (1.68).

Now, taking a $\psi_A \in F$, we know from Eq. (1.60) that this spinor transforms as

$$\psi'_A = M_A{}^B\psi_B \stackrel{(1.70)}{=} \epsilon_{AC}(M^{-1\top})^C{}_D\epsilon^{DB}\psi_B.$$

Multiplying this equation by ϵ^{EA} from the left and again using Eq. (1.68) we arrive at:

$$\epsilon^{EA}\psi'_A = (M^{-1\top})^E{}_D\epsilon^{DB}\psi_B. \tag{1.71}$$

We next define a left-handed Weyl spinor with contravariant spinor index by

$$\psi^A := \epsilon^{AB}\psi_B. \tag{1.72}$$

Then Eq. (1.71) reads:

$$\psi'^A = (M^{-1\top})^A{}_B\psi^B. \tag{1.73}$$

Proposition 1.12:

The representation

$$D(M) =: M^{*-1\top} \tag{1.74}$$

is equivalent to the complex conjugate representation given by Eq. (1.61).

Proof: As in Eq. (1.65) we have to search for a 2×2-matrix $\bar{\epsilon}$ such that

$$\bar{\epsilon}M^*\bar{\epsilon}^{-1} = M^{*-1\top}. \tag{1.75}$$

In the same manner as for Eq. (1.65) one can show that

$$\bar{\epsilon} = \begin{pmatrix} 0 & 1 \\ -1 & 0 \end{pmatrix} =: \left(\epsilon^{\dot{A}\dot{B}}\right) \tag{1.76a}$$

and

$$\bar{\epsilon}^{-1} = \begin{pmatrix} 0 & -1 \\ 1 & 0 \end{pmatrix} =: (\epsilon_{\dot{A}\dot{B}}). \tag{1.76b}$$

In index notation Eq. (1.75) is written

$$\epsilon^{\dot{A}\dot{B}}(M^*)_{\dot{B}}{}^{\dot{C}}\epsilon_{\dot{C}\dot{D}} = (M^{*-1\top})^{\dot{A}}{}_{\dot{D}}. \tag{1.77}$$

Multiplying from the left by $\epsilon_{\dot{E}\dot{A}}$ and from the right by $\epsilon^{\dot{D}\dot{F}}$, and using

$$\epsilon^{\dot{A}\dot{B}}\epsilon_{\dot{B}\dot{C}} = \delta^{\dot{A}}{}_{\dot{C}}, \tag{1.78a}$$

$$\epsilon_{\dot{A}\dot{B}}\epsilon^{\dot{B}\dot{C}} = \delta_{\dot{A}}{}^{\dot{C}}, \tag{1.78b}$$

we obtain:

$$(M^*)_{\dot{A}}{}^{\dot{B}} = \epsilon_{\dot{A}\dot{C}}(M^{*-1\top})^{\dot{C}}{}_{\dot{D}}\epsilon^{\dot{D}\dot{B}}. \tag{1.79}$$

Using Eq. (1.62) we have the transformation property of dotted Weyl spinors (right handed Weyl spinor)

$$\overline{\psi}'_{\dot{A}} = (M^*)_{\dot{A}}{}^{\dot{B}}\overline{\psi}_{\dot{B}} = \epsilon_{\dot{A}\dot{C}}(M^{*-1\top})^{\dot{C}}{}_{\dot{D}}\epsilon^{\dot{D}\dot{B}}\overline{\psi}_{\dot{B}}.$$

Multiplying this equation from the left by $\epsilon^{\dot{E}\dot{A}}$ we obtain

$$\epsilon^{\dot{E}\dot{A}}\overline{\psi}'_{\dot{A}} = \epsilon^{\dot{E}\dot{A}}\epsilon_{\dot{A}\dot{C}}(M^{*-1\top})^{\dot{C}}{}_{\dot{D}}\epsilon^{\dot{D}\dot{B}}\overline{\psi}_{\dot{B}} \stackrel{(1.78a)}{=} (M^{*-1\top})^{\dot{E}}{}_{\dot{D}}\epsilon^{\dot{D}\dot{B}}\overline{\psi}_{\dot{B}}.$$

Defining dotted spinors with contravariant indices by the relation

$$\overline{\psi}^{\dot{A}} := \epsilon^{\dot{A}\dot{B}}\overline{\psi}_{\dot{B}}, \tag{1.80}$$

we conclude that dotted spinors with contravariant indices transform under $M^{*-1\top}$ according to:

$$\overline{\psi}^{\dot{A}} = (M^{*-1\top})^{\dot{A}}{}_{\dot{B}}\overline{\psi}^{\dot{B}}. \tag{1.81}$$

Summarizing we have: For any $M \in SL(2, \mathbb{C})$ the matrix M, its complex conjugate M^*, the inverse of its transpose $(M^\top)^{-1}$, and the inverse of its Hermitian conjugate $(M^\dagger)^{-1}$ all represent the group $SL(2, \mathbb{C})$. Two-component spinors with covariant and contravariant dotted or undotted spinor indices transform under $SL(2, \mathbb{C})$ as follows:

$$\psi'_A = M_A{}^B\psi_B, \tag{1.82a}$$

$$\psi'^A = (M^{-1\top})^A{}_B\psi^B, \tag{1.82b}$$

$$\overline{\psi}'_{\dot{A}} = (M^*)_{\dot{A}}{}^{\dot{B}}\overline{\psi}_{\dot{B}}, \tag{1.82c}$$

$$\overline{\psi}'^{\dot{A}} = (M^{*-1\top})^{\dot{A}}{}_{\dot{B}}\overline{\psi}^{\dot{B}}. \tag{1.82d}$$

The raising and lowering of spinor indices has to be understood in the following way:

(i) Given any spinor which transforms under the special linear group $SL(2,\mathbb{C})$ in the self-representation, ψ_A, we can construct a contravariant spinor ψ^A which transforms under $M^{-1\top}$:

$$\psi^A = \epsilon^{AB}\psi_B = -\psi_B\epsilon^{BA}. \tag{1.83}$$

(ii) For a Weyl spinor which transforms under $M^{-1\top}$, ψ^A, we have

$$\psi_A = \epsilon_{AB}\psi^B. \tag{1.84}$$

(iii) Given any Weyl spinor which transforms under $SL(2,\mathbb{C})$ according to M^*, $\overline{\psi}_{\dot{A}}$

$$\overline{\psi}^{\dot{A}} = \epsilon^{\dot{A}\dot{B}}\overline{\psi}_{\dot{B}} = -\overline{\psi}_{\dot{B}}\epsilon^{\dot{B}\dot{A}}. \tag{1.85}$$

(iv) For a Weyl spinor in the representation $(0, 1/2)$ which transforms under $(M^\dagger)^{-1}$:

$$\overline{\psi}_{\dot{A}} = \epsilon_{\dot{A}\dot{B}}\overline{\psi}^{\dot{B}}. \tag{1.86}$$

The raising and lowering of spinor indices is performed with the help of the matrices ϵ and ϵ^{-1}, which also connect the two equivalent representations M and $M^{-1\top}$. Hence the matrix ϵ plays the role of a *metric in the spinor space* F. Explicitly, the raising and lowering of the spinor indices is given by:

(i) Raising undotted indices:

$$\psi^A = \epsilon^{AB}\psi_B,$$
$$\psi^1 = \epsilon^{1B}\psi_B = \epsilon^{12}\psi_2 = \psi_2,$$

since $\epsilon^{12} = +1$ (see Eq. (1.67)) and

$$\psi^2 = \epsilon^{2B}\psi_B = \epsilon^{21}\psi_1 = -\psi_1.$$

(ii) Lowering undotted indices:

$$\psi_A = \epsilon_{AB}\psi^B,$$
$$\psi_1 = \epsilon_{1B}\psi^B = \epsilon_{12}\psi^2 = -\psi^2,$$

since $\epsilon_{12} = -1$,

$$\psi_2 = \epsilon_{21}\psi^1 = \psi^1.$$

(iii) Raising dotted indices:

$$\overline{\psi}^{\dot{A}} = \epsilon^{\dot{A}\dot{B}}\overline{\psi}_{\dot{B}},$$

$$\overline{\psi}^{\dot{1}} = \epsilon^{\dot{1}\dot{B}}\overline{\psi}_{\dot{B}} = \epsilon^{\dot{1}\dot{2}}\overline{\psi}_{\dot{2}} = \overline{\psi}_{\dot{2}},$$

where $\epsilon^{\dot{1}\dot{2}} = +1$ (see Eq. (1.76a))

$$\overline{\psi}^{\dot{2}} = \epsilon^{\dot{2}\dot{B}}\overline{\psi}_{\dot{B}} = \epsilon^{\dot{2}\dot{1}}\overline{\psi}_{\dot{1}} = -\overline{\psi}_{\dot{1}}.$$

(iv) Lowering dotted indices:

$$\overline{\psi}_{\dot{A}} = \epsilon_{\dot{A}\dot{B}}\overline{\psi}^{\dot{B}},$$

$$\overline{\psi}_{\dot{1}} = \epsilon_{\dot{1}\dot{B}}\overline{\psi}^{\dot{B}} = \epsilon_{\dot{1}\dot{2}}\overline{\psi}^{\dot{2}} = -\overline{\psi}^{\dot{2}},$$

$$\overline{\psi}_{\dot{2}} = \epsilon_{\dot{2}\dot{B}}\overline{\psi}^{\dot{B}} = \epsilon_{\dot{2}\dot{1}}\overline{\psi}^{\dot{1}} = \overline{\psi}^{\dot{1}},$$

since $\epsilon_{\dot{1}\dot{2}} = -1, \epsilon_{\dot{2}\dot{1}} = +1$.

This is a consistent set of equations.

   We now define dual spaces[17] F^* and $\dot{F}^*$ and establish the connection between ψ^* and $\overline{\psi}$. As explained above, we have four types of two-component Weyl spinors transforming according to four different representations of the group $SL(2, \mathbb{C})$, where $M, M^{-1\top}$, and $M^*, M^{*-1\top}$ describe equivalent representations.

(i) Any $\psi \in F$ transforms under the self-representation (1.59) as

$$\psi_A' = M_A{}^B \psi_B.$$

Such spinors are characterized by a lower spinor index.

(ii) From Eq. (1.64) we know that spinors transforming according to

$$\psi'^A = (M^{-1\top})^A{}_B \psi^B$$

form a representation of $SL(2, \mathbb{C})$ that is equivalent to the self-representation. Such spinors carry upper undotted spinor indices and are elements of the dual vector space F^*.

---

[17]Here we adopt the mathematical standard notation for a dual space which is an asterisk. This should not be confused with complex conjugation.

(iii) A spinor transforming according to the complex conjugate self-representation is given by $\overline{\psi}_{\dot{A}} \in \dot{F}$ and is characterized by a lower dotted spinor index,

$$\overline{\psi}'_{\dot{A}} = (M^*)_{\dot{A}}{}^{\dot{B}}\overline{\psi}_{\dot{B}}.$$

(iv) Finally, an equivalent representation of $SL(2,\mathbb{C})$ is given by spinors which transform according to

$$\overline{\psi}'^{\dot{A}} = (M^{*-1\top})^{\dot{A}}{}_{\dot{B}}\overline{\psi}^{\dot{B}}.$$

Such spinors carry upper dotted spinor indices, *i.e.* contravaraint dotted indices and are elements of $\dot{F}^*$.

A mathematical framework for these different types of spinors is given by the following considerations.

Consider F as the vector space of two-component spinors ψ_A, we may construct the dual space F^* in the following way. According to linear algebra,[18] the elements of the dual space F^* are linear maps ϕ from F to a field K, which we take as the field of complex numbers $\mathbb{C}$:

$$\phi : F \longrightarrow \mathbb{C},$$

such that for all $\psi \in F$:

$$\phi(\psi) := \phi^A \psi_A \in \mathbb{C}. \tag{1.87}$$

Hence according to our index convention we may interpret two-component spinors with contravariant undotted indices as elements of the dual space F^*, *i.e.*

$$\psi_A \in F, \quad \psi^A \in F^*.$$

In addition we know from Eq. (1.72) how we can correlate an element of F to the corresponding element of F^*. The ϵ-matrix may be considered as a map

$$(\epsilon^{AB}) : F \longrightarrow F^*,$$
$$\psi_A \longmapsto \psi^A = \epsilon^{AB}\psi_B.$$

The inverse map is, of course, given by the inverse matrix, *i.e.* the ϵ-matrix with covariant spinor indices:

$$(\epsilon_{AB}) : F^* \longrightarrow F,$$
$$\psi^A \longmapsto \psi_A = \epsilon_{AB}\psi^B.$$

---

[18]See the book by S. Lang [64] for a mathematical discussion of these aspects.

Interpreting the composition on the right hand side of Eq. (1.87) as matrix multiplication, spinors with upper undotted indices represent *rows* and those with lower indices represent *columns*.

In the same way one can consider the space $\dot{F}^*$ as the vector space of two-component dotted spinors with upper indices $\overline{\psi}^{\dot{A}} \in \dot{F}^*$ and we have:

$$(\overline{\psi}_{\dot{A}}) \; : \; \dot{F}^* \longrightarrow \mathbb{C} \tag{1.88}$$

with

$$\overline{\psi}(\overline{\phi}) = \overline{\psi}_{\dot{A}} \overline{\phi}^{\dot{A}} \in \mathbb{C}. \tag{1.89}$$

Hence

$$\overline{\psi}_{\dot{A}} \in \dot{F}(\cong \dot{F}^{**}), \quad \text{and} \quad \overline{\phi}^{\dot{A}} \in \dot{F}^*.$$

Thus for dotted spinors we have to assign to $\overline{\psi}_{\dot{A}}$ a *row* and to spinors with upper dotted indices *columns*. We then have:

$$(\epsilon_{\dot{A}\dot{B}}) \; : \; \dot{F}^* \longrightarrow \dot{F},$$
$$\overline{\psi}^{\dot{B}} \longmapsto \overline{\psi}_{\dot{A}} = \epsilon_{\dot{A}\dot{B}} \overline{\psi}^{\dot{B}},$$

and

$$(\epsilon^{\dot{A}\dot{B}}) \; : \; \dot{F} \longrightarrow \dot{F}^*,$$
$$\overline{\psi}_{\dot{B}} \longmapsto \overline{\psi}^{\dot{A}} = \epsilon^{\dot{A}\dot{B}} \overline{\psi}_{\dot{B}}.$$

Furthermore, the correct description of the transition from F to $\dot{F}^*$ is given by complex conjugation and multiplication by the matrix[19] $\overline{\sigma}^0$

$$(\overline{\sigma}^0)^{\dot{A}B} \; : \; F \longrightarrow \dot{F}^*.$$

So we set:

$$(\overline{\sigma}^0)^{\dot{A}A}(\psi_A)^* = \overline{\psi}^{\dot{A}}. \tag{1.90}$$

The inverse map is found to be

$$(\sigma^0)_{A\dot{B}}(\overline{\psi}^{\dot{B}})^* = \psi_A. \tag{1.91}$$

In agreement with Eq. (1.106a) (to be given later) we also have

$$\psi^A = \overline{\psi}^*_{\dot{B}}(\overline{\sigma}^0)^{\dot{B}A},$$

---

[19]See Eq. (1.199) or Sec. 1.3.3 where the σ-matrices are introduced.

and

$$\overline{\psi}_{\dot{A}} = \psi^{B*}(\sigma^0)_{B\dot{A}}.$$

These relations show the connection between ψ_A and $\overline{\psi}^{\dot{A}}$.

Thus the four complex numbers $\psi_A, \overline{\psi}^{\dot{A}}$ where $A = 1, 2; \dot{A} = \dot{1}, \dot{2}$, do in fact, define only four real independent numbers which we can take to be real $\psi_A, \overline{\psi}^{\dot{A}}$ where $A = 1, 2; \dot{A} = \dot{1}, \dot{2}$.

We now consider *dual maps*. From linear algebra we recall the following results. Let V and W be vector spaces over a field K, *i.e.* ℝ or ℂ. For every linear map

$$\phi : V \longrightarrow W$$

we may construct the so-called dual map[20]

$$\phi^* : W^* \longrightarrow V^*$$

by the following prescription: Let $\psi \in W^*$ so that ψ is a map

$$\psi : W \longrightarrow K(= \mathbb{R}, \mathbb{C}).$$

Then we define the dual map by

$$\phi^*(\psi) = \psi \circ \phi.$$

This is to be understood in the following way. Applying the left hand side onto a vector $v \in V$ we have

$$v \in V \xrightarrow{\phi^*(\psi)} K,$$

where $\phi^*(\psi)$ is by construction an element of V^*, and this element maps $v \in V$ onto the real or complex numbers. The right hand side of this equation then gives:

$$v \in V \xrightarrow{\phi : V \longrightarrow W} \phi(v) \in W \xrightarrow{\psi \in W^*} (\psi \circ \phi)(v) \in K.$$

Hence the dual map ϕ^* is defined in such a way that the left and right hand sides of the equation give the same element in the field K when applied to an element of the vector space V. Furthermore, a linear map ϕ then implies a *linear dual map* ϕ^*.

It is important to note that if ϕ is a linear map from a vector space V to a vector space W then the dual map ϕ^* maps W^* onto V^*, *i.e.* dual

---

[20]Here again we use the standard notation of linear algebra for dual spaces and dual maps which should not be confused with complex conjugation.

maps have the 'opposite' direction. Describing ϕ and ϕ^* as matrices, we see that if ϕ corresponds to a matrix A, then the dual map ϕ^* is given by the transposed matrix $A^\top$. For a proof of this statement we refer to books on linear algebra (see for example [64]).

With the help of the mathematical concept of a dual map we may explain the transformation properties of the various types of two-component Weyl spinors; within this formalism we find a natural explanation of Eqs. (1.60), (1.73), (1.62), and (1.81).

We start by considering Eq. (1.60). Formally

$$M \; : \; F \longrightarrow F,$$
$$\psi_B \longmapsto \psi'_A = M_A{}^B \psi_B,$$

i.e. M is a map from the vector space F to the vector space F. According to the above formalism, the dual map is given by

$$\widetilde{M}^{-1} \; : \; F^* \longrightarrow F^*,$$
$$\widetilde{M}^{-1} = M^{-1\top} \quad \text{as stated earlier}$$

(the dual map is represented by the transposed matrix). Here we choose the inverse of the dual map instead of the dual map itself as shown in Fig. 1.1. The figure shows that $M^{-1\top}$ can be re-expressed in terms of a sequence of maps. We start with $\psi^A \in F^*$. Applying the matrix ϵ_{AB} we obtain the corresponding elements $\psi_A = \epsilon_{AB}\psi^B$ of F. Then with $\psi'_A = M_A{}^B \psi_B$ we obtain the transformed element $\psi'_A = M_A{}^B \psi_B = M_A{}^B \epsilon_{BC}\psi^C$ of F. Again applying the metric ϵ^{AB} we finally obtain the transformed element of F^*. The result of this composition of maps has to be the same as applying $M^{-1\top}$ to the spinor $\psi^A \in F^*$. Hence we conclude

$$(M^{-1\top})^A{}_D = \epsilon^{AB} M_B{}^C \epsilon_{CD}.$$

The map $M \; : \; F \longrightarrow F$ induces a corresponding map in the space $\dot{F}$ which can be constructed from M using the general prescription of complex conjugation, multiplication by the matrix $\overline{\sigma}^0$, and contracting with the metric ϵ.

Starting with

$$\psi'_A = M_A{}^B \psi_B, \quad \psi_B \in F,$$

$$\psi_A = \epsilon_{AB}\psi^B \qquad\qquad \psi'_A = M_A{}^B\psi_B = M_A{}^B\epsilon_{BC}\psi^C$$

$$F \xrightarrow{\qquad M \qquad} F$$

$$\epsilon_{AB} \uparrow \qquad\qquad \downarrow \epsilon^{AB}$$

$$F^* \underset{M^\mathsf{T}}{\overset{M^{-1\mathsf{T}}}{\rightleftarrows}} F^*$$

$$\psi^A \qquad\qquad \psi'^A = \epsilon^{AB}\psi'_B$$
$$= \epsilon^{AB}M_B{}^C\epsilon_{CD}\psi^D$$
$$= (M^{-1\mathsf{T}})^A{}_B\psi^B$$

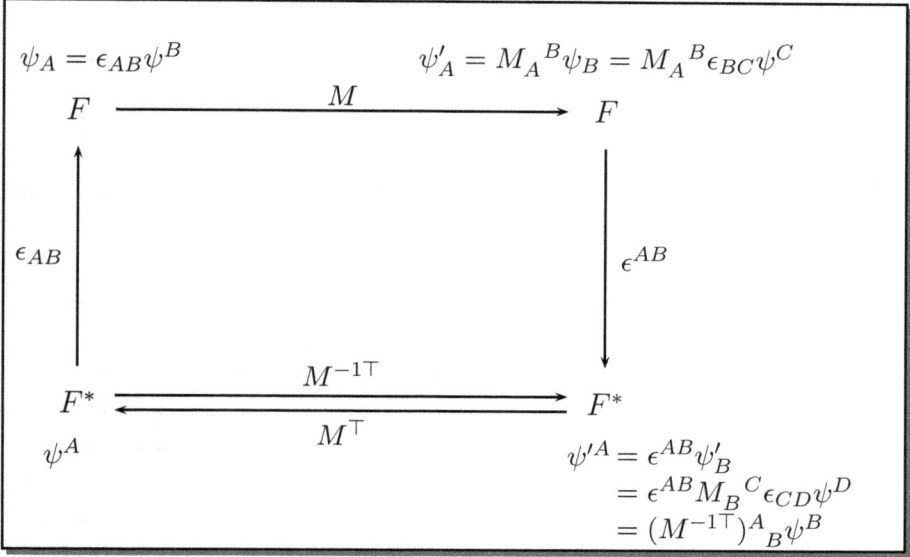

Figure 1.1: Directions of dual maps.

we obtain the corresponding transformation in $\dot{F}^*$:

$$\begin{aligned}
\overline{\psi}'^{\dot{A}} &= (\overline{\sigma}^0)^{\dot{A}A}(\psi'_A)^* \\
&= (\overline{\sigma}^0)^{\dot{A}A}(M_A{}^B\psi_B)^* \\
&= (\overline{\sigma}^0)^{\dot{A}A}(M_A{}^B)^*\psi_B^* \\
&= (\overline{\sigma}^0)^{\dot{A}A}(M_A{}^B)^*(\sigma^0)_{B\dot{C}}\overline{\psi}^{\dot{C}},
\end{aligned}$$

using Eqs. (1.90) and (1.91).

Contracting this equation with the metric $\epsilon_{\dot{D}\dot{A}}$, we obtain the transformation in $\dot{F}$, *i.e.*

$$\begin{aligned}
\overline{\psi}'_{\dot{A}} &= \epsilon_{\dot{A}\dot{B}}\overline{\psi}'^{\dot{B}} \\
&= \epsilon_{\dot{A}\dot{B}}(\overline{\sigma}^0)^{\dot{B}A}(M_A{}^B)^*(\sigma^0)_{B\dot{C}}\delta^{\dot{C}}{}_{\dot{D}}\overline{\psi}^{\dot{D}} \\
&= \epsilon_{\dot{A}\dot{B}}(\overline{\sigma}^0)^{\dot{B}A}(M_A{}^B)^*(\sigma^0)_{B\dot{C}}\epsilon^{\dot{C}\dot{E}}\epsilon_{\dot{E}\dot{D}}\overline{\psi}^{\dot{D}} \\
&= \epsilon_{\dot{A}\dot{B}}(\overline{\sigma}^0)^{\dot{B}A}(M_A{}^B)^*(\sigma^0)_{B\dot{C}}\epsilon^{\dot{C}\dot{E}}\overline{\psi}_{\dot{E}}.
\end{aligned}$$

Defining

$$(M^*)_{\dot{A}}{}^{\dot{B}} := \epsilon_{\dot{A}\dot{C}}(\overline{\sigma}^0)^{\dot{C}A}(M_A{}^B)^*(\sigma^0)_{B\dot{D}}\epsilon^{\dot{D}\dot{B}},$$

then

$$\overline{\psi}'_{\dot{A}} = (M^*)_{\dot{A}}{}^{\dot{B}}\overline{\psi}_{\dot{B}},$$

and

$$M^* : \dot{F} \longrightarrow \dot{F},$$

in agreement with Eq. (1.62). The relation between the various spinors and their respective representation spaces is depicted in Fig. 1.2.

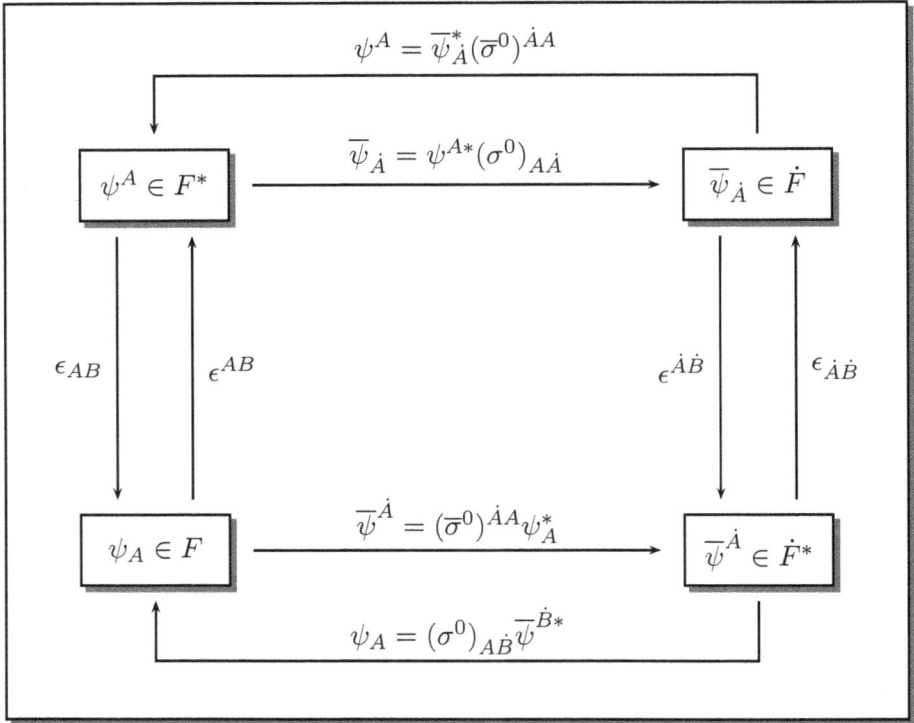

Figure 1.2: Weyl spinors and their respective representation spaces.

Remark: For the matrix

$$(M_A{}^B) = \begin{pmatrix} m_1{}^1 & m_1{}^2 \\ m_2{}^1 & m_2{}^2 \end{pmatrix} \in SL(2,\mathbb{C})$$

the above definition of the matrix $((M^*)_{\dot{A}}{}^{\dot{B}})$ implies:

$$m_{\dot{1}}{}^{\dot{1}} := m^*{}_2{}^2, \qquad m_{\dot{1}}{}^{\dot{2}} := -m^*{}_2{}^1,$$
$$m_{\dot{2}}{}^{\dot{1}} := -m^*{}_1{}^2, \qquad m_{\dot{2}}{}^{\dot{2}} := m^*{}_1{}^1.$$

Note that the above unconventional definition of M^* is due to our index convention; the connection between ψ_A and $\overline{\psi}^{\dot{A}}$ plays a particular role.

Again the transformation in $\dot{F}$ leads to a dual map in the dual space $\dot{F}^*$. This corresponding dual map can be derived by contraction with the metric. Given

$$\overline{\psi}'_{\dot{A}} = (M^*)_{\dot{A}}{}^{\dot{B}}\overline{\psi}_{\dot{B}},$$

and contracting both sides with $\epsilon^{\dot{A}\dot{B}}$, we obtain:

$$\epsilon^{\dot{A}\dot{B}}\overline{\psi}'_{\dot{B}} = \overline{\psi}'^{\dot{A}} = \epsilon^{\dot{A}\dot{B}}(M^*)_{\dot{B}}{}^{\dot{C}}\delta_{\dot{C}}{}^{\dot{D}}\overline{\psi}_{\dot{D}}$$

$$= \epsilon^{\dot{A}\dot{B}}(M^*)_{\dot{B}}{}^{\dot{C}}\epsilon_{\dot{C}\dot{E}}\epsilon^{\dot{E}\dot{D}}\overline{\psi}_{\dot{D}} = \epsilon^{\dot{A}\dot{B}}(M^*)_{\dot{B}}{}^{\dot{C}}\epsilon_{\dot{C}\dot{E}}\overline{\psi}^{\dot{E}}$$

$$=: (M^{*-1\top})^{\dot{A}}{}_{\dot{E}}\overline{\psi}^{\dot{E}},$$

where

$$\left(M^{*-1\top}\right)^{\dot{A}}{}_{\dot{B}} = \epsilon^{\dot{A}\dot{C}}(M^*)_{\dot{C}}{}^{\dot{D}}\epsilon_{\dot{D}\dot{B}},$$

and

$$M^{*-1\top} \;:\; \dot{F}^* \;\longrightarrow\; \dot{F}^*.$$

This is in agreement with Eq. (1.81). Again, for $M \in SL(2, \mathbb{C})$ with

$$(M_A{}^B) = \begin{pmatrix} m_1{}^1 & m_1{}^2 \\ m_2{}^1 & m_2{}^2 \end{pmatrix},$$

the elements of the matrix $M^{*-1\top}$ are given by:

$$m^{\dot{1}}{}_{\dot{1}} := m^*{}_1{}^1, \qquad m^{\dot{1}}{}_{\dot{2}} := m^*{}_1{}^2,$$

$$m^{\dot{2}}{}_{\dot{1}} := m^*{}_2{}^1, \qquad m^{\dot{2}}{}_{\dot{2}} := m^*{}_2{}^2,$$

where

$$\left((M^{*-1\top})^{\dot{A}}{}_{\dot{B}}\right) = \begin{pmatrix} m^{\dot{1}}{}_{\dot{1}} & m^{\dot{1}}{}_{\dot{2}} \\ m^{\dot{2}}{}_{\dot{1}} & m^{\dot{2}}{}_{\dot{2}} \end{pmatrix}.$$

In deriving the matrix M^* we see that we cannot describe complex conjugation as a *linear* map from F to $\dot{F}$. This can, of course, also be shown explicitly by considering a similarity transformation as in the previous equivalence proofs, and showing that an appropriate matrix does not exist.[21] Therefore F and $\dot{F}$ define two inequivalent representation spaces for the special linear group $SL(2, \mathbb{C})$. On the other hand elements of the corresponding dual spaces F^* and $\dot{F}^*$ can be obtained with the metrics ϵ_{AB} and $\epsilon_{\dot{A}\dot{B}}$ respectively. These are matrix representations of linear maps and therefore F and F^* are equivalent representation spaces as well as $\dot{F}$ and $\dot{F}^*$.

---

[21]We shall see later that such a linear transformation exists in the four-dimensional Dirac formalism.

1.3.2   Calculations with Spinors

We now consider a number of explicit calculations with spinors and derive several useful formulas.

Proposition 1.13:

Let $\psi \in F^*, \chi \in F$, F being the representation space of $SL(2,\mathbb{C})$. The quadratic form

$$(\psi\chi) := \psi^A \chi_A \tag{1.92a}$$

is invariant under transformations of $SL(2,\mathbb{C})$.

Proof: We have to show that

$$(\psi'\chi') = (\psi\chi),$$

where

$$\psi' = M^{-1\mathsf{T}}\psi \quad \text{and} \quad \chi' = M\chi.$$

Consider (using Eqs. (1.82a) and (1.82b))

$$(\psi'\chi') = \psi'^A \chi'_A = \left(M^{-1\mathsf{T}}\right)^A{}_B \psi^B M_A{}^C \chi_C = \psi^B \left(M^{-1}\right)_B{}^A M_A{}^C \chi_C$$
$$= \psi^B \delta_B{}^C \chi_C = \psi^B \chi_B = (\psi\chi).$$

Similarly

$$\psi'_A \chi'^A = \psi_A \chi^A.$$

Proposition 1.14:

The quadratic form

$$(\overline{\psi}\overline{\chi}) := \overline{\psi}_{\dot{A}} \overline{\chi}^{\dot{A}} \tag{1.92b}$$

is invariant under $SL(2,\mathbb{C})$.

Proof: We have (using Eqs. (1.82c) and (1.82d))

$$(\overline{\psi}'\overline{\chi}') = \overline{\psi}'_{\dot{A}} \overline{\chi}'^{\dot{A}} = \left(M^*\right)_{\dot{A}}{}^{\dot{B}} \overline{\psi}_{\dot{B}} \left(M^{*-1\mathsf{T}}\right)^{\dot{A}}{}_{\dot{C}} \overline{\chi}^{\dot{C}}$$
$$= \overline{\psi}_{\dot{B}} \left(M^{*\mathsf{T}}\right)^{\dot{B}}{}_{\dot{A}} \left(M^{*-1\mathsf{T}}\right)^{\dot{A}}{}_{\dot{C}} \overline{\chi}^{\dot{C}} = \overline{\psi}_{\dot{B}} \delta^{\dot{B}}{}_{\dot{C}} \overline{\chi}^{\dot{C}} = \overline{\psi}_{\dot{B}} \overline{\chi}^{\dot{B}} \overset{(1.92b)}{=} (\overline{\psi}\overline{\chi}).$$

Remark: For calculations with Weyl spinors it is a useful convention to sum undotted indices from the upper left to the lower right ("north-west to south-east") and dotted indices from the lower left to the upper right ("south-west to north-east"), as indicated in Eqs. (1.92a) and (1.92b). Thus we write

$$(\psi\chi) := \psi^A \chi_A \quad \text{and} \quad (\overline{\psi}\overline{\chi}) := \overline{\psi}_{\dot{A}} \overline{\chi}^{\dot{A}}. \tag{1.93}$$

However, expressions such as Eq. (1.84), *i.e.* $\psi_A = \epsilon_{AB}\psi^B$, show that in general summations in the other direction are also possible.

Postulate

We postulate that the components of spinors are *Grassmann variables* also called *Grassmann numbers* or *Grassmann parameters*. This is not to be confused with anticommuting operators such as the spinor charges Q_A to be defined later. We demand:

$$\{\psi_A, \psi^B\} = \{\psi_A, \psi_B\} = \{\psi^A, \psi^B\} = 0, \qquad (1.94)$$

$$\{\overline{\chi}_{\dot{A}}, \overline{\chi}^{\dot{B}}\} = \{\overline{\chi}_{\dot{A}}, \overline{\chi}_{\dot{B}}\} = \{\overline{\chi}^{\dot{A}}, \overline{\chi}^{\dot{B}}\} = 0. \qquad (1.95)$$

and all mixed anticommutators vanish too, *e.g.*

$$\{\psi_A, \overline{\chi}_{\dot{B}}\} = 0. \qquad (1.96)$$

Here $\{ \quad , \quad \}$ denotes the anticommutator

$$\{\psi_A, \psi^B\} := \psi_A\psi^B + \psi^B\psi_B.$$

If we require the ψ_A's to anticommute, an expression like $(\psi\psi)$ does not vanish. Thus:

$$(\psi\psi) \stackrel{(1.92a)}{=} \psi^A\psi_A \stackrel{(1.83)}{=} \epsilon^{AB}\psi_B\psi_A = \epsilon^{12}\psi_2\psi_1 + \epsilon^{21}\psi_1\psi_2 = \psi_2\psi_1 - \psi_1\psi_2,$$

since $\epsilon^{12} = -\epsilon^{21} = +1$.

If the components ψ_A were required to commute, *i.e.* if $[\psi_1, \psi_2] = 0$, then $(\psi\psi) = 0$. Instead we assume that ψ_A's anticommute with each other and with other Grassmann variables[22] such as fermion fields and spinor charges. This requirement is, of course, nothing but an application of the spin statistic theorem: half-integer spin quantities obey Fermi–Dirac statistics whereas integer spin quantities obey Bose–Einstein statistics.

Proposition 1.15:

For anticommuting ψ's the quadratic form (1.92a) is symmetric on exchange of spinors, *i.e.*

$$(\psi\chi) = (\chi\psi). \qquad (1.97)$$

---

[22]For a readable introduction to this subject see R. Penrose [87], Chap. 11.

Proof: We have

$$(\psi\chi) \overset{(1.92a)}{=} \psi^A\chi_A \overset{(1.94)}{=} -\chi_A\psi^A \overset{\substack{(1.83)\\(1.84)}}{=} -\epsilon_{AB}\chi^B\epsilon^{AC}\psi_C$$

$$= -\chi^B(\epsilon^\top)_{BA}\epsilon^{AC}\psi_C \overset{(1.68)}{=} \chi_B\delta_B{}^C\psi_C = \chi^B\psi_B \overset{(1.92a)}{=} (\chi\psi).$$

Proposition 1.16:

The quadratic form for dotted spinors, *i.e.* Eq. (1.93), is symmetric on exchange, *i.e.*

$$(\overline{\psi}\overline{\chi}) = (\overline{\chi}\overline{\psi}), \tag{1.98}$$

if $\overline{\psi}$ and $\overline{\chi}$ are Grassmann variables.

Proof: We have:

$$(\overline{\psi}\overline{\chi}) \overset{(1.93)}{=} \overline{\psi}_{\dot{A}}\overline{\chi}^{\dot{A}} \overset{(1.95)}{=} -\overline{\chi}^{\dot{A}}\overline{\psi}_{\dot{A}} \overset{(1.85)}{=} -\epsilon^{\dot{A}\dot{B}}\overline{\chi}_{\dot{B}}\epsilon_{\dot{A}\dot{C}}\overline{\psi}^{\dot{C}}$$

$$= -\overline{\chi}_{\dot{B}}(\epsilon^\top)^{\dot{B}\dot{A}}\epsilon_{\dot{A}\dot{C}}\overline{\psi}^{\dot{C}} = \overline{\chi}_{\dot{B}}\epsilon^{\dot{B}\dot{A}}\epsilon_{\dot{A}\dot{C}}\overline{\psi}^{\dot{C}}$$

$$\overset{(1.78a)}{=} \overline{\chi}_{\dot{B}}\delta^{\dot{B}}{}_{\dot{C}}\overline{\psi}^{\dot{C}} = \overline{\chi}_{\dot{B}}\overline{\psi}^{\dot{B}} \overset{(1.93)}{=} (\overline{\chi}\overline{\psi}).$$

Proposition 1.17:

If θ is a Grassmann variable, then

$$\theta^2 := (\theta\theta) = -2\theta^1\theta^2, \tag{1.99a}$$

$$\overline{\theta}^2 := (\overline{\theta}\,\overline{\theta}) = 2\overline{\theta}_{\dot{1}}\overline{\theta}_{\dot{2}}, \tag{1.99b}$$

and hence $(\theta\theta)\theta^A = 0$ *etc.*

Proof: We have[23]

$$\theta^2 = (\theta\theta) \overset{(1.92a)}{=} \theta^A\theta_A \overset{(1.84)}{=} \theta^A\epsilon_{AB}\theta^B$$

$$= \theta^1\epsilon_{12}\theta^2 + \theta^2\epsilon_{21}\theta^1 = -\theta^1\theta^2 + \theta^2\theta^1 \overset{(1.94)}{=} -2\theta^1\theta^2.$$

Similarly:

$$\overline{\theta}_{\dot{A}}\overline{\theta}^{\dot{A}} = \overline{\theta}_{\dot{A}}\epsilon^{\dot{A}\dot{B}}\overline{\theta}_{\dot{B}} = 2\overline{\theta}_{\dot{1}}\overline{\theta}_{\dot{2}}$$

using first Eq. (1.85) and then Eq. (1.76a).

---

[23]Observe: "θ squared" on the left, component "θ two" on the right in our notation.

Proposition 1.18:

For Grassmann variables θ and $\bar{\theta}$ we have

$$\theta^A \theta^B = -\frac{1}{2}\epsilon^{AB}(\theta\theta), \tag{1.100a}$$

$$\theta_A \theta_B = \frac{1}{2}\epsilon_{AB}(\theta\theta), \tag{1.100b}$$

$$\bar{\theta}^{\dot{A}}\bar{\theta}^{\dot{B}} = \frac{1}{2}\epsilon^{\dot{A}\dot{B}}(\bar{\theta}\,\bar{\theta}), \tag{1.100c}$$

$$\bar{\theta}_{\dot{A}}\bar{\theta}_{\dot{B}} = -\frac{1}{2}\epsilon_{\dot{A}\dot{B}}(\bar{\theta}\,\bar{\theta}). \tag{1.100d}$$

Proof: We shall see later (*cf.* Eqs. (1.152c) and (1.152d)) that

$$\epsilon_{AB}\epsilon^{DC} = \delta_A{}^C\delta_B{}^D - \delta_A{}^D\delta_B{}^C$$

and

$$\epsilon_{\dot{A}\dot{B}}\epsilon^{\dot{D}\dot{C}} = \delta_{\dot{A}}{}^{\dot{C}}\delta_{\dot{B}}{}^{\dot{D}} - \delta_{\dot{A}}{}^{\dot{D}}\delta_{\dot{B}}{}^{\dot{C}}.$$

These formulas are needed in proving Eqs. (1.100a) to (1.100d).

(i) Consider the expression on the right hand side of Eq. (1.100a):

$$-\frac{1}{2}\epsilon^{AB}(\theta\theta) = -\frac{1}{2}\epsilon^{AB}\theta^C\theta_C = -\frac{1}{2}\epsilon^{AB}\epsilon_{CD}\theta^C\theta^D$$

$$= -\frac{1}{2}(\delta_C{}^B\delta_D{}^A - \delta_C{}^A\delta_D{}^B)\theta^C\theta^D$$

$$= \frac{1}{2}(\theta^B\theta^A - \theta^A\theta^B) = \theta^A\theta^B.$$

(ii) Consider the expression on the left hand side of Eq. (1.100b):

$$\theta_A\theta_B = \epsilon_{AC}\epsilon_{BD}\theta^C\theta^D = -\frac{1}{2}\epsilon_{AC}\epsilon_{BD}\epsilon^{CD}(\theta\theta) = \frac{1}{2}\epsilon_{AB}(\theta\theta).$$

Multiplying by ϵ^{CA} we obtain another formula:

$$\theta^C\theta_B = \frac{1}{2}\delta^C{}_B(\theta\theta). \tag{1.100e}$$

(iii) In a similar manner we have

$$\frac{1}{2}\epsilon^{\dot{A}\dot{B}}(\bar{\theta}\,\bar{\theta}) = \frac{1}{2}\epsilon^{\dot{A}\dot{B}}\bar{\theta}_{\dot{C}}\bar{\theta}^{\dot{C}} = \frac{1}{2}\epsilon^{\dot{A}\dot{B}}\epsilon_{\dot{C}\dot{D}}\bar{\theta}^{\dot{D}}\bar{\theta}^{\dot{C}}$$

$$= \frac{1}{2}(\delta^{\dot{B}}{}_{\dot{C}}\delta^{\dot{A}}{}_{\dot{D}} - \delta^{\dot{A}}{}_{\dot{C}}\delta^{\dot{B}}{}_{\dot{D}})\bar{\theta}^{\dot{D}}\bar{\theta}^{\dot{C}}$$

$$= \frac{1}{2}(\bar{\theta}^{\dot{A}}\bar{\theta}^{\dot{B}} - \bar{\theta}^{\dot{B}}\bar{\theta}^{\dot{A}}) = \bar{\theta}^{\dot{A}}\bar{\theta}^{\dot{B}}.$$

(iv) Finally:

$$\overline{\theta}_{\dot{A}}\overline{\theta}_{\dot{B}} = \epsilon_{\dot{A}\dot{C}}\epsilon_{\dot{B}\dot{D}}\overline{\theta}^{\dot{C}}\overline{\theta}^{\dot{D}} = \frac{1}{2}\epsilon_{\dot{A}\dot{C}}\epsilon_{\dot{B}\dot{D}}\epsilon^{\dot{C}\dot{D}}\left(\overline{\theta}\,\overline{\theta}\right)$$

$$= \frac{1}{2}\epsilon_{\dot{B}\dot{A}}\left(\overline{\theta}\,\overline{\theta}\right) = -\frac{1}{2}\epsilon_{\dot{A}\dot{B}}\left(\overline{\theta}\,\overline{\theta}\right).$$

Multiplying this result by $\epsilon^{\dot{C}\dot{A}}$ we obtain the formula:

$$\overline{\theta}^{\dot{C}}\overline{\theta}_{\dot{B}} = -\frac{1}{2}\delta^{\dot{C}}_{\ \dot{B}}\left(\overline{\theta}\,\overline{\theta}\right). \tag{1.100f}$$

This completes the proof of Proposition (1.18).

Proposition 1.19:

Let θ, ϕ, ψ be Grassmann variables. Then

$$(\theta\phi)(\theta\psi) = -\frac{1}{2}(\phi\psi)(\theta\theta) = -\frac{1}{2}(\theta\theta)(\phi\psi). \tag{1.101}$$

This result is a *Fierz reordering* formula for Weyl spinors.[24]

Proof: Manipulating the expression on the left hand side we obtain

$$(\theta\phi)(\theta\psi) \overset{(1.97)}{=} (\phi\theta)(\theta\psi)$$

$$\overset{(1.92a)}{=} \phi^A\theta_A\theta^B\psi_B$$

$$\overset{(1.84)}{=} \phi^A\epsilon_{AC}\theta^C\theta^B\psi_B$$

$$\overset{(1.100a)}{=} \phi^A\epsilon_{AC}\left(-\frac{1}{2}\epsilon^{CB}\theta\theta\right)\psi_B$$

$$= -\frac{1}{2}\phi^A\delta_A^{\ B}(\theta\theta)\psi_B$$

$$= -\frac{1}{2}\phi^A\psi_A(\theta\theta)$$

$$= -\frac{1}{2}(\phi\psi)(\theta\theta),$$

applying twice the Grassmann property.

---

[24]For other analogous relations discussed later see Eqs. (1.133), (1.137), (6.89) and (6.92).

1.3.3 Connection between $SL(2, \mathbb{C})$ and $L_+^\uparrow$

We now investigate the connection between the group $SL(2, \mathbb{C})$ and the restricted Lorentz group $L_+^\uparrow$ more closely and study the spinor representations of $SL(2, \mathbb{C})$. We shall show that the restricted Lorentz group $L_+^\uparrow$ is homomorphic to $SL(2, \mathbb{C})$, *i.e.* for any $M \in SL(2, \mathbb{C})$ there is a Lorentz matrix

$$\Lambda = \Lambda(M) \in L_+^\uparrow$$

such that

$$\Lambda(M_1)\Lambda(M_2) = \Lambda(M_1 M_2) \tag{1.102}$$

and

$$\Lambda^{-1}(M) = \Lambda(M^{-1}).$$

In order to obtain a connection between spinor calculus and the Lorentz four-vector calculus, we start by introducing a set of four matrices

$$(\sigma^\mu) = (\mathbb{1}_{2\times2}, \boldsymbol{\sigma}) = (\sigma^0, \boldsymbol{\sigma}), \tag{1.103a}$$

where σ^0 is the 2×2-unit matrix and the $(\sigma^i)_{i=1,2,3}$ are the usual *Pauli matrices*, *i.e.*

$$\sigma^1 = \begin{pmatrix} 0 & 1 \\ 1 & 0 \end{pmatrix}, \quad \sigma^2 = \begin{pmatrix} 0 & -i \\ i & 0 \end{pmatrix}, \quad \sigma^3 = \begin{pmatrix} 1 & 0 \\ 0 & -1 \end{pmatrix}. \tag{1.103b}$$

As is well known[25] and can be checked directly, using the explicit form of the Pauli matrices, *i.e.* Eq. (1.103b), the Pauli matrices obey the following relations:

$$\{\sigma_i, \sigma_j\} = 2\delta_{ij}\,\mathbb{1}_{2\times2}, \quad i, j = 1, 2, 3, \tag{1.104a}$$

$$[\sigma_i, \sigma_j] = 2i\epsilon_{ijk}\sigma_k, \quad i, j, k = 1, 2, 3, \tag{1.104b}$$

$$\mathrm{Tr}\,[\sigma_i] = 0, \quad i = 1, 2, 3, \tag{1.104c}$$

$$\frac{1}{2}\mathrm{Tr}\,[\sigma_i^2] = 1, \quad i = 1, 2, 3. \tag{1.104d}$$

The spinor index structure of the four matrices σ^μ is such that

$$\sigma^\mu := (\sigma^\mu{}_{A\dot{A}}), \tag{1.105}$$

as can be shown from the adjoint representation of $SL(2, \mathbb{C})$ (see Eq. (1.117)). The indices may be raised by application of the ϵ-tensor as in Eqs. (1.83) and (1.85) leading to a new set of matrices, *i.e.*

$$(\bar{\sigma}^\mu)^{\dot{A}A} := \epsilon^{AB}\epsilon^{\dot{A}\dot{B}}(\sigma^\mu)_{B\dot{B}}, \tag{1.106a}$$

---

[25]See *e.g.* the monographs by H. Goldstein [49] or H.J.W. Müller-Kirsten [75], Sec. 5.4.

or

$$(\sigma^\mu)_{A\dot{A}} = \epsilon_{AB}\epsilon_{\dot{A}\dot{B}}(\overline{\sigma}^\mu)^{\dot{B}B}. \tag{1.106b}$$

In another formulation Eq. (1.106a) is

$$(\overline{\sigma}^\mu)^{\dot{A}A} = (\overline{\sigma}^{\mu\top})^{A\dot{A}} = -\epsilon^{AB}(\sigma^\mu)_{B\dot{B}}\epsilon^{\dot{B}\dot{A}}. \tag{1.107}$$

Evaluating Eq. (1.106a) we find that

$$\overline{\sigma}^0 = \sigma^0, \qquad \overline{\sigma}^i = -\sigma^i, \ i = 1,2,3. \tag{1.108}$$

Proof: We demonstrate Eq. (1.108) explicitly for $i = 1$. In this case

$$(\overline{\sigma}^1)^{\dot{A}A} = \epsilon^{AB}\epsilon^{\dot{A}\dot{B}}(\sigma^1)_{B\dot{B}},$$

where

$$(\sigma^1)_{1\dot{1}} = (\sigma^1)_{2\dot{2}} = 0, \qquad (\sigma^1)_{1\dot{2}} = (\sigma^1)_{2\dot{1}} = 1.$$

i.e.

$$(\overline{\sigma}^1)^{\dot{1}1} = \epsilon^{12}\epsilon^{\dot{1}\dot{2}}(\sigma^1)_{2\dot{2}} = 0, \quad (\overline{\sigma}^1)^{\dot{1}2} = \epsilon^{21}\epsilon^{\dot{1}\dot{2}}(\sigma^1)_{1\dot{2}} = -1,$$
$$(\overline{\sigma}^1)^{\dot{2}1} = \epsilon^{12}\epsilon^{\dot{2}\dot{1}}(\sigma^1)_{2\dot{1}} = -1, \quad (\overline{\sigma}^1)^{\dot{2}2} = \epsilon^{21}\epsilon^{\dot{2}\dot{1}}(\sigma^1)_{2\dot{2}} = 0,$$

so that

$$\overline{\sigma}^1 = \begin{pmatrix} 0 & -1 \\ -1 & 0 \end{pmatrix} = -\sigma^1.$$

The cases $i = 2,3$ can be shown in a similar fashion.

Proposition 1.20:

The following formulas can be shown to hold:

$$\mathrm{Tr}\left[\sigma^\mu\overline{\sigma}^\nu\right] = 2\eta^{\mu\nu}, \tag{1.109a}$$

where the metric tensor is $(\eta^{\mu\nu}) = \mathrm{diag}(+1,-1,-1,-1)$, and

$$\sigma^\mu\overline{\sigma}^\nu + \sigma^\nu\overline{\sigma}^\mu = 2\eta^{\mu\nu}\mathbb{1}_{2\times 2}. \tag{1.109b}$$

Proof:

(a) Proof of Eq. (1.109a): Take $\mu = \nu = 0$:

$$\mathrm{Tr}\left[\sigma^0\overline{\sigma}^0\right] = \mathrm{Tr}\left[\mathbb{1}_{2\times 2}\cdot\mathbb{1}_{2\times 2}\right] = 2 = 2\eta^{00}.$$

Next: $\mu = 0, \nu = i = 1,2,3$:

$$\mathrm{Tr}\left[\sigma^0\overline{\sigma}^i\right] = \mathrm{Tr}\left[\mathbb{1}_{2\times 2}\overline{\sigma}^i\right] = -\mathrm{Tr}\left[\sigma^i\right] = 0,$$

since the Pauli matrices are traceless (*cf.* Eq. (1.104c)).
Finally: $\mu = i, \nu = j; \quad i, j = 1, 2, 3$:

$$\begin{aligned}
\text{Tr} \left[\sigma^i \overline{\sigma}^j \right] &= -\text{Tr} \left[\sigma^i \sigma^j \right] \\
&= -\text{Tr} \left[\delta^{ij} \mathbb{1}_{2\times2} + \epsilon^{ijk} \sigma^k \right] \\
&= -\delta^{ij} \text{Tr} \left[\mathbb{1}_{2\times2} \right] - \epsilon^{ijk} \text{Tr} \left[\sigma^k \right] \\
&= 2\eta^{ij}.
\end{aligned}$$

This demonstrates Eq. (1.109a).

(b) To show relation (1.109b) we first consider the case $\mu = \nu$:

$$2\sigma^\mu \overline{\sigma}^\mu = 2\eta^{\mu\mu} \mathbb{1}_{2\times2}.$$

For $\mu \neq \nu$ we get:

(a) $\mu = 0, \nu \neq 0$:

$$\sigma^0 \overline{\sigma}^i + \sigma^i \overline{\sigma}^0 = 0.$$

(b) $\mu = i, \nu = j$:

$$\sigma^i \overline{\sigma}^j + \sigma^j \overline{\sigma}^i = 0.$$

Proposition 1.21:

The matrices $(\sigma^\mu)_{\mu=0,1,2,3}$ form a complete set in the sense that any complex 2×2–matrix can be expressed as a linear combination of them. In particular the completeness relation is:

$$(\sigma^\mu)_{A\dot{A}} (\overline{\sigma}_\mu)^{\dot{B}B} = 2\delta_A{}^B \delta_{\dot{A}}{}^{\dot{B}}. \tag{1.110}$$

Proof: Equation (1.110) has to be shown explicitly.

$$\begin{aligned}
(\sigma^\mu)_{A\dot{A}} (\overline{\sigma}_\mu)^{\dot{B}B} &= \eta_{\mu\nu} (\sigma^\mu)_{A\dot{A}} (\overline{\sigma}^\nu)^{\dot{B}B} \\
&= (\sigma^0)_{A\dot{A}} (\overline{\sigma}^0)^{\dot{B}B} - (\sigma^1)_{A\dot{A}} (\overline{\sigma}^1)^{\dot{B}B} \\
&\quad - (\sigma^2)_{A\dot{A}} (\overline{\sigma}^2)^{\dot{B}B} - (\sigma^3)_{A\dot{A}} (\overline{\sigma}^3)^{\dot{B}B}.
\end{aligned}$$

(i) $A = B = 1, \dot{A} = \dot{B} = \dot{1}$:

$$\begin{aligned}
(\sigma^\mu)_{1\dot{1}} (\overline{\sigma}_\mu)^{\dot{1}1} &= (\sigma^0)_{1\dot{1}} (\overline{\sigma}^0)^{\dot{1}1} - (\sigma^1)_{1\dot{1}} (\overline{\sigma}^1)^{\dot{1}1} \\
&\quad - (\sigma^2)_{1\dot{1}} (\overline{\sigma}^2)^{\dot{1}1} - (\sigma^3)_{1\dot{1}} (\overline{\sigma}^3)^{\dot{1}1} \\
&= 2\delta_1{}^1 \delta_{\dot{1}}{}^{\dot{1}}.
\end{aligned}$$

(ii) $A = B = 2, \dot{A} = \dot{B} = \dot{2}$:

$$
\begin{aligned}
(\sigma^\mu)_{2\dot{2}}(\overline{\sigma}_\mu)^{\dot{2}2} &= (\sigma^0)_{2\dot{2}}(\overline{\sigma}^0)^{\dot{2}2} - (\sigma^1)_{2\dot{2}}(\overline{\sigma}^1)^{\dot{2}2} \\
&\quad -(\sigma^2)_{2\dot{2}}(\overline{\sigma}^2)^{\dot{2}2} - (\sigma^3)_{2\dot{2}}(\overline{\sigma}^3)^{\dot{2}2} \\
&= 2\delta_2{}^2\delta_{\dot{2}}{}^{\dot{2}}.
\end{aligned}
$$

(iii) $A = B = 1, \dot{A} = \dot{B} = \dot{2}$:

$$
\begin{aligned}
(\sigma^\mu)_{1\dot{2}}(\overline{\sigma}_\mu)^{\dot{2}1} &= (\sigma^0)_{1\dot{2}}(\overline{\sigma}^0)^{\dot{2}1} - (\sigma^1)_{1\dot{2}}(\overline{\sigma}^1)^{\dot{2}1} \\
&\quad -(\sigma^2)_{1\dot{2}}(\overline{\sigma}^2)^{\dot{2}1} - (\sigma^3)_{1\dot{2}}(\overline{\sigma}^3)^{\dot{2}1} \\
&= 2\delta_1{}^1\delta_{\dot{2}}{}^{\dot{2}}.
\end{aligned}
$$

(iv) $A = B = 2, \dot{A} = \dot{B} = \dot{1}$:

$$
\begin{aligned}
(\sigma^\mu)_{2\dot{1}}(\overline{\sigma}_\mu)^{\dot{1}2} &= (\sigma^0)_{2\dot{1}}(\overline{\sigma}^0)^{\dot{1}2} - (\sigma^1)_{2\dot{1}}(\overline{\sigma}^1)^{\dot{1}2} \\
&\quad -(\sigma^2)_{2\dot{1}}(\overline{\sigma}^2)^{\dot{1}2} - (\sigma^3)_{2\dot{1}}(\overline{\sigma}^3)^{\dot{1}2} \\
&= 2\delta_2{}^2\delta_{\dot{1}}{}^{\dot{1}}.
\end{aligned}
$$

(v) $A = 1, B = 2, \dot{A} = \dot{B} = \dot{1}$:

$$
\begin{aligned}
(\sigma^\mu)_{1\dot{1}}(\overline{\sigma}_\mu)^{\dot{1}2} &= (\sigma^0)_{1\dot{1}}(\overline{\sigma}^0)^{\dot{1}2} - (\sigma^1)_{1\dot{1}}(\overline{\sigma}^1)^{\dot{1}2} \\
&\quad -(\sigma^2)_{1\dot{1}}(\overline{\sigma}^2)^{\dot{1}2} - (\sigma^3)_{1\dot{1}}(\overline{\sigma}^3)^{\dot{1}2} \\
&= 0.
\end{aligned}
$$

In the same way one can show that all other combinations of the spinor indices vanish identically.

We are now in a position to construct the group homomorphism between $SL(2,\mathbb{C})$ and the restricted Lorentz group $L_+^\uparrow$. First, however, we construct a map from Minkowski space $\mathbb{M}_4$ to the set of Hermitian complex 2×2-matrices, denoted by $\mathbb{H}(2,\mathbb{C})$ ($\mathbb{H}$ for Hermitian). Thus, denoting the map by ρ, we have:

$$
\begin{aligned}
\rho: \quad \mathbb{M}_4 &\longrightarrow \mathbb{H}(2,\mathbb{C}), \\
x^\mu &\longmapsto \rho(x^\mu) = x_\mu \sigma^\mu \\
&= \begin{pmatrix} x^0 - x^3 & x^1 + ix^2 \\ x^1 - ix^2 & x^0 + x^3 \end{pmatrix} \\
&=: X,
\end{aligned}
\tag{1.111}
$$

where the four-vector is

$$\left(x^{\mu}\right) = \left(x^0, \mathbf{x}\right), \quad \left(x_{\mu}\right) = \left(x^0, -\mathbf{x}\right).$$

Proposition 1.22:

The map inverse to Eq. (1.111)

$$\rho^{-1} : \mathbb{H}(2, \mathbb{C}) \longrightarrow \mathbb{M}_4$$

is given by the following trace relation

$$\rho^{-1} : \mathbb{H}(2, \mathbb{C}) \longrightarrow \mathbb{M}_4,$$

$$X \longmapsto \rho^{-1}(X) = x^{\mu} = \frac{1}{2}\mathrm{Tr}\left[X\overline{\sigma}^{\mu}\right]. \tag{1.112}$$

Proof: Using Eqs. (1.111) and (1.109a) we obtain:

$$\frac{1}{2}\mathrm{Tr}\left[X\overline{\sigma}^{\mu}\right] = \frac{1}{2}\mathrm{Tr}\left[x_{\nu}\sigma^{\nu}\overline{\sigma}^{\mu}\right] = \frac{1}{2}\mathrm{Tr}\left[\sigma^{\nu}\overline{\sigma}^{\mu}\right]x_{\nu} = \frac{1}{2} \cdot 2\eta^{\mu\nu}x_{\nu} = x^{\mu}.$$

Proposition 1.23:

The determinant of $\rho(x_{\mu}) \in \mathbb{H}(2, \mathbb{C})$ is given by:

$$\det X = \det \rho(x_{\mu}) = x_{\mu}x^{\mu} = x^2. \tag{1.113}$$

Proof: Using the definition of the map ρ, *i.e.* Eq. (1.111), we have:

$$\det X = \det \rho(x_{\mu}) = \det \begin{pmatrix} x^0 - x^3 & x^1 + ix^2 \\ x^1 - ix^2 & x^0 + x^3 \end{pmatrix}$$

$$= (x^0)^2 - (x^i)^2 = \eta_{\mu\nu}x^{\mu}x^{\nu} = x^2.$$

The Hermiticity of the matrix $X = \rho(x^{\mu})$ is seen immediately from the Hermiticity of the Pauli matrices

$$X = X^{\dagger} = X^{*\top}. \tag{1.114}$$

We now consider the action of $SL(2, \mathbb{C})$ on $\mathbb{H}(2, \mathbb{C})$, which is called the *adjoint representation of the group $SL(2, \mathbb{C})$*. This representation is defined in the following way:

$$\begin{aligned} \mathrm{ad} : \quad SL(2, \mathbb{C}) &\longrightarrow \quad \mathrm{Aut}\big(\mathbb{H}(2, \mathbb{C})\big), \\ M &\longmapsto \quad \mathrm{ad}(M), \end{aligned} \tag{1.115}$$

with

$$M' \equiv \mathrm{ad}M(X) := MXM^\dagger, \quad M, M^\dagger \in SL(2,\mathbb{C}), \tag{1.116}$$

where $\mathrm{Aut}\big(\mathbb{H}(2,\mathbb{C})\big)$ is the automorphism group of $\mathbb{H}(2,\mathbb{C})$, which is isomorphic to $GL(\mathbb{H}(2,\mathbb{C}),\mathbb{C})$ [19]. We have to verify therefore that $\mathrm{ad}M(X) \in \mathbb{H}(2,\mathbb{C})$, *i.e.* we have to show that $\mathrm{ad}M(X)$ is Hermitian. Using Eq. (1.116) we have

$$\big[\mathrm{ad}M(X)\big]^\dagger = \big(MXM^\dagger\big)^\dagger = MX^\dagger M^\dagger = MXM^\dagger = \mathrm{ad}M(X),$$

since $X \in \mathbb{H}(2,\mathbb{C})$.

From the adjoint representation (1.116) we can derive the index structure of the Pauli matrices, as indicated in Eq. (1.105). We had

$$\rho(x^\mu) = X = x_\mu \sigma^\mu,$$

which transforms under the adjoint representation of $SL(2,\mathbb{C})$ as:

$$X' = MXM^\dagger,$$

or

$$x'_\mu \sigma^\mu = M\sigma^\nu x_\nu M^\dagger.$$

Using the index notation of the $SL(2,\mathbb{C})$-matrices M and $M^\dagger$ we conclude

$$x'_\mu(\sigma^\mu)_{A\dot{A}} = M_A{}^B (\sigma^\nu)_{B\dot{B}} x_\nu (M^*)^{\dot{B}}{}_{\dot{A}}, \tag{1.117}$$

and we see that the Pauli matrices must carry a dotted and an undotted index. The matrices σ^μ therefore map $\dot{F}$ into F, and similarly the matrices $\bar{\sigma}^\mu$ map F into $\dot{F}$.

Next we calculate the determinant of a transformed Hermitian matrix. Using Eqs. (1.111) and (1.116) we have

$$\begin{aligned}
x'_\mu x'^\mu &= \det\big(X'\big) = \det\big(MXM^\dagger\big) \\
&= \det(M)\det\big(X\big)\det(M^\dagger) \\
&= \det(X) = x_\mu x^\mu, \tag{1.118}
\end{aligned}$$

since $M \in SL(2,\mathbb{C})$ the determinant of M equals 1. Thus the quadratic form $x_\mu x^\mu = \det(X)$ is left invariant under transformations of the adjoint representation of $SL(2,\mathbb{C})$.

We now have the following construction. Starting from a Minkowski four-vector $x^\mu \in \mathbb{M}_4$ we construct an Hermitian 2×2-matrix $X \in \mathbb{H}(2,\mathbb{C})$. The determinant of this matrix is simply the Minkowski product of x_μ with itself.

We then go to the adjoint representation of $SL(2, \mathbb{C})$ on $\mathbb{H}(2, \mathbb{C})$ and derive the result that the determinant is left invariant under $SL(2, \mathbb{C})$. If we apply the map ρ^{-1} to the transformed Hermitian matrix, we obtain a four-vector x'_μ such that

$$x'_\mu x'^\mu = x_\mu x^\mu.$$

We can demonstrate these steps schematically in the following way

$$\mathbb{M}_4 \xrightarrow{\rho} \mathbb{H}(2, \mathbb{C}) \xrightarrow{\mathrm{ad} M} \mathbb{H}(2, \mathbb{C}) \xrightarrow{\rho^{-1}} \mathbb{M}_4, \qquad (1.119)$$

that is:

$$x_\mu \longrightarrow \begin{aligned} \rho(x_\mu) &= x_\mu \sigma^\mu \\ &= X \end{aligned} \longrightarrow \begin{aligned} \mathrm{ad}\ M \\ &= MXM^\dagger \\ &= X'. \end{aligned} \longrightarrow \rho^{-1}(X') = x'_\mu$$

This transformation from $\mathbb{M}_4$ to $\mathbb{M}_4$ is, of course, simply a Lorentz transformation which transforms four-vectors into four-vectors leaving the Minkowski square, *i.e.* the line-element, invariant. Now from Eq. (1.1) we have

$$x'^\mu = \Lambda^\mu{}_\nu x^\nu.$$

On the other hand we can also write, using Eqs. (1.112), (1.116) and (1.111):

$$x'^\mu = \frac{1}{2}\mathrm{Tr}\left[X'\overline{\sigma}^\mu\right] = \frac{1}{2}\mathrm{Tr}\left[MXM^\dagger\overline{\sigma}^\mu\right] = \frac{1}{2}\mathrm{Tr}\left[Mx_\nu\sigma^\nu M^\dagger\overline{\sigma}^\mu\right]$$
$$= \frac{1}{2}\mathrm{Tr}\left[M\sigma^\nu M^\dagger\overline{\sigma}^\mu\right]x_\nu = \frac{1}{2}\mathrm{Tr}\left[\overline{\sigma}^\mu M\sigma_\nu M^\dagger\right]x^\nu. \qquad (1.120)$$

Comparing this expression with Eq. (1.1) above, we obtain:

$$\boxed{\Lambda^\mu{}_\nu(M) = \frac{1}{2}\mathrm{Tr}\left[\overline{\sigma}^\mu M\sigma_\nu M^\dagger\right].} \qquad (1.121)$$

Equation (1.121) is the explicit form of the group homomorphism

$$SL(2, \mathbb{C}) \longrightarrow L_+^\uparrow.$$

Proposition 1.24:

The following properties hold:

(i) Equation (1.121) is a group homomorphism, *i.e.* $\forall\ M_1, M_2 \in SL(2, \mathbb{C})$ we have:

$$\Lambda^\mu{}_\nu(M_1)\Lambda^\nu{}_\rho(M_2) = \Lambda^\mu{}_\rho(M_1 M_2), \qquad (1.122)$$

and

$$\left(\Lambda^{-1}\right)^\mu{}_\nu(M) = \Lambda^\mu{}_\nu(M^{-1}). \qquad (1.123)$$

(ii) The matrix Λ, defined by Eq. (1.121), is an element of the restricted Lorentz group $L_+^\uparrow$, *i.e.* we have to show:

$$\det\left[\Lambda^\mu{}_\nu(M)\right] = +1 \tag{1.124}$$

and

$$\Lambda^0{}_0(M) \geq 1. \tag{1.125}$$

Proof: We first prove Eq. (1.122). Using Eq. (1.121) we have

$$
\begin{aligned}
\Lambda^\mu{}_\nu(M_1)\Lambda^\nu{}_\rho(M_2) &= \frac{1}{4}\mathrm{Tr}\left[\bar\sigma^\mu M_1 \sigma_\nu M_1^\dagger\right]\mathrm{Tr}\left[\bar\sigma^\nu M_2 \sigma_\rho M_2^\dagger\right]\\
&= \frac{1}{4}\mathrm{Tr}\left[M_1^\dagger\bar\sigma^\mu M_1 \sigma_\nu\right]\mathrm{Tr}\left[\bar\sigma^\nu M_2 \sigma_\rho M_2^\dagger\right]\\
&= \frac{1}{4}\left[M_1^\dagger\bar\sigma^\mu M_1 \sigma_\nu\right]^{\dot B}{}_{\dot B}\left[\bar\sigma^\nu M_2 \sigma_\rho M_2^\dagger\right]^{\dot A}{}_{\dot A}\\
&= \frac{1}{4}\left[M_1^\dagger\bar\sigma^\mu M_1\right]^{\dot B B}(\sigma_\nu)_{B\dot B}(\bar\sigma^\nu)^{\dot A A}\left[M_2 \sigma_\rho M_2^\dagger\right]_{A\dot A}\\
&\overset{(1.110)}{=} \frac{1}{2}\left[M_1^\dagger\bar\sigma^\mu M_1\right]^{\dot B B}\delta_B{}^A\delta_{\dot B}{}^{\dot A}\left[M_2 \sigma_\rho M_2^\dagger\right]_{A\dot A}\\
&= \frac{1}{2}\left[M_1^\dagger\bar\sigma^\mu M_1\right]^{\dot A A}\left[M_2 \sigma_\rho M_2^\dagger\right]_{A\dot A}\\
&= \frac{1}{2}\left[M_1^\dagger\bar\sigma^\mu M_1 M_2 \sigma_\rho M_2^\dagger\right]^{\dot A}{}_{\dot A}\\
&= \frac{1}{2}\mathrm{Tr}\left[M_1^\dagger\bar\sigma^\mu M_1 M_2 \sigma_\rho M_2^\dagger\right]\\
&= \frac{1}{2}\mathrm{Tr}\left[\bar\sigma^\mu (M_1 M_2)\sigma_\rho (M_1 M_2)^\dagger\right]\\
&= \Lambda^\mu{}_\rho(M_1 M_2),
\end{aligned}
$$

again using Eq. (1.121). A direct consequence of Eq. (1.122) is the following result:

$$\left(\Lambda^{-1}\right)^\mu{}_\nu(M) = \Lambda^\mu{}_\nu(M^{-1}), \tag{1.126}$$

which we can prove as follows. Using Eqs. (1.122) and (1.109a):

$$
\begin{aligned}
\Lambda^\mu{}_\nu(M)\Lambda^\nu{}_\rho(M^{-1}) &\overset{(1.122)}{=} \Lambda^\mu{}_\rho(MM^{-1}) = \Lambda^\mu{}_\rho(\mathbb{1}_{SL(2,\mathbb{C})})\\
&= \frac{1}{2}\mathrm{Tr}\left[\bar\sigma^\mu \sigma_\rho\right] \overset{(1.109a)}{=} \frac{1}{2}\cdot 2\eta^\mu{}_\rho = \delta^\mu{}_\rho.
\end{aligned}
$$

In order to prove Eq. (1.124) it suffices to recall that

$$\Lambda^\mu{}_\nu(\mathbb{1}_{SL(2,\mathbb{C})}) = \delta^\mu{}_\nu,$$

and therefore

$$\det\left(\Lambda^\mu{}_\nu(\mathbb{1}_{SL(2,\mathbb{C})})\right) = \det\left(\delta^\mu{}_\nu\right) = 1.$$

Furthermore, det is a continuous function of $\Lambda^\mu{}_\nu$, so that $\det(\Lambda^\mu{}_\nu(M))$ must be $+1$ as M runs through $SL(2,\mathbb{C})$.

Finally, Eq. (1.125) may be shown as follows.[26] Using Eq. (1.121) we have

$$\Lambda^0{}_0(M) = \frac{1}{2}\mathrm{Tr}\left[\bar\sigma^0 M \sigma_0 M^\dagger\right] = \frac{1}{2}\mathrm{Tr}\left[MM^\dagger\right],$$

since

$$\bar\sigma^0 = \sigma^0 = \sigma_0 = \mathbb{1}_{2\times 2}.$$

Now let $M \in SL(2,\mathbb{C})$ and take $U \in U(2,\mathbb{C})$ such that

$$UMU^\dagger = D,$$

where D is diagonal. Then

$$\left(UMU^\dagger\right)^\dagger = D^\dagger,$$

or

$$UM^\dagger U^\dagger = D^\dagger.$$

Now

$$1 = \det M = \det(UMU^\dagger) = \det D = a_1 a_2,$$

where $a_1, a_2 \in \mathbb{C}$ are the eigenvalues of M. Similarly

$$1 = \det M^\dagger = \det(UM^\dagger U^\dagger) = \det D^\dagger = a_1^* a_2^*.$$

Using the fact that a trace is invariant under similarity transformations we can write

$$\mathrm{Tr}\left[MM^\dagger\right] = \mathrm{Tr}\left[UMM^\dagger U^\dagger\right] = \mathrm{Tr}\left[UMU^\dagger UM^\dagger U^\dagger\right]$$
$$= \mathrm{Tr}\left[DD^\dagger\right] = |a_1|^2 + |a_2|^2.$$

Taking into account that $a_1 \cdot a_2 = a_1^* \cdot a_2^* = 1$, we can write

$$\mathrm{Tr}\left[MM^\dagger\right] = |a_1|^2 + \frac{1}{|a_1|^2} = \frac{|a_1|^4 + 1}{|a_1|^2}.$$

We now set

$$c := |a_1|^2 \in \mathbb{R}_+.$$

---

[26] We introduce a unitary matrix U which diagonalizes M.

Then:

$$\mathrm{Tr}\big[MM^\dagger\big] = \frac{c^2 + 1}{c}.$$

With this relation we can show that $\mathrm{Tr}\, MM^\dagger \geq 2$. Suppose we have the opposite, *i.e.*

$$\mathrm{Tr}\big[MM^\dagger\big] < 2.$$

Then

$$\frac{c^2 + 1}{c} < 2$$

which implies

$$(c - 1)^2 < 0.$$

But this is impossible for any $c \in \mathbb{R}_+$. Hence we conclude

$$\mathrm{Tr}\big[MM^\dagger\big] \geq 2,$$

and hence

$$\Lambda^0{}_0(M) = \frac{1}{2}\mathrm{Tr}\big[MM^\dagger\big] \geq 1.$$

Thus, $\Lambda^\mu{}_\nu(M)$ as defined by Eq. (1.121) is, together with Eq. (1.124), a restricted Lorentz transformation.

Our next step is to derive the map inverse to that of Eq. (1.121), *i.e.* given any Lorentz matrix $\Lambda^\mu{}_\nu$ the question is: How can we construct the corresponding $SL(2, \mathbb{C})$-matrix $M = M(\Lambda)$. We start by considering the action of the restricted Lorentz group on Minkowski space $\mathbb{M}_4$:

$$x'^\mu = \Lambda^\mu{}_\nu x^\nu, \quad x^\mu \in \mathbb{M}_4, \quad \Lambda \in L_+^\uparrow. \tag{1.127}$$

However, we know that the action of $SL(2, \mathbb{C})$ on Hermitian 2×2-matrices is given by the adjoint representation:

$$X' = MXM^\dagger, \quad M \in SL(2, \mathbb{C}), \quad X \in \mathbb{H}(2, \mathbb{C}),$$

or using relation (1.111):

$$x'_\mu \sigma^\mu = M x_\nu \sigma^\nu M^\dagger, \qquad x'^\mu \sigma_\mu = M x^\nu \sigma_\nu M^\dagger.$$

Using Eq. (1.127),

$$\Lambda^\mu{}_\nu x^\nu \sigma_\mu = M x^\nu \sigma_\nu M^\dagger,$$

so that

$$\Lambda^\mu{}_\nu \sigma_\mu = M \sigma_\nu M^\dagger. \tag{1.128}$$

Proposition 1.25:

For every $M \in SL(2, \mathbb{C})$

$$\sigma_\mu M \bar{\sigma}^\mu = 2\text{Tr}[M]\,\mathbb{1}_{2\times2} \qquad (1.129)$$

Proof: This can be shown by explicit verification.

Multiplying Eq. (1.128) by $\bar{\sigma}^\nu$ and summing over ν, thereby using Eq. (1.129), we obtain for $M, M^\dagger \in SL(2, \mathbb{C})$:

$$\Lambda^\mu{}_\nu \sigma_\mu \bar{\sigma}^\nu = M\sigma_\nu M^\dagger \bar{\sigma}^\nu = M\big(2\text{Tr}[M^\dagger]\big)\mathbb{1}_{2\times2}, \qquad (1.130)$$

and so

$$M(\Lambda) = \frac{1}{2\text{Tr}[M^\dagger]}\Lambda^\mu{}_\nu \sigma_\mu \bar{\sigma}^\nu.$$

Taking the determinant of Eq. (1.130) we obtain:

$$\det\big(\Lambda^\mu{}_\nu \sigma_\mu \bar{\sigma}^\nu\big) = \det\big[M(2\text{Tr}M^\dagger)\big] = \det M \cdot (2\text{Tr}M^\dagger)^2 \cdot \det\mathbb{1}_{2\times2}.$$

Using $\det M = 1$, since $M \in SL(2, \mathbb{C})$, we can express $\text{Tr}M^\dagger$ as a function of Λ, *i.e.*

$$2\text{Tr}M^\dagger = \pm\Big\{\det\big[\Lambda^\mu{}_\nu \sigma_\mu \bar{\sigma}^\nu\big]\Big\}^{1/2}.$$

The final expression for M as a function of Λ is therefore:

$$M(\Lambda) = \pm\frac{1}{\Big\{\det\big[\Lambda^\mu{}_\nu \sigma_\mu \bar{\sigma}^\nu\big]\Big\}^{1/2}}\Lambda^\mu{}_\nu \sigma_\mu \bar{\sigma}^\nu. \qquad (1.131)$$

In components, Eq. (1.131) reads

$$M_A{}^B(\Lambda) = \pm\frac{1}{\Big\{\det\big[\Lambda^\mu{}_\nu \sigma_\mu \bar{\sigma}^\nu\big]\Big\}^{1/2}}\Lambda^\mu{}_\nu (\sigma_\mu)_{A\dot{B}}(\bar{\sigma}^\nu)^{\dot{B}B}.$$

Thus the Lorentz indices μ, ν are effectively exchanged via $\sigma, \bar{\sigma}$ for spinor indices A, B.

Equations (1.121) and (1.131) show the connection between the two groups $SL(2, \mathbb{C})$ and $L_+^\uparrow$. An important point is that both $SL(2, \mathbb{C})$-matrices M and $-M$ lead to the same Lorentz transformation as can be seen from formula (1.121). In Eq. (1.131) this fact corresponds to the $\pm$ signs. Thus the correspondence $\Lambda \longleftrightarrow \pm M$ defines a two-valued representation of the restriced Lorentz group, which is called the *spinor representation*. This leads us to the identification

$$L_+^\uparrow \cong SL(2, \mathbb{C})/\mathbb{Z}_2, \tag{1.132}$$

and so $SL(2, \mathbb{C})$ is the *universal covering group* of $L_+^\uparrow$, which means the $SL(2, \mathbb{C})$ is simply connected [104].

1.3.4   The Fierz-Reordering Formula

The *Fierz reordering formula* exchanges the order of anticommuting spinors.

Proposition 1.26:

The following relation holds for spinors ψ, ϕ, χ:

$$(\phi\psi)\overline{\chi}_{\dot{A}} = \frac{1}{2}(\phi\sigma^\mu\overline{\chi})(\psi\sigma_\mu)_{\dot{A}}. \tag{1.133}$$

Proof: Using Eqs. (1.92a), (1.84), (1.86) we find:

$$\begin{aligned}
(\phi\psi)\overline{\chi}_{\dot{A}} &= \phi^B\psi_B\overline{\chi}_{\dot{A}} \\
&= \phi^B\epsilon_{BC}\psi^C\epsilon_{\dot{A}\dot{D}}\overline{\chi}^{\dot{D}} \\
&= -\epsilon_{CB}\phi^B\psi^C\epsilon_{\dot{A}\dot{D}}\overline{\chi}^{\dot{D}} \\
&= \epsilon_{CB}\psi^C\phi^B\epsilon_{\dot{A}\dot{D}}\overline{\chi}^{\dot{D}} \\
&= \delta_A{}^C\delta_{\dot{A}}{}^{\dot{E}}\epsilon_{CB}\epsilon_{\dot{E}\dot{D}}\psi^A\phi^B\overline{\chi}^{\dot{D}}.
\end{aligned}$$

Using Eq. (1.110) and finally Eq. (1.106a) we arrive at:

$$\begin{aligned}
(\phi\psi)\overline{\chi}_{\dot{A}} &= \frac{1}{2}(\sigma^\mu)_{A\dot{A}}(\overline{\sigma}^\mu)^{\dot{E}C}\epsilon_{CB}\epsilon_{\dot{E}\dot{D}}\psi^A\phi^B\overline{\chi}^{\dot{D}} \\
&= \frac{1}{2}(\sigma^\mu)_{A\dot{A}}\epsilon^{CF}\epsilon^{\dot{E}\dot{H}}(\sigma_\mu)_{F\dot{H}}\epsilon_{CB}\epsilon_{\dot{E}\dot{D}}\psi^A\phi^B\overline{\chi}^{\dot{D}} \\
&= \frac{1}{2}(\sigma^\mu)_{A\dot{A}}(\sigma_\mu)_{F\dot{H}}\delta^F{}_B\delta^{\dot{H}}{}_{\dot{D}}\psi^A\phi^B\overline{\chi}^{\dot{D}} \\
&= \frac{1}{2}\phi^B(\sigma_\mu)_{B\dot{D}}\overline{\chi}^{\dot{D}}\psi^A(\sigma^\mu)_{A\dot{A}} \\
&= \frac{1}{2}(\phi\sigma_\mu\overline{\chi})(\psi\sigma^\mu)_{\dot{A}}.
\end{aligned}$$

For further Fierz reordering formulas see Eqs. (1.101), (6.90), (6.91), and (6.92).

1.3.5 Further Calculations with Spinors

We begin by deriving some more formulas which are very useful in calculations with spinors.

Proposition 1.27:

Let ϕ and χ be spinors and $\sigma^\mu, \overline{\sigma}^\mu$ given by Eqs. (1.105) and (1.106a), then the following identiy holds:

$$(\phi\sigma^\mu\overline{\chi}) = -(\overline{\chi}\,\overline{\sigma}^\mu\phi) \tag{1.134}$$

Proof: We start with the expression on the left:

$$(\phi\sigma^\mu\overline{\chi}) \overset{(1.92a)}{\underset{(1.93)}{=}} \phi^A(\sigma^\mu)_{A\dot{B}}\overline{\chi}^{\dot{B}} \overset{(1.83)}{\underset{(1.84)}{=}} \epsilon^{AB}\phi_B(\sigma^\mu)_{A\dot{B}}\epsilon^{\dot{B}\dot{C}}\overline{\chi}_{\dot{C}}$$

$$= -\phi_B\epsilon^{BA}(\sigma^\mu)_{A\dot{B}}\epsilon^{\dot{B}\dot{C}}\overline{\chi}_{\dot{C}} \overset{(1.106a)}{=} -\phi_B(-\overline{\sigma}^\mu)^{\dot{C}B}\overline{\chi}_{\dot{C}}$$

$$\overset{(1.94)}{=} -\overline{\chi}_{\dot{C}}(\overline{\sigma}^\mu)^{\dot{C}B}\phi_B = -(\overline{\chi}\,\overline{\sigma}^\mu\phi).$$

Proposition 1.28:

Under Hermitian conjugation the binary spinor expressions $\phi\sigma^\mu\overline{\chi}$ and $(\theta\phi)$ transform according to:

$$(\phi\sigma^\mu\overline{\chi})^\dagger = (\chi\sigma^\mu\overline{\phi}), \tag{1.135a}$$

$$(\theta\phi)^\dagger = \overline{\theta}\,\overline{\phi}. \tag{1.135b}$$

Proof: The first expression, Eq. (1.135a), is $SL(2, \mathbb{C})$ covariant. Therefore, with the definition of the †-operation

$$(\phi\sigma^\mu\overline{\chi})^\dagger = (\overline{\chi}^\dagger\sigma^{\mu\dagger}\phi^\dagger) = \overline{\chi}^{\dagger A}(\sigma^\mu)^\dagger_{A\dot{B}}\phi^{\dagger\dot{B}} = \chi^A(\sigma^\mu)_{A\dot{B}}\overline{\phi}^{\dot{B}} = (\chi\sigma^\mu\overline{\phi}).$$

Here we use:

$$\overline{\chi}^{\dagger A} := \chi^A \quad \text{and} \quad \phi^{\dagger\dot{B}} := \overline{\phi}^{\dot{B}}.$$

It may be observed that the Grassmann property, Eqs. (1.94) and (1.95), has not been used because $(\phi\overline{\chi})^\dagger$ is defined as $\overline{\chi}^\dagger\phi^\dagger$. Similarly, Eq. (1.135b) can be demonstrated as follows:

$$(\theta\phi)^\dagger \overset{(1.97)}{=} (\phi\theta)^\dagger = \theta^\dagger_{\dot{B}}\phi^{\dagger\dot{B}} = \overline{\theta}\,\overline{\phi}.$$

From Eq. (1.135a) we see that $\psi\sigma^\mu\overline{\psi}$ is Hermitian.

Proposition 1.29:

As can be expected, the expression $(\phi\sigma^\mu\overline{\chi})$ transforms as a four-vector under the Lorentz group, *i.e.*

$$(\phi'\sigma^\mu\overline{\chi}') = \Lambda^\mu{}_\nu(M)(\phi\sigma^\nu\overline{\chi}). \tag{1.136}$$

Proof: We have:

$$
\begin{aligned}
(\phi'\sigma^\mu\overline{\chi}') &\overset{(1.134)}{=} -(\overline{\chi}'\overline{\sigma}^\mu\phi') \\
&= -\overline{\chi}'_{\dot{A}}(\overline{\sigma}^\mu)^{\dot{A}B}\phi'_B \\
&\overset{(1.82a)}{=} -(M^*)_{\dot{A}}{}^{\dot{B}}\overline{\chi}_{\dot{B}}(\overline{\sigma}^\mu)^{\dot{A}B}M_B{}^C\phi_C \\
&= -\overline{\chi}_{\dot{B}}(M^\dagger)^{\dot{B}}{}_{\dot{A}}(\overline{\sigma}^\mu)^{\dot{A}B}M_B{}^C\phi_C \\
&= -\overline{\chi}_{\dot{B}}\delta^{\dot{B}}{}_{\dot{D}}(M^\dagger)^{\dot{D}}{}_{\dot{A}}(\overline{\sigma}^\mu)^{\dot{A}B}M_B{}^C\delta_C{}^D\phi_D \\
&\overset{(1.110)}{=} -\frac{1}{2}\overline{\chi}_{\dot{B}}(\overline{\sigma}^\nu)^{\dot{B}D}(\sigma_\nu)_{C\dot{D}}(M^\dagger)^{\dot{D}}{}_{\dot{A}}(\overline{\sigma}^\mu)^{\dot{A}B}M_B{}^C\phi_D \\
&= -\frac{1}{2}\left[(M^\dagger)^{\dot{D}}{}_{\dot{A}}(\overline{\sigma}^\mu)^{\dot{A}B}M_B{}^C(\sigma_\nu)_{C\dot{D}}\right]\left(\overline{\chi}_{\dot{B}}(\overline{\sigma}^\nu)^{\dot{B}D}\phi_D\right) \\
&= -\frac{1}{2}\mathrm{Tr}\left[M^\dagger\overline{\sigma}^\mu M\sigma_\nu\right]\left(\overline{\chi}\,\overline{\sigma}^\nu\phi\right) \\
&= \Lambda^\mu{}_\nu(M)(\phi\sigma^\nu\overline{\chi}),
\end{aligned}
$$

using Eqs. (1.121) and (1.134).

Proposition 1.30:

The following relations can be shown to hold:

$$
\left.
\begin{aligned}
(\sigma^\mu)_{A\dot{A}}\overline{\theta}^{\dot{A}}(\theta\sigma^\nu\overline{\theta}) &= \tfrac{1}{2}\eta^{\mu\nu}\theta_A(\overline{\theta}\,\overline{\theta}), \\
(\theta\sigma^\mu\overline{\theta})(\theta\sigma^\nu\overline{\theta}) &= \tfrac{1}{2}\eta^{\mu\nu}(\theta\theta)(\overline{\theta}\,\overline{\theta}).
\end{aligned}
\right\} \tag{1.137}
$$

Proof: Using Eqs. (1.92a) and (1.92b) we have

$$
\begin{aligned}
(\theta\sigma^\mu\overline{\theta})(\theta\sigma^\nu\overline{\theta}) &= \theta^A(\sigma^\mu)_{A\dot{B}}\overline{\theta}^{\dot{B}}\theta^C(\sigma^\nu)_{C\dot{D}}\overline{\theta}^{\dot{D}} \\
&= -\theta^A(\sigma^\mu)_{A\dot{B}}\theta^C\overline{\theta}^{\dot{B}}(\sigma^\nu)_{C\dot{D}}\overline{\theta}^{\dot{D}} \\
&= -(\sigma^\mu)_{A\dot{B}}(\sigma^\nu)_{C\dot{D}}\theta^A\theta^C\overline{\theta}^{\dot{B}}\overline{\theta}^{\dot{D}} \\
&= \frac{1}{4}(\sigma^\mu)_{A\dot{B}}(\sigma^\nu)_{C\dot{D}}\epsilon^{AC}\epsilon^{\dot{B}\dot{D}}(\theta\theta)(\overline{\theta}\,\overline{\theta}) \\
&\overset{(1.106a)}{=} \frac{1}{4}(\sigma^\mu)_{A\dot{B}}(\overline{\sigma}^\nu)^{\dot{B}A}(\theta\theta)(\overline{\theta}\,\overline{\theta}) \\
&\overset{(1.109a)}{=} \frac{1}{2}\eta^{\mu\nu}(\theta\theta)(\overline{\theta}\,\overline{\theta}).
\end{aligned}
$$

Proposition 1.31:

The generators of the group $SL(2, \mathbb{C})$ in the spinor representations $(1/2, 0)$ and $(0, 1/2)$ are given by (see Eq. (1.131))

$$\sigma^{\mu\nu} = \frac{i}{4}\left(\sigma^{\mu}\overline{\sigma}^{\nu} - \sigma^{\nu}\overline{\sigma}^{\mu}\right), \tag{1.138a}$$

$$\overline{\sigma}^{\mu\nu} = \frac{i}{4}\left(\overline{\sigma}^{\mu}\sigma^{\nu} - \overline{\sigma}^{\nu}\sigma^{\mu}\right), \tag{1.138b}$$

$$(\sigma^{\mu\nu})^{\dagger} = \overline{\sigma}^{\mu\nu}. \tag{1.138c}$$

We will need these two-dimensional spinor representations of the Lorentz generators later for the construction of the complex two-dimensional grading of the Poincaré algebra. In components, Eqs. (1.138a) and (1.138b) read:

$$(\sigma^{\mu\nu})_A{}^B = \frac{i}{4}\left[(\sigma^{\mu})_{A\dot{A}}(\overline{\sigma}^{\nu})^{\dot{A}B} - (\sigma^{\nu})_{A\dot{A}}(\overline{\sigma}^{\mu})^{\dot{A}B}\right],$$

$$(\overline{\sigma}^{\mu\nu})^{\dot{A}}{}_{\dot{B}} = \frac{i}{4}\left[(\overline{\sigma}^{\mu})^{\dot{A}A}(\sigma^{\nu})_{A\dot{B}} - (\overline{\sigma}^{\nu})^{\dot{A}A}(\sigma^{\mu})_{A\dot{B}}\right].$$

Proof of Eq. (1.138a): For an infinitesimal Lorentz transformation (ω_{ij} infinitesimal) we have:

$$\Lambda^{\mu}{}_{\nu}\sigma_{\mu}\overline{\sigma}^{\nu} = \Lambda_{\mu\nu}\sigma^{\mu}\overline{\sigma}^{\nu}$$

$$= \begin{pmatrix} 1 & \omega_{01} & \omega_{02} & \omega_{03} \\ -\omega_{01} & -1 & \omega_{12} & \omega_{13} \\ -\omega_{02} & -\omega_{12} & -1 & \omega_{23} \\ -\omega_{03} & -\omega_{13} & -\omega_{23} & -1 \end{pmatrix}_{\mu\nu} \sigma^{\mu}\overline{\sigma}^{\nu}$$

$$= \begin{pmatrix} 4 - 2(\omega_{03} + i\omega_{12}) & -2(\omega_{01} + i\omega_{23}) \\ & +2i(\omega_{02} + i\omega_{31}) \\ -2(\omega_{01} + i\omega_{23}) & \\ -2i(\omega_{02} + i\omega_{31}) & 4 + 2(\omega_{03} + i\omega_{12}) \end{pmatrix}.$$

Hence:

$$\det\left(\Lambda^{\mu}{}_{\nu}\sigma_{\mu}\overline{\sigma}^{\nu}\right) = 16 + O(\omega^2),$$

and, omitting terms of order ω^2:

$$N = \left[\det\left(\Lambda^{\mu}{}_{\nu}\sigma_{\mu}\overline{\sigma}^{\nu}\right)\right]^{\frac{1}{2}} = 4.$$

The relationship between an element $\Lambda^{\rho}{}_{\tau}$ of the Lie group $SO(1, 3; \mathbb{R})$ and an element $M^{\mu\nu}$ of the Lie algebra $so(1, 3; \mathbb{R})$ is

$$\Lambda^{\rho}{}_{\tau}(\omega) = \left[\exp\left(-\frac{i}{2}\omega_{\mu\nu}M^{\mu\nu}\right)\right]^{\rho}{}_{\tau}.$$

For infinitesimal parameters $\omega_{\mu\nu}$, the element $M(\Lambda) \in SL(2,\mathbb{C})$ is

$$M(\Lambda) = \frac{1}{N}\Lambda^{\rho}{}_{\tau}(\omega)\sigma_{\rho}\bar{\sigma}^{\tau}$$

$$= \frac{1}{N}\left(\eta_{\rho\tau} - \frac{i}{2}\omega_{\mu\nu}(M^{\mu\nu})_{\rho\tau}\right)\sigma^{\rho}\bar{\sigma}^{\tau}$$

$$= \frac{1}{N}\sigma^{\mu}\bar{\sigma}_{\mu} - \frac{i}{2}\omega_{\mu\nu}\sigma^{\mu\nu},$$

where

$$\sigma^{\mu\nu} := \frac{1}{N}\left(M^{\mu\nu}\right)_{\rho\tau}\sigma^{\rho}\bar{\sigma}^{\tau}.$$

Now

$$x'_{\rho} = \Lambda_{\rho\tau}x^{\tau} = (\eta_{\rho\tau} + \omega_{\rho\tau})x^{\tau},$$

and since every homogeneous linear transformation group possesses the trivial self-representation in the four-space spanned by the x_{μ}, we have:

$$-\frac{i}{2}\omega_{\mu\nu}\left(M^{\mu\nu}\right)_{\rho\tau} = \omega_{\rho\tau}.$$

This equation implies for the M's:

$$\left(M^{\mu\nu}\right)^{\rho}{}_{\tau} = i\left(\eta^{\mu\rho}\delta^{\nu}{}_{\tau} - \eta^{\nu\rho}\delta^{\mu}{}_{\tau}\right).$$

Of course, from Eq. (1.11) we know that the generators of the Lorentz group have this form. Hence

$$\sigma^{\mu\nu} = \frac{i}{N}\left(\eta^{\mu\rho}\delta^{\nu}{}_{\tau} - \eta^{\nu\rho}\delta^{\mu}{}_{\tau}\right)\sigma_{\rho}\bar{\sigma}^{\tau} = \frac{i}{4}\left(\sigma^{\mu}\bar{\sigma}^{\nu} - \sigma^{\nu}\bar{\sigma}^{\mu}\right).$$

In a similar way, $\bar{\sigma}^{\mu\nu}$ is obtained by considering M^{*} as a function of the Lorentz transformation Λ. The matrices $\sigma^{\mu\nu}$ and $\bar{\sigma}^{\mu\nu}$ obey the commutation relations of the Lorentz algebra $so(1,3;\mathbb{R})$, *i.e.* Eq. (1.14).

Proposition 1.32:

The generators of rotations $\tilde{J}_i, i = 1,2,3$, and Lorentz boosts $\tilde{K}_j, j = 1,2,3$, defined in Eqs. (1.19) and (1.20), *i.e.*

$$J_i = \frac{1}{2}\epsilon_{ijk}M^{jk}, \quad K_i = M_{0i}$$

can be written:

$$\tilde{J}^i = \frac{1}{2}\epsilon^i{}_{jk}\sigma^{jk} = \frac{1}{2}\sigma^i\bar{\sigma}^0, \tag{1.139a}$$

$$\tilde{K}^i = \sigma^{0i} = -\frac{i}{2}\sigma^i\bar{\sigma}^0. \tag{1.139b}$$

Proof:

$$\tilde{J}_i = \frac{1}{2}\epsilon_{ijk}\sigma^{jk} = \frac{1}{2}\epsilon_{ijk}\frac{i}{4}\left(\sigma^j\overline{\sigma}^k - \sigma^k\overline{\sigma}^j\right) \overset{(1.108)}{=} \frac{i}{8}\epsilon_{ijk}\left(-\sigma^j\sigma^k + \sigma^k\sigma^j\right)$$

$$= -\frac{i}{8}\epsilon_{ijk}\left[\sigma^j, \sigma^k\right] = -\frac{i}{8}\epsilon_{ijk}\cdot 2i\epsilon^{jkl}\sigma_l = \frac{1}{4}\epsilon_{jki}\epsilon^{jkl}\sigma_l = \frac{1}{2}\delta_i{}^l\sigma_l = \frac{1}{2}\sigma_i,$$

where we used

$$\sum_{j,k=1}^{3}\epsilon_{jki}\epsilon^{jkl} = 2\delta_i{}^l.$$

Relation (1.139b) is shown as follows:

$$\tilde{K}_i = \sigma^{0i} = \frac{i}{4}\left(\sigma^0\overline{\sigma}^i - \sigma^i\overline{\sigma}^0\right) = \frac{i}{4}\left(-\sigma^i - \sigma^i\right) = -\frac{i}{2}\sigma^i,$$

where we used again Eq. (1.108).

Proposition 1.33:

The $\sigma^{\mu\nu}$ are selfdual, *i.e.*

$$\sigma^{\mu\nu} = \frac{1}{2i}\epsilon^{\mu\nu\rho\sigma}\sigma_{\rho\sigma}, \tag{1.140a}$$

whereas the $\overline{\sigma}^{\mu\nu}$ are anti-selfdual, *i.e.*

$$\overline{\sigma}^{\mu\nu} = -\frac{1}{2i}\epsilon^{\mu\nu\rho\sigma}\overline{\sigma}_{\rho\sigma}. \tag{1.140b}$$

Proof: In order to prove the first of these relations consider

(i) $\mu = 0, \nu = i; i = 1, 2, 3$:

$$\sigma^{0i} = \frac{i}{4}\left(\sigma^0\overline{\sigma}^i - \sigma^i\overline{\sigma}^0\right) \overset{(1.108)}{=} \frac{i}{4}\left(-\sigma^0\sigma^i - \sigma^i\sigma^0\right) = -\frac{i}{2}\sigma^i = \frac{1}{2i}\sigma^i.$$

On the other hand with $i, j, k = 1, 2, 3$:

$$\frac{1}{2i}\epsilon^{0i\rho\sigma}\sigma_{\rho\sigma} = \frac{1}{2i}\epsilon^{0ijk}\sigma_{jk} = \frac{1}{2i}\epsilon^{0ijk}\frac{i}{4}\left(\sigma_j\overline{\sigma}_k - \sigma_k\overline{\sigma}_j\right)$$

$$= \frac{1}{8}\epsilon^{0ijk}\left(-\sigma_j\sigma_k + \sigma_k\sigma_j\right) = -\frac{1}{8}\epsilon^{0ijk}\left[\sigma_j, \sigma_k\right]$$

$$= -\frac{1}{8}\epsilon^{0ijk}2i\epsilon_{jkl}\sigma^l = \frac{1}{4i}\epsilon^{0ijk}\epsilon_{jkl}\sigma^l$$

$$= \frac{1}{4i}2\delta^i{}_l\sigma^l = \frac{1}{2i}\sigma^i = \sigma^{0i},$$

as shown above. Now consider the case $\mu = i, \nu = j; i, j = 1, 2, 3$:

$$\sigma^{ij} \overset{(1.138a)}{=} \frac{i}{4}\left(\sigma^i\overline{\sigma}^j - \sigma^j\overline{\sigma}^i\right) \overset{(1.108)}{=} -\frac{i}{4}\left(\sigma^i\sigma^j - \sigma^j\sigma^i\right)$$

$$= -\frac{i}{4}\left[\sigma^i, \sigma^j\right] = -\frac{i}{4}2i\epsilon^{ijk}\sigma_k = \frac{1}{2}\epsilon^{ijk}\sigma_k.$$

$$\frac{1}{2i}\epsilon^{ij\rho\sigma}\sigma_{\rho\sigma} = \frac{1}{2i}\left(\epsilon^{ijk0}\sigma_{k0} + \epsilon^{ij0k}\sigma_{0k}\right) = \frac{1}{2i}\left(\epsilon^{ijk0}\sigma_{k0} + \epsilon^{ijk0}\sigma_{k0}\right)$$

$$= \frac{1}{i}\epsilon^{ijk0}\sigma_{k0} = \frac{1}{i}\epsilon^{ijk0}\frac{i}{4}\left(\sigma_k\overline{\sigma}_0 - \sigma_0\overline{\sigma}_k\right)$$

$$= \frac{1}{2}\epsilon^{ijk0}\sigma_k = \sigma^{ij}.$$

(ii) Equation (1.140b) is shown as follows: We start with $\mu = 0, \nu = i = 1, 2, 3$, and obtain from Eq. (1.138b):

$$\overline{\sigma}^{0i} = \frac{i}{4}\left(\overline{\sigma}^0\sigma^i - \overline{\sigma}^i\sigma^0\right) = \frac{i}{4}\left(2\sigma^i\right) = -\frac{1}{2i}\sigma^i.$$

On the other hand:

$$-\frac{1}{2i}\epsilon^{0i\rho\sigma}\overline{\sigma}_{\rho\sigma} = -\frac{1}{2i}\epsilon^{0ijk}\overline{\sigma}_{jk} \overset{(1.138b)}{=} -\frac{1}{2i}\epsilon^{0ijk}\frac{i}{4}\left(\overline{\sigma}_j\sigma_k - \overline{\sigma}_k\sigma_j\right)$$

$$= \frac{1}{8}\epsilon^{0ijk}\left(\sigma_j\sigma_k - \sigma_k\sigma_j\right) = \frac{1}{8}\epsilon^{0ijk}\left[\sigma_j, \sigma_k\right]$$

$$= \frac{1}{8}\epsilon^{0ijk} \cdot 2i\epsilon_{jkl}\sigma^l = -\frac{1}{4i}\epsilon^{0ijk}\epsilon_{jkl}\sigma^l$$

$$= -\frac{1}{4i}2\delta^i{}_l\sigma^l = -\frac{1}{2i}\sigma^i = \overline{\sigma}^{0i}.$$

$\mu = i, \nu = j; \quad i, j = 1, 2, 3$:

$$\overline{\sigma}^{ij} = \frac{i}{4}\left(\overline{\sigma}^i\sigma^j - \overline{\sigma}^j\sigma^i\right) = -\frac{i}{4}\left[\sigma^i, \sigma^j\right] = -\frac{i}{4} \cdot 2i\epsilon^{ijk}\sigma_k = \frac{1}{2}\epsilon^{ijk}\sigma_k.$$

Furthermore:

$$-\frac{1}{2i}\epsilon^{ij\rho\sigma}\overline{\sigma}_{\rho\sigma} = -\frac{1}{2i}\left(\epsilon^{ijk0}\overline{\sigma}_{k0} + \epsilon^{ij0k}\overline{\sigma}_{0k}\right) = -\frac{1}{2i}2\epsilon^{ijk0}\overline{\sigma}_{k0}$$

$$= i\epsilon^{ijk0}\frac{i}{4}\left(\overline{\sigma}_k\sigma_0 - \overline{\sigma}_0\sigma_k\right) = -\frac{1}{4}\epsilon^{ijk0}(-2\sigma_k)$$

$$= \frac{1}{2}\epsilon^{ijk0}\sigma_k = \overline{\sigma}^{ij}.$$

Proposition 1.34:

The matrices σ^μ and $\overline{\sigma}^\mu$, defined by Eqs. (1.105), (1.106a), and (1.106b) obey the following relations:

$$\sigma^\mu \overline{\sigma}^\nu + \sigma^\nu \overline{\sigma}^\mu = 2\eta^{\mu\nu} \mathbb{1}_{2\times 2}, \qquad (1.141a)$$

$$\overline{\sigma}^\mu \sigma^\nu + \overline{\sigma}^\nu \sigma^\mu = 2\eta^{\mu\nu} \mathbb{1}_{2\times 2}. \qquad (1.141b)$$

In components these equations read:

$$(\sigma^\mu)_{A\dot{B}} (\overline{\sigma}^\nu)^{\dot{B}B} + (\sigma^\nu)_{A\dot{B}} (\overline{\sigma}^\mu)^{\dot{B}B} = 2\eta^{\mu\nu} \delta_A{}^B,$$

$$(\overline{\sigma}^\mu)^{\dot{A}B} (\sigma^\nu)_{B\dot{B}} + (\overline{\sigma}^\nu)^{\dot{A}B} (\sigma^\mu)_{B\dot{B}} = 2\eta^{\mu\nu} \delta^{\dot{A}}{}_{\dot{B}}.$$

Proof: From the definitions (1.103a) and (1.108) we have

$$(\sigma^\mu) = (\mathbb{1}_{2\times 2}, \boldsymbol{\sigma}) \qquad \text{and} \qquad (\overline{\sigma}^\mu) = (\mathbb{1}_{2\times 2}, -\boldsymbol{\sigma})$$

We verify Eqs. (1.141a) and (1.141b) by considering the following cases:

(i) $\mu = \nu = 0$:
$$(\sigma^0 \overline{\sigma}^0 + \sigma^0 \overline{\sigma}^0)_A{}^B = 2(\mathbb{1})_A{}^B = 2\eta^{00} \delta_A{}^B.$$

(ii) $\mu = 0, \nu = k = 1, 2, 3$:
$$(\sigma^0 \overline{\sigma}^k + \sigma^k \overline{\sigma}^0)_A{}^B = (\mathbb{1}_{2\times 2}(-\sigma^k) + \sigma^k \mathbb{1}_{2\times 2})_A{}^B = (-\sigma^k + \sigma^k)_A{}^B = 0.$$

(iii) $\mu = i, \nu = j; \; i, j = 1, 2, 3$:
$$(\sigma^i \overline{\sigma}^j + \sigma^j \overline{\sigma}^i)_A{}^B = -(\sigma^i \sigma^j + \sigma^j \sigma^i)_A{}^B = -(\{\sigma^i, \sigma^j\})_A{}^B$$
$$= -2\delta^{ij}(\mathbb{1}_{2\times 2})_A{}^B = -2\delta^{ij} \delta_A{}^B.$$

The second relation (1.141b) can be shown in a similar way.

In later calculations we will frequently use equations (1.141a) and (1.141b) in the form:

$$\sigma^\mu \overline{\sigma}^\nu = 2\eta^{\mu\nu} \mathbb{1}_{2\times 2} - \sigma^\nu \overline{\sigma}^\mu, \qquad (1.142)$$

$$\overline{\sigma}^\mu \sigma^\nu = 2\eta^{\mu\nu} \mathbb{1}_{2\times 2} - \overline{\sigma}^\nu \sigma^\mu. \qquad (1.143)$$

Proposition 1.35:

The generators $\sigma^{\mu\nu}$ of $SL(2,\mathbb{C})$ in the spinor representation obey the following trace relation:

$$\mathrm{Tr}\left[\sigma^{\mu\nu}\sigma^{\rho\sigma}\right] = \frac{1}{2}\left(\eta^{\mu\rho}\eta^{\nu\sigma} - \eta^{\mu\sigma}\eta^{\nu\rho}\right) + \frac{i}{2}\epsilon^{\mu\nu\rho\sigma}, \qquad (1.144)$$

and

$$\mathrm{Tr}\left[\overline{\sigma}^{\mu\nu}\overline{\sigma}^{\rho\sigma}\right] = \frac{1}{2}\left(\eta^{\mu\rho}\eta^{\nu\sigma} - \eta^{\mu\sigma}\eta^{\nu\rho}\right) - \frac{i}{2}\epsilon^{\mu\nu\rho\sigma}. \qquad (1.145)$$

Proof: Using Eq. (1.138a) we find:

$$
\begin{aligned}
\mathrm{Tr}\left[\sigma^{\mu\nu}\sigma^{\rho\sigma}\right] &= -\frac{1}{16}\mathrm{Tr}\left[\left(\sigma^{\mu}\overline{\sigma}^{\nu} - \sigma^{\nu}\overline{\sigma}^{\mu}\right)\left(\sigma^{\rho}\overline{\sigma}^{\sigma} - \sigma^{\sigma}\overline{\sigma}^{\rho}\right)\right] \\
&= -\frac{1}{16}\mathrm{Tr}\left[\sigma^{\mu}\overline{\sigma}^{\nu}\sigma^{\rho}\overline{\sigma}^{\sigma} - \sigma^{\mu}\overline{\sigma}^{\nu}\sigma^{\sigma}\overline{\sigma}^{\rho} - \sigma^{\nu}\overline{\sigma}^{\mu}\sigma^{\rho}\overline{\sigma}^{\sigma} + \sigma^{\nu}\overline{\sigma}^{\mu}\sigma^{\sigma}\overline{\sigma}^{\rho}\right] \\
&= -\frac{1}{16}\mathrm{Tr}\left[\sigma^{\mu}\overline{\sigma}^{\nu}\sigma^{\rho}\overline{\sigma}^{\sigma}\right] + \frac{1}{16}\mathrm{Tr}\left[\sigma^{\mu}\overline{\sigma}^{\nu}\sigma^{\sigma}\overline{\sigma}^{\rho}\right] \\
&\quad + \frac{1}{16}\mathrm{Tr}\left[\sigma^{\nu}\overline{\sigma}^{\mu}\sigma^{\rho}\overline{\sigma}^{\sigma}\right] - \frac{1}{16}\mathrm{Tr}\left[\sigma^{\nu}\overline{\sigma}^{\mu}\sigma^{\sigma}\overline{\sigma}^{\rho}\right].
\end{aligned}
$$

Using now (to be shown below in Eq. (1.146))

$$\frac{1}{2}\mathrm{Tr}\left[\sigma^{\mu}\overline{\sigma}^{\nu}\sigma^{\rho}\overline{\sigma}^{\sigma}\right] = \eta^{\mu\nu}\eta^{\rho\sigma} + \eta^{\nu\rho}\eta^{\mu\sigma} - \eta^{\mu\rho}\eta^{\nu\sigma} - i\epsilon^{\mu\nu\rho\sigma},$$

we obtain:

$$
\begin{aligned}
\mathrm{Tr}\left[\sigma^{\mu\nu}\sigma^{\rho\sigma}\right] &= -\frac{1}{8}\left\{\eta^{\mu\nu}\eta^{\rho\sigma} + \eta^{\nu\rho}\eta^{\mu\sigma} - \eta^{\mu\rho}\eta^{\nu\sigma} - i\epsilon^{\mu\nu\rho\sigma}\right\} \\
&\quad + \frac{1}{8}\left\{\eta^{\mu\nu}\eta^{\sigma\rho} + \eta^{\mu\rho}\eta^{\nu\sigma} - \eta^{\mu\sigma}\eta^{\nu\rho} - i\epsilon^{\mu\nu\sigma\rho}\right\} \\
&\quad + \frac{1}{8}\left\{\eta^{\nu\mu}\eta^{\rho\sigma} + \eta^{\mu\rho}\eta^{\nu\sigma} - \eta^{\nu\rho}\eta^{\mu\sigma} - i\epsilon^{\nu\mu\rho\sigma}\right\} \\
&\quad - \frac{1}{8}\left\{\eta^{\nu\mu}\eta^{\sigma\rho} + \eta^{\mu\sigma}\eta^{\nu\rho} - \eta^{\nu\sigma}\eta^{\mu\rho} - i\epsilon^{\nu\mu\sigma\rho}\right\} \\
&= \frac{1}{2}\eta^{\mu\rho}\eta^{\nu\sigma} - \frac{1}{2}\eta^{\mu\sigma}\eta^{\nu\rho} + \frac{i}{2}\epsilon^{\mu\nu\rho\sigma},
\end{aligned}
$$

taking into account the total antisymmetry of the tensor $\epsilon^{\mu\nu\rho\sigma}$. Hence

$$\mathrm{Tr}\left[\sigma^{\mu\nu}\sigma^{\rho\sigma}\right] = \frac{1}{2}\left(\eta^{\mu\rho}\eta^{\nu\sigma} - \eta^{\mu\sigma}\eta^{\nu\rho}\right) + \frac{i}{2}\epsilon^{\mu\nu\rho\sigma}.$$

as had to be shown. To complete the proof we have to demonstrate the following Proposition.

Proposition 1.36:

The following relation holds

$$\frac{1}{2}\text{Tr}\left[\sigma^\mu\overline{\sigma}^\nu\sigma^\rho\overline{\sigma}^\sigma\right] = \eta^{\mu\nu}\eta^{\rho\sigma} + \eta^{\nu\rho}\eta^{\mu\sigma} - \eta^{\mu\rho}\eta^{\nu\sigma} - i\epsilon^{\mu\nu\rho\sigma}. \tag{1.146}$$

Proof: We decompose the left hand side of Eq. (1.146) in the following way:

$$\frac{1}{2}\text{Tr}\left[\sigma^\mu\overline{\sigma}^\nu\sigma^\rho\overline{\sigma}^\sigma\right] = \frac{1}{2}\text{Tr}\left[\frac{1}{2}\left(\sigma^\mu\overline{\sigma}^\nu\sigma^\rho\overline{\sigma}^\sigma + \sigma^\rho\overline{\sigma}^\nu\sigma^\mu\overline{\sigma}^\sigma\right)\right.$$
$$\left. + \frac{1}{2}\left(\sigma^\mu\overline{\sigma}^\nu\sigma^\rho\overline{\sigma}^\sigma - \sigma^\rho\overline{\sigma}^\nu\sigma^\mu\overline{\sigma}^\sigma\right)\right]. \tag{1.147}$$

The first part of the trace is symmetric in μ and ρ whereas the second part is antisymmetric in μ and ρ. Consider first the symmetric part.

$$\frac{1}{4}\text{Tr}\left[\sigma^\mu\overline{\sigma}^\nu\sigma^\rho\overline{\sigma}^\sigma + \sigma^\rho\overline{\sigma}^\nu\sigma^\mu\overline{\sigma}^\sigma\right]$$

$$\overset{(1.142)}{=} \frac{1}{4}\text{Tr}\left[\left(2\eta^{\mu\nu} - \sigma^\nu\overline{\sigma}^\mu\right)\sigma^\rho\overline{\sigma}^\sigma + \sigma^\rho\overline{\sigma}^\nu\sigma^\mu\overline{\sigma}^\sigma\right]$$

$$= \frac{1}{4}\text{Tr}\left[2\eta^{\mu\nu}\sigma^\rho\overline{\sigma}^\sigma - \sigma^\nu\overline{\sigma}^\mu\sigma^\rho\overline{\sigma}^\sigma + \sigma^\rho\overline{\sigma}^\nu\sigma^\mu\overline{\sigma}^\sigma\right]$$

$$\overset{(1.143)}{=} \frac{1}{2}\eta^{\mu\nu}\text{Tr}\left[\sigma^\rho\overline{\sigma}^\sigma\right] + \frac{1}{4}\text{Tr}\left[-\sigma^\nu\left(2\eta^{\mu\rho} - \overline{\sigma}^\rho\sigma^\mu\right)\overline{\sigma}^\sigma + \sigma^\rho\overline{\sigma}^\nu\sigma^\mu\overline{\sigma}^\sigma\right]$$

$$\overset{(1.109a)}{=} \eta^{\mu\nu}\eta^{\rho\sigma} - \frac{1}{2}\eta^{\mu\rho}\text{Tr}\left[\sigma^\nu\overline{\sigma}^\sigma\right] + \frac{1}{4}\text{Tr}\left[\sigma^\nu\overline{\sigma}^\rho\sigma^\mu\overline{\sigma}^\sigma + \sigma^\rho\overline{\sigma}^\nu\sigma^\mu\overline{\sigma}^\sigma\right]$$

$$\overset{\substack{(1.109a) \\ (1.142)}}{=} \eta^{\mu\nu}\eta^{\rho\sigma} - \eta^{\mu\rho}\eta^{\nu\sigma} + \frac{1}{4}\text{Tr}\left[\left(2\eta^{\nu\rho} - \sigma^\rho\overline{\sigma}^\nu\right)\sigma^\mu\overline{\sigma}^\sigma + \sigma^\rho\overline{\sigma}^\nu\sigma^\mu\overline{\sigma}^\sigma\right]$$

$$= \eta^{\mu\nu}\eta^{\rho\sigma} - \eta^{\mu\rho}\eta^{\nu\sigma} + \eta^{\nu\rho}\eta^{\mu\sigma} + \frac{1}{4}\text{Tr}\left[-\sigma^\rho\overline{\sigma}^\nu\sigma^\mu\overline{\sigma}^\sigma + \sigma^\rho\overline{\sigma}^\nu\sigma^\mu\overline{\sigma}^\sigma\right]$$

$$= \eta^{\mu\nu}\eta^{\rho\sigma} - \eta^{\mu\rho}\eta^{\nu\sigma} + \eta^{\nu\rho}\eta^{\mu\sigma}.$$

Hence, the symmetric part of Eq. (1.147) gives:

$$\frac{1}{2}\text{Tr}\left[\sigma^\mu\overline{\sigma}^\nu\sigma^\rho\overline{\sigma}^\sigma\right] = \eta^{\mu\nu}\eta^{\rho\sigma} + \eta^{\mu\sigma}\eta^{\nu\rho} - \eta^{\mu\rho}\eta^{\nu\sigma}$$
$$+ \frac{1}{4}\text{Tr}\left[\sigma^\mu\overline{\sigma}^\nu\sigma^\rho\overline{\sigma}^\sigma - \sigma^\rho\overline{\sigma}^\nu\sigma^\mu\overline{\sigma}^\sigma\right].$$

Now consider the antisymmetric part of Eq. (1.147). The following result will be needed later.

Proposition 1.37:

The tensor

$$A^{\mu\nu\rho\sigma} := \frac{1}{4}\mathrm{Tr}\left[\sigma^\mu\overline{\sigma}^\nu\sigma^\rho\overline{\sigma}^\sigma - \sigma^\rho\overline{\sigma}^\nu\sigma^\mu\overline{\sigma}^\sigma\right]. \tag{1.148}$$

is totally antisymmetric in its indices μ, ν, ρ, σ and therefore proportional to the totally antisymmetric tensor in 4-dimensional Minkowski space $\epsilon^{\mu\nu\rho\sigma}$, *i.e.*

$$A^{\mu\nu\rho\sigma} = c\epsilon^{\mu\nu\rho\sigma}, \quad c = -i.$$

Proof: We demonstrate the validity of the six possibilities separately.

(i) $A^{\mu\nu\rho\sigma} = -A^{\nu\mu\rho\sigma}$:

$$
\begin{aligned}
A^{\nu\mu\rho\sigma} &= \frac{1}{4}\mathrm{Tr}\left[\sigma^\nu\overline{\sigma}^\mu\sigma^\rho\overline{\sigma}^\sigma - \sigma^\rho\overline{\sigma}^\mu\sigma^\nu\overline{\sigma}^\sigma\right]\\
&\overset{\substack{(1.142)\\(1.143)}}{=} \frac{1}{4}\mathrm{Tr}\left[(2\eta^{\mu\nu} - \sigma^\mu\overline{\sigma}^\nu)\sigma^\rho\overline{\sigma}^\sigma - \sigma^\rho(2\eta^{\mu\nu} - \overline{\sigma}^\nu\sigma^\mu)\overline{\sigma}^\sigma\right]\\
&= \frac{1}{2}\eta^{\mu\nu}\mathrm{Tr}\left[\sigma^\rho\overline{\sigma}^\sigma\right] - \frac{1}{2}\eta^{\mu\nu}\mathrm{Tr}\left[\sigma^\rho\overline{\sigma}^\sigma\right]\\
&\quad -\frac{1}{4}\mathrm{Tr}\left[\sigma^\mu\overline{\sigma}^\nu\sigma^\rho\overline{\sigma}^\sigma - \sigma^\rho\overline{\sigma}^\nu\sigma^\mu\overline{\sigma}^\sigma\right]\\
&= -\frac{1}{4}\mathrm{Tr}\left[\sigma^\mu\overline{\sigma}^\nu\sigma^\rho\overline{\sigma}^\sigma - \sigma^\rho\overline{\sigma}^\nu\sigma^\mu\overline{\sigma}^\sigma\right]\\
&= -A^{\mu\nu\rho\sigma}.
\end{aligned}
$$

(ii) $A^{\mu\nu\rho\sigma} = -A^{\mu\nu\sigma\rho}$:

$$
\begin{aligned}
A^{\mu\nu\sigma\rho} &= \frac{1}{4}\mathrm{Tr}\left[\sigma^\mu\overline{\sigma}^\nu\sigma^\sigma\overline{\sigma}^\rho - \sigma^\sigma\overline{\sigma}^\nu\sigma^\mu\overline{\sigma}^\rho\right]\\
&\overset{(1.142)}{=} \frac{1}{4}\mathrm{Tr}\left[\sigma^\mu\overline{\sigma}^\nu(2\eta^{\sigma\rho} - \sigma^\rho\overline{\sigma}^\sigma) - \overline{\sigma}^\nu\sigma^\mu\overline{\sigma}^\rho\sigma^\sigma\right]\\
&= \frac{1}{2}\eta^{\sigma\rho}\mathrm{Tr}\left[\sigma^\mu\overline{\sigma}^\nu\right] + \frac{1}{4}\mathrm{Tr}\left[-\sigma^\mu\overline{\sigma}^\nu\sigma^\rho\overline{\sigma}^\sigma - \overline{\sigma}^\nu\sigma^\mu(2\eta^{\rho\sigma} - \overline{\sigma}^\sigma\sigma^\rho)\right]\\
&= \frac{1}{2}\eta^{\sigma\rho}\mathrm{Tr}\left[\sigma^\mu\overline{\sigma}^\nu\right] - \frac{1}{2}\eta^{\rho\sigma}\mathrm{Tr}\left[\overline{\sigma}^\nu\sigma^\mu\right]\\
&\quad -\frac{1}{4}\mathrm{Tr}\left[\sigma^\mu\overline{\sigma}^\nu\sigma^\rho\overline{\sigma}^\sigma - \sigma^\rho\overline{\sigma}^\nu\sigma^\mu\overline{\sigma}^\sigma\right]\\
&= -\frac{1}{4}\mathrm{Tr}\left[\sigma^\mu\overline{\sigma}^\nu\sigma^\rho\overline{\sigma}^\sigma - \sigma^\rho\overline{\sigma}^\nu\sigma^\mu\overline{\sigma}^\sigma\right] = -A^{\mu\nu\rho\sigma}.
\end{aligned}
$$

(iii) $A^{\mu\nu\rho\sigma} = -A^{\rho\nu\mu\sigma}$:
This holds by construction.

(iv) $A^{\mu\nu\rho\sigma} = -A^{\sigma\nu\rho\mu}$:

We have (using several times the cyclicity of the trace operation)

$$A^{\sigma\nu\rho\mu} = \frac{1}{4}\mathrm{Tr}\left[\sigma^\sigma\overline{\sigma}^\nu\sigma^\rho\overline{\sigma}^\mu - \sigma^\rho\overline{\sigma}^\nu\sigma^\sigma\overline{\sigma}^\mu\right]$$

$$\overset{(1.142)}{=} \frac{1}{4}\mathrm{Tr}\left[\overline{\sigma}^\mu\sigma^\sigma\overline{\sigma}^\nu\sigma^\rho - \sigma^\rho\overline{\sigma}^\nu\left(2\eta^{\sigma\mu} - \sigma^\mu\overline{\sigma}^\sigma\right)\right]$$

$$\overset{(1.142)}{=} \frac{1}{4}\mathrm{Tr}\left[\left(2\eta^{\mu\sigma} - \overline{\sigma}^\sigma\sigma^\mu\right)\overline{\sigma}^\nu\sigma^\rho + \sigma^\rho\overline{\sigma}^\nu\sigma^\mu\overline{\sigma}^\sigma\right] - \frac{1}{2}\eta^{\sigma\mu}\mathrm{Tr}\left[\sigma^\rho\overline{\sigma}^\nu\right]$$

$$= \frac{1}{2}\eta^{\mu\sigma}\mathrm{Tr}\left[\overline{\sigma}^\nu\sigma^\rho\right] - \frac{1}{2}\eta^{\sigma\mu}\mathrm{Tr}\left[\sigma^\rho\overline{\sigma}^\nu\right]$$

$$\quad - \frac{1}{4}\mathrm{Tr}\left[\sigma^\mu\overline{\sigma}^\nu\sigma^\rho\overline{\sigma}^\sigma - \sigma^\rho\overline{\sigma}^\nu\sigma^\mu\overline{\sigma}^\sigma\right]$$

$$= -\frac{1}{4}\mathrm{Tr}\left[\sigma^\mu\overline{\sigma}^\nu\sigma^\rho\overline{\sigma}^\sigma - \sigma^\rho\overline{\sigma}^\nu\sigma^\mu\overline{\sigma}^\sigma\right]$$

$$= -A^{\mu\nu\rho\sigma}.$$

(v) $A^{\mu\nu\rho\sigma} = -A^{\mu\rho\nu\sigma}$:

We have

$$A^{\mu\rho\nu\sigma} = \frac{1}{4}\mathrm{Tr}\left[\sigma^\mu\overline{\sigma}^\rho\sigma^\nu\overline{\sigma}^\sigma - \sigma^\nu\overline{\sigma}^\rho\sigma^\mu\overline{\sigma}^\sigma\right]$$

$$= \frac{1}{4}\mathrm{Tr}\left[\sigma^\mu\left(2\eta^{\rho\nu} - \overline{\sigma}^\nu\sigma^\rho\right)\overline{\sigma}^\sigma - \left(2\eta^{\nu\rho} - \sigma^\rho\overline{\sigma}^\nu\right)\sigma^\mu\overline{\sigma}^\sigma\right]$$

$$= \frac{1}{2}\eta^{\rho\nu}\mathrm{Tr}\left[\sigma^\mu\overline{\sigma}^\sigma\right] - \frac{1}{2}\eta^{\nu\rho}\mathrm{Tr}\left[\sigma^\mu\overline{\sigma}^\sigma\right]$$

$$\quad - \frac{1}{4}\mathrm{Tr}\left[\sigma^\mu\overline{\sigma}^\nu\sigma^\rho\overline{\sigma}^\sigma - \sigma^\rho\overline{\sigma}^\nu\sigma^\mu\overline{\sigma}^\sigma\right]$$

$$= -A^{\mu\nu\rho\sigma}.$$

(vi) $A^{\mu\nu\rho\sigma} = -A^{\mu\sigma\rho\nu}$:

Using the cyclicity of the trace we immediately get:

$$A^{\mu\sigma\rho\nu} = \frac{1}{4}\mathrm{Tr}\left[\sigma^\mu\overline{\sigma}^\sigma\sigma^\rho\overline{\sigma}^\nu - \sigma^\rho\overline{\sigma}^\sigma\sigma^\mu\overline{\sigma}^\nu\right]$$

$$= -\frac{1}{4}\mathrm{Tr}\left[-\sigma^\rho\overline{\sigma}^\nu\sigma^\mu\overline{\sigma}^\sigma + \sigma^\mu\overline{\sigma}^\nu\sigma^\rho\overline{\sigma}^\sigma\right]$$

$$= -A^{\mu\nu\rho\sigma}.$$

For a quantity with n indices there are $n(n-1)/2$ possibilities of interchanging two indices. The quantity $A^{\mu\nu\rho\sigma}$ has four spacetime indices; hence there are six possible exchanges of any two indices. We have thus shown that $A^{\mu\nu\rho\sigma}$ is antisymmetric in exchanging any two of its indices. Hence, $A^{\mu\nu\rho\sigma}$

is totally antisymmetric. Thus, $A^{\mu\nu\rho\sigma}$ must be proportional to the totally antisymmetric four-tensor,[27] $\epsilon^{\mu\nu\rho\sigma}$ since $\epsilon^{\mu\nu\rho\sigma}$ is the only tensor in 4-space with this property. Hence we conclude

$$A^{\mu\nu\rho\sigma} = c \cdot \epsilon^{\mu\nu\rho\sigma},$$

where c is a constant. This completes the proof of Proposition 1.37.

Proposition 1.38:

The constant c in Eq. (1.148) is given by $c = -i$.

Proof: We use the following complete contraction of the four-dimensional Levi-Civita tensor:

$$\epsilon_{\mu\nu\rho\sigma}\epsilon^{\mu\nu\rho\sigma} = -4! = -24.$$

Then we have

$$\epsilon_{\mu\nu\rho\sigma}A^{\mu\nu\rho\sigma} = c\epsilon_{\mu\nu\rho\sigma}\epsilon^{\mu\nu\rho\sigma} = -24c.$$

On the other hand:

$$
\begin{aligned}
\epsilon_{\mu\nu\rho\sigma}A^{\mu\nu\rho\sigma} &= \frac{1}{4}\epsilon_{\mu\nu\rho\sigma}\operatorname{Tr}\left[\sigma^\mu\overline{\sigma}^\nu\sigma^\rho\overline{\sigma}^\sigma - \sigma^\rho\overline{\sigma}^\nu\sigma^\mu\overline{\sigma}^\sigma\right] \\
&= \frac{1}{4}\operatorname{Tr}\left[\epsilon_{\mu\nu\rho\sigma}\sigma^\mu\overline{\sigma}^\nu\sigma^\rho\overline{\sigma}^\sigma - \epsilon_{\mu\nu\rho\sigma}\sigma^\rho\overline{\sigma}^\nu\sigma^\mu\overline{\sigma}^\sigma\right] \\
&= \frac{1}{4}\operatorname{Tr}\left[\epsilon_{\mu\nu\rho\sigma}\sigma^\mu\overline{\sigma}^\nu\sigma^\rho\overline{\sigma}^\sigma - \epsilon_{\rho\nu\mu\sigma}\sigma^\mu\overline{\sigma}^\nu\sigma^\rho\overline{\sigma}^\sigma\right] \\
&= \frac{1}{4}\operatorname{Tr}\left[\epsilon_{\mu\nu\rho\sigma}\sigma^\mu\overline{\sigma}^\nu\sigma^\rho\overline{\sigma}^\sigma + \epsilon_{\mu\nu\rho\sigma}\sigma^\mu\overline{\sigma}^\nu\sigma^\rho\overline{\sigma}^\sigma\right] \\
&= \frac{1}{2}\operatorname{Tr}\left[\epsilon_{\mu\nu\rho\sigma}\sigma^\mu\overline{\sigma}^\nu\sigma^\rho\overline{\sigma}^\sigma\right].
\end{aligned}
$$

It is not difficult to show by explicit calculation that this expression is simply $24i$. Hence $c = -i$.

Proposition 1.39:

The product

$$\epsilon_{AC}(\sigma^{\mu\nu})_B{}^C$$

is symmetric in the spinor indices A and B. Since

$$\epsilon_{AC}(\sigma^{\mu\nu})_B{}^C = (\sigma^{\mu\nu}\epsilon^\top)_{BA},$$

we demonstrate that

---

[27]This tensor is also called the Levi-Civita tensor.

$$(\sigma^{\mu\nu}\epsilon^\top)_{BA} = (\epsilon\sigma^{\mu\nu\top})_{BA}. \tag{1.149}$$

Similarly

$$(\epsilon\overline{\sigma}^{\mu\nu})_{\dot{A}\dot{C}} = (\epsilon\overline{\sigma}^{\mu\nu})_{\dot{C}\dot{A}}. \tag{1.150}$$

Proof: We have to demonstrate that

$$(\sigma^{\mu\nu}\epsilon^\top)^\top = \sigma^{\mu\nu}\epsilon^\top.$$

Consider the expression on the left hand side:

$$
\begin{aligned}
(\sigma^{\mu\nu}\epsilon^\top)^\top &= \epsilon(\sigma^{\mu\nu})^\top \overset{(1.138a)}{=} \frac{i}{4}\epsilon\big(\sigma^\mu\overline{\sigma}^\nu - \sigma^\nu\overline{\sigma}^\mu\big)^\top \\
&\overset{(1.107)}{=} \frac{i}{4}\epsilon\big(\sigma^\mu\epsilon\sigma^{\nu\top}\epsilon^\top - \sigma^\nu\epsilon\sigma^{\mu\top}\epsilon^\top\big)^\top \\
&= \frac{i}{4}\epsilon\big(\epsilon\sigma^\nu\epsilon^\top\sigma^{\mu\top} - \epsilon\sigma^\mu\epsilon^\top\sigma^{\nu\top}\big) \\
&\overset{(1.66)}{=} \frac{i}{4}\epsilon\big[(-\epsilon^\top)\sigma^\nu(-\epsilon)\sigma^{\mu\top} + \epsilon^\top\sigma^\mu(-\epsilon)\sigma^{\nu\top}\big](-\epsilon^\top)\epsilon^\top \\
&= \frac{i}{4}\big(\sigma^\mu\epsilon\sigma^{\nu\top}\epsilon^\top - \sigma^\nu\epsilon\sigma^{\mu\top}\epsilon^\top\big)\epsilon^\top \\
&\overset{(1.107)}{=} \frac{i}{4}\big(\sigma^\mu\overline{\sigma}^\nu - \sigma^\nu\overline{\sigma}^\mu\big)\epsilon^\top = \sigma^{\mu\nu}\epsilon^\top.
\end{aligned}
$$

Proposition 1.40:

The following relation holds

$$(\phi\sigma^{\mu\nu}\chi) = -(\chi\sigma^{\mu\nu}\phi). \tag{1.151}$$

In particular

$$(\phi\sigma^{\mu\nu}\phi) = 0.$$

Proof: We have:

$$
\begin{aligned}
(\phi\sigma^{\mu\nu}\chi) &= \phi^A(\sigma^{\mu\nu})_A{}^B\chi_B = (\epsilon^{AC}\phi_C)(\sigma^{\mu\nu})_A{}^B(\epsilon_{BD}\chi^D) \\
&\overset{(1.94)}{=} -\chi^D(\sigma^{\mu\nu}\epsilon)_{AD}\epsilon^{AC}\phi_C \overset{(1.67)}{=} \chi^D(\sigma^{\mu\nu}\epsilon^\top)_{AD}\epsilon^{AC}\phi_C \\
&\overset{(1.149)}{=} \chi^D(\sigma^{\mu\nu}\epsilon^\top)_{DA}\epsilon^{AC}\phi_C = \chi^D(\sigma^{\mu\nu})_D{}^B\epsilon^\top_{BA}\epsilon^{AC}\phi_C \\
&\overset{(1.67)}{=} -\chi^D(\sigma^{\mu\nu})_D{}^B\epsilon_{BA}\epsilon^{AC}\phi_C \overset{(1.68)}{=} -\chi^D(\sigma^{\mu\nu})_D{}^B\delta_B{}^C\phi_C \\
&\overset{.}{=} -(\chi\sigma^{\mu\nu}\phi).
\end{aligned}
$$

As a consequence of Eq. (1.151) an expression like $(\phi\sigma^{\mu\nu}\phi)$ vanishes identically.

1.3.6 Higher Order Weyl Spinors and their Representations

Multiplying elementary Weyl spinors by one another, *i.e.* $\psi_A, \psi^B, \overline{\psi}_{\dot{A}}, \overline{\psi}^{\dot{B}}$, one can form higher order spinors or spinors which carry both dotted and undotted indices. In general these spinors transform under $SL(2, \mathbb{C})$ as the product of individual spinors.[28] From previous considerations we know that undotted spinors ψ_A are elements of the $(1/2, 0)$ representation of $SL(2, \mathbb{C})$ and dotted spinors $\overline{\chi}^{\dot{A}}$ transform according to the $(0, 1/2)$ representation. The Kronecker product of these two representations

$$(1/2, 0) \otimes (0, 1/2) = (1/2, 1/2)$$

gives a $(2 \cdot 1/2 + 1)^2 = 4$-dimensional representation of $SL(2, \mathbb{C})$, which can be shown to be equivalent to the four-dimensional self-representation of the group $SO(1, 3; \mathbb{R})$ in the space spanned by the $x_\mu, \mu = 0, 1, 2, 3$, *i.e.* the four components of

$$\psi_A \otimes \overline{\chi}_{\dot{A}} \equiv \psi_A \overline{\chi}_{\dot{A}} = \frac{1}{2} (\psi \sigma_\mu \overline{\chi}) \sigma^\mu_{A\dot{A}} \tag{1.152a}$$

form a four-vector V_μ. We can find the components of this vector by comparing Eq. (1.152a) with Eq. (1.111), *i.e.*

$$(X)_{A\dot{A}} = (x_\mu \sigma^\mu)_{A\dot{A}},$$

where

$$X_{1\dot{1}} = x^0 - x^3, \quad X_{1\dot{2}} = x^1 + ix^2, \quad X_{2\dot{1}} = x^1 - ix^2, \quad X_{2\dot{2}} = x^0 + x^3.$$

Thus

$$V_\mu = \frac{1}{2} (\psi \sigma_\mu \overline{\chi}).$$

An explicit proof of the vector character of $\psi_A \otimes \overline{\chi}_{\dot{A}}$ is given in Proposition 1.33 (see also [97]).

Proof of Eq. (1.152a): Using the summation conventions Eqs. (1.92a), (1.92b), and (1.93) we have:

$$\frac{1}{2} (\psi \sigma^\mu \overline{\chi}) (\sigma_\mu)_{A\dot{A}} = \frac{1}{2} \psi^B (\sigma^\mu)_{B\dot{B}} \overline{\chi}^{\dot{B}} (\sigma_\mu)_{A\dot{A}}$$

$$\overset{(1.83)}{=} \frac{1}{2} \epsilon^{BC} \psi_C (\sigma^\mu)_{B\dot{B}} \epsilon^{\dot{B}\dot{D}} \overline{\chi}_{\dot{D}} (\sigma_\mu)_{A\dot{A}}$$

---

[28]We mention, that of course not every higher order spinor is expressible as the product of elementary spinors.

$$= -\frac{1}{2}\psi_C\epsilon^{BC}\epsilon^{\dot{D}\dot{B}}(\sigma^\mu)_{B\dot{B}}\overline{\chi}_{\dot{D}}(\sigma_\mu)_{A\dot{A}}$$

$$= \frac{1}{2}\psi_C\big(\epsilon^{CB}\epsilon^{\dot{D}\dot{B}}(\sigma^\mu)_{B\dot{B}}\big)\overline{\chi}_{\dot{D}}(\sigma_\mu)_{A\dot{A}}$$

$$\overset{(1.106a)}{=} \frac{1}{2}\psi_C\overline{\chi}_{\dot{D}}(\overline{\sigma}^\mu)^{\dot{D}C}(\sigma_\mu)_{A\dot{A}}$$

$$\overset{(1.110)}{=} \psi_C\overline{\chi}_{\dot{D}}\delta^{\dot{D}}{}_{\dot{A}}\delta^C{}_A$$

$$= \psi_A\overline{\chi}_{\dot{A}}.$$

Equation (1.152a) can also be demonstrated with the help of the Fierz reordering formula, Eq. (1.133).

Proposition 1.41:

The Kronecker product of two left-handed Weyl spinors gives:

$$(1/2, 0) \otimes (1/2, 0) = (0, 0) \oplus (1, 0),$$

that means:

$$\psi_A \otimes \chi_B = \frac{1}{2}\epsilon_{AB}(\psi\chi) + \frac{1}{2}(\sigma^{\mu\nu}\epsilon^\top)_{AB}(\psi\sigma_{\mu\nu}\chi), \qquad (1.152b)$$

corresponding to a scalar $(0, 0)$ and a selfdual second rank tensor ($\sigma_{\mu\nu}$ is selfdual according to Eq. (1.140a)).

Proof: As usual we decompose the tensor product into a symmetric and an antisymmetric part, *i.e.* we write:

$$\psi_A \otimes \chi_B = \frac{1}{2}\big(\psi_A\chi_B + \psi_B\chi_A\big) + \frac{1}{2}\big(\psi_A\chi_B - \psi_B\chi_A\big).$$

Consider first the antisymmetric part. Using Eq. (1.100b) we have:

$$\frac{1}{2}\big(\psi_A\chi_B - \psi_B\chi_A\big) = \frac{1}{2}\Big[\frac{1}{2}(\psi\chi)\epsilon_{AB} - \frac{1}{2}(\psi\chi)\epsilon_{BA}\Big] = \frac{1}{2}\epsilon_{AB}(\psi\chi).$$

Now consider the symmetric part. We show first that

$$(\sigma^{\mu\nu})_A{}^B(\sigma_{\mu\nu})_C{}^D = \epsilon_{AC}\epsilon^{BD} + \delta_A{}^D\delta^B{}_C.$$

We have

$$(\sigma^{\mu\nu})_A{}^B(\sigma_{\mu\nu})_C{}^D \overset{(1.138a)}{=} -\frac{1}{16}\big(\sigma^\mu\overline{\sigma}^\nu - \sigma^\nu\overline{\sigma}^\mu\big)_A{}^B\big(\sigma_\mu\overline{\sigma}_\nu - \sigma_\nu\overline{\sigma}_\mu\big)_C{}^D$$

$$= -\frac{1}{8}\Big[(\sigma^\mu\overline{\sigma}^\nu)_A{}^B(\sigma_\mu\overline{\sigma}_\nu)_C{}^D - (\sigma^\nu\overline{\sigma}^\mu)_A{}^B(\sigma_\mu\overline{\sigma}_\nu)_C{}^D\Big]$$

Now we have

$$
\begin{aligned}
\left(\sigma^\mu \overline{\sigma}^\nu\right)_A{}^B \left(\sigma_\mu \overline{\sigma}_\nu\right)_C{}^D
&= (\sigma^\mu)_{A\dot{A}} (\overline{\sigma}^\nu)^{\dot{A}B} (\sigma_\mu)_{C\dot{B}} (\overline{\sigma}_\nu)^{\dot{B}D} \\
&= (\sigma^\mu)_{A\dot{A}} (\sigma_\mu)_{C\dot{B}} (\overline{\sigma}^\nu)^{\dot{A}B} (\overline{\sigma}_\nu)^{\dot{B}D} \\
&\overset{\underset{(1.106a)}{(1.106b)}}{=} (\sigma^\mu)_{A\dot{A}} \epsilon_{CE} \epsilon_{\dot{B}\dot{C}} (\overline{\sigma}_\mu)^{\dot{C}E} \epsilon^{\dot{A}\dot{F}} \epsilon^{BF} (\sigma^\nu)_{F\dot{F}} (\overline{\sigma}_\nu)^{\dot{B}D} \\
&= (\sigma^\mu)_{A\dot{A}} (\overline{\sigma}_\mu)^{\dot{C}E} (\sigma^\nu)_{F\dot{F}} (\overline{\sigma}_\nu)^{\dot{B}D} \epsilon_{CE} \epsilon_{\dot{B}\dot{C}} \epsilon^{\dot{A}\dot{F}} \epsilon^{BF} \\
&\overset{(1.110)}{=} 4 \delta_A{}^E \delta_{\dot{A}}{}^{\dot{C}} \delta_F{}^D \delta_{\dot{F}}{}^{\dot{B}} \epsilon_{CE} \epsilon_{\dot{B}\dot{C}} \epsilon^{\dot{A}\dot{F}} \epsilon^{BF} \\
&= 4 \epsilon_{CA} \epsilon_{\dot{F}\dot{C}} \epsilon^{\dot{C}\dot{F}} \epsilon^{BD} \\
&\overset{(1.78b)}{=} 4 \epsilon_{CA} \epsilon^{BD} \delta_{\dot{F}}{}^{\dot{F}} \\
&= -8 \epsilon_{AC} \epsilon^{BD}.
\end{aligned}
$$

Furthermore

$$
\begin{aligned}
\left(\sigma^\nu \overline{\sigma}^\mu\right)_A{}^B \left(\sigma_\mu \overline{\sigma}_\nu\right)_C{}^D
&= (\sigma^\nu)_{A\dot{A}} (\overline{\sigma}^\mu)^{\dot{A}B} (\sigma_\mu)_{C\dot{B}} (\overline{\sigma}_\nu)^{\dot{B}D} \\
&= (\sigma^\nu)_{A\dot{A}} (\overline{\sigma}_\nu)^{\dot{B}D} (\sigma_\mu)_{C\dot{B}} (\overline{\sigma}^\mu)^{\dot{A}B} \\
&\overset{(1.110)}{=} 4 \delta_A{}^D \delta_{\dot{A}}{}^{\dot{B}} \delta_C{}^B \delta_{\dot{B}}{}^{\dot{A}} \\
&= 8 \delta_A{}^D \delta_C{}^B.
\end{aligned}
$$

Hence:

$$
\begin{aligned}
(\sigma^{\mu\nu})_A{}^B (\sigma_{\mu\nu})_C{}^D &= -\frac{1}{8}\left\{ (\sigma^\mu \overline{\sigma}^\nu)_A{}^B (\sigma_\mu \overline{\sigma}_\nu)_C{}^D - (\sigma^\nu \overline{\sigma}^\mu)_A{}^B (\sigma_\mu \overline{\sigma}_\nu)_C{}^D \right\} \\
&= \epsilon_{AC} \epsilon^{BD} + \delta_A{}^D \delta^B{}_C.
\end{aligned}
$$

With this result we obtain:

$$
\begin{aligned}
\frac{1}{2} (\sigma^{\mu\nu} \epsilon^\top)_{AB} (\psi \sigma_{\mu\nu} \chi)
&= \frac{1}{2} (\sigma^{\mu\nu})_A{}^C (\epsilon^\top)_{CB} \psi^D (\sigma_{\mu\nu})_D{}^F \chi_F \\
&= \frac{1}{2} \psi^D (\sigma^{\mu\nu})_A{}^C (\sigma_{\mu\nu})_D{}^F (\epsilon^\top)_{CB} \chi_F \\
&= \frac{1}{2} \psi^D \left\{ \epsilon_{AD} \epsilon^{CF} + \delta_A{}^F \delta_D{}^C \right\} (\epsilon^\top)_{CB} \chi_F \\
&= \frac{1}{2} \psi^D \epsilon_{AD} \epsilon^{FC} \epsilon_{CB} \chi_F + \frac{1}{2} \psi^D \delta_A{}^F \delta_D{}^C (\epsilon^\top)_{CB} \chi_F \\
&= \frac{1}{2} \psi_A \delta^F{}_B \chi_F + \frac{1}{2} \psi^C (\epsilon^\top)_{CB} \chi_A \\
&= \frac{1}{2} \left(\psi_A \chi_B + \psi_B \chi_A \right).
\end{aligned}
$$

Hence:

$$\psi_A \otimes \chi_B = \frac{1}{2}(\psi_A \chi_B + \psi_B \chi_A) + \frac{1}{2}(\psi_A \chi_B - \psi_B \chi_A)$$
$$= \frac{1}{2}\epsilon_{AB}(\psi\chi) + \frac{1}{2}(\sigma^{\mu\nu}\epsilon^\top)_{AB}(\psi\sigma_{\mu\nu}\chi).$$

The first term of Eq. (1.152b) transforms as a scalar since ϵ_{AB} transforms as an invariant spinor under transformations of $SL(2, \mathbb{C})$. The second term transforms as a three-vector under $SL(2, \mathbb{C})$, since the representation $(1, 0)$ is equivalent to the space of selfdual tensors of rank two. This second term is just the symmetric part of $\psi_A \otimes \chi_B$ according to Eq. (1.149), and symmetric spinors

$$T_{AB} = \frac{1}{2}(\psi_A \otimes \chi_B + \psi_B \otimes \chi_A) = T_{BA}$$

transform according to the $(1, 0)$-representation as can be shown by an analogous construction following Eq. (1.111). The odd and even parts of this product are recognized as the one-component scalar and three-component vector quantities which are well-known from the construction of the spin part of the wave function of two spin-$\frac{1}{2}$ particles. However, we can see this also as follows. From Eqs. (1.138a) and (1.138b) we infer that

$$\left\{\sigma^{\mu\nu} \mid \mu, \nu = 0, 1, 2, 3\right\}_A^{\ B} \implies \left\{\sigma^i \bar{\sigma}^0 \mid i = 1, 2, 3\right\}_A^{\ B},$$
$$\left\{\bar{\sigma}^{\mu\nu} \mid \mu, \nu = 0, 1, 2, 3\right\}_{\dot{B}}^{\ \dot{A}} \implies \left\{\bar{\sigma}^i \sigma^0 \mid i = 1, 2, 3\right\}_{\dot{B}}^{\ \dot{A}}.$$

Then

$$(\sigma^{\mu\nu}\epsilon^\top)_{AB}(\psi\sigma_{\mu\nu}\chi) \implies \begin{pmatrix} -\psi\sigma^1\chi + i\psi\sigma^2\chi & \psi\sigma^3\chi \\ \psi\sigma^3\chi & \psi\sigma^1\chi + i\psi\sigma^2\chi \end{pmatrix}_{AB},$$

and after rotation the three-component vector is seen to be[29]

$$\psi\boldsymbol{\sigma}\chi.$$

We also observe that

$$\det \begin{pmatrix} -x + iy & z \\ z & x + iy \end{pmatrix} = -(x^2 + y^2 + z^2).$$

Remark: In later calculations we shall need some further relations involving spinors of rank two. It is convenient to derive these relations here.

---

[29] See P. Roman [97], p. 72.

Let Φ_{AB}, $A, B = 1, 2$, be any spinor of rank two which is antisymmetric in its spinor indices A and B, *i.e.* $\Phi_{AB} \in F \oplus F$, and

$$\Phi_{AB} = -\Phi_{BA}.$$

Then this spinor must be proportional to the metric ϵ_{AB}, since Φ_{AB} has only one independent component, *i.e.*

$$(\Phi_{AB}) = \begin{pmatrix} 0 & \phi_{12} \\ \phi_{21} & 0 \end{pmatrix},$$

and since $\Phi_{AB} = -\Phi_{BA}$, we conclude $\phi_{12} = -\phi_{21}$. Therefore

$$\Phi_{AB} = c\epsilon_{AB}, \qquad c = \text{const.}$$

Now considering

$$\epsilon^{BA}\Phi_{AB} = c\epsilon^{BA}\epsilon_{AB} = c\delta^B{}_B = 2c,$$

we have

$$c = \frac{1}{2}\epsilon^{BA}\Phi_{AB} = \frac{1}{2}\Phi^B{}_B,$$

and therefore:

$$\Phi_{AB} = \frac{1}{2}\Phi^C{}_C\epsilon_{AB}.$$

Then we have for an arbitrary $\Phi_{AB} \in F \oplus F$:

$$\Phi_{AB} - \Phi_{BA} = \frac{1}{2}\Phi^D{}_D\epsilon_{AB} - \frac{1}{2}\Phi^D{}_D\epsilon_{BA} = \Phi^D{}_D\epsilon_{AB}.$$

$$\Phi_{AB} - \Phi_{BA} = \delta_A{}^C\delta_B{}^D\Phi_{CD} - \delta_B{}^C\delta_A{}^D\Phi_{CD}$$
$$= (\delta_A{}^C\delta_B{}^D - \delta_B{}^C\delta_A{}^D)\Phi_{CD}$$
$$\overset{!}{=} \Phi^D{}_D\epsilon_{AB} = \epsilon_{AB}\epsilon^{DC}\Phi_{CD}. \qquad (1.152c)$$

Hence

$$\epsilon_{AB}\epsilon^{DC} = \delta_A{}^C\delta_B{}^D - \delta_A{}^D\delta_B{}^C. \qquad (1.152d)$$

Similarly we obtain

$$\epsilon_{\dot{A}\dot{B}}\epsilon^{\dot{D}\dot{C}} = \delta_{\dot{A}}{}^{\dot{C}}\delta_{\dot{B}}{}^{\dot{D}} - \delta_{\dot{A}}{}^{\dot{D}}\delta_{\dot{B}}{}^{\dot{C}}. \qquad (1.152e)$$

Multiplying Eq. (1.152c) by $\epsilon^{CA}\epsilon^{DB}$ and using Eq. (1.83) we obtain:

$$\Phi^{CD} - \Phi^{DC} = \epsilon^{CA}\epsilon^{DB}\Phi^E{}_E\epsilon_{AB}$$
$$= \delta^C{}_B\delta^{DB}\Phi^E{}_E$$
$$= \epsilon^{DC}\Phi^E{}_E$$
$$= -\Phi^E{}_E\epsilon^{CD}. \qquad (1.152f)$$

Corresponding expressions hold for dotted spinors:

$$\Phi_{\dot{A}\dot{B}} - \Phi_{\dot{B}\dot{A}} = \Phi_{\dot{C}}{}^{\dot{C}}\epsilon_{\dot{A}\dot{B}},$$
$$\Phi^{\dot{A}\dot{B}} - \Phi^{\dot{B}\dot{A}} = -\Phi_{\dot{C}}{}^{\dot{C}}\epsilon^{\dot{A}\dot{B}}. \tag{1.152g}$$

These expressions will be needed in later calculations.

1.4 Dirac and Majorana Spinors

In Eqs. (1.22) and (1.23) we defined two sets of operators, *i.e.*

$$N_i := \frac{1}{2}(J_i + iK_i),$$
$$N_i^\dagger := \frac{1}{2}(J_i - iK_i).$$

From these operators we constructed two Casimir operators $N_i N_i$ and $N_i^\dagger N_i^\dagger$, whose eigenvalues $n(n+1)$ and $m(m+1)$ with $n, m = 0, 1/2, 1, \ldots$, determine finite-dimensional, non-unitary repesentations of the Lorentz group, labelled (n, m). Under space reflection, *i.e.* the parity transformation, the generators of rotations, J_i, remain invariant, whereas the boost generators K_i change sign, *i.e.* $K_i \longrightarrow -K_i$. Hence, the parity operation is formally equivalent to the transformation

$$N_i \longrightarrow N_i^\dagger, \quad N_i^\dagger \longrightarrow N_i,$$

i.e. the parity transformation corresponds to complex conjugation. However, we have seen with Eqs. (1.59) and (1.61) that the operation of complex conjugation transforms the $(1/2, 0)$ representation into the $(0, 1/2)$ representation. In order to write such a map as a *linear* transformation it is necessary to double the dimension of the representation space.

Previously we defined:

F as the two-dimensional complex representation space of $SL(2, \mathbb{C})$ whose elements are undotted, left-handed Weyl spinors, and

$\dot{F}^*$ as the two-dimensional complex representation space of the complex conjugate representation of $SL(2, \mathbb{C})$ whose elements are dotted or so-called right-handed Weyl spinors.

We now define the direct sum of the representation spaces F and $\dot{F}^*$ as

$$E := F \oplus \dot{F}^*.$$

E is the four-dimensional complex representation space of Dirac spinors. Let

$$\phi \in F \quad \text{and} \quad \overline{\psi} \in \dot{F}^*.$$

Then

$$\Psi := \begin{pmatrix} \phi \\ \overline{\psi} \end{pmatrix} \in E \tag{1.153}$$

is a Dirac four-spinor. We define a representation of $SL(2, \mathbb{C})$ on E by the map

$$M \in SL(2, \mathbb{C}) \longrightarrow S(M) := \begin{pmatrix} M & 0 \\ 0 & M^{*-1} \end{pmatrix} \in \text{Aut}(E), \tag{1.154}$$

where $\text{Aut}(E)$ is the automorphism group of E. This representation acts on Dirac spinors $\Psi \in E$ as follows. Using Eqs. (1.153) and (1.154):

$$\begin{aligned} \Psi' &= S(M)\Psi \\ &= \begin{pmatrix} M & 0 \\ 0 & M^{*-1} \end{pmatrix} \begin{pmatrix} \phi \\ \overline{\psi} \end{pmatrix} \\ &= \begin{pmatrix} M\phi \\ M^{*-1}\overline{\psi} \end{pmatrix}. \end{aligned} \tag{1.155}$$

From this relation we can obtain the index structure of Dirac spinors as well as that of matrices acting on Dirac spinors. In view of Eqs. (1.60) and (1.81) we obtain:

$$\Psi_a = \begin{pmatrix} \phi_A \\ \overline{\psi}^{\dot{A}} \end{pmatrix}, \tag{1.156}$$

where $A = 1, 2$, and $\dot{A} = \dot{1}, \dot{2}$, and $a = 1, 2, 3, 4$. Furthermore:

$$S_{ab}(M) = \begin{pmatrix} M_A{}^B & 0 \\ 0 & (M^{*-1})^{\dot{A}}{}_{\dot{B}} \end{pmatrix}. \tag{1.157a}$$

Any 4×4-matrix acting on Dirac four-spinors must therefore have the following index structure

$$\Gamma_{ab} = \begin{pmatrix} A_A{}^B & B_{A\dot{B}} \\ C^{\dot{A}B} & D^{\dot{A}}{}_{\dot{B}} \end{pmatrix} \tag{1.157b}$$

with $A, B = 1, 2$, $\dot{A}, \dot{B} = \dot{1}, \dot{2}$, and $a, b = 1, 2, 3, 4$. Here, A, B, C and D are 2×2 complex submatrices. The off-diagonal submatrices B and C must have a dotted as well as an undotted index since by construction B is is a map from $\dot{F}^*$ to F and C is a map from F to $\dot{F}^*$. Applying the matrix Γ to a Dirac spinor we see that

$$\Gamma_{ab}\Psi_b = \begin{pmatrix} A_A{}^B & B_{A\dot{B}} \\ C^{\dot{A}B} & D^{\dot{A}}{}_{\dot{B}} \end{pmatrix} \cdot \begin{pmatrix} \phi_B \\ \overline{\psi}^{\dot{B}} \end{pmatrix} = \begin{pmatrix} A_A{}^B\phi_B + B_{A\dot{B}}\overline{\psi}^{\dot{B}} \\ C^{\dot{A}B}\phi_B + D^{\dot{A}}{}_{\dot{B}}\overline{\psi}^{\dot{B}} \end{pmatrix} \begin{matrix} \in F \\ \in \dot{F}^* \end{matrix}$$

This index picture is consistent with the summation convention for Weyl spinors.

A representation matrix for the parity operation in E is (as can be shown but will not be proved here):

$$S_R = i\begin{pmatrix} 0 & \sigma^0 \\ \overline{\sigma}^0 & 0 \end{pmatrix}.$$

The factor i is inserted as a matter of convention. Thus:

$$S_R\Psi = i\begin{pmatrix} 0 & \sigma^0 \\ \overline{\sigma}^0 & 0 \end{pmatrix}\begin{pmatrix} \phi \\ \psi \end{pmatrix} = i\begin{pmatrix} \sigma^0\overline{\psi} \\ \overline{\sigma}^0\phi \end{pmatrix}. \tag{1.158}$$

In components Eq. (1.158) reads

$$(S_R)_{ab}\Psi_b = i\begin{pmatrix} 0 & (\sigma^0)_{A\dot{B}} \\ (\overline{\sigma}^0)^{\dot{A}B} & 0 \end{pmatrix}\begin{pmatrix} \phi_B \\ \overline{\psi}^{\dot{B}} \end{pmatrix} = i\begin{pmatrix} (\sigma^0)_{A\dot{B}}\overline{\psi}^{\dot{B}} \\ (\overline{\sigma}^0)^{\dot{A}B}\phi_B \end{pmatrix} = \Psi'_a,$$

where $\Psi'_a \in \dot{F}^* \oplus F$ since σ^μ maps $\dot{F}^*$ into F, and the $\overline{\sigma}^\mu$ maps F into $\dot{F}^*$. This transformation of Ψ into Ψ' demonstrates explicitly the irreducibility of the representation space E under the parity transformation S_R. We conclude therefore that if parity is of interest, one has to use the Dirac spinor formalism in a relativistic theory.

We now discuss several representations of Dirac γ-matrices.

1.4.1 The Weyl Basis or Chiral Representations

If we take the following realization of γ-matrices

$$\gamma_W^\mu := \begin{pmatrix} 0 & \sigma^\mu \\ \overline{\sigma}^\mu & 0 \end{pmatrix}, \tag{1.159}$$

we obtain a direct relation between two-component and four-component spinors. The representation in which γ-matrices have the form of Eq. (1.159) is called the *Weyl basis*. As in any other representation of γ-matrices, the γ-matrices have to satisfy the basic Clifford algebra relation

$$\{\gamma_W^\mu, \gamma_W^\nu\} = 2\eta^{\mu\nu}\mathbb{1}_{4\times4}. \tag{1.160}$$

We can use this basic anticommutation relation to verify our two-component index formalism. Thus:

$$
\begin{aligned}
\{\gamma_W^\mu, \gamma_W^\nu\}_{ac} &= (\gamma_W^\mu)_{ab}(\gamma_W^\nu)_{bc} + (\gamma_W^\nu)_{ab}(\gamma_W^\mu)_{bc} \\[2mm]
&= \begin{pmatrix} 0 & (\sigma^\mu)_{A\dot{B}} \\ (\overline{\sigma}^\mu)^{\dot{A}B} & 0 \end{pmatrix} \begin{pmatrix} 0 & (\sigma^\nu)_{B\dot{C}} \\ (\overline{\sigma}^\nu)^{\dot{B}C} & 0 \end{pmatrix} \\[2mm]
&\quad + \begin{pmatrix} 0 & (\sigma^\nu)_{A\dot{B}} \\ (\overline{\sigma}^\nu)^{\dot{A}B} & 0 \end{pmatrix} \begin{pmatrix} 0 & (\sigma^\mu)_{B\dot{C}} \\ (\overline{\sigma}^\mu)^{\dot{B}C} & 0 \end{pmatrix} \\[2mm]
&= \begin{pmatrix} \sigma^\mu_{A\dot{B}}\overline{\sigma}^{\nu\dot{B}C} + \sigma^\nu_{A\dot{B}}\overline{\sigma}^{\mu\dot{B}C} & 0 \\ 0 & \overline{\sigma}^{\mu\dot{A}B}\sigma^\nu_{B\dot{C}} + \overline{\sigma}^{\nu\dot{A}B}\sigma^\mu_{B\dot{C}} \end{pmatrix} \\[2mm]
&= \begin{pmatrix} (\sigma^\mu\overline{\sigma}^\nu + \sigma^\nu\overline{\sigma}^\mu)_A{}^C & 0 \\ 0 & (\overline{\sigma}^\mu\sigma^\nu + \overline{\sigma}^\nu\sigma^\mu)^{\dot{A}}{}_{\dot{C}} \end{pmatrix} \\[2mm]
&= \begin{pmatrix} 2\eta^{\mu\nu}\delta_A{}^C & 0 \\ 0 & 2\eta^{\mu\nu}\delta^{\dot{A}}{}_{\dot{C}} \end{pmatrix} \qquad \text{using Eqs. (1.141a,1.141b)} \\[2mm]
&= 2\eta^{\mu\nu}\begin{pmatrix} \delta_A{}^C & 0 \\ 0 & \delta^{\dot{A}}{}_{\dot{C}} \end{pmatrix} \\[2mm]
&= 2\eta^{\mu\nu}\delta_{ab}.
\end{aligned}
$$

In matrix notation:

$$
\{\gamma_W^\mu, \gamma_W^\nu\} = 2\eta^{\mu\nu}\mathbb{1}_{4\times 4}.
$$

Remark: The 4×4 unit matrix in the representation space E, δ_{ab}, has the index structure

$$
\delta_{ab} = \begin{pmatrix} \delta_A{}^B & 0 \\ 0 & \delta^{\dot{A}}{}_{\dot{B}} \end{pmatrix}, \tag{1.161}
$$

because Eq. (1.157a) reduces to Eq. (1.161) for $M = \mathbb{1}_{SL(2,\mathbb{C})}$, where $\mathbb{1}_{SL(2,\mathbb{C})}$ is the unit element of the group $SL(2, \mathbb{C})$.

We define the γ^5-matrix in the Weyl basis as

$$
\gamma_W^5 := i\gamma_W^0\gamma_W^1\gamma_W^2\gamma_W^3. \tag{1.162a}
$$

Proposition 1.42:

The γ^5-matrix in the Weyl basis, Eq. (1.162a), has the explicit form:

$$
\gamma_W^5 = \begin{pmatrix} -\mathbb{1}_{2\times 2} & 0 \\ 0 & \mathbb{1}_{2\times 2} \end{pmatrix}. \tag{1.162b}
$$

Proof: From the definition, Eq. (1.162a), we obtain:

$$
\begin{aligned}
(\gamma_W^5)_{ab} &= i(\gamma_W^0 \gamma_W^1 \gamma_W^2 \gamma_W^3)_{ab} \\
&= i(\gamma_W^0)_{ac}(\gamma_W^1)_{cd}(\gamma_W^2)_{de}(\gamma_W^3)_{eb} \\
&= i\begin{pmatrix} 0 & \sigma_{A\dot{C}}^0 \\ \overline{\sigma}^{0\dot{A}C} & 0 \end{pmatrix}\begin{pmatrix} 0 & \sigma_{C\dot{D}}^1 \\ \overline{\sigma}^{1\dot{C}D} & 0 \end{pmatrix}\begin{pmatrix} 0 & \sigma_{D\dot{E}}^2 \\ \overline{\sigma}^{2\dot{D}E} & 0 \end{pmatrix}\begin{pmatrix} 0 & \sigma_{E\dot{B}}^3 \\ \overline{\sigma}^{3\dot{E}B} & 0 \end{pmatrix} \\
&= i\begin{pmatrix} \sigma_{A\dot{C}}^0 \overline{\sigma}^{1\dot{C}D} & 0 \\ 0 & \overline{\sigma}^{0\dot{A}C}\sigma_{C\dot{D}}^1 \end{pmatrix}\begin{pmatrix} \sigma_{D\dot{E}}^2 \overline{\sigma}^{3\dot{E}B} & 0 \\ 0 & \overline{\sigma}^{2\dot{D}E}\sigma_{E\dot{B}}^3 \end{pmatrix} \\
&= i\begin{pmatrix} \sigma_{A\dot{C}}^0 \overline{\sigma}^{1\dot{C}D}\sigma_{D\dot{E}}^2 \overline{\sigma}^{3\dot{E}B} & 0 \\ 0 & \overline{\sigma}^{0\dot{A}C}\sigma_{C\dot{D}}^1\overline{\sigma}^{2\dot{D}E}\sigma_{E\dot{B}}^3 \end{pmatrix} \\
&= i\begin{pmatrix} (\sigma^0\overline{\sigma}^1\sigma^2\overline{\sigma}^3)_A{}^B & 0 \\ 0 & (\overline{\sigma}^0\sigma^1\overline{\sigma}^2\sigma^3)^{\dot{A}}{}_{\dot{B}} \end{pmatrix} \\
&= i\begin{pmatrix} (\sigma^0\sigma^1\sigma^2\sigma^3)_A{}^B & 0 \\ 0 & -(\sigma^0\sigma^1\sigma^2\sigma^3)^{\dot{A}}{}_{\dot{B}} \end{pmatrix},
\end{aligned}
$$

using Eq. (1.108). Evaluation of the product of the Pauli matrices yields:

$$
(\gamma_W^5)_{ab} = \begin{pmatrix} -\delta_A{}^B & 0 \\ 0 & \delta^{\dot{A}}{}_{\dot{B}} \end{pmatrix}. \tag{1.162c}
$$

The following properties of the γ^5-matrix are independent of the particular representation:

$$
(\gamma_W^5)^2 = \mathbb{1}_{4\times4}, \tag{1.163a}
$$

$$
\{\gamma_W^5, \gamma_W^\mu\} = 0. \tag{1.163b}
$$

Proposition 1.43:

In the Weyl representation, the γ_W-matrices satisfy the following properties:

$$
\left.\begin{aligned}
\gamma_W^0 &= (\gamma_W^0)^\dagger, & i.e.\ \gamma_W^0 \text{ is Hermitian,} \\
\gamma_W^5 &= (\gamma_W^5)^\dagger, & i.e.\ \gamma_W^5 \text{ is Hermitian,} \\
\gamma_W^i &= -(\gamma_W^i)^\dagger, & i.e.\ \gamma_W^i \text{ is anti-Hermitian for } i = 1,2,3.
\end{aligned}\right\} \tag{1.164}
$$

Proof: We first demonstrate Proposition 1.43 in the matrix notation.

(i) Using Eq. (1.159) we have:

$$
(\gamma_W^0)^\dagger = \begin{pmatrix} 0 & \sigma^0 \\ \overline{\sigma}^0 & 0 \end{pmatrix}^\dagger = \begin{pmatrix} 0 & (\overline{\sigma}^0)^\dagger \\ (\sigma^0)^\dagger & 0 \end{pmatrix} = \begin{pmatrix} 0 & \sigma^0 \\ \overline{\sigma}^0 & 0 \end{pmatrix} = \gamma_W^0,
$$

using the fact that according to Eqs. (1.103a) and (1.108) σ^0 and $\overline{\sigma}^0$ are 2×2 unit matrices.

(ii) By definition of the γ_W^5-matrix, *i.e.* Eq. (1.162a), we can write:

$$
(\gamma_W^5)^\dagger = \left(i\gamma_W^0 \gamma_W^1 \gamma_W^2 \gamma_W^3 \right)^\dagger = -i(\gamma_W^3)^\dagger (\gamma_W^2)^\dagger (\gamma_W^1)^\dagger (\gamma_W^0)^\dagger
$$
$$
\overset{*}{=} i\gamma_W^3 \gamma_W^2 \gamma_W^1 \gamma_W^0 \overset{**}{=} i\gamma_W^0 \gamma_W^1 \gamma_W^2 \gamma_W^3 = \gamma_W^5.
$$

In step * we made use of the Hermiticity of γ_W^0 and the anti-Hermiticity of the γ_W^i (still to be shown). In step ** we used the basic Clifford algebra relation (1.160) to rearrange the γ's.

(iii) Using Eq. (1.159) and Eq. (1.108) we have:

$$
(\gamma_W^i)^\dagger = \begin{pmatrix} 0 & \sigma^i \\ \overline{\sigma}^i & 0 \end{pmatrix}^\dagger = \begin{pmatrix} 0 & (\overline{\sigma}^i)^\dagger \\ (\sigma^i)^\dagger & 0 \end{pmatrix} = \begin{pmatrix} 0 & -(\sigma^i)^\dagger \\ (\sigma^i)^\dagger & 0 \end{pmatrix}
$$
$$
= \begin{pmatrix} 0 & -\sigma^i \\ \sigma^i & 0 \end{pmatrix} = -\begin{pmatrix} 0 & \sigma^i \\ \overline{\sigma}^i & 0 \end{pmatrix} = -\gamma_W^i, \quad i = 1, 2, 3.
$$

We now verify Eq. (1.164) in terms of our index notation. First we have to clarify what we mean by transposition and Hermitian conjugation of γ-matrices in this context. The Hermitian conjugate of $(\gamma_W^\mu)_{ab}$ is given by

$$
(\gamma_W^{\mu\dagger})_{ab} = \begin{pmatrix} 0 & (\overline{\sigma}^\mu)^\dagger_{A\dot{B}} \\ (\sigma^\mu)^{\dagger \dot{A}B} & 0 \end{pmatrix}
$$

with

$$
(\gamma_W^\mu)_{ab} = \begin{pmatrix} 0 & (\sigma^\mu)_{A\dot{B}} \\ (\overline{\sigma}^\mu)^{\dot{A}B} & 0 \end{pmatrix}. \tag{1.165}
$$

This can be verified as follows. From Dirac theory[30] we know that in any representation the matrix γ^0 has the property:

$$
\gamma^0 \gamma^\mu \gamma^0 = \gamma^{\mu\dagger}.
$$

---

[30] See *e.g.* J.D. Bjorken and S.D. Drell [17] or C. Itzykson and J.-B. Zuber [60].

In the Weyl representation this relation implies (see Eq. (1.159))

$$
\begin{aligned}
& \left(\gamma_W^0\right)_{ab}\left(\gamma_W^\mu\right)_{bc}\left(\gamma_W^0\right)_{cd} \\[2mm]
& = \begin{pmatrix} 0 & (\sigma^0)_{A\dot{B}} \\ (\overline{\sigma}^0)^{\dot{A}B} & 0 \end{pmatrix} \begin{pmatrix} 0 & (\sigma^\mu)_{B\dot{C}} \\ (\overline{\sigma}^\mu)^{\dot{B}C} & 0 \end{pmatrix} \begin{pmatrix} 0 & (\sigma^0)_{C\dot{D}} \\ (\overline{\sigma}^0)^{\dot{C}D} & 0 \end{pmatrix} \\[2mm]
& = \begin{pmatrix} 0 & (\sigma^0)_{A\dot{B}}(\overline{\sigma}^\mu)^{\dot{B}C}(\sigma^0)_{C\dot{D}} \\ (\overline{\sigma}^0)^{\dot{A}B}(\sigma^\mu)_{B\dot{C}}(\overline{\sigma}^0)^{\dot{C}D} & 0 \end{pmatrix} \\[2mm]
& = \begin{pmatrix} 0 & (\sigma^0\overline{\sigma}^\mu\sigma^0)_{A\dot{D}} \\ (\overline{\sigma}^0\sigma^\mu\overline{\sigma}^0)^{\dot{A}D} & 0 \end{pmatrix} \\[2mm]
& = \begin{pmatrix} 0 & (\overline{\sigma}^\mu)_{A\dot{D}} \\ (\sigma^\mu)^{\dot{A}D} & 0 \end{pmatrix},
\end{aligned}
$$

where we define the matrices:

$$
\begin{aligned}
\left(\overline{\sigma}^\mu\right)_{A\dot{D}} & := (\sigma^0)_{A\dot{B}}(\overline{\sigma}^\mu)^{\dot{B}C}(\sigma^0)_{C\dot{D}}, \\
\left(\sigma^\mu\right)^{\dot{A}D} & := (\overline{\sigma}^0)^{\dot{A}B}(\sigma^\mu)_{B\dot{C}}(\overline{\sigma}^0)^{\dot{C}D}.
\end{aligned}
$$

This definition is consistent with our discussion following Eq. (1.157b). Using now the Hermiticity of the Pauli matrices we obtain:

$$
\left(\gamma_W^0\right)_{ab}\left(\gamma_W^\mu\right)_{bc}\left(\gamma_W^0\right)_{cd} = \begin{pmatrix} 0 & (\overline{\sigma}^{\mu\dagger})_{A\dot{D}} \\ (\sigma^{\mu\dagger})^{\dot{A}D} & 0 \end{pmatrix} = \left(\gamma_W^{\mu\dagger}\right)_{ad}.
$$

The consistency means that $\gamma_W^{\mu\dagger}$ as γ-matrices in the Weyl representation act on Dirac spinors Eq. (1.156):

$$
\left(\gamma_w^{\mu\dagger}\right)_{ab}\Psi_b = \begin{pmatrix} 0 & (\overline{\sigma}^{\mu\dagger})_{A\dot{B}} \\ (\sigma^{\mu\dagger})^{\dot{A}B} & 0 \end{pmatrix} \begin{pmatrix} \phi_B \\ \overline{\psi}^{\dot{B}} \end{pmatrix} = \begin{pmatrix} (\overline{\sigma}^{\mu\dagger})_{A\dot{B}}\,\overline{\psi}^{\dot{B}} \\ (\sigma^{\mu\dagger})^{\dot{A}B}\,\phi_B \end{pmatrix}.
$$

The upper right 2×2-submatrix must always have the index structure exhibited in Eq. (1.165) since by construction this submatrix maps elements from $\dot{F}^*$ into F, whereas the lower left submatrix maps F into $\dot{F}^*$.

Equation (1.164) can now be demonstrated in our index notation as follows:

(i) For $\mu = 0$ we have:

$$
\begin{aligned}
\left(\gamma_W^{0\dagger}\right)_{ab} & = \begin{pmatrix} 0 & (\overline{\sigma}^{0\dagger})_{A\dot{B}} \\ (\sigma^{0\dagger})^{\dot{A}B} & 0 \end{pmatrix} = \begin{pmatrix} 0 & (\overline{\sigma}^0)_{A\dot{B}} \\ (\sigma^0)^{\dot{A}B} & 0 \end{pmatrix} \\[2mm]
& = \begin{pmatrix} 0 & (\sigma^0)_{A\dot{B}} \\ (\overline{\sigma}^0)^{\dot{A}B} & 0 \end{pmatrix} = \left(\gamma_W^0\right)_{ab}.
\end{aligned}
$$

(ii) For $\mu = i$, $i = 1, 2, 3$, we get:

$$\left(\gamma_W^{i\dagger}\right)_{ab} = \begin{pmatrix} 0 & \left(\bar{\sigma}^{i\dagger}\right)_{A\dot{B}} \\ \left(\sigma^{i\dagger}\right)^{\dot{A}B} & 0 \end{pmatrix} = \begin{pmatrix} 0 & \left(\bar{\sigma}^{i}\right)_{A\dot{B}} \\ \left(\sigma^{i}\right)^{\dot{A}B} & 0 \end{pmatrix}$$

$$= -\begin{pmatrix} 0 & \left(\sigma^{i}\right)_{A\dot{B}} \\ \left(\bar{\sigma}^{i}\right)^{\dot{A}B} & 0 \end{pmatrix} = -\left(\gamma_W^{i}\right)_{ab}.$$

Here we made use of Eqs. (1.108) and (1.165).

This completes the proof of Proposition 1.43.

The Weyl representation is a useful representation when studying the extreme relativistic limit of Dirac theory.[31]

The Dirac equation is

$$\left(i\not{\partial} - m\right)\Psi = \left(i\gamma^{\mu}\partial_{\mu} - m\mathbb{1}_{4\times4}\right)\Psi = 0. \tag{1.166a}$$

In the Weyl representation we have

$$\left(i\gamma_W^{\mu}\partial_{\mu} - m\mathbb{1}_{4\times4}\right)\Psi_W = 0, \tag{1.166b}$$

where Ψ_W is a Dirac four-spinor in the Weyl representation as in Eq. (1.153). In the extreme relativistic limit, *i.e.* when $m \longrightarrow 0$, we obtain:

$$\gamma_W^{\mu}\partial_{\mu}\Psi_W = 0, \tag{1.167a}$$

which can be written, using Eqs. (1.159) and (1.153),

$$\begin{pmatrix} 0 & \left(\sigma^{\mu}\partial_{\mu}\right)_{A\dot{B}} \\ \left(\bar{\sigma}^{\mu}\partial_{\mu}\right)^{\dot{A}B} & 0 \end{pmatrix} \begin{pmatrix} \phi_B \\ \bar{\psi}^{\dot{B}} \end{pmatrix} = 0,$$

and therefore the 4×4 matrix equation (1.166b) decouples into two 2×2 matrix equations:

$$i\left(\sigma^{\mu}\partial_{\mu}\right)\bar{\psi} - m\phi = 0, \tag{1.167b}$$
$$i\left(\bar{\sigma}^{\mu}\partial_{\mu}\right)\phi - m\bar{\psi} = 0, \tag{1.167c}$$

i.e. for $m \longrightarrow 0$ we have:

$$\left.\begin{aligned} \left(\sigma^{\mu}\partial_{\mu}\right)_{A\dot{B}}\bar{\psi}^{\dot{B}} &= 0, \\ \left(\bar{\sigma}^{\mu}\partial_{\mu}\right)^{\dot{A}B}\phi_B &= 0. \end{aligned}\right\} \tag{1.168}$$

---

[31] See *e.g.* M.D. Scadron [103], Chap. 5.E.

These equations can be written in a compact form as

$$\left.\begin{array}{rcl}(i\partial_t + \boldsymbol{\sigma}\cdot\mathbf{p})\overline{\psi} & = & 0, \\ (i\partial_t - \boldsymbol{\sigma}\cdot\mathbf{p})\phi & = & 0. \end{array}\right\} \tag{1.169}$$

Equations (1.169) are wave equations for spin $1/2$ particles and are called *Weyl equations*. From previous considerations it is clear that these equations are not invariant under the parity transformation, whereas the Dirac equation is.

As a result of experiments one usually calls massless neutrinos left-handed and massless antineutrinos right-handed (this is a convention). Furthermore, neutrinos are spin-$1/2$ particles obeying Fermi–Dirac statistics.

We conclude therefore that we can describe massless neutrinos by two-component Weyl spinors transforming under the group $SL(2,\mathbb{C})$ according to the $(1/2,0)$ representation (*left-handed Weyl spinors*), and massless antineutrinos are represented by two-component Weyl spinors transforming under $SL(2,\mathbb{C})$ according to the $(0,1/2)$ representation (*right-handed Weyl spinors*).

In order to link the Weyl equations with helicity eigenstates, we observe that plane wave solutions of the first of Eqs. (1.169) for positive energy eigenstates, proportional to $\exp(ipx)$, where $E = |\mathbf{p}|$ (massless case), satisfy

$$\left(\boldsymbol{\sigma}\cdot\hat{\mathbf{p}}\right)\overline{\psi} = \overline{\psi},$$

where

$$\hat{\mathbf{p}} := \frac{\mathbf{p}}{E} = \frac{\mathbf{p}}{|\mathbf{p}|}.$$

In order to see this consider

$$\overline{\psi} \sim \exp\left(ipx\right).$$

Then

$$i\partial_t\overline{\psi} = -p_0\overline{\psi} = -E\overline{\psi} = -|\mathbf{p}|\overline{\psi}.$$

Inserting this into the first Weyl equation we obtain:

$$-E\overline{\psi} + (\boldsymbol{\sigma}\cdot\mathbf{p})\overline{\psi} = 0,$$

which is equivalent to

$$\frac{1}{2}(\boldsymbol{\sigma}\cdot\hat{\mathbf{p}})\,\overline{\psi} = \frac{1}{2}\overline{\psi}. \tag{1.170}$$

Likewise the plane-wave eigensolution of the second Weyl equation for positive energy eigenstates, proportional to $\exp\{-ipx\}$ where $p^0 = |\mathbf{p}| = E$, obeys

$$\frac{1}{2}(\boldsymbol{\sigma} \cdot \hat{\mathbf{p}}) \, \phi = -\frac{1}{2}\phi. \tag{1.171}$$

In Eq. (1.50) we defined the helicity operator as the projection of total angular momentum onto the momentum direction. The eigenvalues of this operator determine the helicity of the state. As shown in Sec. 1.2, helicity is a Poincaré invariant quantity; its eigenvalues λ determine various representations for the massless case. The sign of the eigenvalue λ determines the polarization state of particles with spin $|\lambda|$. The invariance of the helicity operator under Lorentz transformations implies that left-handed neutrinos are left-handed in any inertial system. Left-handed means $\lambda = -1/2$, right-handed means $\lambda = +1/2$. Hence, we conclude that ϕ is an eigenfunction of the helicity operator with eigenvalue $-1/2$, and $\overline{\psi}$ is an eigenfunction of this operator with eigenvalue $+1/2$.

1.4.2   The Canonical Basis or Dirac Representation

The canonical basis for Dirac matrices is defined by:

$$\gamma_D^0 = \begin{pmatrix} \mathbb{1}_{2\times2} & 0 \\ 0 & -\mathbb{1}_{2\times2} \end{pmatrix}, \quad \gamma_D^i = \begin{pmatrix} 0 & \sigma^i \\ \bar{\sigma}^i & 0 \end{pmatrix}, \quad \gamma_D^5 = \begin{pmatrix} 0 & \sigma^0 \\ \bar{\sigma}^0 & 0 \end{pmatrix}. \tag{1.172}$$

The following Proposition shows that the Weyl and the Dirac representations of the γ-matrices are linked by a similarity transformation.

Proposition 1.44:

The Weyl representation of the γ-matrices, Eq. (1.159), and the Dirac representation, Eq. (1.172), are connected by a similarity transformation, *i.e.*

$$\Gamma_W = X\Gamma_D X^{-1}, \quad X \in GL(4, \mathbb{C}), \tag{1.173a}$$

where Γ_W is any γ-matrix in the Weyl representation and Γ_D is any γ-matrix in the Dirac representation. Furthermore:

$$X = \frac{1}{\sqrt{2}}\begin{pmatrix} -\mathbb{1}_{2\times2} & \sigma^0 \\ -\bar{\sigma}^0 & -\mathbb{1}_{2\times2} \end{pmatrix}, \quad X^{-1} = \frac{1}{\sqrt{2}}\begin{pmatrix} -\mathbb{1}_{2\times2} & -\sigma^0 \\ \bar{\sigma}^0 & -\mathbb{1}_{2\times2} \end{pmatrix}. \tag{1.173b}$$

Proof: Since the γ-matrices in the Weyl representation and also those in the canonical basis obey the same Clifford algebra relation, given by Eq. (1.160), they are related by a similarity transformation. We first verify Eq. (1.173b) as an exercise in the handling of indices. Thus:

$$
\begin{aligned}
X_{ab}X_{bc}^{-1} &= \frac{1}{2}\begin{pmatrix} -\delta_A{}^B & (\sigma^0)_{A\dot{B}} \\ -(\overline{\sigma}^0)^{\dot{A}B} & -\delta^{\dot{A}}{}_{\dot{B}} \end{pmatrix}\begin{pmatrix} -\delta_B{}^C & -(\sigma^0)_{B\dot{C}} \\ (\overline{\sigma}^0)^{\dot{B}C} & -\delta^{\dot{B}}{}_{\dot{C}} \end{pmatrix} \\
&= \frac{1}{2}\begin{pmatrix} \delta_A{}^B\delta_B{}^C + (\sigma^0)_{A\dot{B}}(\overline{\sigma}^0)^{\dot{B}C} & \delta_A{}^B(\sigma^0)_{B\dot{C}} - (\sigma^0)_{A\dot{B}}\delta^{\dot{B}}{}_{\dot{C}} \\ (\overline{\sigma}^0)^{\dot{A}B}\delta_B{}^C - \delta^{\dot{A}}{}_{\dot{B}}(\overline{\sigma}^0)^{\dot{B}C} & (\overline{\sigma}^0)^{\dot{A}B}(\sigma^0)_{B\dot{C}} + \delta^{\dot{A}}{}_{\dot{B}}\delta^{\dot{B}}{}_{\dot{C}} \end{pmatrix} \\
&= \frac{1}{2}\begin{pmatrix} \delta_A{}^C + \delta_A{}^C & (\sigma^0)_{A\dot{C}} - (\sigma^0)_{A\dot{C}} \\ (\overline{\sigma}^0)^{\dot{A}C} - (\overline{\sigma}^0)^{\dot{A}C} & \delta^{\dot{A}}{}_{\dot{C}} + \delta^{\dot{A}}{}_{\dot{C}} \end{pmatrix} = \begin{pmatrix} \delta_A{}^C & 0 \\ 0 & \delta^{\dot{A}}{}_{\dot{C}} \end{pmatrix} \\
&= \left(\mathbb{1}_{4\times 4}\right)_{ac}.
\end{aligned}
$$

We now calculate the Dirac representation of the γ-matrices from those of the Weyl representation (1.173a). Now

$$
\Gamma_W = \gamma_W^0 = \begin{pmatrix} 0 & \sigma^0 \\ \overline{\sigma}^0 & 0 \end{pmatrix},
$$

and so:

$$
\begin{aligned}
\left(\gamma_D^0\right)_{ad} &= \left(X^{-1}\right)_{ab}\left(\gamma_W^0\right)_{bc}(X)_{cd} \\
&= \frac{1}{2}\begin{pmatrix} -\delta_A{}^B & -(\sigma^0)_{A\dot{B}} \\ (\overline{\sigma}^0)^{\dot{A}B} & -\delta^{\dot{A}}{}_{\dot{B}} \end{pmatrix}\begin{pmatrix} 0 & (\sigma^0)_{B\dot{C}} \\ (\overline{\sigma}^0)^{\dot{B}C} & 0 \end{pmatrix}\begin{pmatrix} -\delta_C{}^D & (\sigma^0)_{C\dot{D}} \\ -(\overline{\sigma}^0)^{\dot{C}D} & -\delta^{\dot{C}}{}_{\dot{D}} \end{pmatrix} \\
&= \frac{1}{2}\left(\begin{aligned} &\delta_A{}^B(\sigma^0)_{B\dot{C}}(\overline{\sigma}^0)^{\dot{C}D} + (\sigma^0)_{A\dot{B}}(\overline{\sigma}^0)^{\dot{B}C}\delta_C{}^D \\ &-(\overline{\sigma}^0)^{\dot{A}B}(\sigma^0)_{B\dot{C}}(\overline{\sigma}^0)^{\dot{C}D} + \delta^{\dot{A}}{}_{\dot{B}}(\overline{\sigma}^0)^{\dot{B}C}\delta_C{}^D \end{aligned} \right. \\
&\qquad\qquad \left. \begin{aligned} &\delta_A{}^B(\sigma^0)_{B\dot{C}}\delta^{\dot{C}}{}_{\dot{D}} - (\sigma^0)_{A\dot{B}}(\overline{\sigma}^0)^{\dot{B}C}(\sigma^0)_{C\dot{D}} \\ &-(\overline{\sigma}^0)^{\dot{A}B}(\sigma^0)_{B\dot{C}}\delta^{\dot{C}}{}_{\dot{D}} - \delta^{\dot{A}}{}_{\dot{B}}(\overline{\sigma}^0)^{\dot{B}C}(\sigma^0)_{C\dot{D}} \end{aligned} \right) \\
&= \frac{1}{2}\left(\begin{aligned} &(\sigma^0)_{A\dot{C}}(\overline{\sigma}^0)^{\dot{C}D} + (\sigma^0)_{A\dot{B}}(\overline{\sigma}^0)^{\dot{B}D} \\ &-(\overline{\sigma}^0)^{\dot{A}B}(\sigma^0)_{B\dot{C}}(\overline{\sigma}^0)^{\dot{C}D} + (\overline{\sigma}^0)^{\dot{A}D} \end{aligned} \right. \\
&\qquad\qquad \left. \begin{aligned} &(\sigma^0)_{A\dot{D}} - (\sigma^0)_{A\dot{B}}(\overline{\sigma}^0)^{\dot{B}C}(\sigma^0)_{C\dot{D}} \\ &-(\overline{\sigma}^0)^{\dot{A}B}(\sigma^0)_{B\dot{D}} - (\overline{\sigma}^0)^{\dot{A}C}(\sigma^0)_{C\dot{D}} \end{aligned} \right) \\
&= \frac{1}{2}\begin{pmatrix} \delta_A{}^D + \delta_A{}^D & (\sigma^0)_{A\dot{D}} - (\sigma^0)_{A\dot{B}}\delta^{\dot{B}}{}_{\dot{D}} \\ -(\overline{\sigma}^0)^{\dot{A}B}\delta_B{}^D + (\overline{\sigma}^0)^{\dot{A}D} & -\delta^{\dot{A}}{}_{\dot{D}} - \delta^{\dot{A}}{}_{\dot{D}} \end{pmatrix} \\
&= \begin{pmatrix} \delta_A{}^D & 0 \\ 0 & -\delta^{\dot{A}}{}_{\dot{D}} \end{pmatrix} = \begin{pmatrix} \mathbb{1}_{2\times 2} & 0 \\ 0 & -\mathbb{1}_{2\times 2} \end{pmatrix}_{ad}.
\end{aligned}
$$

Here we used the relations

$$(\overline{\sigma}^0)^{\dot{A}B}(\sigma^0)_{B\dot{C}} = \delta^{\dot{A}}{}_{\dot{C}},$$

$$(\sigma^0)_{A\dot{B}}(\overline{\sigma}^0)^{\dot{B}C} = \delta_A{}^C.$$

These relations can be understood as follows. The matrix $\overline{\sigma}^0$ is a map, the unit map from F to $\dot{F}^*$, and the matrix σ^0 is the unit map from $\dot{F}^*$ to F. Therefore the combined expression $(\overline{\sigma}^0\sigma^0)^{\dot{A}}{}_{\dot{B}}$ can be considered as a map starting from $\dot{F}^*$ and going to F and back to $\dot{F}^*$. Since $\overline{\sigma}^0$ and σ^0 are unit matrices, an expression like $\overline{\sigma}^0\sigma^0$ gives the unit map in $\dot{F}^*$; hence

$$(\overline{\sigma}^0\sigma^0)^{\dot{A}}{}_{\dot{B}} = \delta^{\dot{A}}{}_{\dot{B}} = \mathbb{1}_{\dot{F}^*}.$$

A corresponding consideration applies to $\sigma^0\overline{\sigma}^0 = \mathbb{1}_F$.

Next we transform γ^i_W, $i = 1, 2, 3$, into the Dirac representation. Thus, we have to evaluate

$$\gamma^i_D = X^{-1}\gamma^i_W X.$$

In components:

$$\begin{aligned}
(\gamma^i_D)_{ad} &= (X^{-1})_{ab}(\gamma^i_W)_{bc}(X)_{cd} \\
&= \frac{1}{2}\begin{pmatrix} -\delta_A{}^B & -(\sigma^0)_{A\dot{B}} \\ (\overline{\sigma}^0)^{\dot{A}B} & -\delta^{\dot{A}}{}_{\dot{B}} \end{pmatrix}\begin{pmatrix} 0 & (\sigma^i)_{B\dot{C}} \\ (\overline{\sigma}^i)^{\dot{B}C} & 0 \end{pmatrix}\begin{pmatrix} -\delta_C{}^D & (\sigma^0)_{C\dot{D}} \\ -(\overline{\sigma}^0)^{\dot{C}D} & -\delta^{\dot{C}}{}_{\dot{D}} \end{pmatrix} \\
&= \frac{1}{2}\begin{pmatrix} \delta_A{}^B(\sigma^i)_{B\dot{C}}(\overline{\sigma}^0)^{\dot{C}D} + (\sigma^0)_{A\dot{B}}(\overline{\sigma}^i)^{\dot{B}D} & (\sigma^i)_{A\dot{D}} - (\sigma^0)_{A\dot{B}}(\overline{\sigma}^i)^{\dot{B}C}(\sigma^0)_{C\dot{D}} \\ -(\overline{\sigma}^0)^{\dot{A}B}(\sigma^i)_{B\dot{C}}(\overline{\sigma}^0)^{\dot{C}D} + (\overline{\sigma}^i)^{\dot{A}D} & -(\overline{\sigma}^0)^{\dot{A}B}(\sigma^i)_{B\dot{D}} - (\overline{\sigma}^i)^{\dot{A}C}(\sigma^0)_{C\dot{D}} \end{pmatrix} \\
&= \frac{1}{2}\begin{pmatrix} 0 & (\sigma^i)_{A\dot{D}} - (\overline{\sigma}^i)_{A\dot{D}} \\ -(\sigma^i)^{\dot{A}D} + (\overline{\sigma}^i)^{\dot{A}D} & 0 \end{pmatrix} = \begin{pmatrix} 0 & \sigma^i \\ \overline{\sigma}^i & 0 \end{pmatrix}_{ad},
\end{aligned}$$

where we used (with Eq. (1.108))

$$(\sigma^i)_{A\dot{C}}(\overline{\sigma}^0)^{\dot{C}D} = (\sigma^i)_A{}^D,$$

$$(\sigma^0)_{A\dot{B}}(\overline{\sigma}^i)^{\dot{B}D} = (\overline{\sigma}^i)_A{}^D = -(\sigma^i)_A{}^D,$$

$$-(\overline{\sigma}^0)^{\dot{A}B}(\sigma^i)_{B\dot{C}}(\overline{\sigma}^0)^{\dot{C}D} = -(\sigma^i)^{\dot{A}D} = (\overline{\sigma}^i)^{\dot{A}D},$$

and similar formulas, taking into account that σ^0 and $\overline{\sigma}^0$ are unit matrices which transform dotted indices into undotted and *vice versa*.

Finally we demonstrate that

$$\gamma_D^5 = X^{-1}\gamma_W^5 X.$$

We have:

$$
\begin{aligned}
(\gamma_D^5)_{ad} &= (X^{-1})_{ab}(\gamma_W^5)_{bc}(X)_{cd}\\
&= \frac{1}{2}\begin{pmatrix} -\delta_A{}^B & -(\sigma^0)_{A\dot{B}} \\ (\overline{\sigma}^0)^{\dot{A}B} & -\delta^{\dot{A}}{}_{\dot{B}} \end{pmatrix}\begin{pmatrix} -\delta_B{}^C & 0 \\ 0 & \delta^{\dot{B}}{}_{\dot{C}} \end{pmatrix}\begin{pmatrix} -\delta_C{}^D & (\sigma^0)_{C\dot{D}} \\ -(\overline{\sigma}^0)^{\dot{C}D} & -\delta^{\dot{C}}{}_{\dot{D}} \end{pmatrix}\\
&= \frac{1}{2}\begin{pmatrix} -\delta_A{}^B\delta_B{}^C\delta_C{}^D + (\sigma^0)_{A\dot{B}}\delta^{\dot{B}}{}_{\dot{C}}(\overline{\sigma}^0)^{\dot{C}D} \\ (\overline{\sigma}^0)^{\dot{A}B}\delta_B{}^C\delta_C{}^D + \delta^{\dot{A}}{}_{\dot{B}}\delta^{\dot{B}}{}_{\dot{C}}(\overline{\sigma}^0)^{\dot{C}D} \end{pmatrix.\\
&\qquad\qquad \left. \begin{matrix} \delta_A{}^B\delta_B{}^C(\sigma^0)_{C\dot{D}} + (\sigma^0)_{A\dot{B}}\delta^{\dot{B}}{}_{\dot{C}}\delta^{\dot{C}}{}_{\dot{D}} \\ -(\overline{\sigma}^0)^{\dot{A}B}\delta_B{}^C(\sigma^0)_{C\dot{D}} + \delta^{\dot{A}}{}_{\dot{B}}\delta^{\dot{B}}{}_{\dot{C}}\delta^{\dot{C}}{}_{\dot{D}} \end{matrix} \right)\\
&= \frac{1}{2}\begin{pmatrix} -\delta_A{}^D + \delta_A{}^D & (\sigma^0)_{A\dot{D}} + (\sigma^0)_{A\dot{D}} \\ (\overline{\sigma}^0)^{\dot{A}D} + (\overline{\sigma}^0)^{\dot{A}D} & -\delta^{\dot{A}}{}_{\dot{D}} + \delta^{\dot{A}}{}_{\dot{D}} \end{pmatrix} = \begin{pmatrix} 0 & \sigma^0 \\ \overline{\sigma}^0 & 0 \end{pmatrix}_{ad}.
\end{aligned}
$$

The canonical basis or Dirac representation of γ-matrices has the unique property of all possible representations that it diagonalizes the energy via the matrix

$$\gamma_D^0 = \begin{pmatrix} \mathbb{1}_{2\times 2} & 0 \\ 0 & -\mathbb{1}_{2\times 2} \end{pmatrix}$$

in the nonrelativistic limit.

1.4.3 The Majorana Representation

Of all possible equivalent representations of γ-matrices obtained by a non-singular transformation

$$\gamma^\mu \longrightarrow X\gamma^\mu X^{-1},$$

the *Majorana representation* plays a particular role. It is constructed so as to make the Dirac equation real. This can be seen as follows. In the form originally proposed by Dirac, the Dirac equation is [17, 105]

$$\left(i\frac{\partial}{\partial t} - \frac{1}{i}\boldsymbol{\alpha}\cdot\boldsymbol{\nabla} - \beta m\right)\Psi = 0, \tag{1.174}$$

where $\alpha^i, i = 1, 2, 3$, and β are Hermitian 4×4-matrices. These matrices are related to the γ-matrices by

$$\gamma^0 = \beta, \quad \gamma^i = \beta\alpha^i, \qquad i = 1, 2, 3. \tag{1.175}$$

In the Dirac representation (1.172) we have:

$$\beta = \begin{pmatrix} \mathbb{1}_{2\times2} & 0 \\ 0 & -\mathbb{1}_{2\times2} \end{pmatrix}, \quad \alpha^i = \begin{pmatrix} 0 & \sigma^i \\ -\sigma^i & 0 \end{pmatrix}. \tag{1.176}$$

Hence in order to satisfy the reality property we multiply Eq. (1.174) by $-i$ and obtain

$$\left(\frac{\partial}{\partial t} + \widehat{\boldsymbol{\alpha}} \cdot \boldsymbol{\nabla} + i\widehat{\beta}m \right) \Psi = 0. \tag{1.177}$$

This is a real equation if and only if $\widehat{\boldsymbol{\alpha}}$ are real 4×4-matrices and if $\widehat{\beta}$ is purely imaginary. Hence, if we set:

$$\widehat{\beta} = \alpha^2, \quad \widehat{\alpha}^1 = -\alpha^1, \quad \widehat{\alpha}^2 = \beta, \quad \widehat{\alpha}^3 = -\alpha^3,$$

then Eq. (1.177) is real. Following the prescription (1.175) we obtain the γ-matrices in the Majorana representation:

$$\left.\begin{aligned} \gamma_M^0 = \widehat{\beta} = \begin{pmatrix} 0 & \sigma^2 \\ -\sigma^2 & 0 \end{pmatrix}, \quad \gamma_M^1 = \widehat{\beta}\widehat{\alpha}^1 = \begin{pmatrix} i\sigma^3 & 0 \\ 0 & i\sigma^3 \end{pmatrix}, \\ \gamma_M^2 = \widehat{\beta}\widehat{\alpha}^2 = \begin{pmatrix} 0 & -\sigma^2 \\ -\sigma^2 & 0 \end{pmatrix}, \quad \gamma_M^3 = \widehat{\beta}\widehat{\alpha}^3 = \begin{pmatrix} -i\sigma^1 & 0 \\ 0 & -i\sigma^1 \end{pmatrix}. \end{aligned}\right\} \tag{1.178}$$

The γ^5-matrix in the Majorana representation is:

$$\gamma_M^5 = i\gamma_M^0\gamma_M^1\gamma_M^2\gamma_M^3 = \begin{pmatrix} \sigma^2 & 0 \\ 0 & -\sigma^2 \end{pmatrix}. \tag{1.179}$$

Proposition 1.45:

The Dirac representation and the Majorana representation are connected by the following similarity transformation

$$\Gamma_M = Y\Gamma_D Y^{-1}, \tag{1.180}$$

where Γ_M is any γ-matrix in the Majorana representation and Γ_D is the corresponding γ-matrix in the Dirac representation (1.172). Furthermore

$$Y = \frac{1}{\sqrt{2}} \begin{pmatrix} \mathbb{1}_{2\times2} & \sigma^2 \\ -\sigma^2 & -\mathbb{1}_{2\times2} \end{pmatrix} = Y^{-1}. \tag{1.181}$$

Proof: We first verify that

$$Y_{ab}Y_{bc}^{-1} = \delta_{ac}.$$

Thus

$$
Y_{ab}Y_{bc}^{-1} = \frac{1}{2}\begin{pmatrix} \delta_A{}^B & (\sigma^2)_{A\dot{B}} \\ -(\overline{\sigma}^2)^{\dot{A}B} & -\delta^{\dot{A}}{}_{\dot{B}} \end{pmatrix}\begin{pmatrix} \delta_B{}^C & (\sigma^2)_{B\dot{C}} \\ -(\overline{\sigma}^2)^{\dot{B}C} & -\delta^{\dot{B}}{}_{\dot{C}} \end{pmatrix}
$$

$$
= \frac{1}{2}\begin{pmatrix} \delta_A{}^B\delta_B{}^C - (\sigma^2)_{A\dot{B}}(\overline{\sigma}^2)^{\dot{B}C} & \delta_A{}^B(\sigma^2)_{B\dot{C}} - (\sigma^2)_{A\dot{B}}\delta^{\dot{B}}{}_{\dot{C}} \\ -(\overline{\sigma}^2)^{\dot{A}B}\delta_B{}^C + \delta^{\dot{A}}{}_{\dot{B}}(\overline{\sigma}^2)^{\dot{B}C} & -(\overline{\sigma}^2)^{\dot{A}B}(\sigma^2)_{B\dot{C}} + \delta^{\dot{A}}{}_{\dot{B}}\delta^{\dot{B}}{}_{\dot{C}} \end{pmatrix}
$$

$$
= \frac{1}{2}\begin{pmatrix} \delta_A{}^C - (\sigma^2\overline{\sigma}^2)_A{}^C & (\sigma^2)_{A\dot{C}} - (\sigma^2)_{A\dot{C}} \\ -(\overline{\sigma}^2)^{\dot{A}C} + (\overline{\sigma}^2)^{\dot{A}C} & -(\overline{\sigma}^2\sigma^2)^{\dot{A}}{}_{\dot{C}} + \delta^{\dot{A}}{}_{\dot{C}} \end{pmatrix}.
$$

Using $\overline{\sigma}^2 = -\sigma^2$ as matrices and $(\sigma^2)^2 = \mathbb{1}_{2\times 2}$, we obtain:

$$
Y_{ab}Y_{bc}^{-1} = \frac{1}{2}\begin{pmatrix} \delta_A{}^C + \delta_A{}^C & 0 \\ 0 & \delta^{\dot{A}}{}_{\dot{C}} + \delta^{\dot{A}}{}_{\dot{C}} \end{pmatrix} = \delta_{ac}.
$$

We now investigate the individual γ-matrices.

$\mu = 0$:

$$
(\gamma_M^0)_{ad} = Y_{ab}(\gamma_D^0)_{bc}Y_{cd}^{-1}
$$

$$
= \frac{1}{2}\begin{pmatrix} \delta_A{}^B & (\sigma^2)_{A\dot{B}} \\ -(\overline{\sigma}^2)^{\dot{A}B} & -\delta^{\dot{A}}{}_{\dot{B}} \end{pmatrix}\begin{pmatrix} \delta_B{}^C & 0 \\ 0 & -\delta^{\dot{B}}{}_{\dot{C}} \end{pmatrix}\begin{pmatrix} \delta_C{}^D & (\sigma^2)_{C\dot{D}} \\ -(\overline{\sigma}^2)^{\dot{C}D} & -\delta^{\dot{C}}{}_{\dot{D}} \end{pmatrix}
$$

$$
= \frac{1}{2}\begin{pmatrix} \delta_A{}^D + (\sigma^2)_{A\dot{C}}(\overline{\sigma}^2)^{\dot{C}D} & (\sigma^2)_{A\dot{D}} + (\sigma^2)_{A\dot{D}} \\ -(\overline{\sigma}^2)^{\dot{A}D} - (\overline{\sigma}^2)^{\dot{A}D} & -(\overline{\sigma}^2)^{\dot{A}C}(\sigma^2)_{C\dot{D}} - \delta^{\dot{A}}{}_{\dot{D}} \end{pmatrix}
$$

$$
= \begin{pmatrix} 0 & (\sigma^2)_{A\dot{D}} \\ -(\overline{\sigma}^2)^{\dot{A}D} & 0 \end{pmatrix} = \begin{pmatrix} 0 & \sigma^2 \\ -\overline{\sigma}^2 & 0 \end{pmatrix}_{ad},
$$

again using Eq. (1.108), *i.e.* $\overline{\sigma}^2 = -\sigma^2$ and $(\sigma^2)^2 = \mathbb{1}_{2\times 2}$.

$\mu = 1$:

$$
(\gamma_M^1)_{ad} = Y_{ab}(\gamma_D^1)_{bc}Y_{cd}^{-1}
$$

$$
= \frac{1}{2}\begin{pmatrix} \delta_A{}^B & (\sigma^2)_{A\dot{B}} \\ -(\overline{\sigma}^2)^{\dot{A}B} & -\delta^{\dot{A}}{}_{\dot{B}} \end{pmatrix}\begin{pmatrix} 0 & (\sigma^1)_{B\dot{C}} \\ (\overline{\sigma}^1)^{\dot{B}C} & 0 \end{pmatrix}\begin{pmatrix} \delta_C{}^D & (\sigma^2)_{C\dot{D}} \\ -(\overline{\sigma}^2)^{\dot{C}D} & -\delta^{\dot{C}}{}_{\dot{D}} \end{pmatrix}
$$

$$
= \frac{1}{2}\begin{pmatrix} (\sigma^2)_{A\dot{B}}(\overline{\sigma}^1)^{\dot{B}D} - (\sigma^1)_{A\dot{C}}(\overline{\sigma}^2)^{\dot{C}D} & (\sigma^2)_{A\dot{B}}(\overline{\sigma}^1)^{\dot{B}C}(\sigma^2)_{C\dot{D}} - (\sigma^1)_{A\dot{D}} \\ -(\overline{\sigma}^1)^{\dot{A}D} + (\overline{\sigma}^2)^{\dot{A}B}(\sigma^1)_{B\dot{C}}(\overline{\sigma}^2)^{\dot{C}D} & -(\overline{\sigma}^1)^{\dot{A}C}(\sigma^2)_{C\dot{D}} + (\overline{\sigma}^2)^{\dot{A}B}(\sigma^1)_{B\dot{D}} \end{pmatrix}
$$

$$= \frac{1}{2} \begin{pmatrix} (\sigma^2\overline{\sigma}^1)_A{}^D - (\sigma^1\overline{\sigma}^2)_A{}^D & (\sigma^2\overline{\sigma}^1\sigma^2)_{A\dot{D}} - (\sigma^1)_{A\dot{D}} \\ -(\overline{\sigma}^1)^{\dot{A}D} + (\overline{\sigma}^2\sigma^1\overline{\sigma}^2)^{\dot{A}D} & -(\overline{\sigma}^1\sigma^2)^{\dot{A}}{}_{\dot{D}} + (\overline{\sigma}^2\sigma^1)^{\dot{A}}{}_{\dot{D}} \end{pmatrix}$$

$$= \frac{1}{2} \begin{pmatrix} (\sigma^1\sigma^2 - \sigma^2\sigma^1)_A{}^D & (\sigma^2\overline{\sigma}^1\sigma^2 - \sigma^1)_{A\dot{D}} \\ (-\overline{\sigma}^1 + \overline{\sigma}^2\sigma^1\overline{\sigma}^2)^{\dot{A}D} & (\sigma^1\sigma^2 - \sigma^2\sigma^1)^{\dot{A}}{}_{\dot{D}} \end{pmatrix}$$

$$= \frac{1}{2} \begin{pmatrix} 2i(\sigma^3)_A{}^D & -(\sigma^2\sigma^1\sigma^2 + \sigma^1)_{A\dot{D}} \\ (\sigma^1 + \sigma^2\sigma^1\sigma^2)^{\dot{A}D} & 2i(\sigma^3)^{\dot{A}}{}_{\dot{D}} \end{pmatrix}$$

$$= \frac{1}{2} \begin{pmatrix} 2i(\sigma^3)_A{}^D & (\sigma^1 - \sigma^1)_{A\dot{D}} \\ (\sigma^1 - \sigma^1)^{\dot{A}D} & 2i(\sigma^3)^{\dot{A}}{}_{\dot{D}} \end{pmatrix} = \begin{pmatrix} i\sigma^3 & 0 \\ 0 & i\sigma^3 \end{pmatrix}_{ad}.$$

$\mu = 2$:

$$(\gamma_M^2)_{ad} = Y_{ab}(\gamma_D^2)_{bc}Y_{cd}^{-1}$$

$$= \frac{1}{2} \begin{pmatrix} \delta_A{}^B & (\sigma^2)_{A\dot{B}} \\ -(\overline{\sigma}^2)^{\dot{A}B} & -\delta^{\dot{A}}{}_{\dot{B}} \end{pmatrix} \begin{pmatrix} 0 & (\sigma^2)_{B\dot{C}} \\ (\overline{\sigma}^2)^{\dot{B}C} & 0 \end{pmatrix} \begin{pmatrix} \delta_C{}^D & (\sigma^2)_{C\dot{D}} \\ -(\overline{\sigma}^2)^{\dot{C}D} & -\delta^{\dot{C}}{}_{\dot{D}} \end{pmatrix}$$

$$= \frac{1}{2} \begin{pmatrix} (\sigma^2)_{A\dot{B}}(\overline{\sigma}^2)^{\dot{B}D} - (\sigma^2)_{A\dot{C}}(\overline{\sigma}^2)^{\dot{C}D} \\ -(\overline{\sigma}^2)^{\dot{A}D} + (\overline{\sigma}^2)^{\dot{A}B}(\sigma^2)_{B\dot{C}}(\overline{\sigma}^2)^{\dot{C}D} \end{pmatrix}$$

$$\begin{matrix} (\sigma^2)_{A\dot{B}}(\overline{\sigma}^2)^{\dot{B}C}(\sigma^2)_{C\dot{D}} - (\sigma^2)_{A\dot{D}} \\ -(\overline{\sigma}^2)^{\dot{A}C}(\sigma^2)_{C\dot{D}} + (\overline{\sigma}^2)^{\dot{A}B}(\sigma^2)_{B\dot{D}} \end{matrix}$$

$$= \frac{1}{2} \begin{pmatrix} 0 & -(\sigma^2 - \sigma^2\overline{\sigma}^2\sigma^2)_{A\dot{D}} \\ -(\overline{\sigma}^2 - \overline{\sigma}^2\sigma^2\overline{\sigma}^2)^{\dot{A}D} & 0 \end{pmatrix}.$$

Again using Eq. (1.108), we obtain

$$-\sigma^2 - \sigma^2\overline{\sigma}^2\sigma^2 = \sigma^2(\sigma^2)^2 = \sigma^2,$$

since $(\sigma^2)^2 = \mathbb{1}_{2\times 2}$. Therefore:

$$(\gamma_M^2)_{ad} = \frac{1}{2} \begin{pmatrix} 0 & -2(\sigma^2)_{A\dot{D}} \\ -2(\overline{\sigma}^2)^{\dot{A}D} & 0 \end{pmatrix} = \begin{pmatrix} 0 & -\sigma^2 \\ -\overline{\sigma}^2 & 0 \end{pmatrix}_{ad}.$$

$\mu = 3$:

$$(\gamma_M^3)_{ad} = Y_{ab}(\gamma_D^3)_{bc}Y_{cd}^{-1}$$

$$= \frac{1}{2} \begin{pmatrix} \delta_A{}^B & (\sigma^2)_{A\dot{B}} \\ -(\overline{\sigma}^2)^{\dot{A}B} & -\delta^{\dot{A}}{}_{\dot{B}} \end{pmatrix} \begin{pmatrix} 0 & (\sigma^3)_{B\dot{C}} \\ (\overline{\sigma}^3)^{\dot{B}C} & 0 \end{pmatrix} \begin{pmatrix} \delta_C{}^D & (\sigma^2)_{C\dot{D}} \\ -(\overline{\sigma}^2)^{\dot{C}D} & -\delta^{\dot{C}}{}_{\dot{D}} \end{pmatrix}$$

$$= \frac{1}{2} \begin{pmatrix} (\sigma^2\overline{\sigma}^3)_A{}^D - (\sigma^3\overline{\sigma}^2)_A{}^D & (\sigma^2\overline{\sigma}^3\sigma^2)_{A\dot{D}} - (\sigma^3)_{A\dot{D}} \\ -(\overline{\sigma}^3)^{\dot{A}D} + (\overline{\sigma}^2\sigma^3\overline{\sigma}^2)^{\dot{A}D} & -(\overline{\sigma}^3\sigma^2)^{\dot{A}}{}_{\dot{D}} + (\overline{\sigma}^2\sigma^3)^{\dot{A}}{}_{\dot{D}} \end{pmatrix}$$

$$= \frac{1}{2} \begin{pmatrix} -([\sigma^2,\sigma^3])_A{}^D & 0 \\ 0 & -([\sigma^2,\sigma^3])^{\dot{A}}{}_{\dot{D}} \end{pmatrix}$$

$$= \frac{1}{2} \begin{pmatrix} -2i(\sigma^1)_A{}^D & 0 \\ 0 & -2i(\sigma^1)^{\dot{A}}{}_{\dot{D}} \end{pmatrix}$$

$$= \begin{pmatrix} -i\sigma^1 & 0 \\ 0 & -i\sigma^1 \end{pmatrix}_{ad}.$$

Now the connection between the Weyl representation and the Dirac representation is given by Eq. (1.173a),

$$\Gamma_D = X^{-1}\Gamma_W X.$$

The Dirac representation and the Majorana representation are related via Eq. (1.180),

$$\Gamma_M = Y\Gamma_D Y^{-1}.$$

Thus

$$\Gamma_M = YX^{-1}\Gamma_W XY^{-1},$$

or, using $Y = Y^{-1}$, we obtain the similarity transformation that connects the Weyl representation and the Majorana representation:

$$\Gamma_M = (XY)^{-1}\Gamma_W(XY). \tag{1.182}$$

With

$$X = \frac{1}{\sqrt{2}} \begin{pmatrix} -\mathbb{1}_{2\times 2} & \sigma^0 \\ -\overline{\sigma}^0 & -\mathbb{1}_{2\times 2} \end{pmatrix} = (X^{-1})^\top = (X^{-1})^\dagger,$$

$$Y = \frac{1}{\sqrt{2}} \begin{pmatrix} \mathbb{1}_{2\times 2} & \sigma^2 \\ -\overline{\sigma}^2 & -\mathbb{1}_{2\times 2} \end{pmatrix} = Y^\dagger = Y^{-1},$$

we obtain:

$$X_{ab}Y_{bc} = \frac{1}{2} \begin{pmatrix} -\delta_A{}^B & (\sigma^0)_{A\dot{B}} \\ -(\overline{\sigma}^0)^{\dot{A}B} & -\delta^{\dot{A}}{}_{\dot{B}} \end{pmatrix} \begin{pmatrix} \delta_B{}^C & (\sigma^2)_{B\dot{C}} \\ -(\overline{\sigma}^2)^{\dot{B}C} & -\delta^{\dot{B}}{}_{\dot{C}} \end{pmatrix}$$

$$= \frac{1}{2} \begin{pmatrix} -\delta_A{}^B\delta_B{}^C - (\sigma^0)_{A\dot{B}}(\overline{\sigma}^2)^{\dot{B}C} & -\delta_A{}^B(\sigma^2)_{B\dot{C}} - (\sigma^0)_{A\dot{B}}\delta^{\dot{B}}{}_{\dot{C}} \\ -(\overline{\sigma}^0)^{\dot{A}B}\delta_B{}^C + \delta^{\dot{A}}{}_{\dot{B}}(\overline{\sigma}^2)^{\dot{B}C} & -(\overline{\sigma}^0)^{\dot{A}B}(\sigma^2)_{B\dot{C}} + \delta^{\dot{A}}{}_{\dot{B}}\delta^{\dot{B}}{}_{\dot{C}} \end{pmatrix}$$

$$= \frac{1}{2} \begin{pmatrix} -\delta_A{}^C + (\sigma^2)_A{}^C & -(\sigma^2)_{A\dot{C}} - (\sigma^0)_{A\dot{C}} \\ -(\overline{\sigma}^0)^{\dot{A}C} + (\overline{\sigma}^2)^{\dot{A}C} & -(\sigma^2)^{\dot{A}}{}_{\dot{C}} + \delta^{\dot{A}}{}_{\dot{C}} \end{pmatrix},$$

which is usually written in matrix form as:

$$XY = \frac{1}{2}\begin{pmatrix} -1+\sigma^2 & -\sigma^2-1 \\ -1-\sigma^2 & -\sigma^2+1 \end{pmatrix}. \tag{1.183}$$

The inverse, $(XY)^{-1}$, is

$$(XY)^{-1} = Y^{-1}X^{-1} = \frac{1}{2}\begin{pmatrix} 1 & \sigma^2 \\ -\overline{\sigma}^2 & -1 \end{pmatrix}\begin{pmatrix} -1 & -\sigma^0 \\ \overline{\sigma}^0 & -1 \end{pmatrix}$$

$$= \frac{1}{2}\begin{pmatrix} -1+\sigma^2\overline{\sigma}^0 & -\sigma^0-\sigma^2 \\ \overline{\sigma}^2-\overline{\sigma}^0 & \overline{\sigma}^2\sigma^0+1 \end{pmatrix}, \tag{1.184}$$

which is usually written

$$(XY)^{-1} = \frac{1}{2}\begin{pmatrix} -1+\sigma^2 & -1-\sigma^2 \\ -\sigma^2-1 & -\sigma^2+1 \end{pmatrix} = (XY)^\dagger, \tag{1.185}$$

or

$$(XY)^{-1} = Y^{-1}X^{-1} = Y^\dagger X^\dagger = (XY)^\dagger.$$

However, for a clear index structure it seems to be advantageous to use the explicit form Eq. (1.184). Of course, Eq. (1.185) is the same matrix as Eq. (1.184) since as matrices:

$$\overline{\sigma}^0 = \mathbb{1}_{2\times 2}, \qquad \overline{\sigma}^2 = -\sigma^2,$$

and so on. Next we check that

$$\gamma_M^0 = (XY)^{-1}\gamma_W^0(XY)$$

with

$$\gamma_W^0 = \begin{pmatrix} 0 & \sigma^0 \\ \overline{\sigma}^0 & 0 \end{pmatrix}.$$

We perform this calculation in the compact matrix form of Eq. (1.185):

$$(XY)^{-1}\gamma_W^0(XY) = \frac{1}{4}\begin{pmatrix} -1+\sigma^2 & -1-\sigma^2 \\ -\sigma^2-1 & -\sigma^2+1 \end{pmatrix}\begin{pmatrix} 0 & 1 \\ 1 & 0 \end{pmatrix}\begin{pmatrix} \sigma^2-1 & -\sigma^2-1 \\ -1-\sigma^2 & 1-\sigma^2 \end{pmatrix}$$

$$= \frac{1}{4}\begin{pmatrix} (-1-\sigma^2)(\sigma^2-1)+(-1+\sigma^2)(-1-\sigma^2) \\ (1-\sigma^2)(\sigma^2-1)+(\sigma^2+1)^2 \end{pmatrix}$$

$$\begin{pmatrix} -(1+\sigma^2)(-\sigma^2-1)+(-1+\sigma^2)(1-\sigma^2) \\ (1-\sigma^2)(-\sigma^2-1)+(-\sigma^2-1)(1-\sigma^2) \end{pmatrix}$$

$$= \frac{1}{4}\begin{pmatrix} 0 & 4\sigma^2 \\ 4\sigma^2 & 0 \end{pmatrix} = \begin{pmatrix} 0 & \sigma^2 \\ -\overline{\sigma}^2 & 0 \end{pmatrix} = \gamma_M^0.$$

In the last step we made use of Eq. (1.178). The other γ-matrices can be checked in a similar fashion.

1.4.4   Charge Conjugation, Dirac and Weyl Representations

The *charge conjugation matrix* appears in Dirac theory[32] in the following way. The Dirac theory implies the existence of electrons and positrons, particles with the same mass but opposite charges, which obey the same equation. The Dirac equation must therefore admit a symmetry corresponding to the interchange of particles and antiparticles. We thus seek a transformation

$$\Psi \longrightarrow \Psi^c,$$

which reverses the sign of the charge, so that the Dirac spinor Ψ obeys the Dirac equation

$$\left(i\,\partial\!\!\!/ - e\,A\!\!\!/ - m\right)\Psi = 0, \qquad A\!\!\!/ = A_\mu \gamma^\mu, \tag{1.186}$$

in the presence of the electromagnetic vector potential A_μ, whereas the charge conjugated spinor Ψ^c obeys[33]

$$\left(i\,\partial\!\!\!/ + e\,A\!\!\!/ - m\right)\Psi^c = 0. \tag{1.187}$$

The Dirac equation coupled minimally to the electromagnetic field is

$$\left[\gamma^\mu\left(i\partial_\mu - eA_\mu\right) - m\right]\Psi = 0.$$

Taking the complex conjugate we obtain:

$$\left[\gamma^{\mu*}\left(-i\partial_\mu - eA_\mu\right) - m\right]\Psi^* = 0.$$

Transposition yields

$$\Psi^\dagger\left[\gamma^{\mu\dagger}\left(-i\overleftarrow{\partial}_\mu - eA_\mu\right) - m\right] = 0,$$
$$\Psi^\dagger\gamma^0\left[\gamma^0\gamma^{\mu\dagger}\gamma^0\left(-i\overleftarrow{\partial}_\mu - eA_\mu\right) - m\right] = 0,$$
$$\overline{\Psi}\left[\gamma^\mu\left(-i\overleftarrow{\partial}_\mu - eA_\mu\right) - m\right] = 0,$$

where

$$\overline{\Psi} := \Psi^\dagger\gamma^0$$

is the *Dirac adjoint*, and

$$\gamma^0\gamma^{\mu\dagger}\gamma^0 = \gamma^\mu.$$

Taking again the transpose we obtain

$$\left[-\gamma^{\mu\top}\left(i\partial_\mu + eA_\mu\right) - m\right]\overline{\Psi}^\top = 0.$$

---

[32]See e.g. the discussions in A. Pais [86], p. 381, and J.D. Bjorken and S.D. Drell [17].

[33]For further discussions of these aspects see M.D. Scadron [103], Chap. 3.B.

Multiplying this equation by a 4×4-matrix C from the left and inserting $C^{-1}C$ in front of $\overline{\Psi}^{\top}$, we get:

$$C\left[-\gamma^{\mu\top}\left(i\partial_\mu + eA_\mu\right) - m\right]C^{-1}C\overline{\Psi}^{\top} = 0,$$
$$\left[-C\gamma^{\mu\top}C^{-1}\left(i\partial_\mu + eA_\mu\right) - m\right]C\overline{\Psi}^{\top} = 0.$$

This equation can be identified with Eq. (1.187) provided we set:

$$\Psi^c := C\overline{\Psi}^{\top}, \tag{1.188}$$

except for a phase factor. In addition we have to demand that in any representation of the γ-matrices

$$C\gamma^{\mu\top}C^{-1} = -\gamma^{\mu}. \tag{1.189}$$

The matrix C is called the *charge conjugation matrix*. It suffices to construct the charge conjugation matrix in some particular representation; the unitary transformation which transforms to another representation then gives the matrix C in this new representation. We consider several representations.

The Charge Conjugation Matrix in the Dirac Representation

In the Dirac representation (1.56) the charge conjugation matrix C may be taken as

$$C_D = i\gamma_D^2\gamma_D^0 = i\begin{pmatrix} 0 & -\sigma^2 \\ \overline{\sigma}^2 & 0 \end{pmatrix}. \tag{1.190}$$

Proposition 1.46:

The charge conjugation matrix C_D defined in Eq. (1.190) possesses the following properties:

$$C_D = -C_D^{-1} = -C_D^{\dagger} = -C_D^{\top}. \tag{1.191}$$

Proof: We first verify $C_D = -C_D^{-1}$. Consider

$$
\begin{aligned}
-(C_D)_{ab}(C_D)_{bc} &= \begin{pmatrix} 0 & -(\sigma^2)_{A\dot{B}} \\ (\overline{\sigma}^2)^{\dot{A}B} & 0 \end{pmatrix}\begin{pmatrix} 0 & -(\sigma^2)_{B\dot{C}} \\ (\overline{\sigma}^2)^{\dot{B}C} & 0 \end{pmatrix} \\
&= \begin{pmatrix} -(\sigma^2)_{A\dot{B}}(\overline{\sigma}^2)^{\dot{B}C} & 0 \\ 0 & -(\overline{\sigma}^2)^{\dot{A}B}(\sigma^2)_{B\dot{C}} \end{pmatrix} \\
&= \begin{pmatrix} -(\sigma^2\overline{\sigma}^2)_A{}^C & 0 \\ 0 & -(\overline{\sigma}^2\sigma^2)^{\dot{A}}{}_{\dot{C}} \end{pmatrix}
\end{aligned}
$$

$$\stackrel{(1.108)}{=} \begin{pmatrix} \left((\sigma^2)^2\right)_A{}^C & 0 \\ 0 & \left((\sigma^2)^2\right)^{\dot{A}}{}_{\dot{C}} \end{pmatrix}$$

$$= \begin{pmatrix} \delta_A{}^C & 0 \\ 0 & \delta^{\dot{A}}{}_{\dot{C}} \end{pmatrix} = \left(\mathbb{1}_{4\times4}\right)_{ac}.$$

This shows that $C_D = -C_D^{-1}$.

Next we evaluate the Hermitian conjugate as follows:

$$\left(C_D^\dagger\right)_{ab} = \left[i\begin{pmatrix} 0 & -(\sigma^2)_{A\dot{B}} \\ (\bar{\sigma}^2)^{\dot{A}B} & 0 \end{pmatrix}\right]^\dagger = -i\begin{pmatrix} 0 & (\bar{\sigma}^{2\dagger})_{A\dot{B}} \\ -(\sigma^{2\dagger})^{\dot{A}B} & 0 \end{pmatrix}$$

$$= -i\begin{pmatrix} 0 & -(\sigma^2)_{A\dot{B}} \\ (\bar{\sigma}^2)^{\dot{A}B} & 0 \end{pmatrix} = -\left(C_D\right)_{ab},$$

using the Hermiticity of the Pauli matrices and $\bar{\sigma}^2 = -\sigma^2$ from Eq. (1.108). Finally, the antisymmetry of the charge conjugation matrix C_D is shown as follows:

$$\left(C_D^\top\right)_{ab} = i\begin{pmatrix} 0 & -(\sigma^2)_{A\dot{B}} \\ (\bar{\sigma}^2)^{\dot{A}B} & 0 \end{pmatrix}^\top = i\begin{pmatrix} 0 & (\bar{\sigma}^{2\top})_{A\dot{B}} \\ -(\sigma^{2\top})^{\dot{A}B} & 0 \end{pmatrix}$$

$$= i\begin{pmatrix} 0 & -(\sigma^{2\top})_{A\dot{B}} \\ (\bar{\sigma}^{2\top})^{\dot{A}B} & 0 \end{pmatrix} = i\begin{pmatrix} 0 & (\sigma^2)_{A\dot{B}} \\ -(\bar{\sigma}^2)^{\dot{A}B} & 0 \end{pmatrix}$$

$$= -i\begin{pmatrix} 0 & -(\sigma^2)_{A\dot{B}} \\ (\bar{\sigma}^2)^{\dot{A}B} & 0 \end{pmatrix} = -\left(C_D\right)_{ab}.$$

Here we made use of Eq. (1.108) and the antisymmetry of the σ^2 matrix.

Proposition 1.47:

The charge conjugation matrix C_D, defined in Eq. (1.190), satisfies relation (1.189), *i.e.*

$$C_D\gamma_D^{\mu\top}C_D^{-1} = -\gamma_D^\mu.$$

Proof: First we attach suffixes D to Eq. (1.189) into the following form.

$$C_D\gamma_D^{\mu\top}C_D^{-1} = -\gamma_D^\mu.$$

Using Eq. (1.191) this can be written as

$$-C_D\gamma_D^{\mu\top}C_D = -\gamma_D^\mu.$$

Transposition gives

$$C_D^\top \gamma_D^\mu C_D^\top = \gamma_D^{\mu\top},$$

and using the properties (1.191) again yields

$$C_D \gamma_D^\mu C_D = \gamma_D^{\mu\top}.$$

Consider the case $\mu = 0$, *i.e.*

$$(\gamma_D^0)_{ab} = \begin{pmatrix} \delta_A{}^B & 0 \\ 0 & -\delta^{\dot A}{}_{\dot B} \end{pmatrix}.$$

Then with Eq. (1.190) we have:

$$
\begin{aligned}
(C_D)_{ab}(\gamma_D^0)_{bc}(C_D)_{cd} &= -\begin{pmatrix} 0 & -(\sigma^2)_{A\dot B} \\ (\bar\sigma^2)^{\dot A B} & 0 \end{pmatrix} \begin{pmatrix} \delta_B{}^C & 0 \\ 0 & -\delta^{\dot B}{}_{\dot C} \end{pmatrix} \\
&\quad \times \begin{pmatrix} 0 & -(\sigma^2)_{C\dot D} \\ (\bar\sigma^2)^{\dot C D} & 0 \end{pmatrix} \\
&= -\begin{pmatrix} (\sigma^2)_{A\dot B}(\bar\sigma^2)^{\dot B D} & 0 \\ 0 & -(\bar\sigma^2)^{\dot A B}(\sigma^2)_{B\dot D} \end{pmatrix} \\
&= -\begin{pmatrix} (\sigma^2\bar\sigma^2)_A{}^D & 0 \\ 0 & -(\bar\sigma^2\sigma^2)^{\dot A}{}_{\dot D} \end{pmatrix} \\
&= -\begin{pmatrix} -((\sigma^2)^2)_A{}^D & 0 \\ 0 & ((\sigma^2)^2)^{\dot A}{}_{\dot D} \end{pmatrix} \\
&= -\begin{pmatrix} -\delta_A{}^D & 0 \\ 0 & \delta^{\dot A}{}_{\dot D} \end{pmatrix} \\
&= \begin{pmatrix} (\mathbb{1}_{2\times2}^\top)_A{}^D & 0 \\ 0 & -(\mathbb{1}_{2\times2}^\top)^{\dot A}{}_{\dot D} \end{pmatrix} \\
&= (\gamma^{0\top})_{ad}.
\end{aligned}
$$

Next we work out the case $\mu = 1$:

$$(\gamma_D^1)_{ab} = \begin{pmatrix} 0 & (\sigma^1)_{A\dot B} \\ (\bar\sigma^1)^{\dot A B} & 0 \end{pmatrix}.$$

Using Eq. (1.190) we have:

$$
\begin{aligned}
(C_D)_{ab}(\gamma_D^1)_{bc}(C_D)_{cd} &= -\begin{pmatrix} 0 & -(\sigma^2)_{A\dot{B}} \\ (\overline{\sigma}^2)^{\dot{A}B} & 0 \end{pmatrix} \begin{pmatrix} 0 & (\sigma^1)_{B\dot{C}} \\ (\overline{\sigma}^1)^{\dot{B}C} & 0 \end{pmatrix} \\
&\quad \times \begin{pmatrix} 0 & -(\sigma^2)_{C\dot{D}} \\ (\overline{\sigma}^2)^{\dot{C}D} & 0 \end{pmatrix} \\
&= -\begin{pmatrix} 0 & (\sigma^2)_{A\dot{B}}(\overline{\sigma}^1)^{\dot{B}C}(\sigma^2)_{C\dot{D}} \\ (\overline{\sigma}^2)^{\dot{A}B}(\sigma^1)_{B\dot{C}}(\overline{\sigma}^2)^{\dot{C}D} & 0 \end{pmatrix} \\
&= -\begin{pmatrix} 0 & (\sigma^2\overline{\sigma}^1\sigma^2)_{A\dot{D}} \\ (\overline{\sigma}^2\sigma^1\overline{\sigma}^2)^{\dot{A}D} & 0 \end{pmatrix} \\
&= -\begin{pmatrix} 0 & -(\sigma^2\sigma^1\sigma^2)_{A\dot{D}} \\ (\sigma^2\sigma^1\sigma^2)^{\dot{A}D} & 0 \end{pmatrix} \\
&= -\begin{pmatrix} 0 & (\sigma^1)_{A\dot{D}} \\ -(\sigma^1)^{\dot{A}D} & 0 \end{pmatrix} \qquad (\sigma^2\sigma^i\sigma^2 = -\sigma^i) \\
&= \begin{pmatrix} 0 & -(\sigma^{1\top})_{A\dot{D}} \\ -(\overline{\sigma}^{1\top})^{\dot{A}D} & 0 \end{pmatrix} \qquad (\sigma^1 = \sigma^{1\top}) \\
&= (\gamma_D^{1\top})_{ad},
\end{aligned}
$$

since

$$
\begin{aligned}
(\gamma_D^{1\top})_{ad} &= \begin{pmatrix} 0 & (\sigma^1)_{A\dot{D}} \\ (\overline{\sigma}^1)^{\dot{A}D} & 0 \end{pmatrix}^\top = \begin{pmatrix} 0 & (\overline{\sigma}^{1\top})_{A\dot{D}} \\ (\sigma^{1\top})^{\dot{A}D} & 0 \end{pmatrix} \\
&= -\begin{pmatrix} 0 & (\sigma^{1\top})_{A\dot{D}} \\ (\overline{\sigma}^{1\top})^{\dot{A}D} & 0 \end{pmatrix}.
\end{aligned}
$$

Hence we have demonstrated that

$$
C_D\gamma_D^1 C_D = (\gamma_D^1)^\top.
$$

Next we evaluate the case $\mu = 2$:

$$
(\gamma_D^2)_{ab} = \begin{pmatrix} 0 & (\sigma^2)_{A\dot{B}} \\ (\overline{\sigma}^2)^{\dot{A}B} & 0 \end{pmatrix}.
$$

Using Eq. (1.190):

$$
\begin{aligned}
(C_D)_{ab}(\gamma_D^2)_{bc}(C_D)_{cd} &= -\begin{pmatrix} 0 & -(\sigma^2)_{A\dot{B}} \\ (\overline{\sigma}^2)^{\dot{A}B} & 0 \end{pmatrix} \begin{pmatrix} 0 & (\sigma^2)_{B\dot{C}} \\ (\overline{\sigma}^2)^{\dot{B}C} & 0 \end{pmatrix} \\
&\quad \times \begin{pmatrix} 0 & -(\sigma^2)_{C\dot{D}} \\ (\overline{\sigma}^2)^{\dot{C}D} & 0 \end{pmatrix}
\end{aligned}
$$

$$= -\begin{pmatrix} 0 & (\sigma^2)_{A\dot{B}}(\overline{\sigma}^2)^{\dot{B}C}(\sigma^2)_{C\dot{D}} \\ (\overline{\sigma}^2)^{\dot{A}B}(\sigma^2)_{B\dot{C}}(\overline{\sigma}^2)^{\dot{C}D} & 0 \end{pmatrix}$$

$$= -\begin{pmatrix} 0 & (\sigma^2\overline{\sigma}^2\sigma^2)_{A\dot{D}} \\ (\overline{\sigma}^2\sigma^2\overline{\sigma}^2)^{\dot{A}D} & 0 \end{pmatrix}$$

$$= \begin{pmatrix} 0 & (\sigma^2)_{A\dot{D}} \\ (\overline{\sigma}^2)^{\dot{A}D} & 0 \end{pmatrix}.$$

Now the transpose of γ_D^2 is:

$$(\gamma_D^{2\top})_{ad} = \begin{pmatrix} 0 & (\sigma^2)_{A\dot{D}} \\ (\overline{\sigma}^2)^{\dot{A}D} & 0 \end{pmatrix}^{\top} = \begin{pmatrix} 0 & (\overline{\sigma}^{2\top})_{A\dot{D}} \\ (\sigma^{2\top})^{\dot{A}D} & 0 \end{pmatrix}$$

$$= \begin{pmatrix} 0 & (\sigma^2)_{A\dot{D}} \\ (\overline{\sigma}^2)^{\dot{A}D} & 0 \end{pmatrix},$$

since with Eq. (1.108) $\overline{\sigma}^{2\top} = -\sigma^{2\top} = \sigma^2$. Hence

$$C_D\gamma_D^2 C_D = (\gamma_D^2)^{\top}.$$

Finally we consider $\mu = 3$:

$$(\gamma_D^3)_{ab} = \begin{pmatrix} 0 & (\sigma^3)_{A\dot{B}} \\ (\overline{\sigma}^3)^{\dot{A}B} & 0 \end{pmatrix}.$$

Then:

$$(C_D)_{ab}(\gamma_D^3)_{bc}(C_D)_{cd} = -\begin{pmatrix} 0 & -(\sigma^2)_{A\dot{B}} \\ (\overline{\sigma}^2)^{\dot{A}B} & 0 \end{pmatrix}\begin{pmatrix} 0 & (\sigma^3)_{B\dot{C}} \\ (\overline{\sigma}^3)^{\dot{B}C} & 0 \end{pmatrix}$$

$$\times \begin{pmatrix} 0 & -(\sigma^2)_{C\dot{D}} \\ (\overline{\sigma}^2)^{\dot{C}D} & 0 \end{pmatrix}$$

$$= -\begin{pmatrix} 0 & (\sigma^2)_{A\dot{B}}(\overline{\sigma}^3)^{\dot{B}C}(\sigma^2)_{C\dot{D}} \\ (\overline{\sigma}^2)^{\dot{A}B}(\sigma^3)_{B\dot{C}}(\overline{\sigma}^2)^{\dot{C}D} & 0 \end{pmatrix}$$

$$= -\begin{pmatrix} 0 & (\sigma^2\overline{\sigma}^3\sigma^2)_{A\dot{D}} \\ (\overline{\sigma}^2\sigma^3\overline{\sigma}^2)^{\dot{A}D} & 0 \end{pmatrix}$$

$$= -\begin{pmatrix} 0 & -(\overline{\sigma}^3)_{A\dot{D}} \\ -(\sigma^3)^{\dot{A}D} & 0 \end{pmatrix} = -\begin{pmatrix} 0 & (\sigma^3)_{A\dot{D}} \\ (\overline{\sigma}^3)^{\dot{A}D} & 0 \end{pmatrix}.$$

Now

$$(\gamma_D^{3\top})_{ad} = \begin{pmatrix} 0 & (\sigma^3)_{A\dot{D}} \\ (\overline{\sigma}^3)^{\dot{A}D} & 0 \end{pmatrix}^{\top} = \begin{pmatrix} 0 & (\overline{\sigma}^{3\top})_{A\dot{D}} \\ (\sigma^{3\top})^{\dot{A}D} & 0 \end{pmatrix}$$

$$= -\begin{pmatrix} 0 & -(\sigma^{3\top})_{A\dot{D}} \\ -(\overline{\sigma}^{3\top})^{\dot{A}D} & 0 \end{pmatrix} = -\begin{pmatrix} 0 & (\sigma^3)_{A\dot{D}} \\ (\overline{\sigma}^3)^{\dot{A}D} & 0 \end{pmatrix}$$

using Eq. (1.108) and $\sigma^3 = \sigma^{3\top}$.
Hence

$$C_D \gamma_D^3 C_D = \left(\gamma_D^3\right)^\top,$$

as had to be shown.

The Charge Conjugation Matrix in the Weyl Representation

We transform the charge conjugation matrix Eq. (1.190) from the Dirac representation to the Weyl representation by using Eqs. (1.173a) and (1.173b); thus

$$C_W = X C_D X^{-1} = i X \gamma_D^2 \gamma_D^0 X^{-1}$$

$$= i X \gamma_D^2 X^{-1} X \gamma_D^0 X^{-1} = i \gamma_W^2 \gamma_W^0 = \begin{pmatrix} i\sigma^2 & 0 \\ 0 & -i\sigma^2 \end{pmatrix}. \qquad (1.192)$$

We prove Eq. (1.192) in the submatrix formulation, avoiding cumbersome indices. Thus

$$C_W = \frac{i}{2} \begin{pmatrix} -\mathbb{1} & \sigma^0 \\ -\sigma^0 & -\mathbb{1} \end{pmatrix} \begin{pmatrix} 0 & -\sigma^2 \\ \bar{\sigma}^2 & 0 \end{pmatrix} \begin{pmatrix} -\mathbb{1} & -\sigma^0 \\ \bar{\sigma}^0 & -\mathbb{1} \end{pmatrix}$$

$$= \frac{i}{2} \begin{pmatrix} \sigma^2 \bar{\sigma}^0 - \sigma^0 \bar{\sigma}^2 & -\sigma^2 - \sigma^0 \bar{\sigma}^2 \sigma^0 \\ \bar{\sigma}^0 \sigma^2 \bar{\sigma}^0 + \bar{\sigma}^2 & -\bar{\sigma}^0 \sigma^2 + \bar{\sigma}^2 \sigma^0 \end{pmatrix}$$

$$= \frac{i}{2} \begin{pmatrix} 2\sigma^2 & 0 \\ 0 & -2\sigma^2 \end{pmatrix} = \begin{pmatrix} i\sigma^2 & 0 \\ 0 & -i\sigma^2 \end{pmatrix}.$$

Remark: The correct form of the charge conjugation matrix in the Weyl representation, possessing the correct index structure, is

$$C_W = \begin{pmatrix} i\sigma^2 \bar{\sigma}^0 & 0 \\ 0 & i\bar{\sigma}^2 \sigma^0 \end{pmatrix} \qquad (1.193)$$

with:

$$(C_W)_{ab} = \begin{pmatrix} i\left(\sigma^2 \bar{\sigma}^0\right)_A{}^B & 0 \\ 0 & i\left(\bar{\sigma}^2 \sigma^0\right)^{\dot{A}}{}_{\dot{B}} \end{pmatrix}.$$

Of course, the matrix (1.193) is the same as Eq. (1.192), since $\bar{\sigma}^0$ and σ^0 are unit matrices, which are usually ignored in the literature.

Proposition 1.48:

The charge conjugation matrix C_W in the Weyl representation also satisfies

$$C_W = -C_W^{-1} = -C_W^\top = -C_W^\dagger, \qquad (1.194)$$

and

$$C_W \gamma_W^\mu C_W^{-1} = -\gamma_W^{\mu\top}. \tag{1.195}$$

Proof: Equations (1.194) and (1.195) can be checked directly, using Eq. (1.192) and Eq. (1.159), or alternatively with the help of Eqs. (1.173a) and (1.173b). Thus using first Eq. (1.192) and then Eq. (1.191) we have:

$$C_W = X C_D X^{-1} = -X C_D^{-1} X^{-1} = -\left(X C_D X^{-1}\right)^{-1} = -C_W^{-1}.$$

Also

$$C_W \overset{(1.187)}{=} X C_D X^{-1} \overset{(1.191)}{=} -X C_D^\top X^{-1} = -\left(X^{-1\top} C_D X^\top\right)^\top$$
$$\overset{(1.182)}{=} -\left(X C_D X^{-1}\right)^\top \overset{(1.192)}{=} -C_W^\top.$$

The last property in Eq. (1.194) is shown as follows:

$$C_W = X C_D X^{-1} = -X C_D^\dagger X^{-1}$$
$$= -\left(X^{-1\dagger} C_D X^\dagger\right)^\dagger$$
$$= -\left(X C_D X^{-1}\right)^\dagger = -C_W^\dagger.$$

Finally we verify Eq. (1.195). We know that in the Dirac representation we have:

$$C_D \gamma_D^\mu C_D^{-1} = -\gamma_D^{\mu\top}.$$

Hence:

$$
\begin{aligned}
C_D \gamma_D^\mu C_D^{-1} &= -\gamma_D^{\mu\top}, \\
\Longleftrightarrow \quad X\left(C_D \gamma_D^\mu C_D^{-1}\right) X^{-1} &= -X\left(\gamma_D^{\mu\top}\right) X^{-1}, \\
\Longleftrightarrow \quad X C_D X^{-1} X \gamma_D^\mu X^{-1} X C_D^{-1} X^{-1} &= -X\left(\gamma_D^{\mu\top}\right) X^{-1}, \\
\Longleftrightarrow \quad X C_D X^{-1} X \gamma_D^\mu X^{-1} \left(X C_D X^{-1}\right)^{-1} &= -\left(X \gamma_D^\mu X^{-1}\right)^\top,
\end{aligned}
$$

and so

$$C_W \gamma_W^\mu C_W^{-1} = -\gamma_W^{\mu\top}.$$

Further properties of the C-matrix

Proposition 1.49:

Independently of any representation, the charge conjugation matrix satisfies the following relations:

$$C\gamma^5 C^{-1} = \left(\gamma^5\right)^\top, \tag{1.196}$$

$$C\left(\gamma^5\gamma^\mu\right)C^{-1} = \left(\gamma^5\gamma^\mu\right)^\top. \tag{1.197}$$

Proof: We first show Eq. (1.196). Thus, since Eq. (1.162a) is independent of the particular representation we have:

$$
\begin{aligned}
C\gamma^5 C^{-1} &= iC\gamma^0\gamma^1\gamma^2\gamma^3 C^{-1} \\
&= iC\gamma^0 C^{-1}C\gamma^1 C^{-1}C\gamma^2 C^{-1}C\gamma^3 C^{-1} \\
&= i\gamma^{0\top}\gamma^{1\top}\gamma^{2\top}\gamma^{3\top} \qquad (\text{ with } (1.184)) \\
&= i\left(\gamma^3\gamma^2\gamma^1\gamma^0\right)^\top \\
&= i\left(\gamma^0\gamma^1\gamma^2\gamma^3\right)^\top \\
&= \left(\gamma^5\right)^\top,
\end{aligned}
$$

using the basic Clifford algebra relation (1.160), which is valid in any representation.

Equation (1.197) can be shown in a similar way:

$$C\gamma^5\gamma^\mu C^{-1} = C\gamma^5 C^{-1}C\gamma^\mu C^{-1} = -\left(\gamma^5\right)^\top\left(\gamma^\mu\right)^\top = -\left(\gamma^\mu\gamma^5\right)^\top = \left(\gamma^5\gamma^\mu\right)^\top$$

with Eq. (1.164).

1.4.5 Majorana Spinors

We consider a Dirac four-spinor in the Weyl representation. According to Eq. (1.156) we have:

$$\Psi_a = \left(\frac{\phi}{\overline{\psi}}\right) = \left(\frac{\phi_A}{\overline{\psi}^{\dot A}}\right) \in F \oplus \dot F^*. \tag{1.198}$$

For the index calculus to make sense, rows and columns must have the index structure of

$$\left(\Psi_W^\top\right)_a = \left(\phi^A, \ \overline{\psi}_{\dot A}\right), \qquad \text{and} \qquad \left(\Psi_W\right)_a = \left(\frac{\phi_A}{\overline{\psi}^{\dot A}}\right)$$

respectively. Thus

$$\begin{pmatrix} M_A{}^B & M_{A\dot B} \\ M^{\dot A B} & M^{\dot A}{}_{\dot B} \end{pmatrix} \begin{pmatrix} \phi_B \\ \overline{\psi}^{\dot B} \end{pmatrix} = \begin{pmatrix} M_A{}^B\phi_B + M_{A\dot B}\overline{\psi}^{\dot B} \\ M^{\dot A B}\phi_B + M^{\dot A}{}_{\dot B}\overline{\psi}^{\dot B} \end{pmatrix},$$

and

$$\left(\phi^A, \ \overline{\psi}_{\dot{A}} \right) \begin{pmatrix} M_A{}^B & M_{A\dot{B}} \\ M^{\dot{A}B} & M^{\dot{A}}{}_{\dot{B}} \end{pmatrix} = \left(\phi^A M_A{}^B + \overline{\psi}_{\dot{A}} M^{\dot{A}B}, \ \ \phi^A M_A{}^{\dot{B}} + \overline{\psi}_{\dot{A}} M^{\dot{A}}{}_{\dot{B}} \right).$$

We now observe that the relationship between left-handed and right-handed Weyl spinors may be written

$$\psi^A = \overline{\psi}_{\dot{B}}^*(\overline{\sigma}^0)^{\dot{B}A}, \qquad \overline{\phi}_{\dot{A}} = \phi^{B*}(\sigma^0)_{B\dot{A}}. \tag{1.199}$$

The consistency of these relations with Eq. (1.63), *i.e.*

$$\psi^A = \overline{\psi}_{\dot{A}}^*, \tag{1.200}$$

follows from the fact that the σ^0 and $\overline{\sigma}^0$ matrices are 2×2 unit matrices. It may be noted that ψ_A is a two-component column vector, whereas ψ^A is a two-component row; similarly $\overline{\psi}^{\dot{A}}$ is a column and $\overline{\psi}_{\dot{A}}$ is a row.

The consistency of Eqs. (1.199) and (1.200) can also be seen by calculating the Dirac conjugate of Ψ, *i.e.*

$$\overline{\Psi}_W = \Psi_W^\dagger \gamma_W^0.$$

Using Eq. (1.200) we have:

$$\overline{\Psi}_W = \Psi_W^\dagger \gamma_W^0 = \left(\phi^{A*}, \ \overline{\psi}_{\dot{A}}^* \right) \begin{pmatrix} 0 & \mathbb{1}_{2\times 2} \\ \mathbb{1}_{2\times 2} & 0 \end{pmatrix} = \left(\overline{\psi}_{\dot{A}}^*, \ \phi^{A*} \right) = \left(\psi^A, \ \overline{\phi}_{\dot{A}} \right).$$

Note that since Eq. (1.200) does not preserve a consistent index structure the matrix representation of γ^0 must be used. On the other hand, using Eq. (1.199) we have:

$$\overline{\Psi}_W = \Psi_W^\dagger \gamma_W^0 = \left(\phi^{A*}, \ \overline{\psi}_{\dot{A}}^* \right) \begin{pmatrix} 0 & (\sigma^0)_{A\dot{B}} \\ (\overline{\sigma}^0)^{\dot{A}B} & 0 \end{pmatrix}$$

$$= \left(\overline{\psi}_{\dot{A}}^*(\overline{\sigma}^0)^{\dot{A}B}, \ \phi^{A*}(\sigma^0)_{A\dot{B}} \right) \overset{(1.199)}{=} \left(\psi^B, \ \overline{\phi}_{\dot{B}} \right), \tag{1.201a}$$

in agreement with the previous result. We note that

$$\overline{\Psi}_W^\top = \begin{pmatrix} \psi_A \\ \overline{\phi}^{\dot{A}} \end{pmatrix}. \tag{1.201b}$$

The charge conjugate of a Dirac spinor in the Weyl representation is defined as in Eq. (1.188) with the charge conjugation matrix Eq. (1.192), *i.e.*

$$\Psi_W^c = C_W \overline{\Psi}_W^\top.$$

Thus using Eq. (1.193) and Eqs. (1.201a), (1.201b), we have:

$$(\Psi^c_W)_a = (C_W)_{ab}(\overline{\Psi}^\top_W)_b$$

$$= \begin{pmatrix} (i\sigma^2\overline{\sigma}^0)_A{}^B & 0 \\ 0 & (i\overline{\sigma}^2\sigma^0)^{\dot{A}}{}_{\dot{B}} \end{pmatrix} \begin{pmatrix} \psi_B \\ \overline{\phi}^{\dot{B}} \end{pmatrix} = \begin{pmatrix} (i\sigma^2\overline{\sigma}^0)_A{}^B \psi_B \\ (i\overline{\sigma}^2\sigma^0)^{\dot{A}}{}_{\dot{B}}\overline{\phi}^{\dot{B}} \end{pmatrix}.$$

Now

$$(i\sigma^2\overline{\sigma}^0)^{AB} = \begin{pmatrix} 0 & 1 \\ -1 & 0 \end{pmatrix}^{AB} = (\epsilon^{AB}),$$

and

$$(i\overline{\sigma}^2\sigma^0)_{\dot{A}\dot{B}} = \begin{pmatrix} 0 & -1 \\ 1 & 0 \end{pmatrix}_{\dot{A}\dot{B}} = (\epsilon_{\dot{A}\dot{B}}).$$

Hence

$$(i\sigma^2\overline{\sigma}^0)_A{}^B = \epsilon_{AC}(i\sigma^2\overline{\sigma}^0)^{CB} = \epsilon_{AC}\epsilon^{CB} = \delta_A{}^B,$$

and

$$(i\overline{\sigma}^2\sigma^0)^{\dot{A}}{}_{\dot{B}} = \epsilon^{\dot{A}\dot{C}}(i\overline{\sigma}^2\sigma^0)_{\dot{C}\dot{B}} = \epsilon^{\dot{A}\dot{C}}\epsilon_{\dot{C}\dot{B}} = \delta^{\dot{A}}{}_{\dot{B}}.$$

Hence:

$$(\Psi^c_W)_a = \begin{pmatrix} \psi_A \\ \overline{\phi}^{\dot{A}} \end{pmatrix}. \tag{1.202}$$

Thus, charge conjugation flips ϕ and ψ.

A *Majorana spinor* is a four-component Dirac spinor which satisfies (here in the Weyl representation)

$$\Psi_W = \Psi^c_W. \tag{1.203}$$

Using Eq. (1.202) and the explicit form of a Dirac spinor in the Weyl representation, *i.e.* Eq. (1.156), we obtain

$$\Psi_W = \begin{pmatrix} \phi_A \\ \overline{\psi}^{\dot{A}} \end{pmatrix} \overset{!}{=} \begin{pmatrix} \psi_A \\ \overline{\phi}^{\dot{A}} \end{pmatrix} = \Psi^c_W.$$

Thus for a Majorana spinor in the Weyl representation we can write:

$$\Psi^M_W = \begin{pmatrix} \phi_A \\ \overline{\phi}^{\dot{A}} \end{pmatrix}. \tag{1.204}$$

Thus a Majorana spinor has only two independent complex components and is therefore equivalent to a two-component Weyl spinor or a real Dirac spinor.

1.4.6   Calculations with Dirac Spinors

It is useful to know the connection between the four-component Dirac formalism and the two-component Weyl formalism. The use of the Weyl representation of the four-component Dirac formalism has certain advantages.

We use the following notation for Dirac spinors:

$$\Psi = \begin{pmatrix} \psi_{+A} \\ \overline{\psi}_-^{\dot{A}} \end{pmatrix}, \tag{1.205}$$

where the subscripts $\pm$ distinguish between Weyl spinors which are elements of representation spaces F $(+)$ and $\dot{F}^*$ $(-)$ respectively. Then according to Eq. (1.201a), the Dirac conjugate of Eq. (1.205) is:

$$\overline{\Psi} = \begin{pmatrix} \psi_-^A, & \overline{\psi}_{+\dot{A}} \end{pmatrix}. \tag{1.206}$$

Proposition 1.50:

The following relations hold for Dirac spinors Ψ and χ:

$$(\overline{\Psi}\chi)_4 = (\psi_-\chi_+)_2 + (\overline{\psi}_+\overline{\chi}_-)_2, \tag{1.207a}$$

$$(\overline{\Psi}\gamma^5\chi)_4 = -(\psi_-\chi_+)_2 + (\overline{\psi}_+\overline{\chi}_-)_2, \tag{1.207b}$$

$$(\overline{\Psi}\gamma^\mu\chi)_4 = (\overline{\psi}_+\overline{\sigma}^\mu\chi_+)_2 + (\psi_-\sigma^\mu\overline{\chi}_-)_2, \tag{1.207c}$$

$$(\overline{\Psi}\gamma^\mu\gamma^5\chi)_4 = (\psi_-\sigma^\mu\overline{\chi}_-)_2 - (\overline{\psi}_+\overline{\sigma}^\mu\chi_+)_2, \tag{1.207d}$$

$$(\overline{\Psi}\sigma^{4\mu\nu}\chi)_4 = (\psi_-\sigma_2^{\mu\nu}\chi_+)_2 + (\overline{\psi}_+\overline{\sigma}_2^{\mu\nu}\overline{\chi}_-)_2, \tag{1.207e}$$

where

$$\sigma^{4\mu\nu} := \frac{i}{4}[\gamma^\mu, \gamma^\nu]. \tag{1.207f}$$

Proof:

(a) On using Eqs. (1.92a), (1.92b), and (1.93), we obtain for Eq. (1.207a):

$$(\overline{\Psi}\chi)_4 = \begin{pmatrix} \psi_-^A, & \overline{\psi}_{+\dot{A}} \end{pmatrix} \begin{pmatrix} \chi_{+A} \\ \overline{\chi}_-^{\dot{A}} \end{pmatrix} = \psi_-^A\chi_{+A} + \overline{\psi}_{+\dot{A}}\overline{\chi}_-^{\dot{A}}$$

$$= (\psi_-\chi_+)_2 + (\overline{\psi}_+\overline{\chi}_-)_2.$$

(b) With Eqs. (1.92a), (1.92b), and (1.93), we can prove Eq. (1.207b):

$$(\overline{\Psi}\gamma^5\chi)_4 = \overline{\Psi}_a\gamma^5_{ab}\chi_b = \begin{pmatrix} \psi_-^A, & \overline{\psi}_{+\dot{A}} \end{pmatrix} \begin{pmatrix} -\delta_A{}^B & 0 \\ 0 & \delta^{\dot{A}}{}_{\dot{B}} \end{pmatrix} \begin{pmatrix} \chi_{+B} \\ \overline{\chi}_-^{\dot{B}} \end{pmatrix}$$

$$= \begin{pmatrix} \psi_-^A, & \overline{\psi}_{+\dot{A}} \end{pmatrix} \begin{pmatrix} -\chi_{+A} \\ \overline{\chi}_-^{\dot{A}} \end{pmatrix}$$

$$= -\psi_-^A\chi_{+A} + \overline{\psi}_{+\dot{A}}\overline{\chi}_-^{\dot{A}} = -(\psi_-\chi_+)_2 + (\overline{\psi}_+\overline{\chi}_-)_2.$$

(c) Proof of Eq. (1.207c):

$$\left(\overline{\Psi}\gamma^\mu\chi\right)_4 = \overline{\Psi}_a\gamma^\mu_{ab}\chi_b \stackrel{(1.159)}{=} \left(\psi^A_-, \overline{\psi}_{+\dot A}\right)\begin{pmatrix} 0 & (\sigma^\mu)_{A\dot B} \\ (\overline{\sigma}^\mu)^{\dot A B} & 0 \end{pmatrix}\begin{pmatrix} \chi_{+B} \\ \overline{\chi}^{\dot B}_- \end{pmatrix}$$

$$= \left(\psi^A_-, \overline{\psi}_{+\dot A}\right)\begin{pmatrix} (\sigma^\mu)_{A\dot B}\overline{\chi}^{\dot B}_- \\ (\overline{\sigma}^\mu)^{\dot A B}\chi_{+B} \end{pmatrix}$$

$$= \psi^A_-(\sigma^\mu)_{A\dot B}\overline{\chi}^{\dot B}_- + \overline{\psi}_{+\dot A}(\overline{\sigma}^\mu)^{\dot A B}\chi_{+B}$$

$$= \left(\psi_-\sigma^\mu\overline{\chi}_-\right)_2 + \left(\overline{\psi}_+\overline{\sigma}^\mu\chi_+\right)_2,$$

again using Eqs. (1.92a), (1.92b), and (1.93).

(d) Proof of Eq. (1.207d):

$$\left(\overline{\Psi}\gamma^\mu\gamma^5\chi\right)_4 = \overline{\Psi}_a\gamma^\mu_{ab}\gamma^5_{bc}\chi_c$$

$$= \left(\psi^A_-, \ \overline{\psi}_{+\dot A}\right)\begin{pmatrix} 0 & (\sigma^\mu)_{A\dot B} \\ (\overline{\sigma}^\mu)^{\dot A B} & 0 \end{pmatrix}\begin{pmatrix} -\delta_B{}^C & 0 \\ 0 & \delta^{\dot B}{}_{\dot C} \end{pmatrix}\begin{pmatrix} \chi_{+C} \\ \overline{\chi}^{\dot C}_- \end{pmatrix}$$

$$= \left(\psi^A_-, \ \overline{\psi}_{+\dot A}\right)\begin{pmatrix} 0 & (\sigma^\mu)_{A\dot B} \\ (\overline{\sigma}^\mu)^{\dot A B} & 0 \end{pmatrix}\begin{pmatrix} -\chi_{+B} \\ \overline{\chi}^{\dot B}_- \end{pmatrix}$$

$$= \left(\psi^A_-, \ \overline{\psi}_{+\dot A}\right)\begin{pmatrix} (\sigma^\mu)_{A\dot B}\overline{\chi}^{\dot B}_- \\ -(\overline{\sigma}^\mu)^{\dot A B}\chi_{+B} \end{pmatrix}$$

$$= \psi^A_-(\sigma^\mu)_{A\dot B}\overline{\chi}^{\dot B}_- - \overline{\psi}_{+\dot A}(\overline{\sigma}^\mu)^{\dot A B}\chi_{+B}$$

$$= \left(\psi_-\sigma^\mu\overline{\chi}_-\right)_2 - \left(\overline{\psi}_+\overline{\sigma}^\mu\chi_+\right)_2,$$

again using Eqs. (1.92a), (1.92b), and (1.93).

(e) Proof of Eq. (1.207e): Before we demonstrate Eq. (1.207e), we discuss Eq. (1.207f), *i.e.*

$$\left(\sigma^{4\mu\nu}\right)_{ab} := \frac{i}{4}\left(\left[\gamma^\mu, \gamma^\nu\right]\right)_{ab} = \frac{i}{4}\left(\gamma^\mu_{ac}\gamma^\nu_{cb} - \gamma^\nu_{ac}\gamma^\mu_{cb}\right)$$

$$= \frac{i}{4}\left[\begin{pmatrix} 0 & (\sigma^\mu)_{A\dot C} \\ (\overline{\sigma}^\mu)^{\dot A C} & 0 \end{pmatrix}\begin{pmatrix} 0 & (\sigma^\nu)_{C\dot B} \\ (\overline{\sigma}^\nu)^{\dot C B} & 0 \end{pmatrix}\right.$$

$$\left. - \begin{pmatrix} 0 & (\sigma^\nu)_{A\dot C} \\ (\overline{\sigma}^\nu)^{\dot A C} & 0 \end{pmatrix}\begin{pmatrix} 0 & (\sigma^\mu)_{C\dot B} \\ (\overline{\sigma}^\mu)^{\dot C B} & 0 \end{pmatrix}\right]$$

$$= \frac{i}{4}\begin{pmatrix} \left(\sigma^\mu\overline{\sigma}^\nu - \sigma^\nu\overline{\sigma}^\mu\right)_A{}^B & 0 \\ 0 & \left(\overline{\sigma}^\mu\sigma^\nu - \overline{\sigma}^\nu\sigma^\mu\right)^{\dot A}{}_{\dot B} \end{pmatrix}$$

$$= \begin{pmatrix} (\sigma_2^{\mu\nu})_A{}^B & 0 \\ 0 & (\overline{\sigma}_2^{\mu\nu})^{\dot A}{}_{\dot B} \end{pmatrix},$$

using Eqs. (1.138a) and (1.138b). With this expression one can directly verify Eq. (1.207e):

$$
\begin{aligned}
\left(\overline{\Psi}\sigma^{4\mu\nu}\chi\right)_4 &= \begin{pmatrix} \psi_-^A, & \overline{\psi}_{+\dot A} \end{pmatrix} \begin{pmatrix} (\sigma_2^{\mu\nu})_A{}^B & 0 \\ 0 & (\overline{\sigma}_2^{\mu\nu})^{\dot A}{}_{\dot B} \end{pmatrix} \begin{pmatrix} \chi_{+B} \\ \overline{\chi}_-^{\dot B} \end{pmatrix} \\
&= \begin{pmatrix} \psi_-^A, & \overline{\psi}_{+\dot A} \end{pmatrix} \begin{pmatrix} (\sigma_2^{\mu\nu})_A{}^B \chi_{+B} \\ (\overline{\sigma}_2^{\mu\nu})^{\dot A}{}_{\dot B} \overline{\chi}_-^{\dot B} \end{pmatrix} \\
&= \psi_-^A (\sigma_2^{\mu\nu})_A{}^B \chi_{+B} + \overline{\psi}_{+\dot A} (\overline{\sigma}_2^{\mu\nu})^{\dot A}{}_{\dot B} \overline{\chi}_-^{\dot B} \\
&= \left(\psi_- \sigma_2^{\mu\nu} \chi_+\right)_2 + \left(\overline{\psi}_+ \overline{\sigma}_2^{\mu\nu} \overline{\chi}_-\right)_2,
\end{aligned}
$$

using Eqs. (1.92a), (1.92b), and (1.93).

1.4.7   Calculations with Majorana Spinors

As explained in Sec. 1.4.5, a Majorana spinor has the characteristic property that we can replace the dotted Weyl spinor by the complex conjugate of the undotted one. In the terminology of the previous section this means that we can replace ψ_-^A by ψ_+^A and $\overline{\psi}_-^{\dot A}$ by $\overline{\psi}_+^{\dot A}$. Thus we can drop the suffixes $\pm$ in Eq. (1.205) and write

$$\Psi_M = \begin{pmatrix} \psi_A \\ \overline{\psi}^{\dot A} \end{pmatrix}. \tag{1.208}$$

In this case, Eqs. (1.207a) to (1.207e) read:

$$\left(\overline{\Psi}_M \chi_M\right)_4 = (\psi\chi)_2 + (\overline{\psi}\overline{\chi})_2, \tag{1.209a}$$

$$\left(\overline{\Psi}_M \gamma^5 \chi_M\right)_4 = -(\psi\chi)_2 + (\overline{\psi}\overline{\chi})_2, \tag{1.209b}$$

$$\left(\overline{\Psi}_M \gamma^\mu \chi_M\right)_4 = \left(\overline{\psi}\overline{\sigma}^\mu \chi_+\right)_2 + \left(\psi\sigma^\mu \overline{\chi}\right)_2, \tag{1.209c}$$

$$\left(\overline{\Psi}_M \gamma^\mu \gamma^5 \chi_M\right)_4 = \left(\psi\sigma^\mu \overline{\chi}\right)_2 - \left(\overline{\psi}\overline{\sigma}^\mu \chi\right)_2, \tag{1.209d}$$

$$\left(\overline{\Psi}_M \sigma^{4\mu\nu} \chi_M\right)_4 = \left(\psi\sigma_2^{\mu\nu} \chi\right)_2 + \left(\overline{\psi}\overline{\sigma}_2^{\mu\nu} \overline{\chi}\right)_2. \tag{1.209e}$$

Proposition 1.51:

The following relation holds for Majorana spinors:

$$\left(\overline{\Psi}_M \gamma^\mu \gamma^5 \Psi_M\right)\left(\overline{\Psi}_M \gamma^\nu \gamma^5 \Psi_M\right) = \eta^{\mu\nu} \left(\overline{\Psi}_M \Psi_M\right)_4^2. \tag{1.210}$$

Proof: Using Eq. (1.209d) we have:

$$\left(\overline{\Psi}_M \gamma^\mu \gamma^5 \Psi_M\right)\left(\overline{\Psi}_M \gamma^\nu \gamma^5 \Psi_M\right)$$

$$\overset{(1.209d)}{=} \left[-\left(\overline{\psi}\overline{\sigma}^\mu\psi\right)_2 + \left(\psi\sigma^\mu\overline{\psi}\right)_2\right]\left[-\left(\overline{\psi}\overline{\sigma}^\nu\psi\right)_2 + \left(\psi\sigma^\nu\overline{\psi}\right)_2\right]$$

$$\overset{(1.134)}{=} \left[\left(\psi\sigma^\mu\overline{\psi}\right)_2 + \left(\psi\sigma^\mu\overline{\psi}\right)_2\right]\left[\left(\psi\sigma^\nu\overline{\psi}\right)_2 + \left(\psi\sigma^\nu\overline{\psi}\right)_2\right]$$

$$= 4\left(\psi\sigma^\nu\overline{\psi}\right)_2\left(\psi\sigma^\nu\overline{\psi}\right)_2 \overset{(1.137)}{=} 2\eta^{\mu\nu}\left(\psi\psi\right)_2\left(\overline{\psi}\,\overline{\psi}\right)_2.$$

For the present case Eq. (1.209a) is

$$\left(\overline{\Psi}_M\Psi_M\right)_4 = \left(\psi\psi\right)_2 + \left(\overline{\psi}\,\overline{\psi}\right)_2.$$

Taking the square of this expression we obtain:

$$\left[\left(\overline{\Psi}_M\Psi_M\right)_4\right]^2 = \left[\left(\psi\psi\right)_2 + \left(\overline{\psi}\,\overline{\psi}\right)_2\right]^2$$

$$= \left(\psi\psi\right)_2\left(\psi\psi\right)_2 + \left(\psi\psi\right)_2\left(\overline{\psi}\,\overline{\psi}\right)_2$$

$$+ \left(\overline{\psi}\,\overline{\psi}\right)_2\left(\psi\psi\right)_2 + \left(\overline{\psi}\,\overline{\psi}\right)_2\left(\overline{\psi}\,\overline{\psi}\right)_2$$

$$= 2\left(\psi\psi\right)_2\left(\overline{\psi}\,\overline{\psi}\right)_2,$$

since as will be shown below:

$$\left(\psi\psi\right)_2\left(\psi\psi\right)_2 = \left(\overline{\psi}\,\overline{\psi}\right)_2\left(\overline{\psi}\,\overline{\psi}\right)_2 = 0,$$

and

$$\left(\psi\psi\right)_2\left(\overline{\psi}\,\overline{\psi}\right)_2 = \left(\overline{\psi}\,\overline{\psi}\right)_2\left(\psi\psi\right)_2.$$

Hence it follows that

$$\left(\overline{\Psi}_M\gamma^\mu\gamma^5\Psi_M\right)\left(\overline{\Psi}_M\gamma^\nu\gamma^5\Psi_M\right) = \eta^{\mu\nu}\left(\overline{\Psi}_M\Psi_M\right)_4^2.$$

We now show explicitly that $(\psi\psi)_2^2 = 0$:

$$\left(\psi\psi\right)_2^2 = \left(\psi\psi\right)_2\left(\psi\psi\right)_2$$

$$\overset{(1.92a)}{=} \left(\psi^A\psi_A\right)\left(\psi^B\psi_B\right)$$

$$\overset{(1.83)}{=} \epsilon^{AC}\psi_C\psi_A\epsilon^{BD}\psi_D\psi_B$$

$$= \epsilon^{AC}\epsilon^{BD}\psi_C\psi_A\psi_D\psi_B$$

$$= \epsilon^{12}\epsilon^{12}\psi_2\psi_1\psi_2\psi_1 + \epsilon^{12}\epsilon^{21}\psi_2\psi_1\psi_1\psi_2$$

$$+\epsilon^{21}\epsilon^{12}\psi_1\psi_2\psi_2\psi_1 + \epsilon^{21}\epsilon^{21}\psi_1\psi_2\psi_1\psi_2$$

$$= \psi_2\psi_1\psi_2\psi_1 - \psi_2\psi_1\psi_1\psi_2 - \psi_1\psi_2\psi_2\psi_1 + \psi_1\psi_2\psi_1\psi_2.$$

Taking into account the fact that ψ_A are Grassmann numbers, we obtain (*cf.* discussion following Eq. (1.96))

$$(\psi\psi)_2^2 = -4\psi_1\psi_1\psi_2\psi_2 = 0,$$

since for Grassmann numbers ψ_1

$$\{\psi_1, \psi_1\} = 0 \qquad \text{implies} \qquad \psi_1\psi_1 + \psi_1\psi_1 = 0,$$

i.e. $\psi_1\psi_1 = 0$.

A similar calculation can be carried out for $\left(\overline{\psi}\,\overline{\psi}\right)_2^2$.

We now verify that

$$\left(\overline{\psi}\,\overline{\psi}\right)_2(\psi\psi)_2 = (\psi\psi)_2\left(\overline{\psi}\,\overline{\psi}\right)_2.$$

To this end consider

$$\left(\overline{\psi}\,\overline{\psi}\right)_2(\psi\psi)_2 = \overline{\psi}_{\dot{A}}\overline{\psi}^{\dot{A}}\psi^A\psi_A = -\overline{\psi}_{\dot{A}}\psi^A\overline{\psi}^{\dot{A}}\psi_A$$

$$= \psi^A\overline{\psi}_{\dot{A}}\overline{\psi}^{\dot{A}}\psi_A = \psi^A\psi_A\overline{\psi}_{\dot{A}}\overline{\psi}^{\dot{A}} = (\psi\psi)_2\left(\overline{\psi}\,\overline{\psi}\right)_2.$$

We prove one more formula.

Proposition 1.52:

The following relation holds for Majorana spinors:

$$\left(\overline{\Psi}_M\gamma^5\Psi_M\right)_4^2 = -\left(\overline{\Psi}_M\Psi_M\right)_4^2. \tag{1.211}$$

Proof: Using Eq. (1.209b) we have:

$$\left(\overline{\Psi}_M\gamma^5\Psi_M\right)_4^2 = \left\{-(\psi\psi)_2 + \left(\overline{\psi}\,\overline{\psi}\right)_2\right\}^2$$

$$= (\psi\psi)_2^2 - 2(\psi\psi)_2\left(\overline{\psi}\,\overline{\psi}\right)_2 + \left(\overline{\psi}\,\overline{\psi}\right)_2^2$$

$$= -2(\psi\psi)_2\left(\overline{\psi}\,\overline{\psi}\right)_2$$

$$= -\left\{(\psi\psi)_2^2 + 2(\psi\psi)_2\left(\overline{\psi}\,\overline{\psi}\right)_2 + \left(\overline{\psi}\,\overline{\psi}\right)_2^2\right\}$$

$$= -\left(\overline{\Psi}_M\Psi_M\right)_4^2,$$

where in the last step we used Eq. (1.209a).

Chapter 2

No-Go Theorems and Graded Lie Algebras

2.1 The Theorems of Coleman-Mandula and Haag, Łopuszański, Sohnius

We now discuss the two theorems already referred to in the Introduction.[1]

2.1.1 The Theorem of Coleman–Mandula

The following theorem[2] was established by S. Coleman and J. Mandula [23]:

> Let G be a connected symmetry group of the S-matrix, *i.e.* a group whose generators commute with the S-matrix, and make the following five assumptions:
>
> (i) *Lorentz invariance:* G contains a subgroup which is locally isomorphic to the Poincaré group.
>
> (ii) *Particle finiteness:* All particle types correspond to positive-energy representations of the Poincaré group. For any finite mass M, there is only a finite number of particles with mass less than M.

---

[1]A.O. Barut and R. Rączka [8], p. 43, also give considerations concerning the unification of the Poincaré algebra with internal symmetry algebras. Stronger group theoretical lemmas are given on p. 629, and it is pointed out (p. 630) that the infinite-parameter Lie algebra associated with noncompact dynamical groups (which lead to infinite particle multiplets (*cf.* pp. 411, 609)) does not contradict these theorems.

[2]See also the discussion in S. Weinberg [118].

(iii) *Weak elastic analyticity:* Elastic scattering amplitudes are analytic functions of centre-of-mass energy squared s and invariant momentum transfer squared t in some neighbourhood of the physical region, except at normal thresholds.

(iv) *Occurrence of scattering:* Let $|p\rangle$ and $|p'\rangle$ be any two one-particle momentum eigenstates, and let $|p, p'\rangle$ be the two-particle state constructed from these. Then

$$T|p, p'\rangle \neq 0,$$

where T is the T-matrix defined by

$$S = \mathbb{1} - i(2\pi)^4 \delta^4(p_\mu - p'_\mu)T,$$

except, perhaps, for certain isolated values of S. In simpler terms this assumption means: Two plane waves scatter at almost any energy.

(v) *Technical assumption:* The generators of G, considered as integral operators in momentum space, have distributions for their kernels.

Then the group G is locally isomorphic to the direct product of a compact symmetry group and the Poincaré group.

We recall briefly some basic results of scattering theory. The *Hilbert space* $\mathcal{H}$ is the direct sum of an infinite number of subspaces, *i.e.*

$$\mathcal{H} = \mathcal{H}^{(1)} \oplus \mathcal{H}^{(2)} \oplus \cdots .$$

Here $\mathcal{H}^{(n)}$ denotes the n-particle subspace. It is a subspace of the direct product (symmetric or antisymmetric in accordance with the generalized exclusion principle) of n Hilbert spaces, each being isomorphic to $\mathcal{H}^{(1)}$. The S-matrix is a unitary operator on $\mathcal{H}$. A unitary operator U on $\mathcal{H}$ is said to be a *symmetry transformation* of the S-matrix if

(i) U transforms one-particle states into one-particle states,

(ii) U acts on many-particle states as if they were tensor products of one-particle states,

(iii) U commutes with S.

Thus the theorem of Coleman and Mandula, stated here without proof,[3] demonstrates that the most general Lie algebra of symmetries of the S-matrix contains the energy momentum operator P_μ, the Lorentz rotation generator $M_{\mu\nu}$, and a finite number of Lorentz scalar operators B_l, *i.e.*

$$[P_\mu, B_l] = 0, \qquad [M_{\mu\nu}, B_l] = 0,$$

where the B_l constitute a *Lie algebra*,

$$[B_l, B_m] = i c_{lm}{}^k B_k,$$

and the $c_{lm}{}^k$ are the structure constants of the Lie algebra of the compact internal symmetry[4] group (*e.g. $SU(2, \mathbb{C})$*).

2.1.2   The Theorem of Haag, Łopuszański and Sohnius

Supersymmetries avoid the restrictions of the Coleman-Mandula theorem by relaxing one condition.[5] R. Haag, J.T. Łopuszański and M.F. Sohnius [54] generalize the notion of a Lie algebra to include algebraic systems whose defining relations involve in addition to the usual commutators also anti-commutators. These algebras[6] are called *superalgebras* or *graded Lie algebras*. The generalization of the Poincaré algebra to a superalgebra is obtained in its simplest version by the following procedure. One adds to the Poincaré algebra a Majorana spinor charge with components $Q_a, a = 1, \ldots, 4$. These operators are called *spinor charges*. They have the following properties:

$$\begin{aligned} \{Q_a, \overline{Q}_b\} &= 2(\gamma^\mu)_{ab} P_\mu, \\ [Q_a, P_\mu] &= 0, \\ [Q_a, M^{\mu\nu}] &= (\sigma^{4\mu\nu})_{ab} Q_b. \end{aligned} \qquad (2.1)$$

We shall see later that

$$\{Q_a, Q_b\} = -2(\gamma^\mu C)_{ab} P_\mu,$$

---

[3]For a proof of the theorem see the book of S. Weinberg [118].

[4]In general symmetries are classified as *geometric symmetries* and *internal symmetries*. Geometric symmetry transformations operate on spacetime coordinates (such as Lorentz or Poincaré transformations), whereas internal symmetry transformations do not affect the spacetime manifold but operate on objects (*e.g.* scalar fields, spinor fields), defined on the spacetime manifold.

[5]In this sense, the Haag-Łopuszański-Sohnius theorem is a natural extension of the Coleman-Mandula theorem.

[6]In the mathematical literature, the first who considered groups with commuting and anticommuting parameters were F.A. Berezin and G.I. Kac [11].

where (*cf.* Eq. (1.207f)):

$$\sigma_4^{\mu\nu} = \frac{i}{4}\left[\gamma^\mu, \gamma^\nu\right]$$

and

$$\overline{Q}_a = \left(Q^\dagger \gamma_0\right)_a.$$

The operators P_μ and $M_{\mu\nu}$ are the usual generators of *displacements* and *homogeneous Lorentz transformations* of spacetime. In order to incorporate an internal symmetry in a nontrivial way it is often convenient to rewrite Eqs. (2.1) in terms of two-component *Weyl spinors* Q_A and $\overline{Q}_{\dot{A}}$. Then:

$$\{Q_A, Q_B\} = 0, \qquad\qquad \{\overline{Q}_{\dot{A}}, \overline{Q}_{\dot{B}}\} = 0,$$
$$\{Q_A, \overline{Q}_{\dot{B}}\} = 2(\sigma^\mu)_{A\dot{B}}P_\mu, \qquad [Q_A, M^{\mu\nu}] = i(\sigma_2^{\mu\nu})_A{}^B Q_B,$$
$$[Q_A, P_\mu] = 0, \qquad\qquad [\overline{Q}_{\dot{A}}, P_\mu] = 0, \qquad\qquad (2.2)$$

the $\sigma_2^{\mu\nu}$ being defined by Eq. (1.138a). In Eq. (2.2) the dotted and undotted indices assume values $A = 1,2$ and $\dot{A} = \dot{1}, \dot{2}$ respectively and refer to the $(0, 1/2)$ and $(1/2, 0)$ repesentations of the spinor group $SL(2, \mathbb{C})$.

We now assume that we have a set of generators, *i.e.* spinor charges[7] $Q_A^\alpha, \alpha = 1, \ldots, N$, which transform according to some representation of a compact Lie group G, such as $SU(3, \mathbb{C})$, which represents the internal symmetry group. Then the generators of G are the *Lorentz scalars* B_l. The $\overline{Q}_{\dot{A}}^\alpha$ transform according to the complex conjugate representation of this group. Then the relations (2.2) generalize as follows:

$$\left.\begin{aligned}
\{Q_A^\alpha, Q_B^\beta\} &= 0, \\
\{\overline{Q}_{\dot{A}}^\alpha, \overline{Q}_{\dot{B}}^\beta\} &= 0, \\
\{Q_A^\alpha, \overline{Q}_{\dot{B}}^\beta\} &= 2\delta^{\alpha\beta}(\sigma^\mu)_{A\dot{B}}P_\mu, \\
[Q_A^\alpha, P_\mu] &= 0, \\
[\overline{Q}_{\dot{A}}^\alpha, P_\mu] &= 0, \\
[Q_A^\alpha, B_l] &= iS_l{}^{\alpha\beta}Q_A^\beta, \\
[Q_A^\alpha, M^{\mu\nu}] &= i(\sigma_2^{\mu\nu})_A{}^B Q_B^\alpha, \\
[B_l, B_m] &= ic_{lm}{}^k B_k,
\end{aligned}\right\} \qquad (2.3)$$

where the $S_l{}^{\alpha\beta}$ are the Hermitian representation matrices of the representation containing the charges Q_A^α. As mentioned before, the B_l are the generators of the internal symmetry group. The theorem of Haag, Łopuszański and

---

[7]Here, N denotes the dimension of the chosen representation of the internal symmetry group G.

Sohnius now states that the maximal symmetry of the S-matrix is the direct product of an internal symmetry with the superalgebra given by relations (2.3). The only allowed extension is the possible appearance of so-called *central charges* in the anticommutator of two undotted spinors. Instead of the first relation of Eq. (2.3) one would then have

$$\{Q_A^\alpha, Q_B^\beta\} = \epsilon_{AB} Z^{\alpha\beta}, \quad Z^{\alpha\beta} = -Z^{\beta\alpha},$$

where ϵ_{AB} is given by Eq. (1.66). Furthermore, the Z's commute with the B's, *i.e.*

$$\left[Z^{\alpha\beta}, B_l\right] = 0,$$

which is the reason why the quantities $Z^{\alpha\beta}$ are called *central charges*.

2.2 Graded Lie Algebras

2.2.1 Lie Algebras

Before we consider graded Lie algebras it is worthwhile to recapitulate the definition of a Lie algebra (see *e.g.* [19, Chap. 1], [115, Chap. 2], or [72, Chap. 5]).

Definition (Lie algebra)

A Lie algebra consists of a vector space L over a field (here $\mathbb{R}$ or $\mathbb{C}$) with a composition rule called product, denoted by $\circ$, defined as follows:

$$\circ : L \times L \longrightarrow L.$$

If $v_1, v_2, v_3 \in L$, then the following properties define the Lie algebra:

   (i) $v_1 \circ v_2 \in L$ (closure of L under $\circ$),

  (ii) $v_1 \circ (v_2 + v_3) = v_1 \circ v_2 + v_1 \circ v_3$ (linearity),

 (iii) $v_1 \circ v_2 = -v_2 \circ v_1$ (antisymmetry),

 (iv) $v_1 \circ (v_2 \circ v_3) + v_3 \circ (v_1 \circ v_2) + v_2 \circ (v_3 \circ v_1) = 0$ (Jacobi identity).

Example: The matrix space of complex 2×2-matrices which are traceless and anti-Hermitian forms a Lie algebra, $su(2, \mathbb{C})$, the Lie algebra of the Lie group $SU(2, \mathbb{C})$, provided we define the product as

$$a \circ b := [a, b] = ab - ba, \quad \forall\, a, b \in su(2, \mathbb{C}).$$

A basis of the vector space L is given by the three matrices

$$\tau_i := \frac{i}{2}\sigma_i, \quad \tau_i^\dagger = -\tau_i, \quad i = 1,2,3,$$

where the σ_i are the three *Pauli matrices* (*cf.* Eq. (1.103b)).

We verify that the algebra $su(2,\mathbb{C})$ satisfies items (i) to (iv) of the definition of a Lie algebra.

(i) *Closure.* The product is defined as (see Chapter 1, Eq. (1.104b))

$$\tau_i \circ \tau_i := \left[\tau_i, \tau_k\right] = -\epsilon_{ijk}\tau_k,$$

where ϵ_{ijk} is totally antisymmetric in its indices i,j,k and $\epsilon_{123} = 1$. Hence, the vector space is closed under the product $\circ = [\,,\,]$.

(ii) *Linearity* is demonstrated as follows:

$$\begin{aligned}
\tau_i \circ (\tau_j + \tau_k) &= \left[\tau_i, \tau_j + \tau_k\right] \\
&= \tau_i(\tau_j + \tau_k) - (\tau_j + \tau_k)\tau_i \\
&= \tau_i\tau_j + \tau_i\tau_k - \tau_j\tau_i - \tau_k\tau_i \\
&= \tau_i\tau_j - \tau_j\tau_i + \tau_i\tau_k - \tau_k\tau_i \\
&= \left[\tau_i, \tau_j\right] + \left[\tau_i, \tau_k\right] \\
&= \tau_i \circ \tau_j + \tau_j \circ \tau_k.
\end{aligned}$$

This verifies the linearity of the product $\circ = [\,,\,]$.

(iii) The *antisymmetry* is shown as follows:

$$\begin{aligned}
\tau_i \circ \tau_j = \left[\tau_i, \tau_j\right] &= \tau_i\tau_j - \tau_j\tau_i \\
&= -(\tau_j\tau_i - \tau_i\tau_j) \\
&= -\left[\tau_j, \tau_i\right] \\
&= -\tau_j \circ \tau_i.
\end{aligned}$$

(iv) It is cumbersome to verify the *Jacobi identity*; however, it is well known that the following relation holds which expresses the same property:

$$\left[\tau_1, [\tau_2, \tau_3]\right] + \left[\tau_3, [\tau_1, \tau_2]\right] + \left[\tau_2, [\tau_3, \tau_1]\right] = 0.$$

2.2.2 Graded Algebras

We now define graded algebras. In the simplest case a graded algebra consists of a vector space L which is the direct sum of two subspaces L_0 and L_1; *i.e.*

$$L = L_0 \oplus L_1,$$

and a product $\circ$,

$$\circ \; : \; L \times L \longrightarrow L$$

with the following properties:

(i) $u_1 \circ u_2 \in L_0, \quad \forall \, u_1, u_2 \in L_0,$

(ii) $u \circ v \in L_1, \quad \forall \, u \in L_0, v \in L_1,$

(iii) $v_1 \circ v_2 \in L_0, \quad \forall \, v_1, v_2 \in L_1.$

This algebra is called a $\mathbb{Z}_2$-*graded algebra.*

   More generally, L is the direct sum of $N + 1, N \geq 1$, subspaces L_k, *i.e.*

$$L = \bigoplus_{k=0}^{N} L_k,$$

with a product $\circ$,

$$\circ \; : \; L \times L \longrightarrow L,$$

such that if $u_k \in L_k$, then

$$u_j \circ u_k \in L_{j+k \bmod (N+1)}.$$

A product $\circ$ with such a property is called a *grading*.

2.2.3 Graded Lie Algebras

A graded algebra becomes a *graded Lie algebra* if one modifies the product denoted by $\circ$ in the following way. For simplicity we consider a $\mathbb{Z}_2$-graded algebra.[8] Let L_0 and L_1 be vector spaces and

$$L := L_0 \oplus L_1.$$

Thus, the vector space L can be written as the direct sum of L_0 and L_1. We define the product $\circ$

$$\circ \; : \; L \times L \longrightarrow L$$

with the following properties:

---

[8]An elementary presentation of graded Lie algebras, their classification and some properties of their representations is given in [96]. A mathematical treatment of these topics can be found in the monograph by B. DeWitt [26], Chap. 3. A general classification of super Lie groups can be found in [26], Chap.4. See also the dictionary [43].

(i) *Grading:* For all $x_i \in L_i$, $i = 0, 1$,

$$x_i \circ x_j \in L_{i+j \bmod 2}. \tag{2.4}$$

Then L becomes a graded algebra according to Sec. 2.2.2.

(ii) *Supersymmetrization:* For all $x_i \in L_i$, $x_j \in L_j$, $i, j = 0, 1$,

$$x_i \circ x_j = -(-1)^{ij} x_j \circ x_i. \tag{2.5}$$

(iii) *Generalized Jacobi identities:* For all $x_k \in L_k$, $x_l \in L_l$, $x_m \in L_m$, $k, l, m \in \mathbb{Z}_2$,

$$x_k \circ (x_l \circ x_m)(-1)^{km} + x_l \circ (x_m \circ x_k)(-1)^{lk} + x_m \circ (x_k \circ x_l)(-1)^{ml} = 0. \tag{2.6}$$

With this definition of the product, L as a vector space becomes a *graded Lie algebra*. It is important to note that L is not a Lie algebra, since, as defined in Eq. (2.5), the product is in general not antisymmetric. To see this, it is advantageous to write out Eq. (2.5) explicitly:

(a) $i = j = 0$, i.e. $x_0, y_0 \in L_0$, then:

$$x_0 \circ y_0 = -(-1)^0 y_0 \circ x_0 = -y_0 \circ x_0.$$

Hence in the subspace L_0 the product $\circ$ is antisymmetric.

(b) $i = 0, j = 1$, i.e. $x_0 \in L_0, y_1 \in L_1$, then:

$$x_0 \circ y_1 = -(-1)^0 y_1 \circ x_0 = -y_1 \circ x_0.$$

(c) $i = j = 1$, i.e. $x_1, y_1 \in L_1$, then:

$$x_1 \circ y_1 = -(-1)^1 y_1 \circ x_1 = y_1 \circ x_1.$$

Hence in the subspace L_1 the product $\circ$ is symmetric.

With the above definition of a graded Lie algebra the subspace L_0 spans an ordinary Lie algebra, because the pair $(L_0, \circ)$ satisfies the definition of an ordinary Lie algebra as given in Sec. 2.2.1. The subspace L_1 is not even an algebra, since according to Eq. (2.4) L_1 is not closed under the product $\circ$, i.e. if $x_1, y_1 \in L_1$, then

$$x_1 \circ y_1 \in L_{1+1 \bmod 2} = L_0.$$

2.3   The Graded Lie Algebra of $SU(2, \mathbb{C})$

As an example we discuss the $\mathbb{Z}_2$ grading of $su(2, \mathbb{C})$, the Lie algebra of the group $SU(2, \mathbb{C})$. As stated above, the subspace L_0 in the construction of a graded Lie algebra is an ordinary Lie algebra. It is natural therefore to take L_0 as the Lie algebra of $SU(2, \mathbb{C})$ with generators τ_1, τ_2, τ_3 and

$$\left[\tau_i, \tau_j\right] = -\epsilon_{ijk}\tau_k.$$

We define the product

$$\circ \; : \; L \times L \longrightarrow L$$

on the subspace L_0 as

$$\circ \; : L_0 \times L_0 \longrightarrow L_0,$$
$$(\tau_i, \tau_j) \longmapsto \tau_i \circ \tau_j := \left[\tau_i, \tau_j\right] = -\epsilon_{ijk}\tau_k \in L_0. \tag{2.7}$$

We denote the generators of the subspace L_1 by $Q_a, a = 1, 2, \ldots, N, N = \dim L_1$. We now have to define the product $\circ$ when multiplying any $\tau_i \in L_0$ by Q_a, and multiplying two $Q_a \in L_1$. In the case of the former we have

$$\circ \; : \; L_0 \times L_1 \longrightarrow L_1.$$

Thus if we form the product of any $\tau_i \in L_0$ with a $Q_a \in L_1$, then according to Eq. (2.4) we must obtain an element of the subspace L_1. We define therefore:

$$\circ \; : \; (\tau_i, Q_a) \longmapsto \tau_i \circ Q_a = -(t_i)_{ab}Q_b \in L_1. \tag{2.8}$$

Here the $(t_i)_{ab}$ are coefficients which are restricted by the generalized Jacobi identity (2.6). In the present case we have

$$x_k \in L_k, \quad x_l \in L_l, \quad x_M \in L_m,$$

and

$$\tau_i \in L_0, \quad \tau_j \in L_0, \quad Q_a \in L_1,$$

and so the generalized Jacobi identity is

$$x_k \circ (x_l \circ x_m)(-1)^{km} + x_m \circ (x_k \circ x_l)(-1)^{ml} + x_l \circ (x_m \circ x_k)(-1)^{lk} = 0,$$

so that

$$\tau_i \circ (\tau_j \circ Q_a) + Q_a \circ (\tau_i \circ \tau_j) + \tau_j \circ (Q_a \circ \tau_i) = 0,$$

and, using Eqs. (2.7) and (2.8),

$$\tau_i \circ \left[-(t_j)_{ab}Q_b\right] + Q_a \circ \left[-\epsilon_{ijk}\tau_k\right] + \tau_j \circ \left[(t_i)_{ab}Q_b\right] = 0,$$

that is

$$-(t_j)_{ab}(\tau_i \circ Q_b) - \epsilon_{ijk}(Q_a \circ \tau_k) + (t_i)_{ab}(\tau_j \circ Q_b) = 0.$$

Again using Eq. (2.8),

$$(t_j)_{ab}(t_i)_{bc}Q_c - \epsilon_{ijk}(t_k)_{ab}Q_b - (t_i)_{ab}(t_j)_{bc}Q_c = 0,$$

so that

$$\big[(t_i)_{ab}(t_j)_{bc} - (t_j)_{ab}(t_i)_{bc}\big]Q_c = -\epsilon_{ijk}(t_k)_{ac}Q_c,$$

or since the Q_c's are independent:

$$\big[t_i, t_j\big]_{ac} = -\epsilon_{ijk}(t_k)_{ac}. \tag{2.9}$$

This relation states that the coefficients $(t_i)_{ab}$ of Eq. (2.8) constitute dim $L_1 \times$ dim L_1-matrices which are representation matrices of the algebra L_0. If the dimension of L_1 is two, then $t_i = \tau_i$; if the dimension of L_1 is three then $(t_i)_{ab} = -\epsilon_{iab}$.

Finally we have to define the product $\circ$ on the subspace L_1. According to Eq. (2.4) we have for $Q_a, Q_b \in L_1$:

$$Q_a \circ Q_b \in L_{1+1} = L_2 \cong L_0.$$

According to Eq. (2.5) this product on the subspace L_1 has to be symmetric, i.e.

$$Q_a \circ Q_b = -(-1)^1 Q_b \circ Q_a = Q_b \circ Q_a.$$

Hence we define

$$\circ : L_1 \times L_1 \longrightarrow L_0,$$
$$Q_a, Q_b \longmapsto Q_a \circ Q_b = (h_i)_{ab}\tau_i, \tag{2.10}$$

where $(h_i)_{ab} = (h_i)_{ba}$, and the matrices $h_i, i = 1, 2, 3$, are three symmetric dim $L_1 \times$ dim L_1 matrices. Again we make use of the generalized Jacobi identities to find restrictions for the matrices h_i. An arbitrary multiplicative constant factor can be absorbed in the Q's by a redefinition. Let

$$\tau_i \in L_0 \qquad \text{and} \qquad Q_a, Q_b \in L_1.$$

Then the generalized Jacobi identity reads for this case:

$$\tau_i \circ (Q_a \circ Q_b)(-1)^0 + Q_b \circ (\tau_i \circ Q_a)(-1)^1 + Q_a \circ (Q_b \circ \tau_i)(-1)^0 = 0,$$

so that

$$\tau_i \circ \big((h_j)_{ab}\tau_j\big) - Q_b \circ \big(-(t_i)_{ac}Q_c\big) + Q_a \circ \big((t_i)_{bc}Q_c\big) = 0,$$

or

$$(h_j)_{ab}(\tau_i \circ \tau_j) + (t_i)_{ac}(Q_b \circ Q_c) + (t_i)_{bc}(Q_a \circ Q_c) = 0.$$

Then:

$$-(h_j)_{ab}\epsilon_{ijk}\tau_k + (t_i)_{ac}(h_j)_{bc}\tau_j + (t_i)_{bc}(h_j)_{ac}\tau_j = 0.$$

Rewriting this in the form:

$$\left[(t_i)_{ac}(h_j)_{cb} + (t_i)_{bc}(h_j)_{ca}\right]\tau_j = \epsilon_{ijk}(h_j)_{ab}\tau_k = \epsilon_{ikj}(h_k)_{ab}\tau_j,$$

so that

$$\left[(t_ih_j)_{ab} + (t_ih_j)_{ba}\right]\tau_j = -\epsilon_{ijk}(h_k)_{ab}\tau_j,$$

and

$$t_ih_j + (t_ih_j)^\top = -\epsilon_{ijk}h_k. \tag{2.11}$$

If we take $N = \dim L_1 = 2$, then $t_i = \tau_i, i = 1, 2, 3$, and the most general form of h_i is given by[9]

$$(h_i)_{ab} = a_i\delta_{ab} + b_i(\tau_3)_{ab} + c_i(\tau_1)_{ab} = \begin{pmatrix} a_i + \dfrac{i}{2}b_i & \dfrac{i}{2}c_i \\ \dfrac{i}{2}c_i & a_i - \dfrac{i}{2}b_i \end{pmatrix}_{ab},$$

where $a_i, b_i, c_i \in \mathbb{C}$. The matrix τ_2 does not appear in this expansion of h_i, because, as stated earlier, $(h_i)_{ab}$ has to be symmetric in its indices a and b, and τ_2 is antisymmetric. The nine coefficients a_i, b_i and c_i with $i = 1, 2, 3$, have to be computed from Eq. (2.11), i.e. from the relation

$$\tau_ih_j + (\tau_ih_j)^\top = -\epsilon_{ijk}h_k, \quad i, j, k = 1, 2, 3.$$

Proposition 2.1:

The matrices h_i are given by

$$h_i = 2c_3(\tau_i\tau_2), \quad i = 1, 2, 3. \tag{2.12}$$

This result can be guessed, since h_i is a 2×2-matrix, we expect τ_i on the right hand side; to make this symmetric we must have $\tau_i\tau_2$; fixing the remaining overall multiplicative factor is a matter of convention.

---

[9]Recall that the h_i are symmetric.

Proof: We consider the various cases separately.

The case $i = 1$: With

$$\tau_1 = \frac{i}{2} \begin{pmatrix} 0 & 1 \\ 1 & 0 \end{pmatrix},$$

we get

$$\tau_1 h_j = \frac{i}{2} \begin{pmatrix} 0 & 1 \\ 1 & 0 \end{pmatrix} \begin{pmatrix} a_j + \frac{i}{2} b_j & \frac{i}{2} c_j \\ \frac{i}{2} c_j & a_j - \frac{i}{2} b_j \end{pmatrix} = \frac{i}{2} \begin{pmatrix} \frac{i}{2} c_j & a_j - \frac{i}{2} b_j \\ a_j + \frac{i}{2} b_j & \frac{i}{2} c_j \end{pmatrix}.$$

This leads to:

$$\tau_1 h_j + (\tau_1 h_j)^\top = \frac{i}{2} \begin{pmatrix} \frac{i}{2} c_j & a_j - \frac{i}{2} b_j \\ a_j + \frac{i}{2} b_j & \frac{i}{2} c_j \end{pmatrix} + \frac{i}{2} \begin{pmatrix} \frac{i}{2} c_j & a_j + \frac{i}{2} b_j \\ a_j - \frac{i}{2} b_j & \frac{i}{2} c_j \end{pmatrix}$$

$$= \begin{pmatrix} -\frac{1}{2} c_j & i a_j \\ i a_j & -\frac{1}{2} c_j \end{pmatrix}$$

$$\overset{!}{=} -\epsilon_{1jk} h_k \qquad (\text{ from Eq. (2.11)})$$

$$= -\epsilon_{1jk} \begin{pmatrix} a_k + \frac{i}{2} b_k & \frac{i}{2} c_k \\ \frac{i}{2} c_k & a_k - \frac{i}{2} b_k \end{pmatrix}.$$

Now consider the case $i = 1, j = 2$. In this case we obtain the matrix equation

$$\begin{pmatrix} -\frac{1}{2} c_2 & i a_2 \\ i a_2 & -\frac{1}{2} c_2 \end{pmatrix} = - \begin{pmatrix} a_3 + \frac{i}{2} b_3 & \frac{i}{2} c_3 \\ \frac{i}{2} c_3 & a_3 - \frac{i}{2} b_3 \end{pmatrix},$$

leading to

$$-\frac{1}{2} c_2 = -a_3 - \frac{i}{2} b_3, \quad i a_2 = -\frac{i}{2} c_3, \quad -\frac{1}{2} c_2 = -a_3 + \frac{i}{2} b_3,$$

so that

$$b_3 = 0, \quad a_3 = \frac{1}{2} c_2, \quad a_2 = -\frac{1}{2} c_3.$$

For the case $i = 1, j = 3$ we get

$$\begin{pmatrix} -\frac{1}{2} c_3 & i a_3 \\ i a_3 & -\frac{1}{2} c_3 \end{pmatrix} = \begin{pmatrix} a_2 + \frac{i}{2} b_2 & \frac{i}{2} c_2 \\ \frac{i}{2} c_2 & a_2 - \frac{i}{2} b_2 \end{pmatrix},$$

such that

$$-\frac{1}{2}c_3 = a_2 + \frac{i}{2}b_2, \quad ia_3 = \frac{i}{2}c_2, \quad -\frac{1}{2}c_3 = a_2 - \frac{i}{2}b_2.$$

That means:

$$b_2 = 0, \quad a_3 = \frac{1}{2}c_2, \quad a_2 = -\frac{1}{2}c_3.$$

Next we evaluate the case $i = 2$, using

$$\tau_2 = \frac{i}{2}\begin{pmatrix} 0 & -i \\ i & 0 \end{pmatrix},$$

and obtain

$$\tau_2 h_j = \frac{1}{2}\begin{pmatrix} 0 & 1 \\ -1 & 0 \end{pmatrix}\begin{pmatrix} a_j + \frac{i}{2}b_j & \frac{i}{2}c_j \\ \frac{i}{2}c_j & a_j - \frac{i}{2}b_j \end{pmatrix} = \frac{1}{2}\begin{pmatrix} \frac{i}{2}c_j & a_j - \frac{i}{2}b_j \\ -a_j - \frac{i}{2}b_j & -\frac{i}{2}c_j \end{pmatrix}.$$

Then

$$\tau_2 h_j + \left(\tau_2 h_j\right)^\top = \frac{1}{2}\begin{pmatrix} ic_j & -ib_j \\ -ib_j & -ic_j \end{pmatrix} = \frac{i}{2}\begin{pmatrix} c_j & -b_j \\ -b_j & -c_j \end{pmatrix}.$$

Consider the case $i = 2, j = 1$: Equating the above to $-\epsilon_{213}h_3$ we obtain:

$$\frac{i}{2}\begin{pmatrix} c_1 & -b_1 \\ -b_1 & -c_1 \end{pmatrix} = \begin{pmatrix} a_3 + \frac{i}{2}b_3 & \frac{i}{2}c_3 \\ \frac{i}{2}c_3 & a_3 - \frac{i}{2}b_3 \end{pmatrix}.$$

Hence with $b_3 = 0$ we find

$$\frac{i}{2}c_1 = a_3 + \frac{i}{2}b_3 = a_3, \quad -\frac{i}{2}b_1 = \frac{i}{2}c_3, \quad -\frac{i}{2}c_1 = a_3 - \frac{i}{2}b_3 = a_3.$$

Thus $a_3 = 0, c_1 = 0, b_1 = -c_3$, and since $a_3 = 1/2\,c_2$, we have $c_2 = 0$.

Next we evaluate the case $i = 2, j = 3$:

$$\frac{i}{2}\begin{pmatrix} c_3 & -b_3 \\ -b_3 & -c_3 \end{pmatrix} = -\begin{pmatrix} a_1 + \frac{i}{2}b_1 & \frac{i}{2}c_1 \\ \frac{i}{2}c_1 & a_1 - \frac{i}{2}b_1 \end{pmatrix},$$

leading to:

$$\frac{i}{2}c_3 = -a_1 - \frac{i}{2}b_1, \quad -\frac{i}{2}b_3 = -\frac{i}{2}c_1, \quad -\frac{i}{2}c_3 = -a_1 + \frac{i}{2}b_1,$$

so that $b_3 = 0, c_1 = 0, a_1 = 0, b_1 = -c_3$. Hence, collecting all terms we have:

$$
\begin{array}{lll}
a_1 = 0, & b_1 = -c_3, & c_1 = 0, \\
a_2 = -\frac{1}{2}c_3, & b_2 = 0, & c_2 = 0, \\
a_3 = 0, & b_3 = 0, & c_3 \quad \text{undetermined},
\end{array}
$$

and therefore

$$
h_1 = -\frac{i}{2}c_3 \begin{pmatrix} 1 & 0 \\ 0 & -1 \end{pmatrix} = 2c_3(\tau_1\tau_2), \qquad
h_2 = -\frac{1}{2}c_3 \begin{pmatrix} 1 & 0 \\ 0 & 1 \end{pmatrix} = 2c_3(\tau_2\tau_2),
$$

$$
h_3 = \frac{i}{2}c_3 \begin{pmatrix} 0 & 1 \\ 1 & 0 \end{pmatrix} = 2c_3(\tau_3\tau_2),
$$

i.e.

$$
h_i = 2c_3(\tau_i\tau_2).
$$

The constant c_3 can be absorbed in redefined generators Q_a as mentioned before.

We finally consider the *Jacobi identity* for three operators Q. In this case this relation is:

$$
Q_a \circ (Q_b \circ Q_c)(-1) + Q_c \circ (Q_a \circ Q_b)(-1) + Q_b \circ (Q_c \circ Q_a)(-1) = 0,
$$

and so

$$
Q_a \circ \big((h_i)_{bc}\tau_i\big) + Q_c \circ \big((h_i)_{ab}\tau_i\big) + Q_b \circ \big((h_i)_{ca}\tau_i\big) = 0,
$$

i.e.

$$
(h_i)_{bc}(Q_a \circ \tau_i) + (h_i)_{ab}(Q_c \circ \tau_i) + (h_i)_{ca}(Q_b \circ \tau_i) = 0.
$$

Using Eq. (2.8) with $t_i = \tau_i$,

$$
(h_i)_{bc}(\tau_i)_{ad}Q_d + (h_i)_{ab}(\tau_i)_{cd}Q_d + (h_i)_{ca}(\tau_i)_{bd}Q_d = 0.
$$

Using Eq. (2.12) and the fact that the Q's are independent of one another

$$
(\tau_i\tau_2)_{bc}(\tau_i)_{ad} + (\tau_i\tau_2)_{ab}(\tau_i)_{cd} + (\tau_i\tau_2)_{ca}(\tau_i)_{bd} = 0.
$$

This relation is valid, *i.e.* satisfied by the τ_i's, as can be checked by direct calculation.

In summary we have the following construction. The graded Lie algebra of $su(2, \mathbb{C})$ is:

$$
\begin{array}{ll}
\tau_i \circ \tau_j = -\epsilon_{ijk}\tau_k, & L_0 \times L_0 \longrightarrow L_0, \\
\tau_i \circ Q_a = -(\tau_i)_{ab}Q_b, & L_0 \times L_1 \longrightarrow L_{0+1} = L_1, \\
Q_a \circ Q_b = (\tau_i\tau_2)_{ab}\tau_i, & L_1 \times L_1 \longrightarrow L_{1+1} = L_2 \cong L_0.
\end{array}
$$

The structure of the coefficients of the product in $L_0 \times L_1$ is determined by the generalized Jacobi identities. The only freedom one has is the choice of representation for these coefficients. The important point is that the coefficient matrix of $\tau_i \circ Q_a$ has to be a representation matrix of the algebra given by L_0. This result is an immediate consequence of the generalized Jacobi identity with one Q_a. Thus, the graded Lie algebra $su(2, \mathbb{C})$ has five elements: $\tau_i, i = 1, 2, 3$, and Q_a with $a = 1, 2$, and is written $OSp(1/2)$ ('one slash two') [108]. The dimension of the chosen representation then determines the dimension of the subspace L_1 and therefore the number of operators Q_a; $a = 1, 2, \dots$, dim L_1. Once the representation has been chosen, the coefficient matrices of the product $Q_a \circ Q_b$ may be found from the Jacobi identity with two Q's. For consistency these matrices have to satisfy the generalized Jacobi identity with three Q's; this, however, is not always the case — for example, an extension of $SU(2, \mathbb{C})$ with an n-dimensional L_1 ($n > 2$) is not possible, since the generalized Jacobi identity for three Q's then requires the h_i to be zero.

2.4 $\mathbb{Z}_2$ Graded Lie Algebras

We now discuss $\mathbb{Z}_2$ graded Lie algebras in more detail. We recall first some general properties of $\mathbb{Z}_2$ graded Lie algebras.

Definition: $\mathbb{Z}_2$ graded Lie algebra

A linear algebra $L := \text{Span}\{X_\mu\}$ is given a $\mathbb{Z}_2$ grading if L is the direct sum of two subspaces L_0 and L_1, *i.e.* $L = L_0 \oplus L_1$, where we set:

$$L_0 = \text{Span}\{E_i\}, \; i = 1, \dots, \text{dim } L_0,$$
$$L_1 = \text{Span}\{Q_a\}, \; a = 1, \dots, \text{dim } L_1,$$

on which a composition law:

$$\circ : L \times L \longrightarrow L$$

acts as follows:

$$L_0 \circ L_0 \subset L_0, \qquad L_0 \circ L_1 \subset L_1, \qquad L_1 \circ L_1 \subset L_0.$$

The pair $(L_0, \circ)$ establishes an ordinary Lie algebra.

Definition:

We assign to any $X_\mu \in L$ a *degree* $g \in \mathbb{Z}_2$, by defining:

$$g_\mu := g(X_\mu) = 0 \iff X_\mu \in L_0, \qquad (2.13)$$
$$g_\mu := g(X_\mu) = 1 \iff X_\mu \in L_1. \qquad (2.14)$$

We say the element $X_\mu \in L$ is *even*, if $g_\mu = 0$, and the element $X_\mu \in L$ is *odd*, if $g_\mu = 1$.

With these definitions of degrees of elements we see that the set of generators $E_i, i = 1, 2, 3, \ldots,$ dim L_0, which is a basis of L_0, consists of even elements, whereas the generators Q_a, which span the subspace L_1, are odd.

The subspace L_0 of the graded Lie algebra L, containing the even elements, is called the *bosonic sector*, the subspace L_1 is called the *fermionic sector*.

Definition:

We define the product $\circ$ on L by the assignment:

$$\circ \; : \; L \times L \longrightarrow L$$

where:

$$(X_\mu, X_\nu) \longmapsto X_\mu \circ X_\nu := X_\mu X_\nu - (-1)^{g_\mu g_\nu} X_\nu X_\mu. \qquad (2.15)$$

We now consider this product separately on the two subspaces L_0 and L_1:

(i) $\circ \; : \; L_0 \times L_0 \longrightarrow L_0$
Let $E_i, E_j \in L_0$ and according to Eq. (2.13):

$$g(E_i) = g(E_j) = 0.$$

Then

$$E_i \circ E_j \stackrel{(2.15)}{=} E_i E_j - (-1)^{g_i g_j} E_j E_i = E_i E_j - E_j E_i = \left[E_i, E_j \right]. \quad (2.16)$$

With this construction we are consistent with the basic requirement that the product $\circ$ be antisymmetric on the subspace L_0 (see Eq. (2.5)).

(ii) $\circ \; : \; L_0 \times L_1 \longrightarrow L_1$
Let $E_j \in L_0, Q_a \in L_1$; then we get

$$g(E_i) = 0; \quad g(Q_a) = 1,$$

and we find

$$E_i \circ Q_a \stackrel{(2.15)}{=} E_i Q_a - (-1)^{g_i g_a} Q_a E_i = E_i Q_a - Q_a E_i = \left[E_i, Q_a \right]. \quad (2.17)$$

Again the product is antisymmetric on $L_0 \times L_1$, as demanded by Eq. (2.5).

(iii) $\circ\ :\ L_1 \times L_1 \longrightarrow L_0$

Let $Q_a, Q_b \in L_1$ and according to Eq. (2.14)

$$g(Q_a) = g(Q_b) = 1.$$

Then

$$Q_a \circ Q_b \overset{(2.15)}{=} Q_a Q_b - (-1)^{g_a g_b} Q_a Q_b = Q_a Q_b + Q_b Q_a = \{Q_a, Q_b\}.$$
(2.18)

From Eq. (2.18) we see that the product is symmetric on the subspace L_1.

We next introduce *generalized structure constants*. Since $X_\mu \circ X_\nu \in L$, the product $X_\mu \circ X_\nu$ must be a linear combination of the basis elements X_ω, *i.e.*

$$X_\mu \circ X_\nu = c_{\mu\nu}{}^\omega X_\omega,$$
(2.19)

where

$$c_{\mu\nu}{}^\omega = -(-1)^{g_\mu g_\nu} c_{\nu\mu}{}^\omega$$
(2.20)

are the *generalized structure constants* of the graded Lie algebra L. According to Eq. (2.20) the structure constants are antisymmetric for even-even and even-odd pairs of elements of L, and symmetric for odd-odd pairs. We now prove the relation (2.20).

In order to show Eq. (2.20) we start from Eq. (2.19), *i.e.*

$$X_\mu \circ X_\nu = c_{\mu\nu}{}^\omega X_\omega \qquad \text{and} \qquad X_\nu \circ X_\mu = c_{\nu\mu}{}^\omega X_\omega.$$

On the other hand using Eq. (2.15) we have

$$X_\mu \circ X_\nu = X_\mu X_\nu - (-1)^{g_\mu g_\nu} X_\nu X_\mu,$$
$$X_\nu \circ X_\mu = X_\nu X_\mu - (-1)^{g_\nu g_\mu} X_\mu X_\nu = X_\nu X_\mu - (-1)^{g_\mu g_\nu} X_\mu X_\nu,$$

and again using Eq. (2.15)

$$(-1)^{g_\mu g_\nu} X_\nu \circ X_\mu = (-1)^{g_\mu g_\nu} X_\nu X_\mu - \left[(-1)^{g_\mu g_\nu}\right]^2 X_\mu X_\nu$$
$$= (-1)^{g_\mu g_\nu} X_\nu X_\mu - X_\mu X_\nu$$
$$= -\left[X_\mu X_\nu - (-1)^{g_\mu g_\nu} X_\nu X_\mu\right] = -\left(X_\mu \circ X_\nu\right).$$

Hence

$$X_\mu \circ X_\nu + (-1)^{g_\mu g_\nu} X_\nu \circ X_\mu = 0,$$

and with Eq. (2.19)

$$c_{\mu\nu}{}^\omega X_\omega + (-1)^{g_\mu g_\nu} c_{\nu\mu}{}^\omega X_\omega = 0, \quad \forall X_\omega \in L,$$

i.e.

$$c_{\mu\nu}{}^{\omega} = -(-1)^{g_\mu g_\nu} c_{\nu\mu}{}^{\omega},$$

since the X_ω are independent.

We now consider the generalized Jacobi identities. For $X_\mu, X_\nu, X_\rho \in L$, these are defined by:

$$\left[X_\mu \circ \left(X_\nu \circ X_\rho\right)\right](-1)^{g_\mu g_\rho} + \left[X_\nu \circ \left(X_\rho \circ X_\mu\right)\right](-1)^{g_\nu g_\mu}$$
$$+ \left[X_\rho \circ \left(X_\mu \circ X_\nu\right)\right](-1)^{g_\rho g_\nu} = 0. \qquad (2.21)$$

We demonstrate that the map given in Eq. (2.15) obeys the generalized Jacobi identity (2.21):

$$\left[X_\mu \circ \left(X_\nu \circ X_\rho\right)\right](-1)^{g_\mu g_\rho} + \left[X_\rho \circ \left(X_\mu \circ X_\nu\right)\right](-1)^{g_\rho g_\nu}$$
$$+ \left[X_\nu \circ \left(X_\rho \circ X_\mu\right)\right](-1)^{g_\nu g_\mu}$$

$$= \left[X_\mu \left(X_\nu \circ X_\rho\right) - (-1)^{g_\mu(g_\nu + g_\rho)} \left(X_\nu \circ X_\rho\right)X_\mu\right](-1)^{g_\mu g_\rho}$$
$$+ \left[X_\rho \left(X_\mu \circ X_\nu\right) - (-1)^{g_\rho(g_\mu + g_\nu)} \left(X_\mu \circ X_\nu\right)X_\rho\right](-1)^{g_\rho g_\nu}$$
$$+ \left[X_\nu \left(X_\rho \circ X_\mu\right) - (-1)^{g_\nu(g_\rho + g_\mu)} \left(X_\rho \circ X_\mu\right)X_\nu\right](-1)^{g_\nu g_\mu}$$

$$= \left[X_\mu \left(X_\nu X_\rho - (-1)^{g_\nu g_\rho} X_\rho X_\nu\right)\right.$$
$$\left. - (-1)^{g_\mu(g_\nu + g_\rho)} \left(X_\nu X_\rho - (-1)^{g_\nu g_\rho} X_\rho X_\nu\right)X_\mu\right](-1)^{g_\mu g_\rho}$$
$$+ \left[X_\rho \left(X_\mu X_\nu - (-1)^{g_\mu g_\nu} X_\nu X_\mu\right)\right.$$
$$\left. - (-1)^{g_\rho(g_\mu + g_\nu)} \left(X_\mu X_\nu - (-1)^{g_\mu g_\nu} X_\nu X_\mu\right)X_\rho\right](-1)^{g_\rho g_\nu}$$
$$+ \left[X_\nu \left(X_\rho X_\mu - (-1)^{g_\rho g_\mu} X_\mu X_\rho\right)\right.$$
$$\left. - (-1)^{g_\nu(g_\rho + g_\mu)} \left(X_\rho X_\mu - (-1)^{g_\rho g_\mu} X_\mu X_\rho\right)X_\nu\right](-1)^{g_\nu g_\mu}$$

$$= \left[X_\mu X_\nu X_\rho - (-1)^{g_\nu g_\rho} X_\mu X_\rho X_\nu - (-1)^{g_\mu(g_\nu + g_\rho)} X_\nu X_\rho X_\mu\right.$$
$$\left. + (-1)^{g_\mu g_\nu + g_\mu g_\rho + g_\nu g_\rho} X_\rho X_\nu X_\mu\right](-1)^{g_\mu g_\rho}$$
$$+ \left[X_\rho X_\mu X_\nu - (-1)^{g_\mu g_\nu} X_\rho X_\nu X_\mu - (-1)^{g_\rho g_\mu + g_\rho g_\nu} X_\mu X_\nu X_\rho\right.$$
$$\left. + (-1)^{g_\rho g_\mu + g_\rho g_\nu + g_\mu g_\nu} X_\nu X_\mu X_\rho\right](-1)^{g_\rho g_\nu}$$
$$+ \left[X_\nu X_\rho X_\mu - (-1)^{g_\rho g_\mu} X_\nu X_\mu X_\rho - (-1)^{g_\nu g_\rho + g_\nu g_\mu} X_\rho X_\mu X_\nu\right.$$
$$\left. + (-1)^{g_\nu g_\rho + g_\nu g_\mu + g_\rho g_\mu} X_\mu X_\rho X_\nu\right](-1)^{g_\nu g_\mu}$$

$$\begin{aligned}
&= X_\mu X_\nu X_\rho (-1)^{g_\mu g_\rho} - X_\mu X_\rho X_\nu (-1)^{g_\rho(g_\mu + g_\nu)} \\
&\quad - X_\nu X_\rho X_\mu (-1)^{g_\mu g_\nu} + X_\rho X_\nu X_\mu (-1)^{g_\nu(g_\mu + g_\rho)} \\
&\quad + X_\rho X_\mu X_\nu (-1)^{g_\rho g_\nu} - X_\rho X_\nu X_\mu (-1)^{g_\nu(g_\mu + g_\rho)} \\
&\quad - X_\mu X_\nu X_\rho (-1)^{g_\rho g_\mu} + X_\nu X_\mu X_\rho (-1)^{g_\mu(g_\rho + g_\nu)} \\
&\quad + X_\nu X_\rho X_\mu (-1)^{g_\nu g_\mu} - X_\nu X_\mu X_\rho (-1)^{g_\mu(g_\nu + g_\rho)} \\
&\quad - X_\rho X_\mu X_\nu (-1)^{g_\nu g_\rho} + X_\mu X_\rho X_\nu (-1)^{g_\rho(g_\mu + g_\nu)} \\
&= X_\mu X_\nu X_\rho \Big((-1)^{g_\mu g_\rho} - (-1)^{g_\rho g_\mu} \Big) \\
&\quad + X_\mu X_\rho X_\nu \Big((-1)^{g_\rho(g_\mu + g_\nu)} - (-1)^{g_\rho(g_\nu + g_\mu)} \Big) \\
&\quad + X_\rho X_\nu X_\mu \Big((-1)^{g_\nu(g_\mu + g_\rho)} - (-1)^{g_\nu(g_\mu + g_\rho)} \Big) \\
&\quad + X_\rho X_\mu X_\nu \Big((-1)^{g_\rho g_\nu} - (-1)^{g_\nu g_\rho} \Big) \\
&\quad + X_\nu X_\mu X_\rho \Big((-1)^{g_\mu(g_\rho + g_\nu)} - (-1)^{g_\mu(g_\rho + g_\nu)} \Big) \\
&\quad + X_\nu X_\rho X_\mu \Big((-1)^{g_\nu g_\mu} - (-1)^{g_\mu g_\nu} \Big) = 0.
\end{aligned}$$

Hence the composition law (2.15) is a product which obeys all conditions of the product of a graded Lie algebra as defined by Eqs. (2.4) to (2.6).

We now write out the generalized Jacobi identity for the four different possibilities.

(i) $X_\mu, X_\nu, X_\rho \in L_0$: We take

$$X_\mu = E_i, \quad X_\nu = E_j, \quad X_\rho = E_k,$$

and

$$(-1)^{g_\mu g_\rho} = (-1)^{g_\rho g_\nu} = (-1)^{g_\mu g_\nu} = +1.$$

Then Eq. (2.21) reads in the case of three generators of L_0:

$$\begin{aligned}
\big[X_\mu \circ &(X_\nu \circ X_\rho) \big] (-1)^{g_\mu g_\rho} + \big[X_\rho \circ (X_\mu \circ X_\nu) \big] (-1)^{g_\rho g_\nu} \\
&+ \big[X_\nu \circ (X_\rho \circ X_\mu) \big] (-1)^{g_\nu g_\mu} \\
&= E_i \circ (E_j \circ E_k) + E_k \circ (E_i \circ E_j) + E_j \circ (E_k \circ E_i) \\
&= E_i (E_j \circ E_k) - (E_j \circ E_k) E_i \\
&\quad + E_k (E_i \circ E_j) - (E_i \circ E_j) E_k \\
&\quad + E_j (E_k \circ E_i) - (E_k \circ E_i) E_j \\
&= E_i (E_j E_k - E_k E_j) - (E_j E_k - E_k E_j) E_i \\
&\quad + E_k (E_i E_j - E_j E_i) - (E_i E_j - E_j E_i) E_k \\
&\quad + E_j (E_k E_i - E_i E_k) - (E_k E_i - E_i E_k) E_j
\end{aligned}$$

$$\begin{aligned}
&= E_i[E_j, E_k] - [E_j, E_k]E_i \\
&\quad + E_k[E_i, E_j] - [E_i, E_j]E_k \\
&\quad + E_j[E_k, E_i] - [E_k, E_i]E_j \\
&= \Big[E_i, [E_j, E_k]\Big] + \Big[E_k, [E_i, E_j]\Big] + \Big[E_j, [E_k, E_i]\Big] \\
&= 0.
\end{aligned}$$

Hence the generalized Jacobi identity for three E's reduces to the ususal Jacobi identity for the Lie algebra L_0, *i.e.*

$$\Big[E_i, [E_j, E_k]\Big] + \Big[E_k, [E_i, E_j]\Big] + \Big[E_j, [E_k, E_i]\Big] = 0. \qquad (2.22)$$

(ii) $X_\mu, X_\nu \in L_0, X_\rho \in L_1$: We take

$$X_\mu = E_i, \quad X_\nu = E_j, \quad E_i, E_j \in L_0, \quad \text{and} \quad X_\rho = Q_a \in L_1.$$

In this case Eq. (2.21) becomes

$$\begin{aligned}
\big[X_\mu \circ (X_\nu \circ X_\rho)\big]&(-1)^{g_\mu g_\rho} + \big[X_\rho \circ (X_\mu \circ X_\nu)\big](-1)^{g_\rho g_\nu} \\
&+ \big[X_\nu \circ (X_\rho \circ X_\mu)\big](-1)^{g_\nu g_\mu} \\
&= E_i \circ (E_j \circ Q_a)(-1)^0 + Q_a \circ (E_i \circ E_j)(-1)^0 \\
&\quad + E_j \circ (Q_a \circ E_i)(-1)^0 \\
&= E_i(E_j \circ Q_a) - (-1)^{g_i(g_j + g_a)}(E_j \circ Q_a)E_i \\
&\quad + Q_a(E_i \circ E_j) - (-1)^{g_a(g_i + g_j)}(E_i \circ E_j)Q_a \\
&\quad + E_j(Q_a \circ E_i) - (-1)^{g_j(g_a + g_i)}(Q_a \circ E_i)E_j.
\end{aligned}$$

Using again Eq. (2.15) we find

$$\begin{aligned}
\big[X_\mu \circ (X_\nu \circ X_\rho)\big]&(-1)^{g_\mu g_\rho} + \big[X_\rho \circ (X_\mu \circ X_\nu)\big](-1)^{g_\rho g_\nu} \\
&+ \big[X_\nu \circ (X_\rho \circ X_\mu)\big](-1)^{g_\nu g_\mu} \\
&= E_i(E_j Q_a - (-1)^{g_j g_a} Q_a E_j) - (E_j Q_a - (-1)^{g_j g_a} Q_a E_j)E_i \\
&\quad + Q_a(E_i E_j - (-1)^{g_i g_j} E_j E_i) - (E_i E_j - (-1)^{g_i g_j} E_j E_i)Q_a \\
&\quad + E_j(Q_a E_i - (-1)^{g_a g_i} E_i Q_a) - (Q_a E_i - (-1)^{g_a g_i} E_i Q_a)E_j \\
&= E_i[E_j, Q_a] - [E_j, Q_a]E_i + Q_a[E_i, E_j] - [E_i, E_j]Q_a \\
&\quad + E_j[Q_a, E_i] - [Q_a, E_i]E_j \\
&= \Big[E_i, [E_j, Q_a]\Big] + \Big[Q_a, [E_i, E_j]\Big] + \Big[E_j, [Q_a, E_i]\Big] = 0.
\end{aligned}$$

Hence the generalized Jacobi identity for two E's and one Q is:

$$\Big[E_i, [E_j, Q_a]\Big] + \Big[Q_a, [E_i, E_j]\Big] + \Big[E_j, [Q_a, E_i]\Big] = 0. \qquad (2.23)$$

(iii) $X_\mu \in L_0, X_\nu, X_\rho \in L_1$: Here we set

$$X_\mu = E_i \in L_0, \quad X_\nu = Q_a \in L_1, \quad X_\rho = Q_b \in L_1.$$

Then Eq. (2.21) becomes:

$$
\begin{aligned}
& [X_\mu \circ (X_\nu \circ X_\rho)](-1)^{g_\mu g_\rho} + [X_\rho \circ (X_\mu \circ X_\nu)](-1)^{g_\rho g_\nu} \\
& \qquad + [X_\nu \circ (X_\rho \circ X_\mu)](-1)^{g_\nu g_\mu} \\
&= E_i \circ (Q_a \circ Q_b)(-1)^{g_i g_b} + Q_b \circ (E_i \circ Q_a)(-1)^{g_b g_a} \\
& \qquad + Q_a \circ (Q_b \circ E_i)(-1)^{g_a g_i} \\
&= E_i(Q_a \circ Q_b) - (-1)^{g_i(g_a + g_b)}(Q_a \circ Q_b)E_i \\
& \qquad - Q_b(E_i \circ Q_a) + (-1)^{g_b(g_i + g_a)}(E_i \circ Q_a)Q_b \\
& \qquad + Q_a(Q_b \circ E_i) - (-1)^{g_a(g_b + g_i)}(Q_b \circ E_i)Q_a \\
&= E_i(Q_a Q_b - (-1)^{g_a g_b}Q_b Q_a) - (Q_a Q_b - (-1)^{g_a g_b}Q_b Q_a)E_i \\
& \qquad - Q_b(E_i Q_a - (-1)^{g_i g_a}Q_a E_i) - (E_i Q_a - (-1)^{g_i g_a}Q_a E_i)Q_b \\
& \qquad + Q_a(Q_b E_i - (-1)^{g_i g_b}E_i Q_b) + (Q_b E_i - (-1)^{g_i g_b}E_i Q_b)Q_a \\
&= E_i\{Q_a, Q_b\} - \{Q_a, Q_b\}E_i - Q_b[E_i, Q_a] - [E_i, Q_a]Q_b \\
& \qquad + Q_a[Q_b, E_i] + [Q_b, E_i]Q_a \\
&= \Big[E_i, \{Q_a, Q_b\}\Big] - \Big\{Q_b, [E_i, Q_a]\Big\} + \Big\{Q_a, [Q_b, E_i]\Big\} = 0.
\end{aligned}
$$

Hence, the Jacobi relation for one E and two Q's is:

$$\Big[E_i, \{Q_a, Q_b\}\Big] - \Big\{Q_b, [E_i, Q_a]\Big\} + \Big\{Q_a, [Q_b, E_i]\Big\} = 0. \qquad (2.24)$$

(iv) $X_\mu, X_\nu, X_\rho \in L_1$: Here we set

$$X_\mu = Q_a \in L_1, \quad X_\nu = Q_b \in L_1, \quad X_\rho = Q_c \in L_1.$$

Then Eq. (2.21) becomes for this case of three Q's:

$$
\begin{aligned}
& [X_\mu \circ (X_\nu \circ X_\rho)](-1)^{g_\mu g_\rho} + [X_\rho \circ (X_\mu \circ X_\nu)](-1)^{g_\rho g_\nu} \\
& \qquad + [X_\nu \circ (X_\rho \circ X_\mu)](-1)^{g_\nu g_\mu} \\
&= Q_a \circ (Q_b \circ Q_c)(-1)^{g_a g_c} + Q_c \circ (Q_a \circ Q_b)(-1)^{g_c g_b} \\
& \qquad + Q_b \circ (Q_c \circ Q_a)(-1)^{g_b g_a} \\
&= -Q_a(Q_b \circ Q_c) + (-1)^{g_a(g_b + g_c)}(Q_b \circ Q_c)Q_a \\
& \qquad - Q_c(Q_a \circ Q_b) + (-1)^{g_c(g_a + g_b)}(Q_a \circ Q_b)Q_c \\
& \qquad - Q_b(Q_c \circ Q_a) + (-1)^{g_b(g_c + g_a)}(Q_c \circ Q_a)Q_b
\end{aligned}
$$

$$
\begin{aligned}
&= -Q_a\big(Q_bQ_c - (-1)^{g_b g_c}Q_cQ_b\big) + \big(Q_bQ_c - (-1)^{g_b g_c}Q_cQ_b\big)Q_a \\
&\quad - Q_c\big(Q_aQ_b - (-1)^{g_a g_b}Q_bQ_a\big) + \big(Q_aQ_b - (-1)^{g_a g_b}Q_bQ_a\big)Q_c \\
&\quad - Q_b\big(Q_cQ_a - (-1)^{g_c g_a}Q_aQ_c\big) + \big(Q_cQ_a - (-1)^{g_c g_a}Q_aQ_c\big)Q_b \\
&= -Q_a\{Q_b, Q_c\} + \{Q_b, Q_c\}Q_a \\
&\quad - Q_c\{Q_a, Q_b\} - \{Q_a, Q_b\}Q_c \\
&\quad - Q_b\{Q_c, Q_a\} + \{Q_c, Q_a\}Q_b \\
&= -\Big[Q_a, \{Q_b, Q_c\}\Big] - \Big[Q_c, \{Q_a, Q_b\}\Big] - \Big[Q_b, \{Q_c, Q_a\}\Big] \\
&= 0.
\end{aligned}
$$

Hence, for three Q's the Jacobi identity is:

$$
\Big[Q_a, \{Q_b, Q_c\}\Big] + \Big[Q_c, \{Q_a, Q_b\}\Big] + \Big[Q_b, \{Q_c, Q_a\}\Big] = 0. \qquad (2.25)
$$

Remark: It is important to observe that the commutators and anticommutators in Eqs. (2.22) to (2.25) are consistent with the composition law (2.4) for graded Lie Algebras. Consider for example the generalized Jacobi identity (2.24), *i.e.*

$$
\Big[E_i, \{Q_a, Q_b\}\Big] - \Big\{Q_b, [E_i, Q_a]\Big\} + \Big\{Q_a, [Q_b, E_i]\Big\} = 0.
$$

According to Eq. (2.4) we have $\{Q_a, Q_b\} \in L_0$; but on L_0 the product has to be antisymmetric, so an expression like $\big[E_i, \{Q_a, Q_b\}\big]$ is consistent with Eq. (2.6). Similarly $[E_i, Q_a] \in L_1$ and $\{Q_b, [E_i, Q_a]\}$ has the correct bracket structure.

In conclusion we introduce the so-called *generalized Killing form*. The generalized Killing form on L is defined by:[10]

$$
b_{\mu\nu} = \sum_{\rho, \sigma}(-1)^{g_\sigma} c_{\mu\rho}{}^\sigma c_{\nu\sigma}{}^\rho, \qquad (2.26)
$$

where

$$
g_\sigma := g(X_\sigma), \qquad [X_\mu, X_\nu] = c_{\mu\nu}{}^\sigma X_\sigma,
$$
$$
X_\sigma \in L_0 \implies g_\sigma = 0, \qquad X_\sigma \in L_1 \implies g_\sigma = 1.
$$

Here, $b_{\mu\nu} = B(X_\mu, X_\nu)$ is a symmetric bilinear form when acting on the subspace L_0 and thus reduces to the usual Killing form for the Lie algebra

---

[10]See W. Miller, [72], p. 394, or A.S. Scarrino and P. Sorba, [104].

L_0. We can then apply *Cartan's criterion* [104]: The Lie algebra L_0 is semi-simple if and only if $B(X_\mu, X_\nu)|_{L_0 \times L_0}$ is nondegenerate, *i.e.*

$$\det(b_{\mu\nu}) \neq 0.$$

Furthermore, the Lie algebra L_0 is compact, if and only if the generalized Killing form restricted to L_0 is negative definite. Hence, a compact Lie algebra is always semi-simple. In addition, the generalized Killing form is antisymmetric when acting on the subspace L_1, *i.e.*

$$b_{ab} = B(Q_a, Q_b) = -B(Q_b, Q_a),$$

and

$$B(X_\mu, X_\nu)\Big|_{L_0 \times L_1} = 0.$$

2.5 Graded Matrices

An endomorphism

$$M \ : \ L \longrightarrow L, \qquad L = L_0 \oplus L_1$$

acting on a vector space L [104] can be represented by a *graded matrix* which has the following matrix structure. Let $\dim L_0 = n$, $\dim L_1 = m$. Then

$$M = \begin{pmatrix} A & B \\ C & D \end{pmatrix} \in \text{End}(L) \qquad (2.27)$$

is an $(n + m) \times (n + m)$-matrix where:

A is an $n \times n$ square matrix,

B is an $n \times m$ submatrix,

C is an $m \times n$ submatrix,

D is an $m \times m$ square matrix.

A matrix with such a structure is also called a *supermatrix*. Since L is the direct sum of L_0 and L_1, *i.e.* $L = L_0 \oplus L_1$, a vector $v \in L$ has the structure

$$v = \begin{pmatrix} v_0 \\ v_1 \end{pmatrix}, \qquad v_0 \in L_0, \quad v_1 \in L_1, \qquad (2.28)$$

with $g(v_0) = 0$, $g(v_1) = 1$ according to Eqs. (2.13) and (2.14). Here v_0 is an element of the bosonic sector of L, *i.e.* L_0, and v_1 is an element of the fermionic sector of L, *i.e.* L_1. Introducing the following index notation

$$
\begin{aligned}
A &= \left(A_{ij}\right), & i,j &= 1,2,\ldots,n, \\
B &= \left(B_{ia}\right), & i &= 1,2,\ldots,n;\ a = 1,2,\ldots,m, \\
C &= \left(C_{ai}\right), & a &= 1,2,\ldots,m;\ i = 1,2,\ldots,n, \\
D &= \left(D_{ab}\right), & a,b &= 1,2,\ldots,m, \\
v_0 &= \left(v_{0i}\right), & i &= 1,2,\ldots,n, \\
v_1 &= \left(v_{1a}\right), & a &= 1,2,\ldots,m,
\end{aligned}
\tag{2.29}
$$

we obtain (with $w,z = 1,2,\ldots,m+n$):

$$
M_{wz}v_z = \begin{pmatrix} A_{ij} & B_{ia} \\ C_{bj} & D_{ba} \end{pmatrix}\begin{pmatrix} v_{0j} \\ v_{1a} \end{pmatrix} = \begin{pmatrix} A_{ij}v_{0j} + B_{ia}v_{1a} \\ C_{bj}v_{0j} + D_{ba}v_{1a} \end{pmatrix} \overset{\text{def.}}{=} \begin{pmatrix} v'_{0i} \\ v'_{1b} \end{pmatrix}.
$$

Thus:

$$
v'_0 = Av_0 + Bv_1 \in L_0, \qquad v'_1 = Cv_0 + Dv_1 \in L_1.
$$

Since $v'_0 \in L_0$, we must have $g(v'_0) = 0$. But $g(v_0) = 0$ and $g(v_1) = 1$. Now since $Av_0 \in L_0$ and $Bv_1 \in L_0$, we must have $g(Av_0) = 0$ and $g(Bv_1) = 0 \bmod 2$. Hence

$$
g(Av_0) = g(A) + g(v_0)
$$

implies

$$
g(A) = 0,
$$

and

$$
g(Bv_1) = g(B) + g(v_1)
$$

implies

$$
g(B) = 1.
$$

Similarly since $v'_1 \in L_1$, we have $g(v'_1) = 1$. But $g(v_0) = 0$, $g(v_1) = 1$, so that

$$
g(C) = 1 \qquad \text{and} \qquad g(D) = 0.
$$

This means that A and D must be even submatrices with degree $g(A) = g(D) = 0$, and B and C have to be odd with degree $g(B) = g(C) = 1$. It follows, the elements of B and C are anticommuting variables and therefore behave as Grassmann numbers, *i.e.*

$$
B_{ia}C_{bj} = -C_{bj}B_{ia}.
\tag{2.30}
$$

From this relation we obtain the *transposition rule*

$$BC = -\left(C^\top B^\top\right)^\top \tag{2.31}$$

by setting in Eq. (2.30) $a = b = 1, \ldots, m$, so that

$$(BC)_{ij} = -C_{aj}B_{ia} = -\left(C^\top B^\top\right)_{ji}.$$

Definition:

The *supertrace* of a graded matrix M is defined by:

$$\mathrm{STr}\ M = \mathrm{Tr}\ A - \mathrm{Tr}\ D, \tag{2.32}$$

where A and D are submatrices as in Eq. (2.27).

Proposition 2.2:

Let M_1, M_2 be two graded matrices. The supertrace has the property:

$$\mathrm{STr}\ (M_1 M_2) = \mathrm{STr}\ (M_2 M_1). \tag{2.33}$$

Proof: We consider two graded matrices

$$M_1 = \begin{pmatrix} A_1 & B_1 \\ C_1 & D_1 \end{pmatrix}, \quad M_2 = \begin{pmatrix} A_2 & B_2 \\ C_2 & D_2 \end{pmatrix}.$$

Then:

$$\mathrm{STr}\ (M_1 M_2) = \mathrm{STr}\ \left[\begin{pmatrix} A_1 & B_1 \\ C_1 & D_1 \end{pmatrix} \begin{pmatrix} A_2 & B_2 \\ C_2 & D_2 \end{pmatrix} \right]$$

$$= \mathrm{STr}\ \begin{pmatrix} A_1 A_2 + B_1 C_2 & A_1 B_2 + B_1 D_2 \\ C_1 A_2 + D_1 C_2 & C_1 B_2 + D_1 D_2 \end{pmatrix}.$$

Using Eqs. (2.30) and (2.32) this becomes

$$\begin{aligned}
\mathrm{STr}\ (M_1 M_2) &= \mathrm{Tr}\ (A_1 A_2 + B_1 C_2) - \mathrm{Tr}\ (C_1 B_2 + D_1 D_2) \\
&= \mathrm{Tr}\ (A_1 A_2) + \mathrm{Tr}\ (B_1 C_2) - \mathrm{Tr}\ (C_1 B_2) - \mathrm{Tr}\ (D_1 D_2) \\
&= \mathrm{Tr}\ (A_2 A_1) - \mathrm{Tr}\ (C_2 B_1) + \mathrm{Tr}\ (B_2 C_1) - \mathrm{Tr}\ (D_2 D_1) \\
&= \mathrm{Tr}\ (A_2 A_1 + B_2 C_1) - \mathrm{Tr}\ (C_2 B_1 + D_2 D_1) \\
&= \mathrm{STr}\ \begin{pmatrix} A_2 A_1 + B_2 C_1 & A_2 B_1 + B_2 D_1 \\ C_2 A_1 + D_2 C_1 & C_2 B_1 + D_2 D_1 \end{pmatrix} \\
&= \mathrm{STr}\ \begin{pmatrix} A_2 & B_2 \\ C_2 & D_2 \end{pmatrix} \begin{pmatrix} A_1 & B_1 \\ C_1 & D_1 \end{pmatrix} \\
&= \mathrm{STr}\ (M_2 M_1).
\end{aligned}$$

We observe that the supertrace is defined in such a way that it is cyclic for graded matrices paralleling the analogous property of the trace for ordinary matrices.

Definition:

The determinant of the supermatrix M is the *superdeterminant* Sdet M, defined by

$$\text{Sdet } M := \exp\{\text{STr } \ln M\}. \tag{2.34}$$

If

$$M = \exp X, \tag{2.35}$$

we have, of course,

$$\text{Sdet } M = \exp\{\text{STr } X\}. \tag{2.36}$$

The superdeterminant is defined in analogy to the well-known result

$$\det M = \exp\{\text{ Tr } \ln M\}$$

for ordinary square matrices M. The latter is easily verified by setting $M = 1 + L$. Then:

$$\text{Tr } \ln(1+L) = \text{ Tr } \left(L - \frac{1}{2}L^2 + \frac{1}{3}L^3 - \cdots\right).$$

If U is a matrix with

$$\text{Tr } (L) = \text{ Tr } \left(ULU^{-1}\right) = \text{ Tr } (\Lambda) = \sum_i \lambda_i,$$

where Λ is diagonal, then:

$$\begin{aligned}
\text{Tr } \ln(1+L) &= \sum_i \lambda_i - \frac{1}{2}\sum_i \lambda_i^2 + \frac{1}{3}\sum_i \lambda_i^3 - \cdots \\
&= \sum_i \ln\left(1 + \lambda_i\right) \\
&= \ln \prod_i \left(1 + \lambda_i\right) \\
&= \ln \det\left(1 + \Lambda\right) \\
&= \ln \det\left(1 + L\right),
\end{aligned}$$

so that

$$\text{Tr } \ln M = \ln \det M.$$

It is a natural consequence of Eq. (2.34) to define unimodular graded matrices M by:

$$\text{Sdet } M = 1,$$

and

$$\text{STr } X = 0.$$

Proposition 2.3:

Let M_1 and M_2 be two graded matrices. Then

$$\text{Sdet } (M_1 M_2) = \text{Sdet } (M_1) \cdot \text{Sdet } (M_2). \tag{2.37}$$

Proof: Using the definition of the superdeterminant, *i.e.* Eq. (2.24), we write the left hand side of Eq. (2.37) as

$$\text{Sdet } (M_1 M_2) = \exp\{\text{STr } \ln M_1 M_2\}.$$

We now define two matrices:

$$A := \ln M_1, \quad B := \ln M_2,$$

and use the *Baker–Campbell–Hausdorff* formula[11]

$$e^A e^B = e^{A+B+\frac{1}{2}[A,B]+\frac{1}{12}[A,[A,B]]-\frac{1}{12}[B,[B,A]]+\cdots} = M_1 M_2.$$

Taking the logarithm of both sides we obtain:

$$\ln(M_1 M_2) = \ln(e^A e^B) = A+B+\frac{1}{2}[A,B]+\frac{1}{12}[A,[A,B]]-\frac{1}{12}[B,[B,A]]+\cdots.$$

Next we take the supertrace of both sides so that

$$\begin{aligned} \text{STr } \ln(M_1 M_2) &= \text{STr } A + \text{STr } B + \frac{1}{2}\text{STr }[A,B] + \cdots \\ &= \text{STr } \ln M_1 + \text{STr } \ln M_2 + \frac{1}{2}\text{STr }(AB - BA) + \cdots \\ &= \text{STr } \ln M_1 + \text{STr } \ln M_2. \end{aligned}$$

All commutator terms vanish because the supertrace obeys Eq. (2.33). Hence

$$\begin{aligned} \text{Sdet } (M_1 M_2) &= \exp\{\text{STr } \ln(M_1 M_2)\} \\ &= \exp\{\text{STr } \ln M_1 + \text{STr } \ln M_2\} \\ &= \exp\{\text{STr } \ln M_1\} \cdot \exp\{\text{STr } \ln M_2\} \\ &= \text{Sdet } (M_1) \cdot \text{Sdet } (M_2) \end{aligned}$$

in view of Eq. (2.29).

---

[11]See e.g. W. Miller, [72], p. 161.

Proposition 2.4:

The superdeterminant (2.34) can be expressed in terms of ordinary determinants by means of the following formulas. Let

$$M = \begin{pmatrix} A & B \\ C & D \end{pmatrix}$$

be a supermatrix; then

$$\text{Sdet } M = \frac{\det(A - BD^{-1}C)}{\det D} \tag{2.38}$$

$$= \frac{\det A}{\det(D - CA^{-1}B)}. \tag{2.39}$$

Proof: In order to verify Eq. (2.38), we decompose the matrix M in the following form:

$$M = \begin{pmatrix} E & F \\ 0_{m \times n} & \mathbb{1}_{m \times m} \end{pmatrix} \begin{pmatrix} \mathbb{1}_{n \times n} & 0_{n \times m} \\ G & H \end{pmatrix} = M_1 M_2 = \begin{pmatrix} E + FG & FH \\ G & H \end{pmatrix}.$$

Comparing this decomposition with the standard form above, we obtain:

$$E + FG = A, \quad FH = B, \quad C = G, \quad D = H,$$

or

$$E = A - BD^{-1}C, \quad F = BD^{-1}, \quad G = C, \quad H = D.$$

Then

$$\text{Sdet } M = \text{Sdet } (M_1 M_2)$$
$$\overset{(2.37)}{=} \text{Sdet } M_1 \cdot \text{Sdet } M_2$$
$$= \exp\{\text{STr } \ln M_1\} \cdot \exp\{\text{STr } \ln M_2\}.$$

For the particular form of M_1 under consideration

$$\text{STr } \ln M_1 = \text{Tr } \ln E,$$

as can be shown with the help of the power series expansion of the logarithm. Thus setting $M_1 = 1 + L$, so that

$$L = \begin{pmatrix} E - 1 & F \\ 0 & 0 \end{pmatrix}, \qquad L^2 = \begin{pmatrix} (E-1)^2 & (E-1)F \\ 0 & 0 \end{pmatrix},$$
$$L^3 = \begin{pmatrix} (E-1)^3 & (E-1)^2 F \\ 0 & 0 \end{pmatrix}, \ldots,$$

we have

$$
\begin{aligned}
\text{STr } \ln M_1 &= \text{STr } \ln(1+L) \\
&= \text{STr } \left[L - \frac{1}{2}L^2 + \frac{1}{3}L^3 - \cdots \right] \\
&= \text{STr } \left[\begin{pmatrix} E-1 & F \\ 0 & 0 \end{pmatrix} - \frac{1}{2} \begin{pmatrix} (E-1)^2 & (E-1)F \\ 0 & 0 \end{pmatrix} + \cdots \right] \\
&= \text{Tr } \left[E-1 \right] - \frac{1}{2}\text{Tr } \left[(E-1)^2 \right] + \frac{1}{3}\text{Tr } \left[(E-1)^3 \right] - \cdots \\
&= \text{Tr } \left[\ln E \right].
\end{aligned}
$$

Inserting this in our expression for Sdet M, we obtain:

$$
\begin{aligned}
\text{Sdet } M &= \exp\{ \text{Tr } \left[\ln E \right] \} \cdot \exp\{ -\text{Tr } \left[\ln H \right] \} \\
&= \det E \cdot (\det H)^{-1} \\
&= \frac{\det(A - BD^{-1}C)}{\det D}.
\end{aligned}
$$

Equation (2.39) can be verified by using the decomposition

$$
M = \begin{pmatrix} \mathbb{1}_{n\times n} & 0 \\ P & Q \end{pmatrix} \begin{pmatrix} R & S \\ 0 & \mathbb{1}_{m\times m} \end{pmatrix} = \begin{pmatrix} R & S \\ PR & PS+Q \end{pmatrix} \overset{!}{=} \begin{pmatrix} A & B \\ C & D \end{pmatrix}.
$$

Hence

$$
P = CA^{-1}, \quad Q = D - CA^{-1}B, \quad R = A, \quad B = S.
$$

Then as above

$$
\text{Sdet } M = \det R \cdot (\det Q)^{-1} = \frac{\det A}{\det(D - CA^{-1}B)}.
$$

This completes the proof of Proposition 2.4.

Definition:

Supertransposition of a supermatrix M is defined by

$$
M^{ST} = \begin{pmatrix} A^\top & -C^\top \\ B^\top & D^\top \end{pmatrix}. \tag{2.40}
$$

Proposition 2.5:

Supertransposition as defined in Eq. (2.40) is constructed in such a way to mimic the ordinary law of transposition, *i.e.*

$$\left(M_1 M_2\right)^{ST} = M_2^{ST} M_1^{ST}. \tag{2.41}$$

Proof: We start with:

$$
\begin{aligned}
M_2^{ST} M_1^{ST} &= \begin{pmatrix} A_2 & B_2 \\ C_2 & D_2 \end{pmatrix}^{ST} \begin{pmatrix} A_1 & B_1 \\ C_1 & D_1 \end{pmatrix}^{ST} \\[2mm]
&\overset{(2.40)}{=} \begin{pmatrix} A_2^\top & -C_2^\top \\ B_2^\top & D_2^\top \end{pmatrix} \begin{pmatrix} A_1^\top & -C_1^\top \\ B_1^\top & D_1^\top \end{pmatrix} \\[2mm]
&= \begin{pmatrix} A_2^\top A_1^\top - C_2^\top B_1^\top & -A_2^\top C_1^\top - C_2^\top D_1^\top \\ B_2^\top A_1^\top + D_2^\top B_1^\top & -B_2^\top C_1^\top + D_2^\top D_1^\top \end{pmatrix} \\[2mm]
&\overset{(2.31)}{=} \begin{pmatrix} (A_1 A_2)^\top + (B_1 C_2)^\top & -(C_1 A_2)^\top - (D_1 C_2)^\top \\ (A_1 B_2)^\top + (B_1 D_2)^\top & (C_1 B_2)^\top + (D_1 D_2)^\top \end{pmatrix} \\[2mm]
&= \begin{pmatrix} (A_1 A_2 + B_1 C_2)^\top & -(C_1 A_2 + D_1 C_2)^\top \\ (A_1 B_2 + B_1 D_2)^\top & (C_1 B_2 + D_1 D_2)^\top \end{pmatrix} \\[2mm]
&\overset{(2.40)}{=} \begin{pmatrix} A_1 A_2 + B_1 C_2 & A_1 B_2 + B_1 D_2 \\ C_1 A_2 + D_1 C_2 & C_1 B_2 + D_1 D_2 \end{pmatrix}^{ST} \\[2mm]
&= \left[\begin{pmatrix} A_1 & B_1 \\ C_1 & D_1 \end{pmatrix} \begin{pmatrix} A_2 & B_2 \\ C_2 & D_2 \end{pmatrix} \right]^{ST} = \left(M_1 M_2\right)^{ST}.
\end{aligned}
$$

Finally, we prove the following proposition.

Proposition 2.6:

Let M be a supermatrix. Then

$$\text{Sdet}\left(M^{ST}\right) = \text{Sdet}\left(M\right). \tag{2.42}$$

Proof: Using Eqs. (2.38), (2.30) and again Eq. (2.38), we find:

$$
\begin{aligned}
\text{Sdet}\left(M^{ST}\right) &= \text{Sdet} \begin{pmatrix} A^\top & -C^\top \\ B^\top & D^\top \end{pmatrix} \\[2mm]
&= \frac{\det\left(A^\top + C^\top (D^\top)^{-1} B^\top\right)}{\det(D^\top)} \\[2mm]
&= \frac{\det\left(A^\top - (BD^{-1}C)^\top\right)}{\det D} \\[2mm]
&= \frac{\det\left[(A - BD^{-1}C)^\top\right]}{\det D} \\[2mm]
&= \frac{\det\left(A - BD^{-1}C\right)}{\det D} = \text{Sdet}\left(M\right).
\end{aligned}
$$

Chapter 3

The Supersymmetric Extension of the Poincaré Algebra

3.1 Four-Component Dirac Formulation

With Eqs. (1.34) we obtained the commutation relations of the ten-dimensional Lie algebra of the Poincaré group, *i.e.*

$$
\begin{aligned}
\left[P_\mu, P_\nu\right] &= 0, \\
\left[M_{\mu\nu}, P_\lambda\right] &= i\left(\eta_{\nu\lambda}P_\mu - \eta_{\mu\lambda}P_\nu\right), \\
\left[M_{\mu\nu}, M_{\rho\sigma}\right] &= -i\left(\eta_{\mu\rho}M_{\nu\sigma} - \eta_{\mu\sigma}M_{\nu\rho} - \eta_{\nu\rho}M_{\mu\sigma} + \eta_{\nu\sigma}M_{\mu\rho}\right).
\end{aligned}
\tag{3.1}
$$

The generators P_μ and $M_{\mu\nu}$ span a ten-dimensional vector space with the composition law given by Eqs. (3.1). In order to construct an extension of the Poincaré algebra to a graded Lie algebra it is natural to take the Poincaré algebra as the subspace L_0 of the $\mathbb{Z}_2$ graded Lie algebra L which we want to construct. From previous considerations in Chapter 2 we know that the coefficients of the product on $L_0 \times L_1$ have to form a matrix representation of L_0 with dimension equal to that of L_1. In this section we consider in detail the case when L_1 is four-dimensional. Thus we have four basis elements denoted by Q_a, $a = 1, 2, 3, 4$, which span the subspace L_1. We therefore search for a grading of the Poincaré algebra with the following properties:

(i) L_0: Poincaré algebra (3.1),

(ii) $L_1 = \text{Span } \{Q_a\}, a = 1, 2, 3, 4,$

147

(iii) The product $\circ$: $L_0 \times L_1 \longrightarrow L_1$ is defined by

$$P_\mu \circ Q_a = 0, \tag{3.2}$$

$$M_{\mu\nu} \circ Q_a = -\left(\sigma^4_{\mu\nu}\right)_{ab} Q_b. \tag{3.3}$$

We shall see in a moment that

$$\sigma^4_{\mu\nu} = \frac{i}{4} \left[\gamma_\mu, \gamma_\nu\right].$$

In the above requirements Eq. (3.2) implies that the Q's transform trivially under translations, whereas Eq. (3.3) implies that the Q's transform as spinors under homogeneous Lorentz transformations.

Remarks:

(i) With the above choice of the grading of the Poincaré algebra — in particular Eqs. (3.2) and (3.3) — we satisfy the requirement that the coefficients of the product on $L_0 \times L_1$ form a matrix representation of the Lie algebra L_0. In this particular case, the coefficients are 0 and $\sigma^4_{\mu\nu}$ which together have to form a 4×4-matrix representation of the Poincaré algebra.

(ii) In agreement with the general properties of the product of a graded Lie algebra (*cf.* Eqs. (2.4) to (2.6)), we set (*cf.* also Eq. (2.15))

$$\begin{aligned} P_\mu \circ Q_a &= P_\mu Q_a - (-1)^{g(P_\mu)g(Q_a)} Q_a P_\mu \\ &= P_\mu Q_a - Q_a P_\mu \\ &= \left[P_\mu, Q_a\right] = 0, \end{aligned} \tag{3.4}$$

with

$$\begin{aligned} g(P_\mu) &= 0, & \text{since } P_\mu \in L_0, \\ g(Q_a) &= 1, & \text{since } Q_a \in L_1. \end{aligned}$$

Furthermore we have:

$$\begin{aligned} M_{\mu\nu} \circ Q_a &= M_{\mu\nu} Q_a - (-1)^{g(M_{\mu\nu})g(Q_a)} Q_a M_{\mu\nu} \\ &= M_{\mu\nu} Q_a - Q_a M_{\mu\nu} \\ &= \left[M_{\mu\nu}, Q_a\right] \\ &= -\left(\sigma^4_{\mu\nu}\right)_{ab} Q_b. \end{aligned} \tag{3.5}$$

Here we use

$$\begin{aligned} g(M_{\mu\nu}) &= 0, & \text{since } M_{\mu\nu} \in L_0, \\ g(Q_a) &= 1, & \text{since } Q_a \in L_1. \end{aligned}$$

(iii) The above grading of the Poincaré algebra is by no means unique. There is an infinite number of other possible gradings, for which the Q's transform under other representations of the Poincaré group.

(iv) The graded Poincaré algebra is an example of a graded Lie algebra with more structure than that given in the above definition. If we set

$$L = L_0 \oplus L_1 \oplus L_2$$

with generators

$$M_{\mu\nu} \in L_0, \quad Q_a \in L_1, \quad P_\mu \in L_2,$$

one can show that

$$X_k \circ X_j \in L_{k+j \bmod 3}, \qquad X_k \in L_k, X_j \in L_j.$$

Before we consider the generalized Jacobi identities in detail, we prove the following result.

Proposition 3.1:

Let $\sigma_{\mu\nu}^4$ be defined by

$$\sigma_{\mu\nu}^4 := \frac{i}{4}\left[\gamma_\mu, \gamma_\nu\right]. \tag{3.6}$$

Then

$$\left[\sigma_{\mu\nu}^4, \sigma_{\rho\sigma}^4\right] = -i\left(\eta_{\mu\rho}\sigma_{\nu\sigma}^4 - \eta_{\mu\sigma}\sigma_{\nu\rho}^4 - \eta_{\nu\rho}\sigma_{\mu\sigma}^4 + \eta_{\nu\sigma}\sigma_{\mu\rho}^4\right), \tag{3.7}$$

i.e. the $\sigma_{\mu\nu}$ satisfy the same commutation relation as the generators $M_{\mu\nu}$ of the homogeneous Lorentz transformations, which means that the matrices $\sigma_{\mu\nu}$ form a four-dimensional representation of the Lorentz algebra.

Proof: Consider:

$$\left[\sigma_{\mu\nu}^4, \sigma_{\rho\sigma}^4\right] = -\frac{1}{16}\left[\left[\gamma_\mu, \gamma_\nu\right], \left[\gamma_\rho, \gamma_\sigma\right]\right]$$

$$= -\frac{1}{16}\left[\gamma_\mu\gamma_\nu - \gamma_\nu\gamma_\mu, \gamma_\rho\gamma_\sigma - \gamma_\sigma\gamma_\rho\right]$$

$$= -\frac{1}{16}\left\{\left[\gamma_\mu\gamma_\nu - \gamma_\nu\gamma_\mu, \gamma_\rho\gamma_\sigma\right] - \left[\gamma_\mu\gamma_\nu - \gamma_\nu\gamma_\mu, \gamma_\sigma\gamma_\rho\right]\right\}$$

$$= -\frac{1}{16}\left\{\left[\gamma_\mu\gamma_\nu, \gamma_\rho\gamma_\sigma\right] - \left[\gamma_\nu\gamma_\mu, \gamma_\rho\gamma_\sigma\right]\right.$$

$$\left. - \left[\gamma_\mu\gamma_\nu, \gamma_\sigma\gamma_\rho\right] + \left[\gamma_\nu\gamma_\mu, \gamma_\sigma\gamma_\rho\right]\right\}.$$

Using the relation $[A, BC] = A[B, C] + [A, B]C$, this expression is

$$
\begin{aligned}
[\sigma^4_{\mu\nu}, \sigma^4_{\rho\sigma}] ={}& -\frac{1}{16}\Big\{ \gamma_\mu[\gamma_\nu, \gamma_\rho]\gamma_\sigma + \gamma_\mu\gamma_\rho[\gamma_\nu, \gamma_\sigma] + \gamma_\rho[\gamma_\mu, \gamma_\sigma]\gamma_\nu \\
&+ [\gamma_\mu, \gamma_\rho]\gamma_\sigma\gamma_\nu - \gamma_\nu\gamma_\rho[\gamma_\mu, \gamma_\sigma] - \gamma_\nu[\gamma_\mu, \gamma_\rho]\gamma_\sigma \\
&- \gamma_\rho[\gamma_\nu, \gamma_\sigma]\gamma_\mu - [\gamma_\nu, \gamma_\rho]\gamma_\sigma\gamma_\mu - \gamma_\mu\gamma_\sigma[\gamma_\nu, \gamma_\rho] \\
&- \gamma_\mu[\gamma_\nu, \gamma_\sigma]\gamma_\rho - \gamma_\sigma[\gamma_\mu, \gamma_\rho]\gamma_\nu - [\gamma_\mu, \gamma_\sigma]\gamma_\rho\gamma_\nu \\
&+ \gamma_\nu\gamma_\sigma[\gamma_\mu, \gamma_\rho] + \gamma_\nu[\gamma_\mu, \gamma_\sigma]\gamma_\rho + \gamma_\sigma[\gamma_\nu, \gamma_\rho]\gamma_\mu \\
&+ [\gamma_\nu, \gamma_\sigma]\gamma_\rho\gamma_\mu \Big\} \\
={}& \frac{i}{4}\Big\{ \gamma_\mu\sigma^4_{\nu\rho}\gamma_\sigma + \gamma_\mu\gamma_\rho\sigma^4_{\nu\sigma} + \gamma_\rho\sigma^4_{\mu\sigma}\gamma_\nu + \sigma^4_{\mu\rho}\gamma_\sigma\gamma_\nu \\
&- \gamma_\nu\gamma_\rho\sigma^4_{\mu\sigma} - \gamma_\nu\sigma^4_{\mu\rho}\gamma_\sigma - \gamma_\rho\sigma^4_{\nu\sigma}\gamma_\mu - \sigma^4_{\nu\rho}\gamma_\sigma\gamma_\mu \\
&- \gamma_\mu\gamma_\sigma\sigma^4_{\nu\rho} - \gamma_\mu\sigma^4_{\nu\sigma}\gamma_\rho - \gamma_\sigma\sigma^4_{\mu\rho}\gamma_\nu - \sigma^4_{\mu\sigma}\gamma_\rho\gamma_\nu \\
&+ \gamma_\nu\gamma_\sigma\sigma^4_{\mu\rho} + \gamma_\nu\sigma^4_{\mu\sigma}\gamma_\rho + \gamma_\sigma\sigma^4_{\nu\rho}\gamma_\mu + \sigma^4_{\nu\sigma}\gamma_\rho\gamma_\mu \Big\} \\
={}& \frac{i}{4}\Big\{ \gamma_\mu\gamma_\rho\sigma^4_{\nu\sigma} + \sigma^4_{\nu\sigma}\gamma_\rho\gamma_\mu - \gamma_\rho\sigma^4_{\nu\sigma}\gamma_\mu - \gamma_\mu\sigma^4_{\nu\sigma}\gamma_\rho \\
&+ \sigma^4_{\mu\rho}\gamma_\sigma\gamma_\nu + \gamma_\nu\gamma_\sigma\sigma^4_{\mu\rho} - \gamma_\nu\sigma^4_{\mu\rho}\gamma_\sigma - \gamma_\sigma\sigma^4_{\mu\rho}\gamma_\nu \\
&- \sigma^4_{\mu\sigma}\gamma_\rho\gamma_\nu - \gamma_\nu\gamma_\rho\sigma^4_{\mu\sigma} + \gamma_\nu\sigma^4_{\mu\sigma}\gamma_\rho + \gamma_\rho\sigma^4_{\mu\sigma}\gamma_\nu \\
&- \gamma_\mu\gamma_\sigma\sigma^4_{\nu\rho} - \sigma^4_{\nu\rho}\gamma_\sigma\gamma_\mu + \gamma_\mu\sigma^4_{\nu\rho}\gamma_\sigma + \gamma_\sigma\sigma^4_{\nu\rho}\gamma_\mu \Big\}.
\end{aligned}
$$

Here we made several times use of the definition (3.6).

We now use the following result, which we prove below,

$$
[\sigma^4_{\mu\nu}, \gamma_\rho] = i(\eta_{\nu\rho}\gamma_\mu - \eta_{\mu\rho}\gamma_\nu). \tag{3.8}
$$

This relation shows that the γ-matrices transform as a four-vector under Lorentz-transformations. With Eq. (3.8) it is possible to rearrange various terms so that

$$
\begin{aligned}
[\sigma^4_{\mu\nu}, \sigma^4_{\rho\sigma}] ={}& \frac{i}{4}\Big\{ \gamma_\mu\gamma_\rho\sigma^4_{\nu\sigma} + i\eta_{\sigma\rho}\gamma_\nu\gamma_\mu - i\eta_{\nu\rho}\gamma_\sigma\gamma_\mu + \gamma_\rho\sigma^4_{\nu\sigma}\gamma_\mu - \gamma_\rho\sigma^4_{\nu\sigma}\gamma_\mu \\
&- i\eta_{\sigma\rho}\gamma_\mu\gamma_\nu + i\eta_{\nu\rho}\gamma_\mu\gamma_\sigma - \gamma_\mu\gamma_\rho\sigma^4_{\nu\sigma} + i\eta_{\rho\sigma}\gamma_\mu\gamma_\nu - i\eta_{\mu\sigma}\gamma_\rho\gamma_\nu \\
&+ \gamma_\sigma\sigma^4_{\mu\rho}\gamma_\nu + \gamma_\nu\gamma_\sigma\sigma^4_{\mu\rho} - i\eta_{\rho\sigma}\gamma_\nu\gamma_\mu + i\eta_{\mu\sigma}\gamma_\nu\gamma_\rho - \gamma_\nu\gamma_\sigma\sigma^4_{\mu\rho} \\
&- \gamma_\sigma\sigma^4_{\mu\rho}\gamma_\nu - \sigma^4_{\mu\sigma}\gamma_\rho\gamma_\nu + i\eta_{\sigma\rho}\gamma_\nu\gamma_\mu - i\eta_{\mu\rho}\gamma_\nu\gamma_\sigma - \gamma_\nu\sigma^4_{\mu\sigma}\gamma_\rho \\
&+ \gamma_\nu\sigma^4_{\mu\sigma}\gamma_\rho - i\eta_{\sigma\rho}\gamma_\mu\gamma_\nu + i\eta_{\mu\rho}\gamma_\sigma\gamma_\nu + \sigma^4_{\mu\sigma}\gamma_\rho\gamma_\nu \\
&- \gamma_\mu\gamma_\sigma\sigma^4_{\nu\rho} - i\eta_{\rho\sigma}\gamma_\nu\gamma_\mu + i\eta_{\nu\sigma}\gamma_\rho\gamma_\mu - \gamma_\sigma\sigma^4_{\nu\rho}\gamma_\mu \\
&+ i\eta_{\rho\sigma}\gamma_\mu\gamma_\nu - i\eta_{\nu\sigma}\gamma_\mu\gamma_\rho + \gamma_\mu\gamma_\sigma\sigma^4_{\nu\rho} + \gamma_\sigma\sigma^4_{\nu\rho}\gamma_\mu \Big\}
\end{aligned}
$$

$$= \frac{i}{4}\Big\{i\eta_{\sigma\rho}\big[\gamma_\nu\gamma_\mu - \gamma_\mu\gamma_\nu\big] + i\eta_{\nu\rho}\big[\gamma_\mu\gamma_\sigma - \gamma_\sigma\gamma_\mu\big]$$

$$+ i\eta_{\sigma\rho}\big[\gamma_\mu\gamma_\nu - \gamma_\nu\gamma_\mu\big] + i\eta_{\mu\rho}\big[\gamma_\sigma\gamma_\nu - \gamma_\nu\gamma_\sigma\big]$$

$$+ i\eta_{\mu\sigma}\big[\gamma_\nu\gamma_\rho - \gamma_\rho\gamma_\nu\big] + i\eta_{\nu\sigma}\big[\gamma_\rho\gamma_\mu - \gamma_\mu\gamma_\rho\big]$$

$$+ i\eta_{\sigma\rho}\big[\gamma_\nu\gamma_\mu - \gamma_\mu\gamma_\nu\big] + i\eta_{\sigma\rho}\big[\gamma_\mu\gamma_\nu - \gamma_\nu\gamma_\mu\big]\Big\}$$

$$= \frac{i}{4}\Big\{i\eta_{\nu\rho}\big[\gamma_\mu,\gamma_\sigma\big] + i\eta_{\mu\rho}\big[\gamma_\sigma,\gamma_\nu\big] + i\eta_{\mu\sigma}\big[\gamma_\nu,\gamma_\rho\big] + i\eta_{\nu\sigma}\big[\gamma_\rho,\gamma_\mu\big]\Big\}$$

$$= i\eta_{\nu\rho}\sigma^4_{\mu\sigma} + i\eta_{\mu\rho}\sigma^4_{\sigma\nu} + i\eta_{\mu\sigma}\sigma^4_{\nu\rho} + i\eta_{\nu\sigma}\sigma^4_{\rho\mu}$$

$$= -i\big(\eta_{\mu\rho}\sigma^4_{\nu\sigma} - \eta_{\mu\sigma}\sigma^4_{\nu\rho} - \eta_{\nu\rho}\sigma^4_{\mu\sigma} + \eta_{\nu\sigma}\sigma^4_{\mu\rho}\big).$$

We now prove the result (3.8). To this end we consider

$$\big[\sigma^4_{\mu\nu},\gamma_\rho\big] = \frac{i}{4}\Big[[\gamma_\mu,\gamma_\nu],\gamma_\rho\Big] = \frac{i}{4}\Big[\gamma_\mu\gamma_\nu - \gamma_\nu\gamma_\mu,\gamma_\rho\Big]$$

$$= \frac{i}{4}\Big(\gamma_\mu\gamma_\nu\gamma_\rho - \gamma_\nu\gamma_\mu\gamma_\rho - \gamma_\rho\gamma_\mu\gamma_\nu + \gamma_\rho\gamma_\nu\gamma_\mu\Big).$$

With the help of the basic Clifford algebra relation

$$\big\{\gamma_\mu,\gamma_\nu\big\} = 2\eta_{\mu\nu}\mathbb{1}_{4\times4},$$

we can rearrange various terms so that

$$\big[\sigma^4_{\mu\nu},\gamma_\rho\big] = \frac{i}{4}\Big\{\gamma_\mu\gamma_\nu\gamma_\rho - 2\eta_{\mu\nu}\gamma_\rho + \gamma_\mu\gamma_\nu\gamma_\rho - 2\eta_{\mu\rho}\gamma_\nu$$

$$+ \gamma_\mu\gamma_\rho\gamma_\nu + 2\eta_{\mu\nu}\gamma_\rho - \gamma_\rho\gamma_\mu\gamma_\nu\Big\}$$

$$= \frac{i}{4}\Big\{2\gamma_\mu\gamma_\nu\gamma_\rho - 2\eta_{\mu\rho}\gamma_\nu + 2\eta_{\rho\nu}\gamma_\mu - \gamma_\mu\gamma_\nu\gamma_\rho - 2\eta_{\mu\rho}\gamma_\nu + \gamma_\mu\gamma_\rho\gamma_\nu\Big\}$$

$$= \frac{i}{4}\Big\{\gamma_\mu\gamma_\nu\gamma_\rho - 4\eta_{\mu\rho}\gamma_\nu + 2\eta_{\rho\nu}\gamma_\mu + 2\eta_{\rho\nu}\gamma_\mu - \gamma_\mu\gamma_\nu\gamma_\rho\Big\}$$

$$= i\big(\eta_{\rho\nu}\gamma_\mu - \eta_{\mu\rho}\gamma_\nu\big).$$

We now consider the generalized Jacobi identities. According to Eq. (2.21) these are defined by:

$$X_\mu \circ (X_\nu \circ X_\rho)(-1)^{g_\mu g_\rho} + X_\rho \circ (X_\mu \circ X_\nu)(-1)^{g_\rho g_\nu} + X_\nu \circ (X_\rho \circ X_\mu)(-1)^{g_\mu g_\nu} = 0,$$

where $X_\mu, X_\nu, X_\rho \in L$. We investigate these relations for the case where the subspace L_0 is spanned by the Poincaré algebra (3.1). For $X_\mu, X_\nu, X_\rho \in L_0$ Eq. (2.21) reduces to the Jacobi identity of the Poincaré algebra which is automatically satisfied since L_0 is a Lie algebra.

For $X_\mu, X_\nu \in L_0$, *i.e.* $X_\mu, X_\nu \in \{M_{\rho\sigma}, P_\lambda\}$, and $X_\rho \in L_1$, *i.e.* $X_\rho = Q_a$, we have

$$g(X_\mu) = g_\mu = 0, \qquad g(X_\nu) = g_\nu = 0, \qquad g(Q_a) = g_a = 1,$$

and Eq. (2.21) becomes (using Eqs. (3.4) and (3.5))

$$X_\mu \circ (X_\nu \circ Q_a)(-1)^{g_\mu g_a} + Q_a \circ (X_\mu \circ X_\nu)(-1)^{g_a g_\nu} + X_\nu \circ (Q_a \circ X_\mu)(-1)^{g_\mu g_\nu}$$
$$= \Big[X_\mu, [X_\nu, Q_a]\Big] + \Big[Q_a, [X_\mu, X_\nu]\Big] + \Big[X_\nu, [Q_a, X_\mu]\Big]$$
$$= 0. \tag{3.9}$$

For reasons of consistency we have to check that our choice of grading for the Poincaré algebra, *i.e.* Eqs. (3.2) and (3.3), satisfies the Jacobi identities. We therefore consider these for the various possible cases.

(i) $X_\mu = P_\mu, X_\nu = P_\nu, X_\rho = Q_a$. In this case the expression on the left hand side of Eq. (3.9) vanishes so that the following Jacobi identity is satisfied:

$$\Big[P_\mu, [P_\nu, Q_a]\Big] + \Big[Q_a, [P_\mu, P_\nu]\Big] + \Big[P_\nu, [Q_a, P_\mu]\Big] = 0. \tag{3.10}$$

In verifying this relation one uses, of course, the relations (3.1) and (3.4). The verification of the other Jacobi identites is handled in a similar way.

(ii) $X_\mu = M_{\mu\nu}, X_\nu = P_\rho, X_\rho = Q_a$. In this case we have

$$\Big[M_{\mu\nu}, [P_\rho, Q_a]\Big] + \Big[Q_a, [M_{\mu\nu}, P_\rho]\Big] + \Big[P_\rho, [Q_a, M_{\mu\nu}]\Big]$$
$$\overset{(*)}{=} [M_{\mu\nu}, 0] + [Q_a, i\eta_{\nu\rho}P_\mu - i\eta_{\mu\rho}P_\nu] + [P_\rho, (\sigma^4_{\mu\nu})_{ab}Q_b]$$
$$= i\eta_{\nu\rho}[Q_a, P_\mu] - i\eta_{\mu\rho}[Q_a, P_\nu] + (\sigma^4_{\mu\nu})_{ab}[P_\rho, Q_b]$$
$$= 0.$$

In step (*) we used Eqs. (3.4), (3.1) and (3.5), and the last step is performed by applying the commutator (3.4). Hence:

$$\Big[M_{\mu\nu}, [P_\rho, Q_a]\Big] + \Big[Q_a, [M_{\mu\nu}, P_\rho]\Big] + \Big[P_\rho, [Q_a, M_{\mu\nu}]\Big] = 0. \tag{3.11}$$

(iii) $X_\mu = M_{\mu\nu}, X_\nu = M_{\rho\sigma}, X_\rho = Q_a$. In this case we have

$$\left[M_{\mu\nu}, [M_{\rho\sigma}, Q_a]\right] + \left[Q_a, [M_{\mu\nu}, M_{\rho\sigma}]\right] + \left[M_{\rho\sigma}, [Q_a, M_{\mu\nu}]\right]$$

$$\overset{(3.1)}{\underset{(3.5)}{=}} \left[M_{\mu\nu}, (-\sigma^4_{\rho\sigma})_{ab}Q_b\right] + \left[M_{\rho\sigma}, (\sigma^4_{\mu\nu})_{ab}Q_b\right]$$

$$+ \left[Q_a, \{-i(\eta_{\mu\rho}M_{\nu\sigma} - \eta_{\mu\sigma}M_{\nu\rho}\eta_{\nu\rho}M_{\mu\sigma} + \eta_{\nu\sigma}M_{\mu\rho})\}\right]$$

$$= -(\sigma^4_{\rho\sigma})_{ab}\left[M_{\mu\nu}, Q_b\right] - i\eta_{\mu\rho}\left[Q_a, M_{\nu\sigma}\right]$$

$$+ i\eta_{\mu\sigma}\left[Q_a, M_{\nu\rho}\right] + i\eta_{\nu\rho}\left[Q_a, M_{\mu\sigma}\right]$$

$$- i\eta_{\nu\sigma}\left[Q_a, M_{\mu\rho}\right] + (\sigma^4_{\mu\nu})_{ab}\left[M_{\rho\sigma}, Q_b\right]$$

$$\overset{(3.5)}{=} (\sigma^4_{\rho\sigma})_{ab}(\sigma^4_{\mu\nu})_{bc}Q_c - i\eta_{\mu\rho}(\sigma^4_{\nu\sigma})_{ac}Q_c$$

$$+ i\eta_{\mu\sigma}(\sigma^4_{\nu\rho})_{ac}Q_c + i\eta_{\nu\rho}(\sigma^4_{\mu\sigma})_{ac}Q_c$$

$$- i\eta_{\nu\sigma}(\sigma^4_{\mu\rho})_{ac}Q_c - (\sigma^4_{\mu\nu})_{ab}(\sigma^4_{\rho\sigma})_{bc}Q_c$$

$$= -\left([\sigma^4_{\mu\nu}, \sigma^4_{\rho\sigma}]\right)_{ac}Q_c - i\left[\eta_{\mu\rho}(\sigma^4_{\nu\sigma})_{ac} - \eta_{\mu\sigma}(\sigma^4_{\nu\rho})_{ac}\right.$$

$$\left. - \eta_{\nu\rho}(\sigma^4_{\mu\sigma})_{ac} + \eta_{\nu\sigma}(\sigma^4_{\mu\rho})_{ac}\right]Q_c$$

$$\overset{(3.7)}{=} -\left([\sigma^4_{\mu\nu}, \sigma^4_{\rho\sigma}]\right)_{ac}Q_c + \left([\sigma^4_{\mu\nu}, \sigma^4_{\rho\sigma}]\right)_{ac}Q_c = 0.$$

Hence:

$$\left[M_{\mu\nu}, [M_{\rho\sigma}, Q_a]\right] + \left[Q_a, [M_{\mu\nu}, M_{\rho\sigma}]\right] + \left[M_{\rho\sigma}, [Q_a, M_{\mu\nu}]\right] = 0. \quad (3.12)$$

From the derivation of Eq. (3.12) we see that the essential point is the fact that the matrices $\sigma^4_{\mu\nu}$ form a representation of the Lorentz algebra, *i.e.* the validity of Eq. (3.7), where the $\sigma^4_{\mu\nu}$ are the coefficients of the product $M_{\mu\nu} \circ Q_a$ (see Eq. (3.3)). One can also argue the other way round as was done in Chapter 2 for the group $SU(2, \mathbb{C})$: Given the generalized Jacobi identity (3.12) (and also, of course, relations (3.10) and (3.11)) one can ask, what are the properties which the coefficients of the products $P_\mu \circ Q_a$ and $M_{\mu\nu} \circ Q_a$ must obey? The conclusion is — as was demonstrated in Chapter 2 — that these matrices must be representation matrices of the Lie algebra L_0.

We have not yet defined the product on L_1. With Eqs. (2.4) and (2.15) we have:

$$\begin{aligned} \circ \ : \ L_1 \times L_1 &\longrightarrow L_0, \\ Q_a, Q_b &\longmapsto Q_a \circ Q_b = Q_aQ_b - (-1)^{g_a g_b}Q_bQ_a \\ &= Q_aQ_b + Q_bQ_a \\ &=: \{Q_a, Q_b\}. \end{aligned} \quad (3.13)$$

According to the general theory of graded Lie algebras this product has to be symmetric as indicated in Eq. (3.13), and it must close into L_0. Hence, the most general form of Eq. (3.13) is

$$\{Q_a, Q_b\} = (h^\mu)_{ab} P_\mu + (k^{\mu\nu})_{ab} M_{\mu\nu}, \tag{3.14}$$

where (h^μ) and $(k^{\mu\nu})$ are symmetric 4×4-matrices, since the product, *i.e.* the left hand side of Eq. (3.14), is symmetric on exchange. Furthermore, $(k^{\mu\nu})$ is antisymmetric in μ and ν.

In order to obtain more information about the matrices h^μ and $k^{\mu\nu}$, we choose as a basis for the 4×4-matrices the following set of γ-matrices:

$$\{\mathbb{1}_{4\times4}, \gamma^\mu, \sigma^4_{\mu\nu}, \gamma^5\gamma^\mu, \gamma^5\}, \tag{3.15}$$

where γ^5 is defined by Eq. (1.162a). From Sec. 1.4 we know that there exists a matrix C such that

$$C\gamma^{\mu\top}C^{-1} = -\gamma^\mu,$$

or equivalently

$$\gamma^\mu C = -(\gamma^\mu C^\top)^\top. \tag{3.16}$$

Furthermore, the matrix C is antisymmetric (*cf.* Eq. (1.191)) and therefore Eq. (3.16) tells us that $(\gamma^\mu C) = (\gamma^\mu C)^\top$, *i.e.* $\gamma^\mu C$ is symmetric. We now establish the following result.

Proposition 3.2:

Provided

$$\gamma^\mu C = (\gamma^\mu C)^\top, \tag{3.17}$$

then

$$\sigma^4_{\mu\nu} C = (\sigma^4_{\mu\nu} C)^\top, \tag{3.18}$$

i.e. $\sigma^4_{\mu\nu} C$ is symmetric.

Proof: We have

$$
\begin{aligned}
(\sigma^4_{\mu\nu} C)^\top &= \frac{i}{4}([\gamma_\mu, \gamma_\nu]C)^\top = \frac{i}{4}(\gamma_\mu\gamma_\nu C - \gamma_\nu\gamma_\mu C)^\top \\
&= \frac{i}{4}(C^\top\gamma_\nu^\top\gamma_\mu^\top - C^\top\gamma_\mu^\top\gamma_\nu^\top) = \frac{i}{4}((\gamma_\nu C)^\top\gamma_\mu^\top - (\gamma_\mu C)^\top\gamma_\nu^\top) \\
&\overset{(3.17)}{=} \frac{i}{4}(\gamma_\nu C\gamma_\mu^\top - \gamma_\mu C\gamma_\nu^\top) = \frac{i}{4}(-\gamma_\nu\gamma_\mu C + \gamma_\mu\gamma_\nu C) \\
&= \frac{i}{4}[\gamma_\mu, \gamma_\nu]C = \sigma^4_{\mu\nu} C.
\end{aligned}
$$

Furthermore, the γ-matrices satisfy:

$$(\gamma^5 C)^\top = -\gamma^5 C, \tag{3.19a}$$

$$(\gamma^5 \gamma^\mu C)^\top = -\gamma^5 \gamma^\mu C. \tag{3.19b}$$

Thus, from the set of basis elements for the 4×4-matrices, Eq. (3.15), we can construct two symmetric matrices, $\gamma^\mu C$ and $\sigma^4_{\mu\nu} C$. In addition, the symmetric 4×4-matrix $\sigma^4_{\mu\nu} C$ is antisymmetric in μ and ν. This is exactly what we need for h^μ and $k^{\mu\nu}$, which appear in the general expansion (3.14). We conclude therefore, that

$$k^\mu = a\gamma^\mu C, \qquad k^{\mu\nu} = b\sigma^{4\mu\nu} C, \tag{3.20}$$

where a and b are constants.

We obtain further restrictions on the h^μ and $k^{\mu\nu}$ — that means the right hand side of Eq. (3.14) — by studying the generalized Jacobi identity for two generators Q_a, Q_b. In the case of $X_\mu \in L_0$ and $X_\nu, X_\rho \in L_1$, the degrees are

$$g(X_\mu) = 0, \quad X_\mu \in \{P_\mu, M_{\rho\sigma}\},$$
$$g(X_\nu) = g(X_\rho) = 1, \qquad X_\nu = Q_a, \qquad X_\rho = Q_b.$$

The generalized Jacobi identity (2.21) becomes:

$$X_\mu \circ (Q_a \circ Q_b)(-1)^{g_\mu g_b} + Q_b \circ (X_\mu \circ Q_a)(-1)^{g_a g_b} + Q_a \circ (Q_b \circ X_\mu)(-1)^{g_a g_\mu}$$
$$= [X_\mu, \{Q_a, Q_b\}] - \{Q_b, [X_\mu, Q_a]\} + \{Q_a, [Q_b, X_\mu]\}$$
$$\overset{!}{=} 0. \tag{3.21}$$

Now we have two possibilities.

(i) $X_\mu = P_\mu$: In this case the generalized Jacobi identity (3.21) becomes

$$\begin{aligned}
0 &\overset{!}{=} [P_\mu, \{Q_a, Q_b\}] - \{Q_b, [P_\mu, Q_a]\} + \{Q_a, [Q_b, P_\mu]\} \\
&= [P_\mu, a(\gamma^\nu C)_{ab} P_\nu + b(\sigma^{4\rho\sigma} C)_{ab} M_{\rho\sigma}] \\
&= a(\gamma^\nu C)_{ab}[P_\mu, P_\nu] + b(\sigma^{4\rho\sigma} C)_{ab}[P_\mu, M_{\rho\sigma}] \\
&= b(\sigma^{4\rho\sigma} C)_{ab}\{i\eta_{\rho\mu} P_\sigma - i\eta_{\sigma\mu} P_\rho\} \\
&= ib\Big((\sigma^{4\rho\sigma}\eta_{\rho\mu} P_\sigma - \sigma^{4\rho\sigma}\eta_{\sigma\mu} P_\rho)C\Big)_{ab} \\
&= ib\Big((\sigma^{4\sigma}_\mu P_\sigma - \sigma^{4\rho}_\mu P_\rho)C\Big)_{ab} \\
&= ib\Big((\sigma^4_{\mu\nu} - \sigma^4_{\nu\mu})C\Big)_{ab} P^\nu \\
&= 2ib(\sigma^4_{\mu\nu} C)_{ab} P^\nu.
\end{aligned}$$

In the last step we used the antisymmetry of $\sigma^4_{\mu\nu}$ in μ and ν. Hence, in order that the Jacobi identity (3.21) be satisfied the coefficient b must be zero. This result has an important consequence: The anticommutator of two Q's is a linear combination of the momentum operators P_μ. Hence, the requirement that the Jacobi identity for P_μ and two operators Q be satisfied leads to the relation

$$\boxed{\{Q_a, Q_b\} = a(\gamma^\mu C)_{ab} P_\mu.}$$
(3.22)

We now consider the other case.

(ii) $X_\mu = M_{\mu\nu}$: Inserting the generator of Lorentz transformations $M_{\mu\nu}$ into Eq. (3.21) we obtain:

$$
\begin{aligned}
0 &\overset{!}{=} \big[M_{\mu\nu}, \{Q_a, Q_b\}\big] - \{Q_b, [M_{\mu\nu}, Q_a]\} + \{Q_a, [Q_b, M_{\mu\nu}]\}\\
&= \Big[M_{\mu\nu}, a(\gamma^\rho C)_{ab} P_\rho\Big] - \Big\{Q_b, (-\sigma^4_{\mu\nu})_{ac} Q_c\Big\} + \Big\{Q_a, (\sigma^4_{\mu\nu})_{bc} Q_c\Big\}\\
&= a(\gamma^\rho C)_{ab}[M_{\mu\nu}, P_\rho] + (\sigma^4_{\mu\nu})_{ac}\{Q_b, Q_c\} + (\sigma^4_{\mu\nu})_{bc}\{Q_a, Q_c\}\\
&= a(\gamma^\rho C)_{ab} i(\eta_{\nu\rho}P_\mu - \eta_{\mu\rho}P_\nu) + (\sigma^4_{\mu\nu})_{ac} a(\gamma^\rho C)_{bc} P_\rho\\
&\quad + (\sigma^4_{\mu\nu})_{bc} a(\gamma^\rho C)_{ac} P_\rho.
\end{aligned}
$$

Since $a \neq 0$, we can write

$$-i(\gamma^\rho C)_{ab}(\eta_{\nu\rho}P_\mu - \eta_{\mu\rho}P_\nu) = \Big[(\sigma^4_{\mu\nu}\gamma^\rho C)_{ab} + (\sigma^4_{\mu\nu}\gamma^\rho C)_{ba}\Big]P_\rho,$$
(3.23)

using Eq. (3.17). Now, since

$$
\begin{aligned}
\frac{1}{2}\{\sigma^4_{\mu\nu}, \gamma_\rho\}C + \frac{1}{2}[\sigma^4_{\mu\nu}, \gamma_\rho]C &= \frac{1}{2}\sigma^4_{\mu\nu}\gamma_\rho C + \frac{1}{2}\gamma_\rho\sigma^4_{\mu\nu}C\\
&\quad + \frac{1}{2}\sigma^4_{\mu\nu}\gamma_\rho C - \frac{1}{2}\gamma_\rho\sigma^4_{\mu\nu}C\\
&= \sigma^4_{\mu\nu}\gamma_\rho C,
\end{aligned}
$$

we can write

$$\sigma^4_{\mu\nu}\gamma_\rho C = \frac{1}{2}\{\sigma^4_{\mu\nu}, \gamma_\rho\}C + \frac{1}{2}[\sigma^4_{\mu\nu}, \gamma_\rho]C.$$
(3.24)

The anticommutator and the commutator in Eq. (3.24) can be rewritten as:

$$\{\sigma^4_{\mu\nu}, \gamma_\rho\} = -i\epsilon_{\mu\nu\rho\sigma}\gamma^5\gamma^\sigma,$$
(3.25)

$$[\sigma^4_{\mu\nu}, \gamma_\rho] = i(\eta_{\nu\rho}\gamma_\mu - \eta_{\mu\rho}\gamma_\nu).$$
(3.26)

(see Eq. (3.8)). Then

$$-i\left(\gamma^{\rho}C\right)_{ab}\left(\eta_{\nu\rho}P_{\mu}-\eta_{\mu\rho}P_{\nu}\right)$$

$$= \left[\left(\sigma^{4}_{\mu\nu}\gamma^{\rho}C\right)_{ab}+\left(\sigma^{4}_{\mu\nu}\gamma^{\rho}C\right)_{ba}\right]P_{\rho}$$

$$\overset{(3.24)}{=} \left[\frac{1}{2}\left(\{\sigma^{4}_{\mu\nu},\gamma_{\rho}\}C\right)_{ab}+\frac{1}{2}\left(\left[\sigma^{4}_{\mu\nu},\gamma_{\rho}\right]C\right)_{ab}\right.$$

$$\left.+\frac{1}{2}\left(\{\sigma^{4}_{\mu\nu},\gamma_{\rho}\}C\right)_{ba}+\frac{1}{2}\left(\left[\sigma^{4}_{\mu\nu},\gamma_{\rho}\right]C\right)_{ba}\right]P^{\rho}$$

$$\overset{(3.25)}{\underset{(3.26)}{=}} \left[-\frac{i}{2}\epsilon_{\mu\nu\rho\sigma}\left(\gamma^{5}\gamma^{\sigma}\right)_{ab}+\frac{i}{2}\eta_{\nu\rho}\left(\gamma_{\mu}C\right)_{ab}\right.$$

$$-\frac{i}{2}\eta_{\mu\rho}\left(\gamma_{\nu}C\right)_{ab}-\frac{i}{2}\epsilon_{\mu\nu\rho\sigma}\left(\gamma^{5}\gamma^{\sigma}\right)_{ba}$$

$$\left.+\frac{i}{2}\eta_{\nu\rho}\left(\gamma_{\mu}C\right)_{ba}-\frac{i}{2}\eta_{\mu\rho}\left(\gamma_{\nu}C\right)_{ba}\right]P^{\rho}$$

$$\overset{(3.17)}{\underset{(3.19b)}{=}} \left[i\eta_{\nu\rho}\left(\gamma_{\mu}C\right)_{ab}-i\eta_{\mu\rho}\left(\gamma_{\nu}C\right)_{ab}\right]P^{\rho}$$

$$= i\left(\gamma_{\mu}C\right)_{ab}P_{\nu}-i\left(\gamma_{\nu}C\right)_{ab}P_{\mu}.$$

As can be seen from Eq. (3.23), the left hand side is

$$-i\left(\gamma^{\rho}C\right)_{ab}\left(\eta_{\nu\rho}P_{\mu}-\eta_{\mu\rho}P_{\nu}\right)=i\left(\gamma_{\mu}C\right)_{ab}P_{\nu}-i\left(\gamma_{\nu}C\right)_{ab}P_{\mu}.$$

Thus, we see that the right hand side of Eq. (3.23) cancels exactly the left hand side. Hence the generalized Jacobi relation for the generators $M_{\mu\nu}$, Q_{a} and Q_{b} is satisfied.

Next we demonstrate that the Jacobi identity for three supercharges Q is satisfied. For three Q's, the relation (2.21) becomes:

$$Q_{a}\circ\left(Q_{b}\circ Q_{c}\right)(-1)^{g_{a}g_{c}}+Q_{c}\circ\left(Q_{a}\circ Q_{b}\right)(-1)^{g_{c}g_{b}}+Q_{b}\circ\left(Q_{c}\circ Q_{a}\right)(-1)^{g_{b}g_{a}}=0.$$

Using Eqs. (3.13) and (3.22), the left hand side can be written

$$\left[Q_{a},\{Q_{b},Q_{c}\}\right]+\left[Q_{c},\{Q_{a},Q_{b}\}\right]+\left[Q_{b},\{Q_{c},Q_{a}\}\right]$$
$$= \left[Q_{a},a\left(\gamma^{\mu}C\right)_{bc}P_{\mu}\right]+\left[Q_{c},a\left(\gamma^{\mu}C\right)_{ab}P_{\mu}\right]+\left[Q_{b},a\left(\gamma^{\mu}C\right)_{ca}P_{\mu}\right]$$
$$= a\left(\gamma^{\mu}C\right)_{bc}\left[Q_{a},P_{\mu}\right]+a\left(\gamma^{\mu}C\right)_{ab}\left[Q_{c},P_{\mu}\right]+a\left(\gamma^{\mu}C\right)_{ca}\left[Q_{b},P_{\mu}\right]$$
$$= 0,$$

using the commutator (3.4). Hence the Jacob identity for three supercharges is satisfied.

We have therefore shown that the chosen grading of the Poincaré algebra is a correct grading which is consistent with all possible Jacobi identities, and we have seen that the product on $L_1 \times L_1$ is given by the anticommutator of the supercharges

$$\{Q_a, Q_b\} = a(\gamma^\mu C)_{ab} P_\mu.$$

The coefficient a which appears in this anticommutator is arbitrary and can be absorbed in the Q's. We can impose one further restriction on the supercharges Q which does not change the structure of the underlying algebra. From the chosen grading and from previous considerations concerning the subspace L_1 of a $\mathbb{Z}_2$ graded Lie algebra, in this case spanned by the Q's, it is clear that we have four complex operators Q_a, $a = 1, 2, 3, 4$, and hence eight independent components. Thus, besides Eq. (3.22) we would also have to specify the products

$$Q_a \circ \overline{Q}_b = \{Q_a, \overline{Q}_b\} \qquad \text{and} \qquad \overline{Q}_a \circ \overline{Q}_b = \{\overline{Q}_a, \overline{Q}_b\}.$$

Here, however, we restrict our considerations to the particular case where the supercharge Q is a *Majorana spinor*. In this case we know from Sec. 1.5 that Q and $\overline{Q}$ are no longer independent.

Proposition 3.3:

If we impose the Majorana condition on the generators Q, *i.e.* condition (1.203), we obtain from Eq. (3.22) the relation

$$\{Q_a, \overline{Q}_b\} = -a(\gamma^\mu)_{ab} P_\mu. \tag{3.27}$$

Corollary:

For P^0 to be positive definite the constant a has to be negative, conventionally chosen to be -2.

Proof: The Majorana condition (1.203) is

$$Q_a = Q_a^c = (C\overline{Q}^\top)_a = C_{ab}(\overline{Q}^\top)_b,$$

where

$$\overline{Q} = Q^\dagger \gamma_0.$$

Multiplying the last equation from the left by the charge conjugation matrix C_{ca} we obtain:

$$C_{ca} Q_a = C_{ca} C_{ab}(\overline{Q}^\top)_b = (C^2)_{cb}(\overline{Q}^\top)_b \overset{(1.191)}{=} -\delta_{cb}(\overline{Q}^\top)_b = -(\overline{Q}^\top)_c,$$

from which we conclude:

$$(CQ)_a = -\overline{Q}_a^\top,$$
$$(CQ)_a^\top = -\overline{Q}_a,$$
$$(Q^\top C^\top)_a = -\overline{Q}_a.$$

Again using Eq. (1.191)

$$(Q^\top C)_a = \overline{Q}_a. \tag{3.28a}$$

For later reference purposes we derive two further relations. From the Majorana condition for the Q's we obtain

$$\frac{\partial}{\partial \overline{Q}_b^\top} Q_a = C_{ab}. \tag{3.28b}$$

Now from Eq. (3.28a):

$$\overline{Q}_b^\top = C_{bc}^\top Q_c.$$

We can verify that these relations agree with each other. Thus:

$$\frac{\partial}{\partial \overline{Q}_b^\top} Q_a = \frac{\partial}{\partial \overline{Q}_b^\top} \left((C^\top)^{-1} C^\top Q \right)_a$$

$$= \frac{\partial}{\partial \overline{Q}_b^\top} (C^\top)_{ac}^{-1} \overline{Q}_c^\top$$

$$= (C^\top)_{ab}^{-1}$$

$$= C_{ab} \qquad (\text{since } C_W^\top = C_W^{-1}),$$

in agreement with Eq. (3.28b). Similarly we obtain

$$\frac{\partial}{\partial Q_b} \overline{Q}_a = \frac{\partial}{\partial Q_b} Q_c^\top C_{ca} = C_{ba}. \tag{3.28c}$$

These relations will be needed at a later stage.

Now returning to our original problem to show the Corollary and starting from Eq. (3.22) we obtain:

$$Q_a Q_b + Q_b Q_a = a \left(\gamma^\mu \right)_{ac} C_{cb} P_\mu.$$

Multiplying by the charge conjugation matrix C:

$$Q_a Q_b C_{bd} + Q_b C_{bd} Q_a = a \left(\gamma^\mu \right)_{ac} C_{cb} C_{bd} P_\mu,$$

so that

$$Q_a (Q^\top C)_d + (Q^\top C)_d Q_a = -a \left(\gamma^\mu \right)_{ad} P_\mu.$$

Using Eq. (3.28a) this becomes:

$$Q_a \overline{Q}_b + \overline{Q}_b Q_a = -a (\gamma^\mu)_{ab} P_\mu,$$

and so

$$\{Q_a, \overline{Q}_b\} = -a (\gamma^\mu)_{ab} P_\mu.$$

This completes the proof of Proposition 3.3. In a similar fashion one can show that

$$\{\overline{Q}_a, \overline{Q}_b\} = -a (C^{-1} \gamma^\mu)_{ab} P_\mu.$$

In order to prove the Corollary we go to the rest frame in which the three-momentum vanishes, i.e. $\mathbf{P} = 0$, so that the anticommutator (3.27) reduces to:

$$\{Q_a, \overline{Q}_b\} = -a (\gamma^0)_{ab} P_0,$$

or

$$\{Q_a, \overline{Q}_b\} (\gamma^0)_{ba} = -4a P_0,$$

since $(\gamma^0)^2 = \mathbb{1}_{4\times4}$. Thus

$$
\begin{aligned}
P_0 &= -\frac{1}{4a} \left(Q_a \gamma^{0\top}_{ab} \overline{Q}_b + \overline{Q}_b \gamma^0_{ba} Q_a \right) \\
&= -\frac{1}{4a} \left(Q_a \gamma^{0\top}_{ab} Q^\dagger_c \gamma^0_{cb} + Q^\dagger_c \gamma^0_{cb} \gamma^0_{ba} Q_a \right) \\
&= -\frac{1}{4a} \left(Q_a Q^\dagger_b + Q^\dagger_b Q_a \right).
\end{aligned}
$$

This expression is positive definite provided a is negative. In the following we choose $a = -2$. Further, if we go to the Majorana representation, the Majorana spinors are real. Then:

$$H \equiv P_0 = \frac{1}{4} \sum_b Q^2_b \geq 0.$$

Hence for any state $|\psi\rangle$ of the Hilbert space over which H is defined

$$\langle \psi | H | \psi \rangle \geq 0,$$

and just like

$$P|\Omega\rangle = 0, \quad P = -i \frac{\partial}{\partial x},$$

for a state which is invariant under translations in x, so a state $|\Omega\rangle$ is supersymmetric if

$$Q_a |\Omega\rangle = 0.$$

Hence, supersymmetry is unbroken if and only if

$$\langle\Omega|H|\Omega\rangle = 0 = \min\ \langle\psi|H|\psi\rangle.$$

It follows that the supersymmetric state $|\Omega\rangle$ is the ground state. Further discussion of these aspects is given in Chapter 9.

3.2 Two-Component Weyl Formulation

Starting with Eq. (3.27) it is possible to express the anticommuting part of the Super–Poincaré algebra in terms of two-component Weyl spinors, taking into account the close connection between four-component Majorana spinors and two-component Weyl spinors. Writing Eq. (3.27) in the Weyl representation and using Eqs. (1.159) and (1.204), we see that Eq. (3.27) becomes a set of four anticommutators with the constant a in Eqs. (3.22) and (3.27) chosen as -2, *i.e.*

$$Q_a\overline{Q}_b + \overline{Q}_b Q_a = 2\gamma^\mu_{ab}P_\mu,$$

so that

$$\begin{pmatrix} Q_A \\ \overline{Q}^{\dot{A}} \end{pmatrix} (Q^B,\ \overline{Q}_{\dot{B}}) + (Q^B,\ \overline{Q}_{\dot{B}}) \begin{pmatrix} Q_A \\ \overline{Q}^{\dot{A}} \end{pmatrix} = 2 \begin{pmatrix} 0 & (\sigma^\mu)_{A\dot{B}} \\ (\overline{\sigma}^\mu)^{\dot{A}B} & 0 \end{pmatrix} P_\mu,$$

explicitly:

$$\begin{pmatrix} Q_A Q^B + Q^B Q_a & Q_A\overline{Q}_{\dot{B}} + \overline{Q}_{\dot{B}}Q_A \\ \overline{Q}^{\dot{A}}Q^B + Q^B\overline{Q}^{\dot{A}} & \overline{Q}^{\dot{A}}\overline{Q}_{\dot{B}} + \overline{Q}_{\dot{B}}\overline{Q}^{\dot{A}} \end{pmatrix} = 2 \begin{pmatrix} 0 & (\sigma^\mu)_{A\dot{B}} \\ (\overline{\sigma}^\mu)^{\dot{A}B} & 0 \end{pmatrix} P_\mu,$$

so that

$$\left.\begin{aligned} \{Q_A, Q^B\} &= 0, \\ \{Q_A, \overline{Q}_{\dot{B}}\} &= 2(\sigma^\mu)_{A\dot{B}}P_\mu, \\ \{\overline{Q}^{\dot{A}}, Q^B\} &= 2(\overline{\sigma}^\mu)^{\dot{A}B}P_\mu, \\ \{\overline{Q}^{\dot{A}}, \overline{Q}_{\dot{B}}\} &= 0. \end{aligned}\right\} \tag{3.29}$$

Furthermore, using

$$\sigma^4_{\mu\nu} = \begin{pmatrix} (\sigma^2_{\mu\nu})_A{}^B & 0 \\ 0 & (\overline{\sigma}^2_{\mu\nu})^{\dot{A}}{}_{\dot{B}} \end{pmatrix},$$

the commutator (3.5) becomes:

$$\left.\begin{aligned} [M_{\mu\nu}, Q_A] &= -(\sigma^2_{\mu\nu})_A{}^B Q_B, \\ [M_{\mu\nu}, \overline{Q}^{\dot{A}}] &= -(\overline{\sigma}^2_{\mu\nu})^{\dot{A}}{}_{\dot{B}}\overline{Q}^{\dot{B}}. \end{aligned}\right\} \tag{3.30}$$

Chapter 4

Representations of the Super–Poincaré Algebra

4.1 Casimir Operators

In order to classify representations of the Super–Poincaré algebra we use a method similar to that described in Chapter 1. We recall first the structure of the Super–Poincaré algebra as developed in the previous chapter.

In summary, the Super–Poincaré algebra is given by the following set of commutators and anticommutators:

$$
\begin{aligned}
&\left[P_\mu, P_\nu\right] = 0, \\
&\left[M_{\mu\nu}, P_\rho\right] = -i\left(\eta_{\mu\rho}P_\nu - \eta_{\nu\rho}P_\mu\right), \\
&\left[M_{\mu\nu}, M_{\rho\sigma}\right] = -i\left(\eta_{\mu\rho}M_{\nu\sigma} - \eta_{\mu\sigma}M_{\nu\rho} - \eta_{\nu\rho}M_{\mu\sigma} + \eta_{\nu\sigma}M_{\mu\rho}\right), \\
&\left[P_\mu, Q_a\right] = 0, \\
&\left[M_{\mu\nu}, Q_a\right] = -\left(\sigma^4_{\mu\nu}\right)_{ab}Q_b, \\
&\left\{Q_a, \overline{Q}_b\right\} = 2\left(\gamma^\mu\right)_{ab}P_\mu, \\
&\left\{Q_a, Q_b\right\} = -2\left(\gamma^\mu C\right)_{ab}P_\mu, \\
&\left\{\overline{Q}_a, \overline{Q}_b\right\} = 2\left(C^{-1}\gamma^\mu\right)_{ab}P_\mu, \quad\quad\quad\quad (4.1)
\end{aligned}
$$

$(a, b = 1, 2, 3, 4)$, where

$$
\sigma^4_{\mu\nu} = \frac{i}{4}\left[\gamma_\mu, \gamma_\nu\right].
$$

The algebra (4.1) satisfies all relevant generalized Jacobi identities as was demonstrated in Chapter 3. The chosen algebra has 14 generators, *i.e.*

- 4 generators of translations P_μ, $\mu = 0, 1, 2, 3$,

- 6 generators of Lorentz transformations $M_{\mu\nu}$,

- 4 spinor charges $Q_a, a = 1, 2, 3, 4$ (Majorana spinors), also called *super charges*.

In view of our discussion at the end of Chapter 3 we recall the following steps in adding an internal symmetry. As explained in Sec. 2.1, the theorem of Haag, Łopuszański and Sohnius allows the presence of an internal symmetry group with generators B_l [54]. According to the theorem of Coleman and Mandula these generators B_l commute with P_μ and $M_{\mu\nu}$ [23]. Furthermore we saw that it is necessary in this case to introduce a set of N spinor charges

$$Q_a^\alpha, \quad \alpha = 1, 2, \ldots, N; \quad a = 1, 2, 3, 4,$$

such that the commutator of Q_a^α and B_l closes into the subspace L_1, *i.e.*

$$[Q_a^\alpha, B_l] = iS_l{}^{\alpha\beta} Q_a^\beta, \tag{4.2}$$

where the matrices S_l have to be a representation of the algebra of the internal symmetry group, *i.e.* of

$$[B_l, B_m] = ic_{lm}{}^k B_k.$$

The coefficients S_l appearing in Eq. (4.2) are $N \times N$-dimensional matrices where N depends on the representation one has chosen. The product of two supercharges Q_a^α is a simple extension of Eq. (3.27), *i.e.*

$$\{Q_a^\alpha, \overline{Q}_b^\beta\} = 2\delta^{\alpha\beta} (\gamma^\mu)_{ab} P_\mu. \tag{4.3}$$

The algebra with $N = 1$ is called the *supersymmetry algebra*, and algebras with $N > 1$ are called *extended supersymmetry algebras*.

As mentioned before, in order to classify representations of the Super–Poincaré algebra (4.1) we use the method of Casimir operators as in Chapter 1. The basic assumption is that the operators $M_{\mu\nu}, P_\mu$ and Q_a can be realized as linear operators acting on an infinite dimensional Hilbert space of state vectors. In order to classify representations, we construct Casimir operators whose eigenvalues determine the representation.[1]

From the basic commutation and anticommutation relations of the Super-Poincaré algebra (4.1) we conclude that $P^2 = P_\mu P^\mu$ is again an invariant

---

[1]The complete classification of supersymmetries in more that $1+1$ dimensions and their representations is evaluated in W. Nahm [76].

operator. It is important to note that this fact, *i.e.* that P^2 is a Casimir operator of the Super–Poincaré algebra (4.1), depends crucially on the grading chosen. For example, if we take L_1 five-dimensional — *i.e.* five generators Q — and take as five-dimensional representation of L_0 the matrix representation

$$g(a, \Lambda) = \begin{pmatrix} \Lambda & a \\ 0 & 1 \end{pmatrix}, \qquad \Lambda \in L_+^\uparrow, \ a \in \mathbb{T}_4,$$

then the commutator of P and Q will be nonzero. In this case P^2 will not be an invariant operator.

Proposition 4.1:

The square of the Pauli–Ljubanski vector

$$W^\mu = \frac{1}{2} \epsilon^{\mu\nu\rho\sigma} M_{\nu\rho} P_\sigma \tag{4.4}$$

is not a Casimir operator of the Super–Poincaré algebra (4.1).

Proof: We saw in Chapter 1 that $W^2 = W_\mu W^\mu$ commutes with all generators of the Poincaré algebra, which span the L_0 space of the Super–Poincaré algebra. Therefore the only commutator of interest is

$$[W^2, Q_a].$$

First we derive the commutation relation of the Pauli-Ljubanski vector W_μ with the supercharge Q_a, *i.e.*

$$
\begin{aligned}
[W_\mu, Q_a] &= \frac{1}{2} [\epsilon_{\mu\nu\rho\sigma} M^{\nu\rho} P^\sigma, Q_a] \\
&= \frac{1}{2} \epsilon_{\mu\nu\rho\sigma} \left\{ M^{\nu\rho} [P^\sigma, Q_a] + [M^{\nu\rho}, Q_a] P^\sigma \right\} \\
&\overset{(4.1)}{=} \frac{1}{2} \epsilon_{\mu\nu\rho\sigma} [M^{\nu\rho}, Q_a] P^\sigma \\
&\overset{(4.1)}{=} -\frac{1}{2} \epsilon_{\mu\nu\rho\sigma} (\sigma^{4\nu\rho})_{ab} Q_b P^\sigma \\
&= \frac{1}{2} \epsilon_{\sigma\mu\nu\rho} (\sigma^{4\nu\rho})_{ab} Q_b P^\sigma \\
&= -\frac{1}{2} \epsilon_{\mu\sigma\nu\rho} (\sigma^{4\nu\rho})_{ab} Q_b P^\sigma.
\end{aligned}
$$

Using[2]

$$\epsilon_{\mu\nu\rho\sigma} \sigma^{4\rho\sigma} = -2i\gamma^5 \sigma_{\mu\nu}^4, \tag{4.5}$$

---

[2]This relation can be verified by using $\sigma_2^{\mu\nu}, \overline{\sigma}_2^{\mu\nu}$ of Eqs. (1.140a) and (1.140b) and the expression for $\sigma^{4\mu\nu}$ in the proof of Eq. (1.207e).

we obtain

$$[W_\mu, Q_a] = i(\gamma^5 \sigma^4_{\mu\sigma} Q)_a P^\sigma = i(\sigma^4_{\mu\sigma} \gamma^5 Q)_a P^\sigma,$$

since

$$[\gamma^5, \sigma^4_{\mu\nu}] = 0.$$

Now

$$\gamma_\sigma \gamma_\mu = \frac{1}{2}\{\gamma_\sigma, \gamma_\mu\} + \frac{1}{2}[\gamma_\sigma, \gamma_\mu] = \eta_{\sigma\mu} - 2i\sigma^4_{\sigma\mu},$$

so that

$$\sigma^4_{\mu\sigma} = -\frac{i}{2}(\gamma_\sigma \gamma_\mu - \eta_{\mu\sigma}),$$

taking into account the antisymmetry of $\sigma^4_{\mu\nu}$. Hence

$$\begin{aligned}
[W_\mu, Q_a] &= i(\sigma^4_{\mu\sigma} \gamma^5 Q)_a P^\sigma \\
&= \frac{1}{2}[(\gamma_\sigma \gamma_\mu - \eta_{\mu\sigma})\gamma^5 Q]_a P^\sigma \\
&= \frac{1}{2}(\gamma_\sigma \gamma_\mu \gamma^5 Q)_a P^\sigma - \frac{1}{2}\eta_{\mu\sigma}(\gamma^5 Q)_a P^\sigma \\
&= \frac{1}{2}(\slashed{P}\gamma_\mu \gamma^5 Q)_a - \frac{1}{2}P_\mu (\gamma^5 Q)_a,
\end{aligned} \qquad (4.6)$$

since

$$[P_\mu, Q_a] = 0,$$

and where we defined the Feynman–slash:

$$\slashed{P} := P_\mu \gamma^\mu. \qquad (4.7)$$

We can now calculate the commutator

$$\begin{aligned}
[W^2, Q_a] &= [W_\mu W^\mu, Q_a] \\
&= W^\mu [W_\mu, Q_a] + [W_\mu, Q_a] W^\mu \\
&\overset{(4.6)}{=} W^\mu \left\{ -\frac{1}{2}P_\mu(\gamma^5 Q)_a + \frac{1}{2}(\slashed{P}\gamma_\mu \gamma^5 Q)_a \right\} \\
&\quad + \left\{ -\frac{1}{2}P_\mu(\gamma^5 Q)_a + \frac{1}{2}(\slashed{P}\gamma_\mu \gamma^5 Q)_a \right\} W^\mu \\
&= -\frac{1}{2}W^\mu P_\mu(\gamma^5 Q)_a + \frac{1}{2}W^\mu(\slashed{P}\gamma_\mu \gamma^5 Q)_a \\
&\quad - \frac{1}{2}P_\mu(\gamma^5 Q)_a W^\mu + \frac{1}{2}(\slashed{P}\gamma_\mu \gamma^5 Q)_a W^\mu.
\end{aligned}$$

Due to Proposition 1.10, the $W_\mu P^\mu$-terms vanish. Hence

$$
\begin{aligned}
[W^2, Q_a] &= \frac{1}{2} W^\mu \big(\not{P} \gamma_\mu \gamma^5 Q \big)_a + \frac{1}{2} \big(\not{P} \gamma_\mu \gamma^5 Q \big)_a W^\mu \\
&= \frac{1}{2} W^\mu \big(\not{P} \gamma_\mu \gamma^5 Q \big)_a + \frac{1}{4} \epsilon^{\mu\nu\rho\sigma} \big(\not{P} \gamma_\mu \gamma^5 Q \big)_a M_{\nu\rho} P_\sigma \\
&= \frac{1}{2} W^\mu \big(\not{P} \gamma_\mu \gamma^5 Q \big)_a + \frac{1}{4} \epsilon^{\mu\nu\rho\sigma} \big(\not{P} \gamma_\mu \gamma^5 \big)_{ab} Q_b M_{\nu\rho} P_\sigma \\
&\overset{(4.1)}{=} \frac{1}{2} W^\mu \big(\not{P} \gamma_\mu \gamma^5 Q \big)_a + \frac{1}{4} \epsilon^{\mu\nu\rho\sigma} \big(\not{P} \gamma_\mu \gamma^5 \big)_{ab} \big(M_{\nu\rho} Q_b + (\sigma^4_{\nu\rho})_{bc} Q_c \big) P_\sigma \\
&= \frac{1}{2} W^\mu \big(\not{P} \gamma_\mu \gamma^5 Q \big)_a + \frac{1}{4} \epsilon^{\mu\nu\rho\sigma} \big(P_\lambda \gamma^\lambda \gamma_\mu \gamma^5 \big)_{ab} M_{\nu\rho} Q_b P_\sigma \\
&\quad + \frac{1}{4} \epsilon^{\mu\nu\rho\sigma} \big(P_\lambda \gamma^\lambda \gamma_\mu \gamma^5 \big)_{ab} (\sigma^4_{\nu\rho})_{bc} Q_c P_\sigma \\
&= \frac{1}{2} W^\mu \big(\not{P} \gamma_\mu \gamma^5 Q \big)_a + \frac{1}{4} \epsilon^{\mu\nu\rho\sigma} P_\lambda M_{\nu\rho} P_\sigma \big(\gamma^\lambda \gamma_\mu \gamma^5 Q \big)_a \\
&\quad + \frac{1}{4} \epsilon^{\mu\nu\rho\sigma} P_\lambda \big(\gamma^\lambda \gamma_\mu \gamma^5 \sigma^4_{\nu\rho} Q \big)_a P_\sigma \\
&= \frac{1}{2} W^\mu \big(\not{P} \gamma_\mu \gamma^5 Q \big)_a \\
&\quad + \frac{1}{4} \epsilon^{\mu\nu\rho\sigma} \big(M_{\nu\rho} P_\lambda + i \eta_{\nu\lambda} P_\rho - i \eta_{\rho\lambda} P_\nu \big) P_\sigma \big(\gamma^\lambda \gamma_\mu \gamma^5 Q \big)_a \\
&\quad + \frac{1}{4} \epsilon^{\mu\nu\rho\sigma} P^\lambda \big(\gamma_\lambda \gamma_\mu \gamma^5 \sigma^4_{\nu\rho} Q \big)_a P_\sigma \\
&= \frac{1}{2} W^\mu \big(\not{P} \gamma_\mu \gamma^5 Q \big)_a + \frac{1}{4} \epsilon^{\mu\nu\rho\sigma} M_{\nu\rho} P_\sigma P_\lambda \big(\gamma^\lambda \gamma_\mu \gamma^5 Q \big)_a \\
&\quad + \frac{i}{4} \epsilon^{\mu\nu\rho\sigma} \eta_{\nu\lambda} P_\rho P_\sigma \big(\gamma^\lambda \gamma_\mu \gamma^5 Q \big)_a \\
&\quad - \frac{i}{4} \epsilon^{\mu\nu\rho\sigma} \eta_{\rho\lambda} P_\nu P_\sigma \big(\gamma^\lambda \gamma_\mu \gamma^5 Q \big)_a \\
&\quad + \frac{1}{4} \epsilon^{\mu\nu\rho\sigma} P^\lambda \big(\gamma_\lambda \gamma_\mu \gamma^5 \sigma^4_{\nu\rho} Q \big)_a P_\sigma \\
&= \frac{1}{2} W^\mu \big(\not{P} \gamma_\mu \gamma^5 Q \big)_a + \frac{1}{2} W^\mu \big(\not{P} \gamma_\mu \gamma^5 Q \big)_a \\
&\quad + \frac{1}{4} \epsilon^{\mu\nu\rho\sigma} P^\lambda \big(\gamma_\lambda \gamma_\mu \gamma^5 \sigma^4_{\nu\rho} Q \big)_a P_\sigma \\
&= W^\mu \big(\not{P} \gamma_\mu \gamma^5 Q \big)_a + \frac{1}{4} P^\lambda \big(\gamma_\lambda \gamma_\mu \gamma^5 \epsilon^{\mu\nu\rho\sigma} \sigma^4_{\nu\rho} Q \big)_a P_\sigma \\
&= W^\mu \big(\not{P} \gamma_\mu \gamma^5 Q \big)_a - \frac{i}{2} P^\lambda \big(\gamma_\lambda \gamma_\mu (\gamma^5)^2 \sigma^{4\mu\sigma} Q \big)_a P_\sigma \\
&= W^\mu \big(\not{P} \gamma_\mu \gamma^5 Q \big)_a - \frac{i}{2} \big(\not{P} \gamma_\mu \sigma^{4\mu\sigma} Q \big)_a P_\sigma .
\end{aligned}
$$

Now using the following contraction of γ-matrices:

$$\gamma_\mu \sigma^{4\mu\sigma} = \frac{i}{4}\gamma_\mu\left[\gamma^\mu, \gamma^\sigma\right] = \frac{i}{4}\gamma_\mu\left(\gamma^\mu\gamma^\sigma - \gamma^\sigma\gamma^\mu\right)$$

$$= \frac{i}{4}\gamma_\mu\left(\gamma^\mu\gamma^\sigma - 2\eta^{\mu\sigma} + \gamma^\mu\gamma^\sigma\right) = \frac{i}{4}\left(\gamma_\mu\gamma^\mu\gamma^\sigma - \gamma^\sigma\right)$$

$$= \frac{i}{4}\left(4\gamma^\sigma - \gamma^\sigma\right) = \frac{3i}{4}\gamma^\sigma,$$

we find

$$\left[W^2, Q_a\right] = W^\mu\left(\not{P}\gamma_\mu\gamma^5 Q\right)_a + \frac{3}{4}\left(\not{P}\gamma^\sigma Q\right)_a P_\sigma = W^\mu\left(\not{P}\gamma_\mu\gamma^5 Q\right)_a + \frac{3}{4}\left(\not{P}\not{P}Q\right)_a,$$

or

$$\left[W^2, Q_a\right] = W^\mu\left(\not{P}\gamma_\mu\gamma^5 Q\right)_a + \frac{3}{4}P^2 Q_a. \tag{4.8}$$

Thus, the commutator of the square of the Pauli–Ljubanski polarization vector W^2 and the supercharge Q_a does not vanish. This completes the proof of Proposition 4.1.

Our next step is the explicit construction of an invariant operator. To this end, we first define the following pseudovector

$$X_\mu := \frac{1}{2}\overline{Q}\gamma_\mu\gamma^5 Q. \tag{4.9}$$

Proposition 4.2:

The pseudovector (4.9) and the supercharge Q_a satisfy:

$$\left[X_\mu, Q_a\right] = -2\left(\not{P}\gamma_\mu\gamma^5 Q\right)_a. \tag{4.10}$$

Proof: A straightforward calculation using Eq. (4.9) gives:

$$\begin{aligned}
\left[X_\mu, Q_a\right] &= \frac{1}{2}\left[\overline{Q}\gamma_\mu\gamma^5 Q, Q_a\right] \\
&= \frac{1}{2}\left(\overline{Q}\gamma_\mu\gamma^5\right)_b Q_b Q_a - \frac{1}{2}Q_a\left(\overline{Q}\gamma_\mu\gamma^5 Q\right) \\
&\overset{(3.22)}{=} -\frac{1}{2}\left(\overline{Q}\gamma_\mu\gamma^5\right)_b Q_a Q_b - \left(\overline{Q}\gamma_\mu\gamma^5\right)_b\left(\gamma_\lambda C\right)_{ba}P^\lambda - \frac{1}{2}Q_a\left(\overline{Q}\gamma_\mu\gamma^5 Q\right) \\
&= -\frac{1}{2}\overline{Q}_b Q_a\left(\gamma_\mu\gamma^5 Q\right)_b - \left(\overline{Q}\gamma_\mu\gamma^5\gamma_\lambda C\right)_a P^\lambda - \frac{1}{2}Q_a\overline{Q}_b\left(\gamma_\mu\gamma^5 Q\right)_b \\
&\overset{(3.27)}{=} -\not{P}_{ab}\left(\gamma_\mu\gamma^5 Q\right)_b - \left(\overline{Q}\gamma_\mu\gamma^5\gamma_\lambda C\right)_a P^\lambda \\
&= -\left(\not{P}\gamma_\mu\gamma^5 Q\right)_a - \overline{Q}_c\left(\gamma_\mu\gamma^5\gamma_\lambda C\right)_{ca}P^\lambda \\
&\overset{(3.28a)}{=} -\left(\not{P}\gamma_\mu\gamma^5 Q\right)_a - Q_b^\top C_{bc}\left(\gamma_\mu\gamma^5\gamma_\lambda C\right)_{ca}P^\lambda \\
&= -\left(\not{P}\gamma_\mu\gamma^5 Q\right)_a - \left(Q^\top C\gamma_\mu\gamma^5\gamma_\lambda C\right)_a P^\lambda.
\end{aligned}$$

In order to simplify the last term, we use the following γ-matrix identities:

$$C\gamma_\mu = -\gamma_\mu^\top C, \qquad C\gamma^5 = \gamma^{5\top} C.$$

Then, since $C^2 = -\mathbb{1}_{4\times4}$:

$$
\begin{aligned}
\left[X_\mu, Q_a\right] &= -\left(\slashed{P}\gamma_\mu\gamma^5 Q\right)_a - \left(Q^\top\gamma_\mu^\top\gamma^{5\top}\gamma_\lambda^\top C^2\right)_a P^\lambda \\
&= -\left(\slashed{P}\gamma_\mu\gamma^5 Q\right)_a + \left(Q^\top\gamma_\mu^\top\gamma^{5\top}\gamma_\lambda^\top\right)_a P^\lambda \qquad \text{since } C^2 = -\mathbb{1}_{4\times4} \\
&= -\left(\slashed{P}\gamma_\mu\gamma^5 Q\right)_a + \left(P_\lambda\gamma^\lambda\gamma^5\gamma_\mu Q\right)_a^\top \\
&= -\left(\slashed{P}\gamma_\mu\gamma^5 Q\right)_a - \left(\slashed{P}\gamma_\mu\gamma^5 Q\right)_a \\
&= -2\left(\slashed{P}\gamma_\mu\gamma^5 5 Q\right)_a.
\end{aligned}
$$

This completes the proof of Proposition 4.2.

We define a new vector B_μ by

$$B_\mu := W_\mu + \frac{1}{4}X_\mu, \tag{4.11}$$

where W_μ is the Pauli–Ljubanski polarization vector and X_μ is given by Eq. (4.9). Then:

$$
\begin{aligned}
\left[B_\mu, Q_a\right] &= \left[W_\mu + \frac{1}{4}X_\mu, Q_a\right] = \left[W_\mu, Q_a\right] + \frac{1}{4}\left[X_\mu, Q_a\right] \\
&= \frac{1}{2}\left(\slashed{P}\gamma_\mu\gamma^5 Q\right)_a - \frac{1}{2}P_\mu\left(\gamma^5 Q\right)_a - \frac{1}{2}\left(\slashed{P}\gamma_\mu\gamma^5 Q\right)_a,
\end{aligned}
$$

so that

$$\left[B_\mu, Q_a\right] = -\frac{1}{2}P_\mu\left(\gamma^5 Q\right)_a. \tag{4.12}$$

We now define the following second rank tensor

$$C_{\mu\nu} := B_\mu P_\nu - B_\nu P_\mu. \tag{4.13}$$

Then the following result can be shown to hold.

Proposition 4.3:

The second rank tensor, defined in Eq. (4.13), commutes with the supercharge Q_a, i.e.

$$\left[C_{\mu\nu}, Q_a\right] = 0. \tag{4.14}$$

Proof: We consider the commutator on the left hand side, *i.e.*

$$
\begin{aligned}
[C_{\mu\nu}, Q_a] &= [B_\mu P_\nu - B_\nu P_\mu, Q_a] = [B_\mu P_\nu, Q_a] - [B_\nu P_\mu, Q_a] \\
&= B_\mu [P_\nu, Q_a] + [B_\mu, Q_a] P_\nu - B_\nu [P_\mu, Q_a] - [B_\nu, Q_a] P_\mu \\
&\overset{(4.1)}{=} [B_\mu, Q_a] P_\nu - [B_\nu, Q_a] P_\mu \\
&\overset{(4.12)}{=} -\frac{1}{2} P_\mu (\gamma^5 Q)_a P_\nu + \frac{1}{2} P_\nu (\gamma^5 Q)_a P_\mu \\
&\overset{(4.1)}{=} -\frac{1}{2} (\gamma^5 Q_a)_a P_\mu P_\nu + \frac{1}{2} (\gamma^5 Q)_a P_\nu P_\mu \\
&= -\frac{1}{2} (\gamma^5 Q_a)_a [P_\mu, P_\nu] = 0.
\end{aligned}
$$

From the tensor $C_{\mu\nu}$ we can construct the Casimir operator

$$C^2 = C_{\mu\nu} C^{\mu\nu}. \tag{4.15}$$

Proposition 4.4:

The operator C^2 commutes with all generators of the Super–Poincaré group, *i.e.*

$$[C^2, Q_a] = 0, \tag{4.16}$$
$$[C^2, P_\mu] = 0, \tag{4.17}$$
$$[C^2, M_{\mu\nu}] = 0. \tag{4.18}$$

The three equations (4.16) to (4.18) establish that C^2, defined by Eq. (4.15), is a Casimir operator of the Super–Poincaré algebra.

Proof: The proof of Eq. (4.16) is trivial, since

$$[C^2, Q_a] = [C_{\mu\nu} C^{\mu\nu}, Q_a] = C_{\mu\nu} [C^{\mu\nu}, Q_a] + [C_{\mu\nu}, Q_a] C^{\mu\nu} = 0$$

due to Eq. (4.14).

Next consider the commutator of C^2 with the energy-momentum vector:

$$
\begin{aligned}
[C^2, P_\rho] &= C_{\mu\nu} [C^{\mu\nu}, P_\rho] + [C_{\mu\nu}, P_\rho] C^{\mu\nu} \\
&= C_{\mu\nu} [B^\mu P^\nu - B^\nu P^\mu, P_\rho] + [B_\mu P_\nu - B_\nu P_\mu, P_\rho] C^{\mu\nu}
\end{aligned}
$$

$$= C_{\mu\nu}\Big\{ B^\mu \big[P^\nu, P_\rho\big] + \big[B^\mu, P_\rho\big]P^\nu - B^\nu\big[P^\mu, P_\rho\big] - \big[B^\nu, P_\rho\big]P^\mu \Big\}$$
$$+ \Big\{ B_\mu \big[P_\nu, P_\rho\big] + \big[B_\mu, P_\rho\big]P_\nu - B_\nu\big[P_\mu, P_\rho\big] - \big[B_\nu, P_\rho\big]P_\mu \Big\}C^{\mu\nu}$$
$$= C_{\mu\nu}\Big\{ \big[B^\mu, P_\rho\big]P^\nu - \big[B^\nu, P_\rho\big]P^\mu \Big\}$$
$$+ \Big\{ \big[B_\mu, P_\rho\big]P_\nu - \big[B_\nu, P_\rho\big]P_\mu \Big\}C^{\mu\nu}$$
$$= C_{\mu\nu}\Big\{ \big[W^\mu + \tfrac{1}{4}X^\mu, P_\rho\big]P^\nu - \big[W^\nu + \tfrac{1}{4}X^\nu, P_\rho\big]P^\mu \Big\}$$
$$+ \Big\{ \big[W_\mu + \tfrac{1}{4}X_\mu, P_\rho\big]P_\nu - \big[W_\nu + \tfrac{1}{4}X_\nu, P_\rho\big]P_\mu \Big\}C^{\mu\nu}$$
$$= C_{\mu\nu}\Big\{ \big[W^\mu, P_\rho\big]P^\nu + \tfrac{1}{4}\big[X^\mu, P_\rho\big]P^\nu$$
$$- \big[W^\nu, P_\rho\big]P^\mu - \tfrac{1}{4}\big[X^\nu, P_\rho\big]P^\mu \Big\}$$
$$+ \Big\{ \big[W_\mu, P_\rho\big]P_\nu + \tfrac{1}{4}\big[X_\mu, P_\rho\big]P_\nu$$
$$- \big[W_\nu, P_\rho\big]P_\mu - \tfrac{1}{4}\big[X_\nu, P_\rho\big]P_\mu \Big\}C^{\mu\nu}$$
$$= \tfrac{1}{4}C_{\mu\nu}\Big\{ \big[X^\mu, P_\rho\big]P^\nu - \big[X^\nu, P_\rho\big]P^\mu \Big\}$$
$$+ \tfrac{1}{4}\Big\{ \big[X_\mu, P_\rho\big]P_\nu - \big[X_\nu, P_\rho\big]P_\mu \Big\}C^{\mu\nu} = 0,$$

where in the next to last step we use Eq. (1.38a). The commutator of the X_μ and the P_μ is zero, because

$$[X_\mu, P_\rho] \overset{(4.9)}{=} \frac{1}{2}\Big[\overline{Q}\gamma_\mu\gamma^5 Q, P_\rho\Big]$$
$$= \frac{1}{2}\big(\overline{Q}\gamma_\mu\gamma^5\big)_a [Q_a, P_\mu] + \frac{1}{2}\big[\overline{Q}_a, P_\rho\big]\big(\gamma_\mu\gamma^5 Q\big)_a \overset{(4.1)}{=} 0.$$

Finally, we observe that the operator $C^2 = C_{\mu\nu}C^{\mu\nu}$ commutes with the generators $M_{\rho\sigma}$, because by construction it is a scalar under Lorentz transformations. This completes the proof of Proposition 4.4.

Hence we have shown that there are two Casimir operators of the Super–Poincaré algebra: P^2 and C^2.

In the following we use the eigenvalues of the Casimir operators to label the irreducible representations — and thus particle multiplets — of the Super–Poincaré algebra. The calculations will be performed in the rest frame. States boosted to an arbitrary moment P_μ are then defined by the particle's wave equation.

4.2   Classification of Irreducible Representations

4.2.1 $N = 1$ Supersymmetry

The irreducible representations of the Super–Poinaré algebra are character-ized by the eigenvalues of the operators P^2 and C^2. Let $P^2 = m^2 > 0$, and for simplicity we choose the rest frame

$$P_\mu = (m, \mathbf{0}).$$

Then in the rest frame the Casimir operator C^2 is

$$
\begin{aligned}
C^2 = C_{\mu\nu}C^{\mu\nu} &= \big(B_\mu P_\nu - B_\nu P_\mu\big)\big(B^\mu P^\nu - B^\nu P^\mu\big) \\
&= 2B_\mu P_\nu B^\mu P^\nu - 2B_\mu P_\nu B^\nu P^\mu \\
&= 2B_\mu B^\mu m^2 - 2B_0^2 m^2 \\
&= 2m^2 B_k B^k,
\end{aligned}
\tag{4.19}
$$

where $k = 1, 2, 3$. Now from Eq. (4.11) with Eq. (1.45) we obtain:

$$B_k = W_k + \frac{1}{4}X_k = mS_k + \frac{1}{8}\overline{Q}\gamma_k\gamma^5 Q =: mJ_k. \tag{4.20}$$

The operator J_k is defined by this relation. Subsequently we will use the Weyl formalism. It is therefore desirable to know the definition of the operator J_k in terms of Weyl operators. For this purpose we use Eqs. (1.159) and (1.162c) in rewriting Eq. (4.20). Then:

$$
\begin{aligned}
\overline{Q}\gamma_k\gamma^5 Q &= (Q^A, \ \overline{Q}_{\dot A}) \begin{pmatrix} 0 & (\sigma_k)_{A\dot B} \\ (\overline{\sigma}_k)^{\dot A B} & 0 \end{pmatrix} \begin{pmatrix} -\delta_B{}^C & 0 \\ 0 & \delta^{\dot B}{}_{\dot C} \end{pmatrix} \begin{pmatrix} Q_C \\ \overline{Q}{}^{\dot C} \end{pmatrix} \\
&= \left(\overline{Q}_{\dot A}(\overline{\sigma}_k)^{\dot A B}, \ Q^A(\sigma_k)_{A\dot B}\right) \begin{pmatrix} -\delta_B{}^C Q_C \\ \delta^{\dot B}{}_{\dot C}\overline{Q}{}^{\dot C} \end{pmatrix} \\
&= -\overline{Q}_{\dot A}(\overline{\sigma}_k)^{\dot A B} Q_B + Q^A(\sigma_k)_{A\dot B}\overline{Q}{}^{\dot B} \\
&= -(\overline{Q}\overline{\sigma}_k Q_B) + (Q\sigma_k\overline{Q}) \\
&= 2(Q\sigma_k\overline{Q}) = -2(\overline{Q}\overline{\sigma}_k Q_B),
\end{aligned}
$$

where we used Eq. (1.134). Hence, Eq. (4.20) can be rewritten

$$mJ_k = mS_k - \frac{1}{4}\big(\overline{Q}\overline{\sigma}_k Q\big). \tag{4.21}$$

We can now rewrite C^2 in terms of J_k. Then Eq. (4.19) becomes

$$C^2 = 2m^4 J_k J^k, \tag{4.22}$$

and this implies that J_k is an angular momentum, *i.e.*

$$\left[J_k, J_l\right] = i\epsilon_{klm} J_m. \tag{4.23}$$

One can check that this relation holds for J_k defined by Eq. (4.20). Furthermore

$$\left[J_k, Q_a\right] \overset{(4.20)}{=} \frac{1}{m}\left[B_k, Q_a\right] \overset{(4.12)}{=} -\frac{1}{2m}P_k\left(\gamma^5 Q\right)_a = 0, \tag{4.24a}$$

since in the rest frame we have $P_k = 0$. Correspondingly

$$\left[J_k, Q_A\right] = 0, \quad \left[J_k, \overline{Q}_{\dot{A}}\right] = 0 \tag{4.24b}$$

by decomposing the Majorana spinor Q_a into its Weyl components.

Now according to Eq. (4.22) $J^2 = J_k J^k$ is an invariant operator with eigenvalues of the form $j(j+1)$ as in the case of ordinary angular momentum; here j is either integral or half-integral.[3] The irreducible representations of the Super–Poincaré algebra are specified by the values of m^2 and $j(j+1)$, the eigenvalues of the Casimir operators. The states of the representation can then be given the labels m^2, j and j_3, the latter being the eigenvalue of J_3 which assumes the values $-j, -j+1, \ldots, j-1, j$. However, these states are not, in general, eigenstates of $\mathbf{S}^2$ and S_3 of ordinary spin $\mathbf{S}$. In order to find the spin content it is convenient to work in terms of the two-component Weyl formalism with commutation relations (3.29) and (3.30). We have:

$$\left[M_{\mu\nu}, Q_A\right] = -\left(\sigma_{\mu\nu}^2\right)_A{}^B Q_B,$$

so that

$$\left[M_{ij}, Q_A\right] = -\left(\sigma_{ij}^2\right)_A{}^B Q_B.$$

In the rest frame the spin operator is given by

$$S_k = \frac{1}{2}\epsilon_{kij} M_{ij}.$$

Hence the commutator of the two operators S_k and Q_a is

$$\left[S_k, Q_A\right] = -\frac{1}{2}\epsilon_{kij}\left(\sigma_{ij}^2\right)_A{}^B Q_B = -\frac{1}{2}\left(\sigma_k \overline{\sigma}^0\right)_A{}^B Q_B, \tag{4.25}$$

where we used Eq. (1.139a), and the unit matrix $\overline{\sigma}^0$ has been inserted to provide the correct index structure for σ_k in agreement with Eq. (1.105). Similarly

$$\left[S_k, \overline{Q}^{\dot{A}}\right] = -\frac{1}{2}\left(\overline{\sigma}_k \sigma^0\right)^{\dot{A}}{}_{\dot{B}} \overline{Q}^{\dot{B}}. \tag{4.26}$$

---

[3]This means, there exist appropriate representations of J_k which satisfy for integral or half-integral spin the relation (4.23).

From Eq. (3.29) we obtain for the anticommutator of two supercharges in the rest frame:

$$\{Q_A, \overline{Q}_{\dot{B}}\} = 2(\sigma^0)_{A\dot{B}} P_0 = 2m(\sigma^0)_{A\dot{B}}.$$

Hence

$$
\left.
\begin{aligned}
\{Q_1, \overline{Q}_{\dot{1}}\} &= \{Q_2, \overline{Q}_{\dot{2}}\} &= 2m, \\
\{Q_1, \overline{Q}_{\dot{2}}\} &= \{Q_2, \overline{Q}_{\dot{1}}\} &= 0, \\
\{Q_A, Q_B\} &= \{\overline{Q}_{\dot{A}}, \overline{Q}_{\dot{B}}\} &= 0.
\end{aligned}
\right\}
\qquad (4.27)
$$

These relations define a *Clifford algebra*[4] for operators which we call *creation* ($\overline{Q}$) and *annihilation* (Q) *operators*.

We now consider irreducible spinor representations of the Super–Poincaré algebra characterized by m and j, and consider the states with fixed value of j_3. Among these states is a state $|\Omega\rangle$ which is annihilated by the Q_A, $A = 1, 2$, *i.e.*

$$Q_A|\Omega\rangle = 0. \qquad (4.28)$$

This state, called the *Clifford vacuum*, is the vacuum state of spinor representations and can be either bosonic or fermionic. This state must not be confused with the quantum mechanical ground state or vacuum defined as the state of lowest energy. Instead the Clifford vacuum is an eigenstate of the Casimir operator $P_\mu P^\mu$, *i.e.*

$$P_\mu P^\mu |\Omega\rangle = m^2 |\Omega\rangle. \qquad (4.29)$$

Thus the Clifford vacuum is that particular state in the irreducible representation of the Super–Poincaré algebra specified by the numbers m, j and j_3 which is such that Eq. (4.28) holds. Since by construction P^2 and C^2 are the only Casimir operators in the present case, the Clifford vacuum $|\Omega\rangle$ is nondegenerate (ignoring internal symmetries).

We now consider another state in the chosen representation, which we write $|\beta\rangle$, with fixed value of j_3. Then (see below)

$$|\Omega\rangle = Q_1 Q_2 |\beta\rangle,$$

and

$$Q_1|\Omega\rangle = Q_1 Q_1 Q_2 |\beta\rangle = 0,$$

since Q_1 anticommutes with itself, and similarly

$$Q_2|\Omega\rangle = Q_2 Q_1 Q_2 |\beta\rangle = 0.$$

---

[4]See *e.g.* I.M. Benn and R.W. Tucker [9], Chap. 2.

Furthermore, the Clifford vacuum $|\Omega\rangle$ will have the same eigenvalue of J_3 as $|\beta\rangle$ since

$$
\begin{aligned}
J_3|\Omega\rangle &= J_3 Q_1 Q_2 |\beta\rangle \stackrel{(4.24a)}{=} Q_1 J_3 Q_2 |\beta\rangle \stackrel{(4.24a)}{=} Q_1 Q_2 J_3 |\beta\rangle \\
&= j_3 Q_1 Q_2 |\beta\rangle = j_3|\Omega\rangle.
\end{aligned}
$$

This shows that we must have a Clifford vacuum in any representation of the Super–Poincaré algebra. From Eq. (4.21) we obtain

$$
J_k = S_k - \frac{1}{4m}(\overline{Q}\overline{\sigma}_k Q),
$$

which implies that the Clifford vacuum is an eigenstate of the operators $\mathbf{S}^2$ and S_3 of ordinary spin angular momentum with eigenvalues $s(s+1)$ and s_3. Hence we write

$$
|\Omega\rangle = |m, s, s_3\rangle. \tag{4.30}
$$

We consider the action of the creation operators $\overline{Q}^{\dot{A}}$ on the Clifford vacuum, *i.e.* we consider the states

$$
\overline{Q}^{\dot{1}}|m, s, s_3\rangle \qquad \text{and} \qquad \overline{Q}^{\dot{2}}|m, s, s_3\rangle. \tag{4.31}
$$

Proposition 4.5:

The states (4.31) are eigenstates of the J_3 operator, *i.e.*

$$
J_3 \overline{Q}^{\dot{A}}|m, s, s_3\rangle = j_3 \overline{Q}^{\dot{A}}|m, s, s_3\rangle. \tag{4.32}
$$

Proof: Using Eqs. (4.24b) and (4.30) we have:

$$
\begin{aligned}
J_3 \overline{Q}^{\dot{A}}|m, s, s_3\rangle &= \overline{Q}^{\dot{A}} J_3 |m, s, s_3\rangle = \overline{Q}^{\dot{A}} J_3 |\Omega\rangle \\
&= \overline{Q}^{\dot{A}} j_3 |\Omega\rangle = j_3 \overline{Q}^{\dot{A}} |\Omega\rangle = j_3 \overline{Q}^{\dot{A}}|m, s, s_3\rangle.
\end{aligned}
$$

Equation (4.32) shows that the states (4.31) are states in the representation of the Super–Poincaré algebra with eigenvalues j, m of the Casimir operators and eigenvalue j_3 of the operator J_3. In the following we use the identities:

$$
(\overline{\sigma}_3 \sigma^0)^{\dot{1}}{}_{\dot{1}} = -1, \qquad (\overline{\sigma}_3 \sigma^0)^{\dot{2}}{}_{\dot{2}} = 1.
$$

Now, using Eq. (4.26) we have:

$$
\begin{aligned}
S_3 \overline{Q}^{\dot{1}}|\Omega\rangle &= \overline{Q}^{\dot{1}}S_3|\Omega\rangle - \frac{1}{2}\left(\overline{\sigma}_3\sigma^0\right)^{\dot{1}}{}_{\dot{B}}\overline{Q}^{\dot{B}}|\Omega\rangle \\
&\stackrel{(4.21)}{=} \overline{Q}^{\dot{1}}J_3|\Omega\rangle - \frac{1}{2}\left(\overline{\sigma}_3\sigma^0\right)^{\dot{1}}{}_{\dot{B}}\overline{Q}^{\dot{B}}|\Omega\rangle \\
&= \overline{Q}^{\dot{1}}J_3|\Omega\rangle - \frac{1}{2}\left(\overline{\sigma}_3\sigma^0\right)^{\dot{1}}{}_{\dot{1}}\overline{Q}^{\dot{1}}|\Omega\rangle \\
&= \overline{Q}^{\dot{1}}j_3|\Omega\rangle + \frac{1}{2}\overline{Q}^{\dot{1}}|\Omega\rangle \\
&= \left(j_3 + \frac{1}{2}\right)\overline{Q}^{\dot{1}}|\Omega\rangle .
\end{aligned}
\tag{4.33}
$$

In the same way one obtains for the other creation operator:

$$
\begin{aligned}
S_3 \overline{Q}^{\dot{2}}|\Omega\rangle &= \overline{Q}^{\dot{2}}S_3|\Omega\rangle - \frac{1}{2}\left(\overline{\sigma}_3\sigma^0\right)^{\dot{2}}{}_{\dot{B}}\overline{Q}^{\dot{B}}|\Omega\rangle \\
&\stackrel{(4.21)}{=} \overline{Q}^{\dot{2}}J_3|\Omega\rangle - \frac{1}{2}\left(\overline{\sigma}_3\sigma^0\right)^{\dot{2}}{}_{\dot{B}}\overline{Q}^{\dot{B}}|\Omega\rangle \\
&= \overline{Q}^{\dot{2}}J_3|\Omega\rangle - \frac{1}{2}\left(\overline{\sigma}_3\sigma^0\right)^{\dot{2}}{}_{\dot{2}}\overline{Q}^{\dot{2}}|\Omega\rangle \\
&= \overline{Q}^{\dot{2}}j_3|\Omega\rangle - \frac{1}{2}\overline{Q}^{\dot{2}}|\Omega\rangle \\
&= \left(j_3 - \frac{1}{2}\right)\overline{Q}^{\dot{2}}|\Omega\rangle .
\end{aligned}
\tag{4.34}
$$

Equations (4.33) and (4.34) imply that the operator $\overline{Q}^{\dot{1}}$ raises the value of j_3 by $1/2$ and the operator $\overline{Q}^{\dot{2}}$ lowers it by the same amount. Applying $\overline{Q}^{\dot{1}}\overline{Q}^{\dot{2}}$ to the Clifford vacuum we obtain yet another state:

$$
\begin{aligned}
S_3 \overline{Q}^{\dot{1}}\overline{Q}^{\dot{2}}|\Omega\rangle &\stackrel{(4.26)}{=} \overline{Q}^{\dot{1}}S_3\overline{Q}^{\dot{2}}|\Omega\rangle - \frac{1}{2}\left(\overline{\sigma}_3\sigma^0\right)^{\dot{1}}{}_{\dot{B}}\overline{Q}^{\dot{B}}\overline{Q}^{\dot{2}}|\Omega\rangle \\
&= \overline{Q}^{\dot{1}}S_3\overline{Q}^{\dot{2}}|\Omega\rangle + \frac{1}{2}\overline{Q}^{\dot{1}}\overline{Q}^{\dot{2}}|\Omega\rangle \\
&= \overline{Q}^{\dot{1}}\overline{Q}^{\dot{2}}S_3|\Omega\rangle + \frac{1}{2}\overline{Q}^{\dot{1}}\overline{Q}^{\dot{2}}|\Omega\rangle - \frac{1}{2}\overline{Q}^{\dot{1}}\left(\overline{\sigma}_3\sigma^0\right)^{\dot{2}}{}_{\dot{B}}\overline{Q}^{\dot{B}}|\Omega\rangle \\
&= \overline{Q}^{\dot{1}}\overline{Q}^{\dot{2}}J_3|\Omega\rangle + \frac{1}{2}\overline{Q}^{\dot{1}}\overline{Q}^{\dot{2}}|\Omega\rangle - \frac{1}{2}\overline{Q}^{\dot{1}}\overline{Q}^{\dot{2}}|\Omega\rangle \\
&= j_3\overline{Q}^{\dot{1}}\overline{Q}^{\dot{2}}|\Omega\rangle .
\end{aligned}
\tag{4.35}
$$

This result shows that $\overline{Q}^{\dot{1}}\overline{Q}^{\dot{2}}|\Omega\rangle$ is a state whose eigenvalue is j_3. Applying $\overline{Q}^{\dot{A}}$ once more gives zero and so ends the construction of eigenstates of S_3.

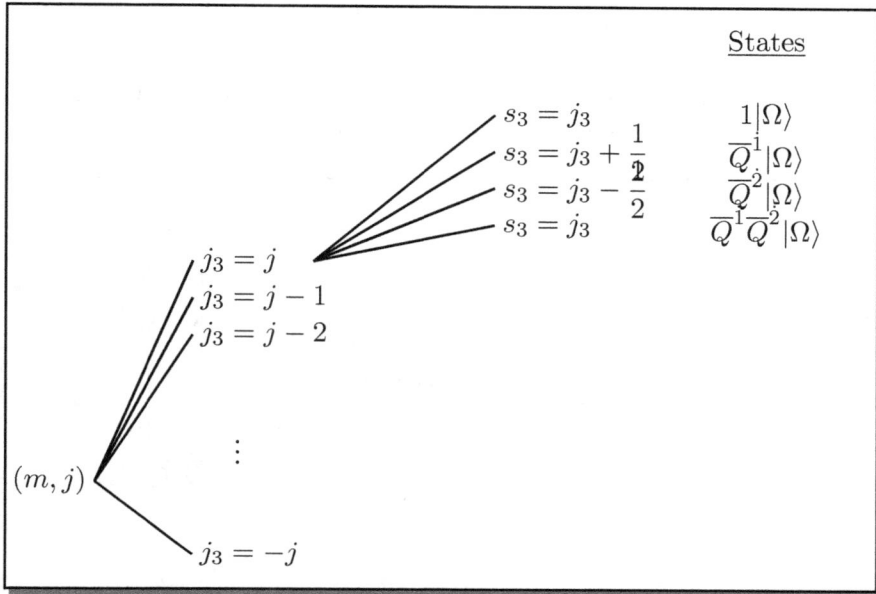

Figure 4.1: Representation states of the Super–Poincaré algebra.

We now summarize our findings. For each pair of values (m, j) of the Casimir operators we obtain an irreducible representation of the Super–Poincaré algebra (4.1). The states corresponding to specific values of m and j contain $2j+1$ subspaces according to the possible values of j_3. For fixed j_3 each subspace contains four eigenstates of S_3 with eigenvalues $s_3 = j_3, j_3+1/2, j_3-1/2$ and again j_3, corresponding to the operators

$$1, \overline{Q}^{\dot{1}}, \overline{Q}^{\dot{2}} \text{ and } \overline{Q}^{\dot{1}}\overline{Q}^{\dot{2}}$$

respectively. These states span an irreducible $4(2j + 1)$-dimensional representation of the little algebra given by Eq. (4.27). This scenario is illustrated in Fig. 4.1. From Eq. (3.29) we deduce that any product of two Q's has negative parity. Thus if $|\Omega\rangle$ is a scalar state, $\overline{Q}\,\overline{Q}|\Omega\rangle$ is pseudoscalar[5] and *vice versa*.

We consider two examples. The lowest-dimensional representation is the representation with $j = 0$ (in this case, the Clifford vacuum $|\Omega\rangle$ is a bosonic state). For this representation there are only three spins: $s = 0, s = 1/2$, and again $s = 0$, corresponding to a scalar particle, a spin-1/2 particle (with $s_3 = \pm 1/2$) and a pseudoscalar particle.[6] We shall see in Chapter 5 that this

---

[5] A pseudoscalar quantity changes sign under the parity transformation.

[6] This is an exception to the general rule that the number of fermions equals the number

case is realized in the *Wess–Zumino model.*

As a second example we consider $j = 1/2$ for fixed m. In this case the Clifford vacuum $|\Omega\rangle$ is a fermionic state. In this example we have a set of states which is given schematically in Fig. 4.2.

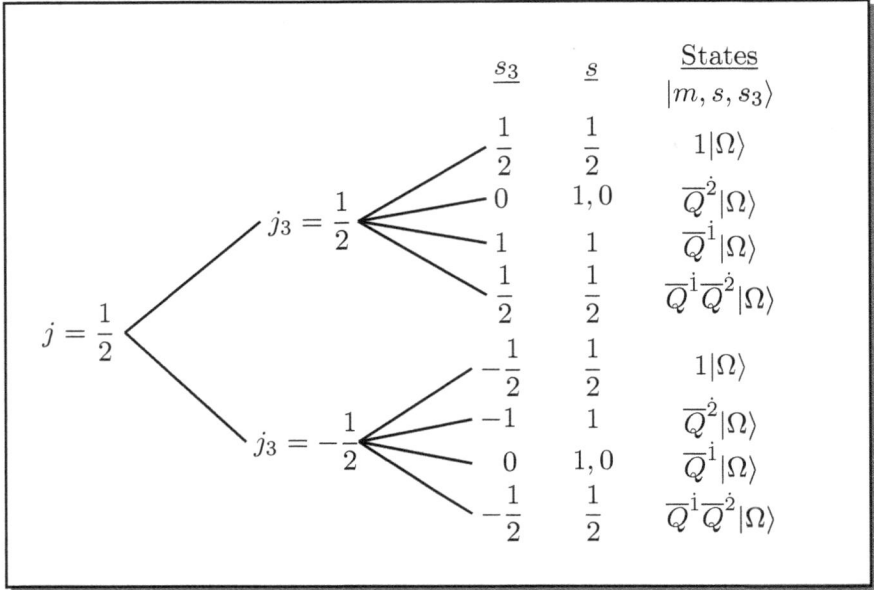

Figure 4.2: Representation of the Super–Poincaré algebra for $j = 1/2$.

The particles described by these states are

- a pseudoscalar particle, represented by the state $\overline{Q}|\Omega\rangle$,

- a vector particle, represented by the states $\overline{Q}^{\dot{1}}|\Omega\rangle, \overline{Q}^{\dot{2}}|\Omega\rangle$,

- two spin-1/2 particles, one is described by the Clifford vacuum $1|\Omega\rangle$, the other by the state $\overline{Q}^{\dot{1}}\overline{Q}^{\dot{2}}|\Omega\rangle$.

Of course, the pseudoscalar state and the associated vector state arise from the Kronecker product of two Weyl spinors, *i.e.*

$$(0, 1/2) \otimes (0, 1/2) = (0, 0) \oplus (0, 1),$$

in much the same way as in Eq. (1.152b). Thus a fermionic state $|\Omega\rangle$ carries effectively an index such as $\dot{A}$ and then transforms like $\overline{Q}^{\dot{A}}$. Thus

of bosons in a supersymmetric theory; however, the number of bosonic states equals the number of fermionic states.

$\overline{Q}^{\dot{2}}|\Omega\rangle \sim \overline{Q}^{\dot{2}}\overline{Q}^{\dot{1}}|\ \rangle$, where $|\ \rangle$ transforms as a bosonic state. Clearly there is only one such pseudoscalar state.

The above scheme reveals an important feature of supersymmetric theories: An irreducible representation of the Super–Poincaré algebra has the number of fermionic states equal to the number of bosonic states and thus the number of bosons equal to the number of fermions.[7] Before we proceed to representations of the $N > 1$ Super–Poincaré algebra, we prove the following result.

Proposition 4.6:

> Every representation of the supersymmetry algebra (4.1) contains as many bosonic states as fermionic states.

Proof: We first introduce a *fermion number operator* N_F. This operator has the property that $(-1)^{N_F}$ has eigenvalue $+1$ when applied to a bosonic state (*i.e.* states containing an even number of fermions n_F). Thus

$$(-1)^{N_F}|\ \rangle = (-1)^{n_F}|\ \rangle$$

when the state $|\ \rangle$ contains n_F fermions. Since Q_A and $\overline{Q}^{\dot{A}}$ annihilate or create fermions, we have

$$Q_A(-1)^{N_F}|\ \rangle = (-1)^{N_F-1}Q_A|\ \rangle,$$

i.e.

$$(-1)^{N_F}Q_A = -Q_A(-1)^{N_F}.$$

For any finite-dimensional representation of the Super–Poincaré algebra (4.1) which is such that the trace operation is well-defined, we have:

$$\mathrm{Tr}\left[(-1)^{N_F}\{Q_A,\overline{Q}_{\dot{B}}\}\right] = \mathrm{Tr}\left[(-1)^{N_F}(Q_A\overline{Q}_{\dot{B}} + \overline{Q}_{\dot{B}}Q_A)\right]$$
$$= \mathrm{Tr}\left[(-1)^{N_F}Q_A\overline{Q}_{\dot{B}}\right] + \mathrm{Tr}\left[Q_A(-1)^{N_F}\overline{Q}_{\dot{B}}\right]$$
$$= 0.$$

This is a result of the above relation. On the other hand, using Eq. (3.29), we obtain:

$$0 = \mathrm{Tr}\left[(-1)^{N_F}\{Q_A,\overline{Q}_{\dot{B}}\}\right] = \mathrm{Tr}\left[(-1)^{N_F}2(\sigma^\mu)_{A\dot{B}}P_\mu\right]$$
$$= 2(\sigma^\mu)_{A\dot{B}}P_\mu\mathrm{Tr}\left[(-1)^{N_F}\right].$$

---

[7]With the exception mentioned above.

For fixed nonzero momentum P_μ this reduces to

$$\text{Tr}\left[(-1)^{N_F}\right] = 0,$$

i.e.

$$\sum_B \langle B|(-1)^{N_F}|B\rangle + \sum_F \langle F|(-1)^{N_F}|F\rangle = N_B - N_F = 0.$$

This demonstrates that supersymmetric representations contain equal numbers of bosonic and fermionic states.

Remark: The factor $(-1)^{N_F}$ is in fact the operator used by E. Witten [131] to discuss qualitatively features of supersymmetry breaking. The trace of this operator is a particular example of the much more general concept of the *index of an operator*. Considerations of this type lead to a proof of the Atiyah–Singer Index Theorem in the context of supersymmetric quantum mechanics. For details we refer to D. Friedan and P. Windey [45], L. Alvarez–Gaumé [3] and E. Witten [131].

4.2.2 $N > 1$ **Supersymmetry**

We remarked at the beginning of this chapter that if we want to include an internal symmetry in the Super–Poincaré algebra in a nontrivial way, we have to extend the algebra (4.1) by introducing a set of N spinor charges $Q_A^\alpha, \overline{Q}_{\dot{A}}^\alpha$ where the index α runs from 1 to N. The number N is the dimension of a representation of the internal symmetry group. For the Super–Poincaré algebra we have in the rest frame according to Eq. (3.29):

$$\{Q_A^\alpha, Q_B^\beta\} = \{\overline{Q}_{\dot{A}}^\alpha, \overline{Q}_{\dot{B}}^\beta\} = 0, \tag{4.36}$$

$$\{Q_A^\alpha, \overline{Q}_{\dot{B}}^\beta\} = 2m(\sigma^0)_{A\dot{B}}\delta^{\alpha\beta}. \tag{4.37}$$

Here the indices α and β run from 1 to N. Rescaling the supercharges Q by writing

$$a_A^\alpha := \frac{1}{\sqrt{2m}}Q_A^\alpha, \tag{4.38a}$$

$$(a_A^\alpha)^\dagger := \frac{1}{\sqrt{2m}}\overline{Q}_{\dot{A}}^\alpha, \tag{4.38b}$$

we obtain:

$$\left.\begin{aligned}
\{a_A^\alpha, a_B^\beta\} &= 0, \\
\{(a_A^\alpha)^\dagger, (a_B^\beta)^\dagger\} &= 0, \\
\{a_A^\alpha, (a_B^\beta)^\dagger\} &= \delta_{AB}\delta^{\alpha\beta}.
\end{aligned}\right\} \tag{4.39}$$

This is the algebra of $2N$ fermionic annihilation and creation operators a_A^α and $\left(a_A^\alpha\right)^\dagger$ respectively. The representations of the algebra (4.39) can be constructed with the help of a Clifford vacuum $|\Omega\rangle$ defined by

$$a_A^\alpha|\Omega\rangle = 0, \quad \alpha = 1, 2, \ldots, N, \quad A = 1, 2. \tag{4.40}$$

As before, this Clifford vacuum state is an eigenstate of the Casimir operator P^2 with eigenvalue m^2, *i.e.*

$$P^2|\Omega\rangle = m^2|\Omega\rangle. \tag{4.41}$$

Furthermore, the Clifford vacuum $|\Omega\rangle$ transforms according to an irreducible representation of the little group[8] $SU(2, \mathbb{C})$ of the momentum operator in the rest frame. In other words, if the Clifford vacuum $|\Omega\rangle$ has spin s, it belongs to a $2s + 1$-dimensional representation of $SU(2, \mathbb{C})$.

The states of the representation of the algebra (4.39) are constructed by applying creation operators $\left(a_A^\alpha\right)^\dagger$ to $|\Omega\rangle$. Then an n-particle state is given by

$$\Phi^{(n)\alpha_1\ldots\alpha_n}_{A_1\ldots A_n} = \frac{1}{\sqrt{n!}}\left(a_{A_1}^{\alpha_1}\right)^\dagger \cdots \left(a_{A_n}^{\alpha_n}\right)^\dagger|\Omega\rangle. \tag{4.42}$$

Now, from the anticommutators (4.39) we find for the present case:

$$\left\{\left(a_{A_i}^{\alpha_i}\right)^\dagger, \left(a_{A_j}^{\alpha_j}\right)^\dagger\right\} = 0, \tag{4.43}$$

so that

$$\Phi^{(n)\alpha_1\ldots\alpha_i\ldots\alpha_j\ldots\alpha_n}_{A_1\ldots A_i\ldots A_j\ldots A_n} = -\Phi^{(n)\alpha_1\ldots\alpha_j\ldots\alpha_i\ldots\alpha_n}_{A_1\ldots A_j\ldots A_i\ldots A_n}.$$

Thus, the state $\Phi^{(n)}$ is antisymmetric under exchange of two pairs of indices (α_i, A_i) and (α_j, A_j). Here, each index pair takes $2N$ different values, since the spinor index is $A = 1, 2$, and the index $\alpha_i = 1, \ldots, N$. Thus n must be less than or equal to $2N$.

As a first example we consider the case $N = 1$, the case treated explicitly before. According to the above, the only states which can be constructed are:

$$\Phi^{(0)} = |\Omega\rangle,$$
$$\Phi^{(1)} \sim \overline{Q}_{\dot{A}}|\Omega\rangle,$$
$$\Phi^{(2)} \sim \overline{Q}_{\dot{A}}\overline{Q}_{\dot{B}}|\Omega\rangle.$$

These are the four states which span the subspace for a given value of j_3 as in Sec. 4.2.1.

---

[8]See footnote 12, Chap. 1.

In view of the anticommutator (4.43) we can construct $\binom{2N}{n}$ different states for any given value of n. In the example of $N = 1$ this means for $n = 0$ we have $\binom{2}{0} = 1$ state, i.e. $\Phi^{(0)} = |\Omega\rangle$, for $n = 1$ we have $\binom{2}{1} = 2$ different states, namely $\overline{Q}_{\dot{1}}|\Omega\rangle$ and $\overline{Q}_{\dot{2}}|\Omega\rangle$; finally for $n = 2$ there is only one possible state $\Phi^{(2)}$.

Summing over all possible values of n gives the dimension d of the representation, i.e.

$$d = \sum_{d=0}^{2N} \binom{2N}{n} = (1+1)^{2N} = 2^{2N}. \tag{4.44}$$

The case $N = 1$ has the dimension $d = 2^2 = 4$, corresponding to the four states which span the subspace for a given value of j_3.

If the Clifford vacuum $|\Omega\rangle$ is not degenerate, Eq. (4.42) is called the *fundamental irreducible massive multiplet*. This multiplet — with n assuming values from 0 to $2N$ — has dimension 2^{2N} with 2^{2N-1} bosonic and the same number of fermionic states. In the case $N = 1$ the fundamental multiplet is given by

$$\Phi^{(0)} = |\Omega\rangle,$$
$$\Phi^{(1)}{}_A = (a_A)^\dagger |\Omega\rangle,$$
$$\Phi^{(2)}{}_{AB} = \frac{1}{2^{1/2}} (a_A)^\dagger (a_B)^\dagger |\Omega\rangle$$
$$= -\frac{1}{2^{3/2}} \epsilon_{AB} (a_C)^\dagger (a^C)^\dagger |\Omega\rangle,$$

(with Eq. (1.100d)) and thus yields as many bosonic as fermionic states as was demonstrated in Sec. 4.2.1.

In the following we consider in detail the case of $N = 2$ supersymmetry. This means we take spinor charges $Q_A^\alpha, \overline{Q}_A^\alpha$ in the fundamental representation of the internal symmetry group $SU(2, \mathbb{C})$ which describes, for instance, isospin. According to Eq. (4.44) the dimension of the representation of the algebra (4.39) is $d = 2^4 = 16$. Furthermore the fundamental multiplet is given by a 16-plet of states.

- For $n = 0$ there is exactly one state, given by

$$\Phi^{(0)} = |\Omega\rangle.$$

- For $n = 1$ there are $\binom{2N}{1} = \binom{4}{1} = 4$ states

$$\Phi^{(1)}{}^\alpha{}_A = (a_A^\alpha)^\dagger |\Omega\rangle,$$

or explicitly:

$$\Phi^{(1)\,1}_{\quad\ 1} = \left(a_1^1\right)^\dagger |\Omega\rangle,$$

$$\Phi^{(1)\,2}_{\quad\ 1} = \left(a_1^2\right)^\dagger |\Omega\rangle,$$

$$\Phi^{(1)\,1}_{\quad\ 2} = \left(a_2^1\right)^\dagger |\Omega\rangle,$$

$$\Phi^{(1)\,2}_{\quad\ 2} = \left(a_2^2\right)^\dagger |\Omega\rangle.$$

- For $n = 2$ there are $\binom{2N}{2} = \binom{4}{2} = 6$ possible states. These are given by:

$$\Phi^{(2)\,11}_{\quad\ 11} = \frac{1}{2^{1/2}}\left(a_1^1\right)^\dagger\left(a_1^1\right)^\dagger |\Omega\rangle = 0,$$

$$\Phi^{(2)\,21}_{\quad\ 11} = \frac{1}{2^{1/2}}\left(a_1^2\right)^\dagger\left(a_1^1\right)^\dagger |\Omega\rangle = -\Phi^{(2)\,12}_{\quad\ 11},$$

$$\Phi^{(2)\,22}_{\quad\ 11} = \frac{1}{2^{1/2}}\left(a_1^2\right)^\dagger\left(a_1^2\right)^\dagger |\Omega\rangle = 0,$$

$$\Phi^{(2)\,11}_{\quad\ 21} = \frac{1}{2^{1/2}}\left(a_2^1\right)^\dagger\left(a_1^1\right)^\dagger |\Omega\rangle = -\Phi^{(2)\,11}_{\quad\ 12},$$

$$\Phi^{(2)\,11}_{\quad\ 22} = \frac{1}{2^{1/2}}\left(a_2^1\right)^\dagger\left(a_2^1\right)^\dagger |\Omega\rangle = 0,$$

$$\Phi^{(2)\,12}_{\quad\ 22} = \frac{1}{2^{1/2}}\left(a_2^1\right)^\dagger\left(a_2^2\right)^\dagger |\Omega\rangle = -\Phi^{(2)\,21}_{\quad\ 22},$$

$$\Phi^{(2)\,22}_{\quad\ 22} = \frac{1}{2^{1/2}}\left(a_2^2\right)^\dagger\left(a_2^2\right)^\dagger |\Omega\rangle = 0,$$

$$\Phi^{(2)\,21}_{\quad\ 21} = \frac{1}{2^{1/2}}\left(a_2^2\right)^\dagger\left(a_1^1\right)^\dagger |\Omega\rangle = -\Phi^{(2)\,12}_{\quad\ 12},$$

$$\Phi^{(2)\,21}_{\quad\ 12} = \frac{1}{2^{1/2}}\left(a_1^2\right)^\dagger\left(a_2^1\right)^\dagger |\Omega\rangle = -\Phi^{(2)\,12}_{\quad\ 21},$$

$$\Phi^{(2)\,22}_{\quad\ 12} = \frac{1}{2^{1/2}}\left(a_1^2\right)^\dagger\left(a_2^2\right)^\dagger |\Omega\rangle = -\Phi^{(2)\,21}_{\quad\ 22}.$$

The vanishing of various terms is due to the anticommutator (4.43) which is for $\Phi^{(2)\,22}_{\quad\ 22}$:

$$\left\{\left(a_2^2\right)^\dagger, \left(a_2^2\right)^\dagger\right\} = 2\left(a_2^2\right)^\dagger\left(a_2^2\right)^\dagger = 0,$$

and therefore

$$\left(a_2^2\right)^\dagger\left(a_2^2\right)^\dagger |\Omega\rangle = 0.$$

- For $n = 3$ there are $\binom{2N}{3} = \binom{4}{3} = 4$ possible states; the nonvanishing ones are:

$$\Phi^{(3)\,211}_{\quad\,121} = \frac{1}{(3!)^{1/2}} \left(a_1^2\right)^\dagger \left(a_2^1\right)^\dagger \left(a_1^1\right)^\dagger |\Omega\rangle,$$

$$\Phi^{(3)\,211}_{\quad\,221} = \frac{1}{(3!)^{1/2}} \left(a_2^2\right)^\dagger \left(a_2^1\right)^\dagger \left(a_1^1\right)^\dagger |\Omega\rangle,$$

$$\Phi^{(3)\,221}_{\quad\,121} = \frac{1}{(3!)^{1/2}} \left(a_1^2\right)^\dagger \left(a_2^2\right)^\dagger \left(a_1^1\right)^\dagger |\Omega\rangle,$$

$$\Phi^{(3)\,122}_{\quad\,221} = \frac{1}{(3!)^{1/2}} \left(a_2^1\right)^\dagger \left(a_2^2\right)^\dagger \left(a_1^2\right)^\dagger |\Omega\rangle.$$

- Finally, for $n = 4$, we have one nonvanishing state given by

$$\Phi^{(4)\,2211}_{\quad\,2121} = \frac{1}{(4!)^{1/2}} \left(a_2^2\right)^\dagger \left(a_1^2\right)^\dagger \left(a_2^1\right)^\dagger \left(a_1^1\right)^\dagger |\Omega\rangle.$$

Chapter 5

The Wess–Zumino Model

5.1 The Lagrangian and the Equations of Motion

We now construct a field theoretical realization of the Super–Poincaré algebra (4.1), first considered by J. Wess and B. Zumino [124]. In this application we choose the representation corresponding to nonvanishing mass $(m^2 \neq 0)$ and $j = 0$. According to the considerations of Sec. 4.2.1 this model describes two spin-0 fields A, B, one scalar and the other pseudoscalar, and one spin-1/2 Majorana field ψ. According to the theory of representations of the Super–Poincaré algebra, all three fields belong to the same mass multiplet. The case to be discussed here is the so-called *on-shell* case, in which the algebra of generators is treated on the basis of the equations of motion of the fields. The *off-shell* case does not require the validity of the equations of motion; we shall see later that this necessitates the introduction of further fields, called *auxiliary fields*, as additional degrees of freedom.

Our starting point is the following Lagrangian density

$$
\mathcal{L} = \frac{1}{2}(\partial_\mu A)(\partial^\mu A) - \frac{1}{2}m^2 A^2 + \frac{1}{2}(\partial_\mu B)(\partial^\mu B) - \frac{1}{2}m^2 B^2
$$
$$
+ \frac{1}{2}\overline{\psi}(i\slashed{\partial} - m)\psi - mgA(A^2 + B^2)
$$
$$
- g(\overline{\psi}\psi A + i\overline{\psi}\gamma^5\psi B) - \frac{1}{2}g^2(A^2 + B^2)^2. \tag{5.1}
$$

In this Wess–Zumino Lagrangian density, we have $A = A^\dagger, B = B^\dagger$, and the spinor field ψ is a Majorana field, satisfying $\psi = \psi^c$. We see that in contrast to non-supersymmetric field theories all fields have the same mass m and couple with the same strength g. As explained in Chapter 4 this is due to the fact that states of a particular representation of the Super–Poincaré algebra are characterized by the eigenvalue m^2 of the Casimir operator P^2 and different values of spin s.

For reasons of simplicity we consider here only the free case, *i.e.*

$$\mathcal{L}\Big|_{g=0} = \mathcal{L}_{\text{free}}.$$

The equations of motion are obtained by applying to the Lagrangian density (5.1) the Euler–Langrange equation

$$\frac{\partial\mathcal{L}}{\partial\Phi_i} - \partial_\mu \frac{\partial\mathcal{L}}{\partial\partial_\mu\Phi_i} = 0, \tag{5.2}$$

where

$$\Phi_i \in \{A, B, \psi, \overline{\psi}\}.$$

(i) Scalar field $\Phi_i = A$: The equation of motion for the scalar field $A(x)$ is:

$$\frac{\partial\mathcal{L}}{\partial A} - \partial_\mu \frac{\partial\mathcal{L}}{\partial\partial_\mu A} = 0,$$

where

$$\frac{\partial\mathcal{L}}{\partial A} = -m^2 A - g\overline{\psi}\psi - mg(3A^2 + B^2) - 2g^2 A(A^2 + B^2), \frac{\partial\mathcal{L}}{\partial\partial_\mu A} = \partial^\mu A.$$

We obtain therefore:

$$(\Box + m^2)A = -g\big[\,\overline{\psi}\psi + m(3A^2 + B^2) + 2gA(A^2 + B^2)\big]. \tag{5.3}$$

(ii) Pseudoscalar field $\Phi_i = B$: The equation of motion for the pseudoscalar field $B(x)$ is:

$$\frac{\partial\mathcal{L}}{\partial B} - \partial_\mu \frac{\partial\mathcal{L}}{\partial\partial_\mu B} = 0,$$

with

$$\frac{\partial\mathcal{L}}{\partial B} = -m^2 B - ig\overline{\psi}\gamma^5\psi - 2mgAB - 2g^2 B(A^2 + B^2), \frac{\partial\mathcal{L}}{\partial\partial_\mu B} = \partial^\mu B.$$

Hence

$$(\Box + m^2)B = -g\big[i\,\overline{\psi}\gamma^5\psi + 2mAB + 2gB(A^2 + B^2)\big]. \tag{5.4}$$

(iii) Majorana field $\Phi_i = \overline{\psi}$: Here we have to evaluate:

$$\frac{\partial \mathcal{L}}{\partial \overline{\psi}} - \partial_\mu \frac{\partial \mathcal{L}}{\partial \partial_\mu \overline{\psi}} = 0,$$

with

$$\frac{\partial \mathcal{L}}{\partial \overline{\psi}} = \frac{1}{2}\big(i\slashed{\partial} - m\big)\psi - g\psi A - ig\gamma^5 \psi B, \qquad \frac{\partial \mathcal{L}}{\partial \partial_\mu \overline{\psi}} = 0.$$

Hence the equation of motion for $\overline{\psi}$ reads:

$$\big(i\slashed{\partial} - m\big)\psi = 2g\big(A + i\gamma^5 B\big)\psi. \tag{5.5}$$

(iv) Majorana field $\Phi_i = \psi$: In this case we have to work out:

$$\frac{\partial \mathcal{L}}{\partial \psi} - \partial_\mu \frac{\partial \mathcal{L}}{\partial \partial_\mu \psi} = 0,$$

with

$$\frac{\partial \mathcal{L}}{\partial \psi} = -\frac{1}{2}m\overline{\psi} - g\overline{\psi}A - ig\overline{\psi}\gamma^5 B, \qquad \frac{\partial \mathcal{L}}{\partial \partial_\mu \psi} = \frac{i}{2}\overline{\psi}\gamma^\mu,$$

and the equation of motion for the spinor field ψ is seen to be

$$i\partial_\mu \overline{\psi}\gamma^\mu + m\overline{\psi} = -2g\big(\overline{\psi}A + i\overline{\psi}\gamma^5 B\big). \tag{5.6}$$

5.2 Symmetries

We now investigate continuous transformations of the fields A, B and ψ, which are symmetry transformations of the Wess–Zumino model defined by the Lagrangian density (5.1). Such transformations are *symmetry transformations* of the theory if the action

$$S = \int \mathcal{L}\, d^4x \tag{5.7}$$

is left invariant. Under supersymmetry transformations — as defined below — the variation of $\mathcal{L}$ is a total derivative, *i.e.*

$$\mathcal{L}' - \mathcal{L} = \delta\mathcal{L} = \partial_\mu V^\mu \neq 0. \tag{5.8}$$

In the case of *e.g.* Lorentz transformations $\mathcal{L} = \mathcal{L}'$ and $\delta\mathcal{L} = 0$. However, for S to be invariant it suffices to require $\delta\mathcal{L} = \partial_\mu V^\mu$, where $\int d^4x \partial_\mu V^\mu$ can be converted into a surface integral with the help of Gauss's theorem, and for a surface sufficiently far away from where the fields are nonzero this integral will vanish. This is, in fact, the situation in the case of the supersymmetry transformations. We shall see later explicitly that so-called "superfields" (to be defined later) integrated in the Grassmann sense with respect to "supercoordinates" — which together with Minkowski coordinates comprise a "superspace" — yield precisely a Minkowski total derivative, and thus permit the construction of manifestly supersymmetric action integrals. The variation of the Lagrangian density under an arbitrary infinitesimal variation of the fields

$$A \longrightarrow A' = A + \delta A,$$
$$B \longrightarrow B' = B + \delta B,$$
$$\psi \longrightarrow \psi' = \psi + \delta\psi,$$

is (with $\Phi_i = A, B, \psi$)

$$
\begin{aligned}
\delta\mathcal{L} &= \mathcal{L}(\Phi_i', \partial_\mu\Phi_i') - \mathcal{L}(\Phi_i, \partial_\mu\Phi_i) = \frac{\partial\mathcal{L}}{\partial\Phi_i}\delta\Phi_i + \frac{\partial\mathcal{L}}{\partial\partial_\mu\Phi_i}\delta\partial_\mu\Phi_i \\
&= \left\{\frac{\partial\mathcal{L}}{\partial\Phi_i} - \partial_\mu\frac{\partial\mathcal{L}}{\partial\partial_\mu\Phi_i}\right\}\delta\Phi_i + \partial_\mu\left(\frac{\partial\mathcal{L}}{\partial\partial_\mu\Phi_i}\delta\Phi_i\right) \\
&\overset{(5.2)}{=} \partial_\mu\left(\frac{\partial\mathcal{L}}{\partial\partial_\mu\Phi_i}\delta\Phi_i\right) \overset{(5.8)}{=} \partial_\mu V^\mu.
\end{aligned}
$$

From this equality we see that the equations of motion define a conserved current density, given by:

$$j^\mu = V^\mu - \sum_i \frac{\partial\mathcal{L}}{\partial\partial_\mu\Phi_i}\delta\Phi_i. \tag{5.9}$$

From now on we consider only the free part of the Wess–Zumino Lagrangian density (5.1), i.e.

$$\mathcal{L}_{\text{free}} = \frac{1}{2}(\partial_\mu A)(\partial^\mu A) - \frac{1}{2}m^2 A^2 + \frac{1}{2}(\partial_\mu B)(\partial^\mu B) - \frac{1}{2}m^2 B^2 + \frac{1}{2}\overline{\psi}(i\slashed{\partial} - m)\psi. \tag{5.10}$$

We consider the following variations of the fields which transform fermions and bosons into each other[1]

$$\delta A(x) = \overline{\epsilon}\psi(x), \tag{5.11a}$$
$$\delta B(x) = -i\overline{\epsilon}\gamma^5\psi(x), \tag{5.11b}$$
$$\delta\psi(x) = -(i\slashed{\partial} + m)(A - i\gamma^5 B)\epsilon, \tag{5.11c}$$
$$\delta\overline{\psi}(x) = \overline{\epsilon}(A - i\gamma^5 B)(i\overset{\leftarrow}{\slashed{\partial}} - m). \tag{5.11d}$$

---

[1]These transformations will be derived later.

Here ϵ is an x-independent Grassmann variable. Equations (5.11a) to (5.11d) constitute the infinitesimal supersymmetry transformations of the set of fields A, B and ψ. For later reference and comparison purposes we rewrite the variations (5.11a) to (5.11d) in the Weyl formalism. Thus, using our previous considerations we have:

$$\delta A(x) = \bar{\epsilon}\psi(x) = \begin{pmatrix} \epsilon^A, & \bar{\epsilon}_{\dot{A}} \end{pmatrix} \begin{pmatrix} \psi_A \\ \overline{\psi}^{\dot{A}} \end{pmatrix} = \epsilon^A \psi_A + \bar{\epsilon}_{\dot{A}} \overline{\psi}^{\dot{A}} = (\epsilon\psi)_2 + (\bar{\epsilon}\overline{\psi})_2.$$

For the B-field we obtain:

$$\delta B(x) = -i\bar{\epsilon}\gamma_W^5 \psi(x) \overset{(1.162c)}{=} i \begin{pmatrix} \epsilon^A, & \bar{\epsilon}_{\dot{A}} \end{pmatrix} \begin{pmatrix} \mathbb{1}_{2\times2} & 0 \\ 0 & -\mathbb{1}_{2\times2} \end{pmatrix} \begin{pmatrix} \psi_A \\ \overline{\psi}^{\dot{A}} \end{pmatrix}$$

$$= i\big(\epsilon^A \psi_A - \bar{\epsilon}_{\dot{A}} \overline{\psi}^{\dot{A}}\big) = i\big[(\epsilon\psi)_2 - (\bar{\epsilon}\overline{\psi})_2\big].$$

The supersymmetry transformation of the Majorana field ψ becomes:

$$
\begin{aligned}
\delta\psi(x) &= -(i\slashed{\partial}_W + m)(A - i\gamma_W^5 B)\epsilon \\
&\overset{(1.162c)}{=} -\Big[i \begin{pmatrix} 0 & (\sigma^\mu)_{A\dot{A}} \\ (\overline{\sigma}^\mu)^{\dot{A}A} & 0 \end{pmatrix} \partial_\mu + m\mathbb{1}_{4\times4}\Big] \\
&\qquad \times \Big[A(x)\mathbb{1}_{4\times4} + i \begin{pmatrix} \mathbb{1}_{2\times2} & 0 \\ 0 & \mathbb{1}_{2\times2} \end{pmatrix} B(x)\Big] \begin{pmatrix} \epsilon_A \\ \bar{\epsilon}^{\dot{A}} \end{pmatrix} \\
&= -\begin{pmatrix} m & i\sigma^\mu\partial_\mu \\ i\overline{\sigma}^\mu\partial_\mu & m \end{pmatrix} \begin{pmatrix} A + iB & 0 \\ 0 & A - iB \end{pmatrix} \begin{pmatrix} \epsilon_A \\ \bar{\epsilon}^{\dot{A}} \end{pmatrix} \\
&= -\begin{pmatrix} m(A + iB)\epsilon_A + i(\sigma^\mu)_{A\dot{A}}\partial_\mu(A - iB)\bar{\epsilon}^{\dot{A}} \\ i(\overline{\sigma}^\mu)^{\dot{A}A}\partial_\mu(A + iB)\epsilon_A + m(A - iB)\bar{\epsilon}^{\dot{A}} \end{pmatrix} \\
&= -\begin{pmatrix} m(A + iB)\epsilon_A + i(\sigma^\mu\bar{\epsilon})_A\partial_\mu(A - iB) \\ i(\overline{\sigma}^\mu\epsilon)^{\dot{A}}\partial_\mu(A + iB) + m(A - iB)\bar{\epsilon}^{\dot{A}} \end{pmatrix}.
\end{aligned}
$$

Hence, setting

$$f(x) = \frac{1}{2}(A + iB),$$

we find, using the transformation properties of the A- and B-fields in the two-component Weyl formalism:

$$\delta f(x) = (\bar{\epsilon}\overline{\psi})_2, \qquad \delta f^*(x) = (\epsilon\psi)_2,$$

and

$$
\left.
\begin{aligned}
\delta\psi_A(x) &= -2mf(x)\epsilon_A + 2i(\sigma^\mu\bar{\epsilon})_A\partial_\mu f^*(x), \\
\delta\overline{\psi}^{\dot{A}}(x) &= -2mf^*(x)\bar{\epsilon}^{\dot{A}} - 2i(\overline{\sigma}^\mu\epsilon)^{\dot{A}}\partial_\mu f(x).
\end{aligned}
\right\} \tag{5.12}
$$

Since $A(x)$ is a scalar field and $B(x)$ a pseudoscalar field, the complex combination $f(x)$ transforms under the parity transformation like complex conjugation. Thus, $f(x)$ does not have a well defined parity — as expected, since the Weyl formulation does not preserve a definite parity. Thus the Wess–Zumino theory can be looked at as the field theory of a single complex scalar field and a Majorana spinor field in the Weyl formalism — and in fact, we shall recover it later in precisely this form (see the discussion following Eq. (8.36)).

As stated earlier, $\psi(x)$ is a Majorana spinor field, so that ψ and $\overline{\psi}$ are not independent, as was demonstrated in Sec. 1.4.5. Hence the transformations (5.11c) and (5.11d) are not independent, *i.e.* the transformation (5.11d) follows from Eq. (5.11c). We demonstrate this connection as an exercise. Since ψ is a Majorana spinor we have according to Eq. (1.203)

$$\psi = \psi^c = C\overline{\psi}^{\top},$$

where C is the charge conjugation matrix and $\overline{\psi} = \psi^\dagger \gamma_0$. Then

$$\overline{\psi} = -\psi^{\top} C^{-1}, \tag{5.13}$$

using the antisymmetry of the charge conjugation matrix C. Then we find for the variation:

$$
\begin{aligned}
\delta\overline{\psi}(x) &\overset{(5.13)}{=} -(\delta\psi)^{\top} C^{-1} \\
&\overset{(5.11c)}{=} \left[(i\slashed{\partial} + m)(A - i\gamma^5 B)\epsilon\right]^{\top} C^{-1} \\
&= \epsilon^{\top}(A - i\gamma^5 B)^{\top}\left(i\overleftarrow{\partial}_\mu \gamma^\mu + m\right)^{\top} C^{-1} \\
&= \epsilon^{\top} C^{-1} C(A - i\gamma^{5\top} B)\left(i\overleftarrow{\partial}_\mu \gamma^{\mu\top} + m\right) C^{-1} \\
&= -\overline{\epsilon} C(A - i\gamma^{5\top} B)\left(i\overleftarrow{\partial}_\mu \gamma^{\mu\top} + m\right) C^{-1} \\
&\overset{(1.196)}{=} -\overline{\epsilon}(A - iC\gamma^{5\top} B)\left(i\overleftarrow{\partial}_\mu \gamma^{\mu\top} C^{-1} + m\right) \\
&\overset{(1.189)}{=} \overline{\epsilon}(A - i\gamma^5 B)\left(i\overleftarrow{\partial}_\mu \gamma^\mu - m\right).
\end{aligned}
$$

Next we show that the variation $\delta\mathcal{L}$ caused by the supersymmetry transformations (5.11a) to (5.11d) is a total derivative. Our starting point is the free Lagrangian density (5.10).

$$
\begin{aligned}
\delta\mathcal{L}_{\text{free}} = &(\partial_\mu A)\delta(\partial^\mu A) - m^2 A\delta A + (\partial_\mu B)\delta(\partial^\mu B) - m^2 B\delta B \\
&+ \frac{1}{2}\delta\overline{\psi}(i\slashed{\partial} - m)\psi + \frac{1}{2}\overline{\psi}(i\slashed{\partial} - m)\delta\psi
\end{aligned}
$$

$$= \bar{\epsilon}(\partial_\mu A)(\partial^\mu \psi) - m^2 A \bar{\epsilon}\psi - i\bar{\epsilon}\gamma^5(\partial_\mu B)\partial^\mu \psi + i\bar{\epsilon}\gamma^5 m^2 B\psi$$
$$+ \frac{1}{2}\bar{\epsilon}(A - i\gamma^5 B)(i\overleftarrow{\partial} - m)(i\partial\!\!\!/ - m)\psi$$
$$- \frac{1}{2}\overline{\psi}(i\partial\!\!\!/ - m)(i\partial\!\!\!/ + m)(A - i\gamma^5 B)\epsilon$$
$$= \bar{\epsilon}(\partial_\mu A)(\partial^\mu \psi) - m^2 A(x)\bar{\epsilon}\psi - i\bar{\epsilon}\gamma^5(\partial_\mu B)\partial^\mu \psi + i\bar{\epsilon}\gamma^5 m^2 B(x)\psi$$
$$+ \frac{1}{2}\overline{\psi}(\Box + m^2)(A - i\gamma^5 B)\epsilon.$$

In the last step we made use of

$$\partial\!\!\!/\partial\!\!\!/ = \partial_\mu \gamma^\mu \partial_\nu \gamma^\nu = \frac{1}{2}\partial_\mu \partial_\nu (\gamma^\mu \gamma^\nu + \gamma^\nu \gamma^\mu) = \partial_\mu \partial_\nu \eta^{\mu\nu} = \partial_\mu \partial^\mu = \Box.$$

Using the equations of motion of the A- and B-fields, we obtain

$$\delta\mathcal{L} = \bar{\epsilon}(\partial_\mu A)(\partial^\mu \psi) - m^2 A \bar{\epsilon}\psi - i\bar{\epsilon}\gamma^5(\partial_\mu B)(\partial^\mu \psi) + i\bar{\epsilon}\gamma^5 m^2 B\psi$$
$$= \bar{\epsilon}(\partial_\mu A)(\partial^\mu \psi) + \Box A\, \bar{\epsilon}\psi - i\bar{\epsilon}\gamma^5(\partial_\mu B)(\partial^\mu \psi) - i\bar{\epsilon}\gamma^5 \Box B\psi$$
$$= \partial_\mu\left\{\bar{\epsilon}(\partial^\mu A)\psi - i\bar{\epsilon}\gamma^5(\partial^\mu B)\psi\right\} = \partial_\mu\left\{\bar{\epsilon}[\partial^\mu(A - i\gamma^5 B)]\psi\right\} \overset{!}{=} \partial_\mu V^\mu.$$

Hence

$$V^\mu = \bar{\epsilon}[\partial^\mu(A - i\gamma^5 B)]\psi. \tag{5.14}$$

For the other contributions to the conserved current density (5.9) we consider:

$$\sum_i \frac{\partial \mathcal{L}}{\partial \partial_\mu \Phi_i}\delta\Phi_i = \frac{\partial \mathcal{L}}{\partial \partial_\mu A}\delta A + \frac{\partial \mathcal{L}}{\partial \partial_\mu B}\delta B + \frac{\partial \mathcal{L}}{\partial \partial_\mu \psi}\delta\psi$$

$$= \partial_\mu A \delta A + \partial_\mu B \delta B + \frac{i}{2}\overline{\psi}\gamma_\mu \delta\psi$$

$$= \partial_\mu A \bar{\epsilon}\psi - i\partial_\mu B \bar{\epsilon}\gamma^5 \psi - \frac{i}{2}\overline{\psi}\gamma_\mu(i\partial\!\!\!/ + m)(A - i\gamma^5 B)\epsilon$$

$$= \bar{\epsilon}[\partial_\mu(A - i\gamma^5 B)]\psi - \frac{i}{2}\overline{\psi}\gamma_\mu(i\partial\!\!\!/ + m)(A - i\gamma^5 B)\epsilon$$

$$= \bar{\epsilon}[\partial_\mu(A - i\gamma^5 B)]\psi + \frac{i}{2}\epsilon^\top(A - i\gamma^5 B)^\top(i\overleftarrow{\partial} + m)^\top \gamma_\mu^\top \overline{\psi}^\top$$

$$= \bar{\epsilon}[\partial_\mu(A - i\gamma^5 B)]\psi$$
$$+ \frac{i}{2}\epsilon^\top C^{-1} C[A - i\gamma^5 B]^\top(i\overleftarrow{\partial} + m)^\top \gamma_\mu^\top \overline{\psi}^\top$$

$$= \bar{\epsilon}\big[\partial_\mu(A - i\gamma^5 B)\big]\psi$$

$$-\frac{i}{2}\bar{\epsilon}\big[AC - iC\gamma^{5\top}B\big]\big(i\overleftarrow{\partial}_\rho \gamma^{\rho\top} + m\big)\gamma_\mu^\top\bar{\psi}^\top$$

$$= \bar{\epsilon}\big[\partial_\mu(A - i\gamma^5 B)\big]\psi$$

$$+\frac{i}{2}\bar{\epsilon}\big[A - i\gamma^5 B\big]\big(i\overleftarrow{\partial}_\rho C\gamma^{\rho\top} + mC\big)C^{-1}\gamma_\mu^\top C\bar{\psi}^\top$$

$$= \bar{\epsilon}\big[\partial_\mu(A - i\gamma^5 B)\big]\psi$$

$$+\frac{i}{2}\bar{\epsilon}\big[A - i\gamma^5 B\big]\big(i\overleftarrow{\partial}_\rho C\gamma^{\rho\top}C^{-1} + m\big)\gamma_\mu C\bar{\psi}^\top$$

$$= \bar{\epsilon}\big[\partial_\mu(A - i\gamma^5 B)\big]\psi - \frac{i}{2}\bar{\epsilon}\big[A - i\gamma^5 B\big]\big(i\overleftarrow{\partial}_\rho \gamma^\rho - m\big)\gamma_\mu\psi. \qquad (5.15)$$

This is an algebraic relation derived without explicit use of the equations of motion. Inserting this result into Eq. (5.9), we obtain for the conserved current (using Eq. (5.14)):

$$j^\mu = V^\mu - \sum_i \frac{\partial\mathcal{L}}{\partial\partial_\mu\Phi_i}\delta\Phi_i = \frac{i}{2}\bar{\epsilon}\big[A - i\gamma^5 B\big]\big(i\overleftarrow{\partial}_\rho \gamma^\rho - m\big)\gamma_\mu\psi.$$

Hence:

$$\boxed{j^\mu = \frac{i}{2}\bar{\epsilon}\big[A - i\gamma^5 B\big]\big(i\overleftarrow{\partial}_\rho \gamma^\rho - m\big)\gamma_\mu\psi.} \qquad (5.16)$$

The vector j^μ is called the *supercurrent*. It is useful to define a slightly modified current, the conserved *Majorana spinor current* k^μ, by the relation

$$j^\mu = \frac{1}{\lambda}\bar{\epsilon}k^\mu, \qquad (5.17)$$

where λ is a multiplicative real constant which has to be determined such that the spinor charge

$$Q_a := \int k_a^0\, d^3x$$

satisfies the anticommutation relation

$$\{Q_a, Q_b\} = -2\big(\gamma^\mu C\big)_{ab}P_\mu.$$

Thus, in the Wess–Zumino model, the *spinor current density* is:

$$k^\mu = \frac{i}{2}\lambda\big[A - i\gamma^5 B\big]\big(i\overleftarrow{\partial}_\rho \gamma^\rho - m\big)\gamma^\mu\psi. \qquad (5.18)$$

We can verify explicitly that this current is conserved, *i.e.*

$$\partial_\mu k^\mu = 0. \qquad (5.19)$$

This must, of course, be valid since by construction Eq. (5.17) is a conserved current. We first rewrite Eq. (5.18) in the form:

$$
\begin{aligned}
k^\mu &= \frac{i}{2}\lambda\big[A - i\gamma^5 B\big]\big(i\,\overleftarrow{\partial}_\rho\,\gamma^\rho - m\big)\gamma^\mu\psi \\
&= \frac{i}{2}\lambda\big[A - i\gamma^5 B\big]i\,\overleftarrow{\partial}_\rho\,\gamma^\rho\gamma^\mu\psi - \frac{i}{2}\lambda m\big[A - i\gamma^5 B\big]\gamma^\mu\psi \\
&= -\frac{i}{2}\lambda m\gamma^\mu(A + i\gamma^5 B)\psi - \frac{1}{2}\lambda\big[\slashed{\partial}(A + i\gamma^5 B)\big]\gamma^\mu\psi
\end{aligned}
$$

using Eq. (1.164). Then:

$$
\begin{aligned}
\partial_\mu k^\mu &= -\frac{i}{2}\lambda m\slashed{\partial}\big[(A + i\gamma^5 B)\psi\big] - \frac{1}{2}\lambda\big[\slashed{\partial}\slashed{\partial}(A - i\gamma^5 B)\big]\psi \\
&\quad - \frac{1}{2}\lambda\big[\slashed{\partial}(A + i\gamma^5 B)\big]\slashed{\partial}\psi \\
&= -\frac{i}{2}m\lambda\big(\slashed{\partial}A - i\gamma^5\slashed{\partial}B\big)\psi - \frac{i}{2}m\lambda(A - i\gamma^5 B)\slashed{\partial}\psi \\
&\quad - \frac{1}{2}\lambda\big[\slashed{\partial}\slashed{\partial}A - i\gamma^5\slashed{\partial}\slashed{\partial}B\big]\psi - \frac{1}{2}\lambda\big[\slashed{\partial}(A + i\gamma^5 B)\big]\slashed{\partial}\psi \\
&= -\frac{1}{2}\lambda\big(\slashed{\partial}A - i\gamma^5\slashed{\partial}B\big)\big(\slashed{\partial}\psi + im\psi\big) \\
&\quad - \frac{1}{2}\big[\partial_\mu\partial^\mu A - i\gamma^5\partial_\mu\partial^\mu B\big]\psi - \frac{i}{2}m\lambda(A - i\gamma^5 B)\slashed{\partial}\psi \\
&= -\frac{1}{2}\lambda\big(\slashed{\partial}A - i\gamma^5\slashed{\partial}B\big)(-i)\big(i\slashed{\partial}\psi - m\psi\big) \\
&\quad - \frac{1}{2}\lambda\big[\square A - i\gamma^5\square B\big]\psi - \frac{1}{2}m^2\lambda(A - i\gamma^5 B)\psi \\
&= -\frac{1}{2}\lambda\Big\{\big(\square + m^2\big)A - i\gamma^5\big(\square + m^2\big)B\Big\}\psi = 0.
\end{aligned}
$$

As mentioned above, the spinor charges Q_a are defined by

$$
Q_a := \int d^3x\, k_a^0 \tag{5.20}
$$

with the *spinor charge density* of the Wess–Zumino model:

$$
k_a^0 = \frac{i}{2}\lambda\Big[\big[A(x) - i\gamma^5 B(x)\big]\big(i\,\overleftarrow{\slashed{\partial}} - m\big)\gamma^0\psi(x)\Big]_a \tag{5.21}
$$

with $a = 1, 2, 3, 4$. The crucial point is the following. The Wess–Zumino model is a supersymmetric field theory, if the spinor charges (5.20) with the charge density given by Eq. (5.21) satisfy the commutation and anticommutation relations of the Super–Poincaré algebra (4.1). In particular, we have to show with a suitably chosen factor λ that the spinor charges (5.20) satisfy the relation:

$$\{Q_a, \overline{Q}_b\} = 2P_\mu (\gamma^\mu)_{ab}.$$

If we can verify this, we have demonstrated that the Wess–Zumino model is a field theoretical realization of the supersymmetry algebra (4.1). Thus, we have to search for a representation of the spinor charges (5.20) as linear operators acting on Fock space.

5.3   Plane Wave Expansions

The equation of motion of a free Dirac field $\psi(x)$ is given by the Dirac equation

$$(i\slashed{\partial} - m)\psi(x) = 0. \tag{5.22}$$

Thus a plane wave solution

$$\psi(x) = \exp\{-ipx\}u(\mathbf{p}) \tag{5.23}$$

satisfies

$$(\slashed{p} - m)u(\mathbf{p}) = 0. \tag{5.24}$$

This equation has four independent solutions:

- Two solutions for

$$p^0 := \omega_p = +(\mathbf{p}^2 + m^2)^{1/2},$$

- and two solutions for

$$p^0 := -\omega_p = -(\mathbf{p}^2 + m^2)^{1/2}.$$

In the rest frame the energy-momentum four vector has the form $(p_\mu) = (p_0, \mathbf{0}) = (m, \mathbf{0})$ and Eq. (5.24) becomes

$$(\gamma^0 - \mathbb{1}_{4\times4})u(\mathbf{0}) = 0, \tag{5.25}$$

so that $u(\mathbf{0})$ is an eigenstate of the matrix γ^0. Taking the γ-matrices in the Dirac representation (1.172), *i.e.*

$$\gamma^0 = \begin{pmatrix} \mathbb{1}_{2\times2} & 0 \\ 0 & \mathbb{1}_{2\times2} \end{pmatrix}, \quad \gamma^i = \begin{pmatrix} 0 & \sigma^i \\ \bar{\sigma}^i & 0 \end{pmatrix}, \quad \gamma^5 = \begin{pmatrix} 0 & \sigma^0 \\ \bar{\sigma}^0 & 0 \end{pmatrix},$$

we can define two eigenvectors of γ^0 in the form:

$$u(\mathbf{0},1) = \begin{pmatrix} 1 \\ 0 \\ 0 \\ 0 \end{pmatrix}, \qquad u(\mathbf{0},2) = \begin{pmatrix} 0 \\ 1 \\ 0 \\ 0 \end{pmatrix}. \tag{5.26}$$

In an analogous way we construct eigenvectors for

$$p_0 < 0, \quad \mathbf{p} = \mathbf{0}, \qquad p_0 = -m,$$

so that Eq. (5.24) becomes:

$$\left(\gamma^0 + \mathbb{1}_{4\times4}\right)v(\mathbf{0}) = 0, \quad \gamma^0 v(\mathbf{0}) = -v(\mathbf{0}),$$

and we define

$$v(\mathbf{0},1) = \begin{pmatrix} 0 \\ 0 \\ 0 \\ 1 \end{pmatrix}, \qquad v(\mathbf{0},2) = \begin{pmatrix} 0 \\ 0 \\ -1 \\ 0 \end{pmatrix}. \tag{5.27}$$

We could now boost these solutions (5.26) and (5.27) from the rest frame to an arbitrary inertial system moving with velocity $v = |\mathbf{p}|/p_0$ by a pure Lorentz transformation. However, it is easier to observe that

$$\left(\not{p} - m\right)\left(\not{p} + m\right) = p^2 - m^2 = 0,$$

and therefore (with $s = 1, 2$)

$$u(\mathbf{p},s) = N(p)\left(\not{p} + m\right)u(\mathbf{0},s), \tag{5.28}$$

and we find

$$\left(\not{p} - m\right)u(\mathbf{p},s) = N(p)\left(\not{p} - m\right)\left(\not{p} + m\right)u(\mathbf{0},s) = 0,$$

such that the momentum space equation (5.24) is satisfied. The factor $N(p)$ is a normalization factor. Furthermore

$$v(\mathbf{p},s) = \left(-\not{p} + m\right)N(p)v(\mathbf{0},s), \tag{5.29}$$

such that

$$\begin{aligned}\left(\not{p} + m\right)v(\mathbf{p},s) &= \left(\not{p} + m\right)\left(-\not{p} + m\right)N(p)v(\mathbf{0},s) \\ &= -(p^2 + m^2)N(p)v(\mathbf{0},s) = 0.\end{aligned}$$

For the conjugate spinors we find:

$$\overline{u}(\mathbf{p}, s) = \overline{u}(\mathbf{0}, s)N(p)(\not{p} + m),$$
$$\overline{v}(\mathbf{p}, s) = \overline{v}(\mathbf{0}, s)N(p)(-\not{p} + m). \tag{5.30}$$

The normalization factor $N(p)$ has to be determined from the requirement of the orthogonality of different solutions:

$$\left.\begin{array}{rcl} \overline{u}(\mathbf{p}, s)u(\mathbf{p}, r) & = & \delta_{rs}, \\ \overline{v}(\mathbf{p}, s)v(\mathbf{p}, r) & = & -\delta_{rs}. \end{array}\right\} \tag{5.31}$$

It is easily seen that also

$$\left.\begin{array}{rcl} \overline{u}(\mathbf{p}, s)v(\mathbf{p}, r) & = & 0, \\ \overline{v}(\mathbf{p}, s)u(\mathbf{p}, r) & = & 0. \end{array}\right\} \tag{5.32}$$

Proposition 5.1:

The normalization factor $N(p)$ is given by

$$N(p) = \left[2m(m + \omega_p)\right]^{-1/2}. \tag{5.33}$$

Proof: We consider:

$$\overline{u}(\mathbf{p}, 1)u(\mathbf{p}, 1) = 1.$$

Using Eqs. (5.28) and (5.30) this can be written in the form:

$$\overline{u}(\mathbf{0}, 1)N(p)(\not{p} + m)N(p)(\not{p} + m)u(\mathbf{0}, 1) = 1,$$

i.e.

$$\overline{u}(\mathbf{0}, 1)(\not{p} + m)(\not{p} + m)u(\mathbf{0}, 1) = \frac{1}{N^2(p)}.$$

This implies

$$\overline{u}(\mathbf{0}, 1)(p^2 + 2m\not{p} + m^2)u(\mathbf{0}, 1) = \frac{1}{N^2(p)}.$$

Using Eq. (5.26) we can write this

$$\frac{1}{N^2(p)} = u^\dagger(\mathbf{0},1)\gamma^0\left(2m^2 + 2mp_\mu\gamma^\mu\right)u(\mathbf{0},1)$$

$$= \begin{pmatrix} 1 & 0 & 0 & 0 \end{pmatrix}\begin{pmatrix} \mathbb{1}_{2\times2} & 0 \\ 0 & -\mathbb{1}_{2\times2} \end{pmatrix}\left(2m^2\mathbb{1}_{4\times4} + 2mp_\mu\gamma^\mu\right)\begin{pmatrix} 1 \\ 0 \\ 0 \\ 0 \end{pmatrix}$$

$$= \begin{pmatrix} 1 & 0 & 0 & 0 \end{pmatrix}\left[2m^2\mathbb{1}_{4\times4} \right.$$

$$\left. + 2m\begin{pmatrix} p_0 & 0 & p_3 & p_1 - ip_2 \\ 0 & p_0 & p_1 + ip_2 & -p_3 \\ -p_3 & -p_1 + ip_2 & -p_0 & 0 \\ -p_1 - ip_2 & p_3 & 0 & -p_0 \end{pmatrix}\right]\begin{pmatrix} 1 \\ 0 \\ 0 \\ 0 \end{pmatrix}$$

$$= \begin{pmatrix} 1 & 0 & 0 & 0 \end{pmatrix}\left[2m^2\begin{pmatrix} 1 \\ 0 \\ 0 \\ 0 \end{pmatrix} + 2m\begin{pmatrix} p_0 \\ 0 \\ -p_3 \\ -p_1 - ip_2 \end{pmatrix}\right]$$

$$= 2m^2 + 2mp_0,$$

so that

$$\frac{1}{N^2(p)} = 2m(m + \omega_p).$$

This proves Proposition 5.1.

With this normalization factor the momentum space representations of the spinors u and v have the form:

$$u(\mathbf{p},s) = \frac{(m + \not{p})u(\mathbf{0},s)}{[2m(m+\omega_p)]^{1/2}}, \tag{5.34}$$

$$v(\mathbf{p},s) = \frac{(m - \not{p})v(\mathbf{0},s)}{[2m(m+\omega_p)]^{1/2}}. \tag{5.35}$$

Second quantization is achieved by imposing equal-time anticommutation relations on the spinor fields $\psi(x)$, *i.e.*

$$\left.\begin{array}{rcl} \{\psi_a(\mathbf{x},t), \psi_b(\mathbf{x}',t)\} &=& 0, \\ \{\psi_a^\dagger(\mathbf{x},t), \psi_b^\dagger(\mathbf{x}',t)\} &=& 0, \\ \{\psi_a(\mathbf{x},t), \psi_b^\dagger(\mathbf{x}',t)\} &=& \delta_{ab}\delta(\mathbf{x} - \mathbf{y}). \end{array}\right\} \tag{5.36}$$

The field $\psi(x)$ is now a linear operator acting on Fock space. Fourier decomposition of $\psi(x)$ leads to:

$$\psi(x) = \frac{1}{(2\pi)^{3/2}} \sum_s \int d^3p \left(\frac{m}{\omega_p}\right)^{1/2} \left[b(\mathbf{p}, s)u(\mathbf{p}, s)e^{-ipx} + d^\dagger(\mathbf{p}, s)v(\mathbf{p}, s)e^{ipx} \right].$$

(5.37)

In this Fourier expansion of this field, in the exponents of the exponentials we have $px = p_\mu x^\mu$, the sum $\sum_s$ indicates summation over spin states, and $b^\dagger(\mathbf{p}, s)$ and $d(\mathbf{p}, s)$ are momentum dependent creation and annihilation operators. Equation (5.37) is the *plane wave expansion* of the Dirac spinor field. We obtain some restrictions on the operators $b(\mathbf{p}, s)$ and $d^\dagger(\mathbf{p}, s)$ if we impose the Majorana property on the spinor field $\psi(x)$. The Majorana condition is

$$\psi(x) = \psi^c(x) = C\overline{\psi}(x)^\top,$$

where C is the charge conjugation matrix. Taking the Hermitian conjugate of Eq. (5.37) we obtain:

$$\psi^\dagger(x) = \frac{1}{(2\pi)^{3/2}} \sum_s \int d^3p \left[\frac{m}{\omega_p}\right]^{1/2} \left[b^\dagger(\mathbf{p}, s)u^\dagger(\mathbf{p}, s)e^{ipx} + d(\mathbf{p}, s)v^\dagger(\mathbf{p}, s)e^{-ipx} \right].$$

Then

$$\begin{aligned}
\overline{\psi}^\top(x) &= \left(\psi^\dagger \gamma^0\right)^\top \\
&= \frac{1}{(2\pi)^{3/2}} \sum_s \int d^3p \left(\frac{m}{\omega_p}\right)^{1/2} \left[b^\dagger(\mathbf{p}, s)\overline{u}^\top(\mathbf{p}, s)e^{ipx} \right. \\
&\quad \left. + d(\mathbf{p}, s)\overline{v}^\top(\mathbf{p}, s)e^{-ipx} \right].
\end{aligned}$$

Then the charge conjugated spinor has the expansion

$$\begin{aligned}
\psi^c(x) &= C\overline{\psi}^\top(x) \\
&= \frac{1}{(2\pi)^{3/2}} \sum_s \int d^3p \left(\frac{m}{\omega_p}\right)^{1/2} \left[b^\dagger(\mathbf{p}, s)C\overline{u}^\top(\mathbf{p}, s)e^{ipx} \right. \\
&\quad \left. + d(\mathbf{p}, s)C\overline{v}^\top(\mathbf{p}, s)e^{-ipx} \right].
\end{aligned}$$

(5.38)

Proposition 5.2:

The following relations hold

$$u^c(\mathbf{0}, s) \equiv C\overline{u}^\top(\mathbf{0}, s) = v(\mathbf{0}, s), \tag{5.39}$$

$$v^c(\mathbf{0}, s) \equiv C\overline{v}^\top(\mathbf{0}, s) = u(\mathbf{0}, s). \tag{5.40}$$

Proof: In order to demonstrate Eq. (5.39) we start with:

$$
u(\mathbf{p}, s) = \begin{pmatrix} u_1(\mathbf{p}) \\ u_2(\mathbf{p}) \\ u_3(\mathbf{p}) \\ u_4(\mathbf{p}) \end{pmatrix},
$$

where

$$
u_2(\mathbf{0}) = u_3(\mathbf{0}) = u_4(\mathbf{0}) = 0,
$$

and $u_1(\mathbf{0}) = 1$, for $s = 1$, see Eq. (5.26). Then

$$
\bar{u}^\top = \left(u^\dagger \gamma^0 \right)^\top = \begin{pmatrix} u_1^* \\ u_2^* \\ -u_3^* \\ -u_4^* \end{pmatrix}, \quad
\gamma^0 = \begin{pmatrix} \mathbb{1}_{2\times 2} & 0 \\ 0 & -\mathbb{1}_{2\times 2} \end{pmatrix}.
$$

Now the charge conjugation matrix is given by

$$
C = i\gamma^2\gamma^0 = -i \begin{pmatrix} 0 & \sigma^2 \\ \sigma^2 & 0 \end{pmatrix}
$$

(see Eq. (1.190)), so that

$$
u^c(\mathbf{p}, s) = C\bar{u}^\top(\mathbf{p}, s) = - \begin{pmatrix} 0 & 0 & 0 & 1 \\ 0 & 0 & -1 & 0 \\ 0 & 1 & 0 & 0 \\ -1 & 0 & 0 & 0 \end{pmatrix} \begin{pmatrix} u_1^* \\ u_2^* \\ -u_3^* \\ -u_4^* \end{pmatrix} = - \begin{pmatrix} -u_4^* \\ u_3^* \\ u_2^* \\ -u_1^* \end{pmatrix}.
$$

Then we find, using Eq. (5.27):

$$
u^c(\mathbf{0}, 1) = \begin{pmatrix} 0 \\ 0 \\ 0 \\ 1 \end{pmatrix} = v(\mathbf{0}, 1), \quad
u^c(\mathbf{0}, 2) = \begin{pmatrix} 0 \\ 0 \\ -1 \\ 0 \end{pmatrix} = v(\mathbf{0}, 2).
$$

Equation (5.40) can be shown in a similar way. This completes the proof of Proposition 5.2.

We need some more results.

Proposition 5.3:

From Eqs. (5.39) and (5.40) we can obtain

$$u^c(\mathbf{p}, s) = v(\mathbf{p}, s), \tag{5.41}$$

$$v^c(\mathbf{p}, s) = u(\mathbf{p}, s). \tag{5.42}$$

Proof: In order to demonstrate Eq. (5.41), we start with:

$$u^c(\mathbf{p}, s) \overset{(1.188)}{=} C\big[\overline{u}(\mathbf{p}, s)\big]^{\top} \overset{(5.30)}{=} N(p) C\big[\overline{u}(\mathbf{0}, s)(\not{p} + m)\big]^{\top}$$

$$= N(p) C\big[(\not{p} + m)^{\top}\overline{u}^{\top}(\mathbf{0}, s)\big] = N(p)\big[p_\mu C\gamma^{\mu\top} + mC\big]\overline{u}^{\top}(\mathbf{0}, s)$$

$$\overset{(1.189)}{=} N(p)\big[-p_\mu\gamma^{\mu} + m\big]C\overline{u}^{\top}(\mathbf{0}, s) = N(p)\big[-\not{p} + m\big]v(\mathbf{0}, s)$$

$$\overset{(5.35)}{=} v(\mathbf{p}, s).$$

In a similar way one shows the second relation, Eq. (5.42):

$$v^c(\mathbf{p}, s) \overset{(1.188)}{=} C\big[\overline{v}(\mathbf{p}, s)\big]^{\top} \overset{(5.30)}{=} N(p) C\big[\overline{v}(\mathbf{0}, s)(-\not{p} + m)\big]^{\top}$$

$$= N(p) C\big[(-\not{p} + m)^{\top}\overline{v}^{\top}(\mathbf{0}, s)\big]$$

$$= N(p)\big[-p_\mu C\gamma^{\mu\top} + mC\big]\overline{v}^{\top}(\mathbf{0}, s)$$

$$\overset{(1.189)}{=} N(p)\big[p_\mu\gamma^{\mu} + m\big]C\overline{v}^{\top}(\mathbf{0}, s) \overset{(5.40)}{=} N(p)\big[\not{p} + m\big]u(\mathbf{0}, s)$$

$$\overset{(5.34)}{=} u(\mathbf{p}, s).$$

This completes the proof of Proposition 5.3.

Inserting Eqs. (5.41) and (5.42) into Eq. (5.38) we obtain:

$$\psi^c(x) = \frac{1}{(2\pi)^{3/2}} \sum_s \int d^3 p \left(\frac{m}{\omega_p}\right)^{1/2} \big[b^\dagger(\mathbf{p}, s)v(\mathbf{p}, s)e^{ipx} + d(\mathbf{p}, s)u(\mathbf{p}, s)e^{-ipx}\big].$$

Imposing the Majorana condition this has to be equal to the expansion (*cf.* Eq. (5.37))

$$\psi(x) = \frac{1}{(2\pi)^{3/2}} \sum_s \int d^3 p \left(\frac{m}{\omega_p}\right)^{1/2} \big[d^\dagger(\mathbf{p}, s)v(\mathbf{p}, s)e^{ipx} + b(\mathbf{p}, s)u(\mathbf{p}, s)e^{-ipx}\big].$$

Thus for Majorana spinors the momentum dependent creation and annihilation operators must obey:

$$b(\mathbf{p}, s) = d(\mathbf{p}, s) \quad \text{and} \quad b^\dagger(\mathbf{p}, s) = d^\dagger(\mathbf{p}, s). \tag{5.43}$$

Hence, a Majorana spinor has the following plane wave expansion:

$$\psi^M(x) = \frac{1}{(2\pi)^{3/2}} \sum_s \int d^3 p \left(\frac{m}{\omega_p}\right)^{1/2} \big[d(\mathbf{p}, s)u(\mathbf{p}, s)e^{-ipx} + d^\dagger(\mathbf{p}, s)v(\mathbf{p}, s)e^{ipx}\big]$$

$$\tag{5.44}$$

satisfying the Majorana condition $\psi^M(x) = \psi^{Mc}(x)$. Thus a Majorana spinor involves creation and annihilation operators $d^\dagger(\mathbf{p}, s)$ and $d(\mathbf{p}, s)$ instead of $d^\dagger(\mathbf{p}, s)$ and $b(\mathbf{p}, s)$ as in the expansion of a Dirac spinor.

In order to obtain the anticommutation relations for the creation and annihilation operators $d(\mathbf{p}, s)$ and $d^\dagger(\mathbf{p}, s)$ it is advantageous to solve Eq. (5.44) for $d(\mathbf{p}, s)$. Then:

$$d(\mathbf{p}, s) = \frac{1}{(2\pi)^{3/2}} \left(\frac{m}{\omega_p}\right)^{1/2} \int d^3x \, e^{ipx} u^\dagger(\mathbf{p}, s) \psi(x). \tag{5.45}$$

Using the orthogonality relations (5.31) and (5.32) it can be shown that

$$\left. \begin{aligned} \{d(\mathbf{p}, r), d(\mathbf{p}', s)\} &= 0, \\ \{d^\dagger(\mathbf{p}, r), d^\dagger(\mathbf{p}', s)\} &= 0, \\ \{d(\mathbf{p}, r), d^\dagger(\mathbf{p}', s)\} &= \delta_{rs}\delta(\mathbf{p} - \mathbf{p}'). \end{aligned} \right\} \tag{5.46}$$

This completes our considerations of the plane wave expansion of a Majorana field. For the scalar fields $A(x)$ and $B(x)$ appearing in the Wess–Zumino model we use the following plane wave expansions:

$$A(x) = \frac{1}{(2\pi)^{3/2}} \int d^3p \, (2\omega_p)^{-1/2} \left[a(\mathbf{p})e^{-ipx} + a^\dagger(\mathbf{p})e^{ipx} \right], \tag{5.47}$$

and

$$B(x) = \frac{1}{(2\pi)^{3/2}} \int d^3p \, (2\omega_p)^{-1/2} \left[b(\mathbf{p})e^{-ipx} + b^\dagger(\mathbf{p})e^{ipx} \right]. \tag{5.48}$$

Our next step is the calculation of the spinor charge Q_a in the Wess–Zumino model.

Proposition 5.4:

The supercharge of the Wess–Zumino model, defined by

$$Q_a = \int d^3x \, k_a^0$$

with the spinor current k_a^0 given by Eq. (5.21) is given by

$$Q_a = \frac{i\lambda m^{1/2}}{\sqrt{2}} \sum_s \int d^3p \left\{ C(\mathbf{p})d^\dagger(\mathbf{p})v(\mathbf{p}, s) - D(\mathbf{p})d(\mathbf{p}, s)u(\mathbf{p}, s) \right\}_a, \tag{5.49}$$

where

$$\left. \begin{aligned} C(\mathbf{p}) &= a(\mathbf{p})\mathbb{1}_{4\times4} - i\gamma^5 b(\mathbf{p}), \\ D(\mathbf{p}) &= a^\dagger(\mathbf{p})\mathbb{1}_{4\times4} - i\gamma^5 b^\dagger(\mathbf{p}). \end{aligned} \right\} \tag{5.50}$$

Proof: Using Eqs. (5.20), (5.21) and the momentum space expansions of the fields, Eqs. (5.44), (5.47) and (5.48), we obtain:

$$Q_a = \int d^3x\, k_a^0 = \frac{i}{2}\lambda \int d^3x \left[\left[A(x) - i\gamma^5 B(x) \right] \left(i\overleftarrow{\partial\!\!\!/} - m \right) \gamma^0 \psi(x) \right]_a$$

$$= \frac{i}{2}\lambda \int d^3x \left[\left\{ \frac{1}{(2\pi)^{3/2}} \int d^3p \frac{1}{(2\omega_p)^{1/2}} \left(a(\mathbf{p})e^{-ipx} + a^\dagger(\mathbf{p})e^{ipx} \right) \right. \right.$$

$$\left. - i\gamma^5 \frac{1}{(2\pi)^{3/2}} \int d^3p \frac{1}{(2\omega_p)^{1/2}} \left(b(\mathbf{p})e^{-ipx} + b^\dagger(\mathbf{p})e^{ipx} \right) \right\}$$

$$\left(i\overleftarrow{\partial\!\!\!/} - m \right) \gamma^0 \frac{1}{(2\pi)^{3/2}} \sum_s \int d^3k \left(\frac{m}{\omega_k} \right)^{1/2}$$

$$\left. \left\{ d(\mathbf{k},s)u(\mathbf{k},s)e^{-ikx} + d^\dagger(\mathbf{k},s)v(\mathbf{k},s)e^{ikx} \right\} \right]_a$$

$$= \frac{i\lambda}{2(2\pi)^3} \sum_s \int d^3x \int d^3p \int d^3k \left(\frac{m}{2\omega_p\omega_k} \right)^{1/2} \left\{ \left[C(\mathbf{p})e^{-ipx} + D(\mathbf{p})e^{ipx} \right] \right.$$

$$\left. \left(i\overleftarrow{\partial\!\!\!/} - m \right) \gamma^0 \left[d(\mathbf{k},s)u(\mathbf{k},s)e^{-ikx} + d^\dagger(\mathbf{k},s)v(\mathbf{k},s)e^{ikx} \right] \right\}_a$$

$$= \frac{i\lambda}{2(2\pi)^3} \sum_s \int d^3x \int d^3p \int d^3k \left(\frac{m}{2\omega_p\omega_k} \right)^{1/2}$$

$$\left\{ \left[C(\mathbf{p})e^{-ipx}(p\!\!\!/ - m) + D(\mathbf{p})e^{ipx}(-p\!\!\!/ - m) \right] \gamma^0 \right.$$

$$\left. \left[d(\mathbf{k},s)u(\mathbf{k},s)e^{-ikx} + d^\dagger(\mathbf{k},s)v(\mathbf{k},s)e^{ikx} \right] \right\}_a$$

$$= \frac{i\lambda}{2(2\pi)^3} \sum_s \int d^3x \int d^3p \int d^3k \left(\frac{m}{2\omega_p\omega_k} \right)^{1/2}$$

$$\left\{ C(\mathbf{p})(p\!\!\!/ - m)\gamma^0 d(\mathbf{k},s)u(\mathbf{k},s)e^{-i(p+k)x} \right.$$

$$+ C(\mathbf{p})(p\!\!\!/ - m)\gamma^0 d^\dagger(\mathbf{k},s)v(\mathbf{k},s)e^{-i(p-k)x}$$

$$- D(\mathbf{p})(p\!\!\!/ + m)\gamma^0 d(\mathbf{k},s)u(\mathbf{k},s)e^{i(p-k)x}$$

$$\left. - D(\mathbf{p})(p\!\!\!/ + m)\gamma^0 d^\dagger(\mathbf{k},s)v(\mathbf{k},s)e^{i(p+k)x} \right\}_a .$$

Now

$$\frac{1}{(2\pi)^3} \int d^3x\, e^{i(p+k)x} = \frac{1}{(2\pi)^3} \int d^3x\, e^{i(\omega_p+\omega_k)t - i(\mathbf{p}+\mathbf{k})\mathbf{x}}$$

$$= \frac{e^{i(\omega_p+\omega_k)t}}{(2\pi)^3} \int d^3x\, e^{-i(\mathbf{p}+\mathbf{k})\mathbf{x}}$$

$$= e^{i(\omega_p + \omega_k)t}\delta(\mathbf{p} + \mathbf{k}) = e^{2i\omega_p t}\delta(\mathbf{p} + \mathbf{k}).$$

Using the momentum space representation of the Dirac delta function

$$(2\pi)^{-3}\int d^3x\, e^{i(p-k)x} = \delta(\mathbf{p} - \mathbf{k}),$$

we obtain:

$$Q_a = \frac{i\lambda}{2}\sum_s \int d^3p \int d^3k \left(\frac{m}{2\omega_p\omega_k}\right)^{1/2}$$

$$\Big\{ C(\mathbf{p})(\not{p} - m)\gamma^0 d(\mathbf{k}, s)u(\mathbf{k}, s)e^{-2i\omega_p t}\delta(\mathbf{p} + \mathbf{k})$$
$$+ C(\mathbf{p})(\not{p} - m)\gamma^0 d^\dagger(\mathbf{k}, s)v(\mathbf{k}, s)\delta(\mathbf{p} - \mathbf{k})$$
$$- D(\mathbf{p})(\not{p} + m)\gamma^0 d(\mathbf{k}, s)u(\mathbf{k}, s)\delta(\mathbf{p} - \mathbf{k})$$
$$- D(\mathbf{p})(\not{p} + m)\gamma^0 d^\dagger(\mathbf{k}, s)v(\mathbf{k}, s)e^{2i\omega_p t}\delta(\mathbf{p} + \mathbf{k})\Big\}_a$$

$$= \frac{i\lambda}{2}\sum_s \int d^3p \left(\frac{m}{2}\right)^{1/2}\frac{1}{\omega_p}$$

$$\Big\{ C(\mathbf{p})(\not{p} - m)\gamma^0 d(-\mathbf{p}, s)u(-\mathbf{p}, s)e^{-2i\omega_p t}$$
$$+ C(\mathbf{p})(\not{p} - m)\gamma^0 d^\dagger(\mathbf{p}, s)v(\mathbf{p}, s)$$
$$- D(\mathbf{p})(\not{p} + m)\gamma^0 d(\mathbf{p}, s)u(\mathbf{p}, s)$$
$$- D(\mathbf{p})(\not{p} + m)\gamma^0 d^\dagger(-\mathbf{p}, s)v(-\mathbf{p}, s)e^{2i\omega_p t}\Big\}_a. \tag{5.51}$$

Now according to Eq. (5.24) we can evaluate the Dirac equation in momentum space

$$(\not{p} - m)u(\mathbf{p}, s) = 0,$$
$$(p_0\gamma_0 - \mathbf{p}\cdot\boldsymbol{\gamma} - m)u(\mathbf{p}, s) = 0,$$
$$\gamma_0(p_0\gamma_0 - \mathbf{p}\cdot\boldsymbol{\gamma} - m)u(\mathbf{p}, s) = 0,$$
$$(p_0\gamma_0^2 - \mathbf{p}\gamma^0\cdot\boldsymbol{\gamma} - m\gamma_0)u(\mathbf{p}, s) = 0.$$

Using the basic Clifford algebra relation

$$\{\gamma^\mu, \gamma^\nu\} = 2\eta^{\mu\nu}\mathbb{1}_{4\times 4},$$

we have

$$\gamma^0\gamma^i = -\gamma^i\gamma^0,$$

and

$$(p_0\gamma_0^2 + \mathbf{p}\cdot\boldsymbol{\gamma}\gamma^0 - m\gamma^0)u(\mathbf{p}, s) = 0$$
$$(p_0\gamma_0 + \mathbf{p}\cdot\boldsymbol{\gamma} - m)\gamma^0 u(\mathbf{p}, s) = 0.$$

Replacing $\mathbf{p}$ by $-\mathbf{p}$ we have:

$$(\not{p} - m)\gamma^0 u(-\mathbf{p}, s) = 0. \tag{5.52}$$

Furthermore:

$$(\not{p} + m)v(\mathbf{p}, s) = 0,$$
$$(p_0\gamma_0 - \mathbf{p} \cdot \boldsymbol{\gamma} + m)v(\mathbf{p}, s) = 0,$$
$$\gamma_0(p_0\gamma_0 - \mathbf{p} \cdot \boldsymbol{\gamma} + m)v(\mathbf{p}, s) = 0,$$
$$(p_0\gamma_0^2 - \mathbf{p}\gamma^0 \cdot \boldsymbol{\gamma} + m\gamma_0)v(\mathbf{p}, s) = 0,$$
$$(p_0\gamma_0 + \mathbf{p} \cdot \boldsymbol{\gamma} + m)\gamma_0 v(\mathbf{p}, s) = 0,$$
$$(\not{p} + m)\gamma_0 v(-\mathbf{p}, s) = 0. \tag{5.53}$$

Inserting Eqs. (5.52) and (5.53) into Eq. (5.51) we see that the time-dependent terms vanish, so that

$$Q_a = \frac{i\lambda}{2} \sum_s \int d^3p \left(\frac{m}{2}\right)^{1/2} \frac{1}{\omega_p} \Big\{ C(\mathbf{p})(\not{p} - m)\gamma^0 d^\dagger(\mathbf{p}, s)v(\mathbf{p}, s)$$
$$- D(\mathbf{p})(\not{p} + m)\gamma^0 d(\mathbf{p}, s)u(\mathbf{p}, s) \Big\}_a. \tag{5.54}$$

Now

$$(\not{p} - m)\gamma^0 v(\mathbf{p}, s) = (p_0\gamma_0 - \mathbf{p} \cdot \boldsymbol{\gamma} - m)\gamma^0 v(\mathbf{p}, s)$$
$$= (p_0\gamma_0^2 - \mathbf{p} \cdot \boldsymbol{\gamma}\gamma^0 - m\gamma^0)v(\mathbf{p}, s), \tag{5.55}$$

and from Eq. (5.53)

$$(p_0\gamma_0^2 + \mathbf{p} \cdot \boldsymbol{\gamma}\gamma^0 + m\gamma^0)v(\mathbf{p}, s) = 0,$$

i.e.

$$p_0\gamma_0^2 v(\mathbf{p}, s) = (-\mathbf{p} \cdot \boldsymbol{\gamma}\gamma^0 - m\gamma^0)\gamma^0 v(\mathbf{p}, s),$$

so that Eq. (5.55) becomes

$$(\not{p} - m)\gamma^0 v(\mathbf{p}, s) = 2p_0(\gamma^0)^2 v(\mathbf{p}, s) = 2p_0 v(\mathbf{p}, s). \tag{5.56}$$

We also have the following result for the momentum space solution $u(\mathbf{p}, s)$.

Proposition 5.5:

The plane wave solution $u(\mathbf{p}, s)$ satisfies the equation

$$(\not{p} + m)\gamma^0 u(\mathbf{p}, s) = 2p_0 u(\mathbf{p}, s). \tag{5.57}$$

Proof: A direct calculation yields:

$$
\begin{aligned}
(\not{p} + m)\gamma^0 u(\mathbf{p}, s) &= (p_0\gamma_0 - \mathbf{p} \cdot \boldsymbol{\gamma} + m)\gamma^0 u(\mathbf{p}, s) \\
&= \left[p_0(\gamma_0)^2 - \mathbf{p} \cdot \boldsymbol{\gamma}\gamma_0 + m\gamma_0 \right] u(\mathbf{p}, s).
\end{aligned}
$$

From Eq. (5.52) we obtain:

$$p_0(\gamma_0)^2 u(\mathbf{p}, s) = \left(-\mathbf{p} \cdot \boldsymbol{\gamma}\gamma_0 + m\gamma_0 \right) u(\mathbf{p}, s),$$

so that

$$(\not{p} + m)\gamma^0 u(\mathbf{p}, s) = 2p_0(\gamma_0)^2 u(\mathbf{p}, s) = 2p_0 u(\mathbf{p}, s).$$

Inserting Eqs. (5.56) and (5.57) into Eq. (5.54) we obtain

$$
\begin{aligned}
Q_a &= \frac{i\lambda}{2}\left(\frac{m}{2}\right)^{1/2} \sum_s \int d^3p \frac{1}{\omega_p} \Big\{ C(\mathbf{p})d^\dagger(\mathbf{p}, s)2p_0 v(\mathbf{p}, s) \\
&\quad - D(\mathbf{p})d(\mathbf{p}, s)2p_0 u(\mathbf{p}, s) \Big\}_a \\
&= i\lambda\left(\frac{m}{2}\right)^{1/2} \sum_s \int d^3p \Big\{ C(\mathbf{p})d^\dagger(\mathbf{p}, s)v(\mathbf{p}, s) - D(\mathbf{p})d(\mathbf{p}, s)u(\mathbf{p}, s \Big\}_a.
\end{aligned}
$$

This completes the proof of Proposition 5.4.

5.4   Projection Operators

We define the operators

$$\Lambda_+ := \frac{\not{p} + m}{2m}, \tag{5.58}$$

and

$$\Lambda_- := \frac{-\not{p} + m}{2m}. \tag{5.59}$$

Proposition 5.6:

The operators $\Lambda_\pm$, defined in Eqs. (5.58) and (5.59) are projection operators, *i.e.*

$$
\left.
\begin{aligned}
(\Lambda_\pm)^2 &= \Lambda_\pm, \\
\Lambda_\pm \Lambda_\mp &= 0, \\
\Lambda_+ + \Lambda_- &= \mathbb{1}.
\end{aligned}
\right\} \tag{5.60}
$$

Proof: Evaluating the squares of the operators $\Lambda_\pm$ as defined in Eqs. (5.58) and (5.59) yields:

$$
\left(\Lambda_\pm\right)^2 = \frac{1}{4m^2}\left(\not{p}\pm m\right)\left(\not{p}\pm m\right) = \frac{1}{4m^2}\left(\not{p}\not{p}\pm 2m\not{p}+m^2\right)
$$

$$
= \frac{1}{4m^2}\left(p^2\pm 2m\not{p}+m^2\right) = \frac{1}{4m^2}\left(2m^2\pm 2m\not{p}\right) = \Lambda_\pm.
$$

Next:

$$
\Lambda_\pm\Lambda_\mp = \frac{1}{4m^2}\left(\pm\not{p}+m\right)\left(\mp\not{p}+m\right) = \frac{1}{4m^2}\left(-\not{p}\not{p}+m^2\right)
$$

$$
= \frac{1}{4m^2}\left(-p^2+m^2\right) = \frac{1}{4m^2}\left(-m^2+m^2\right) = 0.
$$

Finally:

$$
\Lambda_+ + \Lambda_- = \frac{\not{p}+m}{2m} + \frac{-\not{p}+m}{2m} = \mathbb{1}.
$$

This completes the proof of Proposition 5.6.

Proposition 5.7:

The operators Λ_+ and Λ_- project out the positive and negative energy-states respectively, *i.e.*

$$
\left.\begin{array}{rcl}
\Lambda_+ u(\mathbf{p},s) &=& u(\mathbf{p},s), \\
\Lambda_- u(\mathbf{p},s) &=& 0, \\
\Lambda_+ v(\mathbf{p},s) &=& 0, \\
\Lambda_- v(\mathbf{p},s) &=& v(\mathbf{p},s).
\end{array}\right\}
\tag{5.61}
$$

Proof: The second and third of Eqs. (5.61) follow immediately from Eq. (5.24). In order to demonstrate the first relation, we have

$$
\Lambda_+ u(\mathbf{p},s) = \frac{1}{2m}\left(\not{p}+m\right)u(\mathbf{p},s)
$$

$$
= \frac{1}{2m}\left(p_0\gamma_0 - \mathbf{p}\cdot\boldsymbol{\gamma} + m\right)u(\mathbf{p},s)
$$

Now the positive energy eigenstate $u(\mathbf{p},s)$ satisfies

$$
\left(\not{p}-m\right)u(\mathbf{p},s) = 0,
$$

so that

$$
p_0\gamma_0 u(\mathbf{p},s) = \left(\mathbf{p}\cdot\boldsymbol{\gamma}+m\right)u(\mathbf{p},s).
$$

Then

$$
\Lambda_+ u(\mathbf{p},s) = \frac{1}{2m}2m u(\mathbf{p},s) = u(\mathbf{p},s).
$$

The last of Eqs. (5.61) can be shown in a similar fashion.

Proposition 5.8:

The following completeness relations hold:

$$(\Lambda_+)_{ab}(\mathbf{p}) = \sum_s u_a(\mathbf{p}, s) \otimes \bar{u}_b(\mathbf{p}, s), \tag{5.62}$$

$$(\Lambda_-)_{ab}(\mathbf{p}) = -\sum_s v_a(\mathbf{p}, s) \otimes \bar{v}_b(\mathbf{p}, s). \tag{5.63}$$

Proof: Consider the right hand side of Eq. (5.62):

$$\sum_s u(\mathbf{p}, s) \otimes \bar{u}(\mathbf{p}, s) \overset{\underset{(5.30)}{=}}{\underset{(5.34)}{}} (\not{p} + m) \frac{\sum_s u(\mathbf{0}, s) \otimes \bar{u}(\mathbf{0}, s)}{2m(m + \omega_p)} (\not{p} + m)$$

$$= \frac{(\not{p} + m)}{2m(m + \omega_p)} \Big[u(\mathbf{0}, 1) \otimes \bar{u}(\mathbf{0}, 1)$$

$$+ u(\mathbf{0}, 2) \otimes \bar{u}(\mathbf{0}, 2) \Big] (\not{p} + m)$$

$$= \frac{1}{2m(m + \omega_p)} (\not{p} + m) \frac{1}{2} (\mathbb{1}_{4 \times 4} + \gamma^0) (\not{p} + m)$$

$$\overset{\underset{(5.26)}{=}}{\underset{(5.27)}{}} \frac{1}{4m(m + \omega_p)} \Big[(\not{p} + m)(\not{p} + m)$$

$$+ (\not{p} + m)\gamma_0 (\not{p} + m) \Big]$$

$$\overset{(**)}{=} \frac{1}{4m(m + \omega_p)} \left\{ p^2 + 2m\not{p} + m^2 + 2\omega_p (\not{p} + m) \right\}$$

$$\overset{(*)}{=} \frac{1}{4m(m + \omega_p)} \left\{ (2m + 2\omega_p)(\not{p} + m) \right\}$$

$$= \frac{\not{p} + m}{2m}$$

$$= \Lambda_+(p).$$

In step (*) we used $p^2 - m^2 = 0$. In step (**) we made use of the following relation:

$$(\not{p} + m)\gamma_0 (\not{p} + m) = 2\omega_p (\not{p} + m). \tag{5.64}$$

This can be shown as follows:

$$(\not{p} + m)\gamma_0 (\not{p} + m)$$

$$= \not{p}\gamma_0\not{p} + m^2\gamma_0 + m(\not{p}\gamma_0 + \gamma_0\not{p})$$

$$= \not{p}\gamma_0 (p_0\gamma_0 - \mathbf{p} \cdot \boldsymbol{\gamma}) + m^2\gamma_0 + m\Big[p_\mu\gamma^\mu\gamma_0 + \gamma_0 (p_0\gamma_0 - \mathbf{p} \cdot \boldsymbol{\gamma}) \Big]$$

$$= \not{p}(p_0\gamma_0^2 + \mathbf{p} \cdot \boldsymbol{\gamma}\gamma_0) + p^2\gamma_0 + m\Big[p_0\gamma^0 - \mathbf{p} \cdot \boldsymbol{\gamma} + p_0\gamma_0 + \mathbf{p} \cdot \boldsymbol{\gamma} \Big]\gamma_0$$

$$= (p_0\gamma^0 - \mathbf{p} \cdot \boldsymbol{\gamma})(p_0\gamma_0 + \mathbf{p} \cdot \boldsymbol{\gamma})\gamma_0 + p^2\gamma_0 + 2mp_0\gamma_0^2$$

$$= \left\{ (p_0\gamma^0)^2 + p_0\gamma^0 \mathbf{p}\cdot\boldsymbol{\gamma} - \mathbf{p}\cdot\boldsymbol{\gamma}p_0\gamma_0 - \mathbf{p}\cdot\boldsymbol{\gamma}\mathbf{p}\cdot\boldsymbol{\gamma} \right\}\gamma_0$$

$$+ \left(p_0\gamma_0 - \mathbf{p}\cdot\boldsymbol{\gamma} \right)^2\gamma_0 + 2mp_0$$

$$= \left\{ p_0^2(\gamma^0)^2 + p_0\gamma^0 \mathbf{p}\cdot\boldsymbol{\gamma} - \mathbf{p}\cdot\boldsymbol{\gamma}p_0\gamma_0 - \mathbf{p}\cdot\boldsymbol{\gamma}\mathbf{p}\cdot\boldsymbol{\gamma} \right\}\gamma_0$$

$$+ \left\{ p_0^2(\gamma^0)^2 - p_0\gamma_0 \mathbf{p}\cdot\boldsymbol{\gamma} - \mathbf{p}\cdot\boldsymbol{\gamma}p_0\gamma_0 + \mathbf{p}\cdot\boldsymbol{\gamma}\mathbf{p}\cdot\boldsymbol{\gamma} \right\}\gamma_0 + 2mp_0$$

$$= 2p_0^2\gamma^0 - 2\mathbf{p}\cdot\boldsymbol{\gamma}p_0\gamma_0^2 + 2mp_0 = 2p_0\left(p_0\gamma^0 - \mathbf{p}\cdot\boldsymbol{\gamma} \right) + 2mp_0$$

$$= 2p_0\not{p} + 2mp_0 = 2\omega_p\left(\not{p} + m \right).$$

Here we made use of

$$p_0 = \omega_p = \sqrt{\mathbf{p}^2 + m^2}.$$

In a similar way one can show that

$$(\not{p} - m)\gamma_0(\not{p} - m) = 2\omega_p(\not{p} - m). \qquad (5.65)$$

We have

$$(\not{p} - m)\gamma_0(\not{p} - m) = \not{p}\gamma_0\not{p} + m^2\gamma_0 - m(\not{p}\gamma_0 + \gamma_0\not{p}) = 2p_0\not{p} - 2mp_0.$$

Here the last expression results in the same way as above so that

$$(\not{p} - m)\gamma_0(\not{p} - m) = p_0(\not{p} - m),$$

which had to be shown.

We now come to the proof of Eq. (5.63). We have, using Eqs. (5.30) and (5.35),

$$\sum_s v(\mathbf{p}, s) \otimes \overline{v}(\mathbf{p}, s) = (m - \not{p})\frac{\sum_s v(\mathbf{0}, s) \otimes \overline{v}(\mathbf{0}, s)}{2m(m + \omega_p)}(m - \not{p})$$

$$= \frac{1}{2m(m + \omega_p)}\Big\{ (m - \not{p})\left[v(\mathbf{0}, 1) \otimes \overline{v}(\mathbf{0}, 1) \right.$$

$$\left. + v(\mathbf{0}, 2) \otimes \overline{v}(\mathbf{0}, 2) \right](m - \not{p}) \Big\}.$$

Using Eq. (5.27), we obtain

$$v(\mathbf{0}, 1) \otimes \overline{v}(\mathbf{0}, 1) + v(\mathbf{0}, 2) \otimes \overline{v}(\mathbf{0}, 2)$$

$$= \begin{pmatrix} 0 \\ 0 \\ 0 \\ 1 \end{pmatrix} \otimes (0 \ \ 0 \ \ 0 \ \ -1) \begin{pmatrix} 0 \\ 0 \\ -1 \\ 0 \end{pmatrix} \otimes (0 \ \ 0 \ \ 1 \ \ 0)$$

$$= \begin{pmatrix} 0 & 0 & 0 & 0 \\ 0 & 0 & 0 & 0 \\ 0 & 0 & -1 & 0 \\ 0 & 0 & 0 & -1 \end{pmatrix} = -\frac{1}{2}\left(\mathbb{1}_{4\times 4} - \gamma^0 \right).$$

Hence

$$\sum_s v(\mathbf{p}, s) \otimes \overline{v}(\mathbf{p}, s)$$

$$= -\frac{1}{2m(m + \omega_p)} \left\{ (\not{p} - m) \frac{\mathbf{1}_{4 \times 4} - \gamma^0}{2} (\not{p} - m) \right\}$$

$$= -\frac{1}{4m(m + \omega_p)} \left\{ (\not{p} - m)(\not{p} - m) - (\not{p} - m)\gamma^0(\not{p} - m) \right\}.$$

Applying Eq. (5.65) we find

$$\sum_s v(\mathbf{p}, s) \otimes \overline{v}(\mathbf{p}, s) = -\frac{1}{4m(m + \omega_p)} \left\{ 2m^2 - m\not{p} - 2\omega_p(\not{p} - m) \right\}$$

$$= -\frac{1}{2m(m + \omega_p)} \left\{ m(m - \not{p}) + \omega_p(m - \not{p}) \right\}$$

$$= -\frac{1}{2m(m + \omega_p)} (m + \omega_p)(m - \not{p})$$

$$= -\frac{m - \not{p}}{2m} = -\Lambda_-.$$

This completes the proof of Proposition 5.8.

Some other formulas which will be needed later follow immediately from Eq. (5.63), *e.g.*

$$\sum_a \overline{u}_a(\mathbf{p}, s) \Lambda_{-ab}(\mathbf{p}) = -\sum_a \overline{u}_a(\mathbf{p}, s) \sum_t v_a(\mathbf{p}, t) \overline{v}_b(\mathbf{p}, t) = 0,$$

where the last step is a consequence of Eq. (5.32). Thus

$$\overline{u}_a(\mathbf{p}, s) \Lambda_{-ab}(\mathbf{p}) = 0 \tag{5.66}$$

and similarly

$$\overline{v}_a(\mathbf{p}, s) \Lambda_{+ab}(\mathbf{p}) = 0. \tag{5.67}$$

5.5 Anticommutation Relations

We mentioned earlier that we want to show that the Wess–Zumino model is a supersymmetric field theory. This means that we have to show that the spinor charges for that model, given by Eq. (5.49), obey the anticommutation relation

$$\{Q_a, \overline{Q}_b\} = 2\not{P}_{ab}.$$

In order to evaluate this anticommutator, we first have to derive the Dirac adjoint of the Majorana spinor charge, *i.e.* $\overline{Q}_a$. Now, according to Eq. (5.49), the spinor charge of the Wess–Zumino model is given by

$$Q_a = \frac{i\lambda m^{1/2}}{\sqrt{2}} \sum_s \int d^3p \Big[C(\mathbf{p})d^\dagger(\mathbf{p},s)v(\mathbf{p},s) - D(\mathbf{p})d(\mathbf{p},s)u(\mathbf{p},s) \Big]_a.$$

Then the Hermitian conjugate of this charge operator is:

$$Q_a^\dagger = -\frac{i\lambda m^{1/2}}{\sqrt{2}} \sum_s \int d^3p \Big[v^\dagger(\mathbf{p},s)d(\mathbf{p},s)C^\dagger(\mathbf{p}) - u^\dagger(\mathbf{p})d^\dagger(\mathbf{p},s)D^\dagger(\mathbf{p}) \Big]_a,$$

and the Dirac adjoint is

$$\begin{aligned}
\overline{Q}_a &= \left(Q^\dagger \gamma_0 \right)_a \\
&= -\frac{i\lambda m^{1/2}}{\sqrt{2}} \sum_s \int d^3p \Big[v^\dagger(\mathbf{p},s)d(\mathbf{p},s)C^\dagger(\mathbf{p})\gamma_0 - u^\dagger(\mathbf{p})d^\dagger(\mathbf{p},s)D^\dagger(\mathbf{p})\gamma_0 \Big]_a.
\end{aligned}$$

$$(5.68)$$

Now, using Eq. (5.50) we have:

$$\begin{aligned}
C^\dagger(\mathbf{p})\gamma_0 &= \big(a^\dagger(\mathbf{p})\mathbb{1}_{4\times4} + i\gamma_5^\dagger b^\dagger(\mathbf{p}) \big)\gamma_0 \\
&= a^\dagger(\mathbf{p})\gamma_0 - i\gamma_0\gamma_5^\dagger b^\dagger(\mathbf{p}) \\
&= \gamma_0 \big(a^\dagger(\mathbf{p})\mathbb{1}_{4\times4} - i\gamma_5 b^\dagger(\mathbf{p}) \big) \\
&= \gamma_0 D(\mathbf{p}),
\end{aligned}$$

where we used Eq. (5.50). Hence

$$C^\dagger(\mathbf{p})\gamma_0 = \gamma_0 D(\mathbf{p}). \qquad (5.69)$$

From Eq. (5.69) it follows

$$D^\dagger(\mathbf{p})\gamma_0 = \gamma_0 C(\mathbf{p}). \qquad (5.70)$$

Inserting Eqs. (5.69) and (5.70) into the momentum space expansion of the supercharge, *i.e.* Eq. (5.68), we obtain the Dirac adjoint of the supercharge:

$$\begin{aligned}
\overline{Q}_a &= -\frac{i\lambda m^{1/2}}{\sqrt{2}} \sum_s \int d^3p \Big[v^\dagger(\mathbf{p},s)\gamma_0 d(\mathbf{p},s)D(\mathbf{p}) - u^\dagger(\mathbf{p},s)\gamma_0 d^\dagger(\mathbf{p},s)C(\mathbf{p}) \Big]_a \\
&= -\frac{i\lambda m^{1/2}}{\sqrt{2}} \sum_s \int d^3p \Big[\overline{v}(\mathbf{p},s)d(\mathbf{p},s)D(\mathbf{p}) - \overline{u}(\mathbf{p},s)d^\dagger(\mathbf{p},s)C(\mathbf{p}) \Big]_a.
\end{aligned}$$

$$(5.71)$$

Before we start the calculation of the anticommutator of the spinor charges Q and $\overline{Q}$, it is advantageous to derive a number of identities, which will be helpful later on.

Proposition 5.9:

The following relations hold:

$$\left[C_{ac}(\mathbf{p}), D_{bd}(\mathbf{k})\right] = \left(\delta_{ac}\delta_{bd} - \gamma^5_{ac}\gamma^5_{bd}\right)\delta(\mathbf{p} - \mathbf{k}), \tag{5.72}$$

and

$$\left.\begin{aligned}
\left[C_{ac}(\mathbf{p}), C_{bd}(\mathbf{k})\right] &= 0, \\
\left[C^\dagger_{ac}(\mathbf{p}), C^\dagger_{bd}(\mathbf{k})\right] &= 0, \\
\left[D_{ac}(\mathbf{p}), D_{bd}(\mathbf{k})\right] &= 0, \\
\left[D^\dagger_{ac}(\mathbf{p}), D^\dagger_{bd}(\mathbf{k})\right] &= 0.
\end{aligned}\right\} \tag{5.73}$$

Proof: In order to prove Eq. (5.72) we make use of the definitions of $C(\mathbf{p})$ and $D(\mathbf{p})$ given in Eq. (5.50), *i.e.*

$$\begin{aligned}
\left[C_{ac}(\mathbf{p}), D_{bd}(\mathbf{k})\right] &= \left[a(\mathbf{p})\delta_{ac} - i\gamma^5_{ac}b(\mathbf{p}), a^\dagger(\mathbf{k})\delta_{bd} - i\gamma^5_{bd}b^\dagger(\mathbf{k})\right] \\
&= \left[a(\mathbf{p}), a^\dagger(\mathbf{k})\right]\delta_{ac}\delta_{bd} \\
&\quad - i\gamma^5_{ac}\delta_{bd}\left[b(\mathbf{p}), a^\dagger(\mathbf{k})\right] \\
&\quad - i\delta_{ac}\gamma^5_{bd}\left[a(\mathbf{p}), b^\dagger(\mathbf{k})\right] \\
&\quad - \gamma^5_{ac}\gamma^5_{bd}\left[b(\mathbf{p}), b^\dagger(\mathbf{k})\right] \\
&= \delta_{ac}\delta_{bd}\delta(\mathbf{p} - \mathbf{k}) - \gamma^5_{ac}\gamma^5_{bd}\delta(\mathbf{p} - \mathbf{k}) \\
&= \left(\delta_{ac}\delta_{bd} - \gamma^5_{ac}\gamma^5_{bd}\right)\delta(\mathbf{p} - \mathbf{k}).
\end{aligned}$$

Here we used the commutation relations of the bosonic creation and annihilation operators $a(\mathbf{p}), a^\dagger(\mathbf{p}), b(\mathbf{p})$ and $b^\dagger(\mathbf{p})$ of the scalar fields $A(x)$ and $B(x)$ in the Wess–Zumino model. These operators obey:

$$\begin{aligned}
\left[a(\mathbf{p}), a^\dagger(\mathbf{k})\right] &= \delta(\mathbf{p} - \mathbf{k}), \\
\left[b(\mathbf{p}), b^\dagger(\mathbf{k})\right] &= \delta(\mathbf{p} - \mathbf{k}),
\end{aligned}$$

all other commutators of these operators vanish. Using these commutators of the bosonic creation and annihilation operators, it is easy to see that all the commutators of Eq. (5.73) vanish. This completes the proof of Proposition 5.9.

Now we turn our attention to the computation of the anticommutator of the supercharges Q and $\overline{Q}$. Using Eqs (5.49) and (5.71) we have

$$\{Q_a, \overline{Q}_b\}$$

$$= \left\{ i\lambda \left(\frac{m}{2}\right)^{1/2} \sum_s \int d^3p \left(C(\mathbf{p})d^\dagger(\mathbf{p}, s)v(\mathbf{p}, s) - D(\mathbf{p})d(\mathbf{p}, s)u(\mathbf{p}, s) \right)_a, \right.$$

$$\left. - i\lambda \left(\frac{m}{2}\right)^{1/2} \sum_r \int d^3k \left(\overline{v}(\mathbf{k}, r)d(\mathbf{k}, r)D(\mathbf{k}) - \overline{u}(\mathbf{k}, r)d^\dagger(\mathbf{k}, r)C(\mathbf{k}) \right)_b \right\}$$

$$= \frac{m\lambda^2}{2} \sum_{s,r} \int d^3p \int d^3k \left\{ C_{ac}(\mathbf{p})d^\dagger(\mathbf{p}, s)v_c(\mathbf{p}, s) - D_{ac}(\mathbf{p})d(\mathbf{p}, s)u_c(\mathbf{p}, s), \right.$$

$$\left. \overline{v}_d(\mathbf{k}, r)d(\mathbf{k}, r)D_{db}(\mathbf{k}) - \overline{u}_d(\mathbf{k}, r)d^\dagger(\mathbf{k}, r)C_{db}(\mathbf{k}) \right\}$$

$$= \frac{m\lambda^2}{2} \sum_{s,r} \int d^3p\, d^3k \left(\left[C_{ac}(\mathbf{p})d^\dagger(\mathbf{p}, s)v_c(\mathbf{p}, s) - D_{ac}(\mathbf{p})d(\mathbf{p}, s)u_c(\mathbf{p}, s) \right] \right.$$

$$\times \left[D_{db}(\mathbf{k})d(\mathbf{k}, r)\overline{v}_d(\mathbf{k}, r) - C_{db}(\mathbf{k})d^\dagger(\mathbf{k}, r)\overline{u}_d(\mathbf{k}, r) \right]$$

$$+ \left[D_{db}(\mathbf{k})d(\mathbf{k}, r)\overline{v}_d(\mathbf{k}, r) - C_{db}(\mathbf{k})d^\dagger(\mathbf{k}, r)\overline{u}_d(\mathbf{k}, r) \right] \times$$

$$\left. \left[C_{ac}(\mathbf{p})d^\dagger(\mathbf{p}, s)v_c(\mathbf{p}, s) - D_{ac}(\mathbf{p})d(\mathbf{p}, s)u_c(\mathbf{p}, s) \right] \right)$$

$$= \frac{m\lambda^2}{2} \sum_{s,r} \int d^3p\, d^3k \left(C_{ac}(\mathbf{p})d^\dagger(\mathbf{p}, s)v_c(\mathbf{p}, s)D_{db}(\mathbf{k})d(\mathbf{k}, r)\overline{v}_d(\mathbf{k}, r) \right.$$

$$- C_{ac}(\mathbf{p})d^\dagger(\mathbf{p}, s)v_c(\mathbf{p}, s)C_{db}(\mathbf{k})d^\dagger(\mathbf{k}, r)\overline{u}_d(\mathbf{k}, r)$$

$$- D_{ac}(\mathbf{p})d(\mathbf{p}, s)u_c(\mathbf{p}, s)D_{db}(\mathbf{k})d(\mathbf{k}, r)\overline{v}_d(\mathbf{k}, r)$$

$$+ D_{ac}(\mathbf{p})d(\mathbf{p}, s)u_c(\mathbf{p}, s)C_{db}(\mathbf{k})d^\dagger(\mathbf{k}, r)\overline{u}_d(\mathbf{k}, r)$$

$$+ D_{db}(\mathbf{k})d(\mathbf{k}, r)\overline{v}_d(\mathbf{k}, r)C_{ac}(\mathbf{p})d^\dagger(\mathbf{p}, s)v_c(\mathbf{p}, s)$$

$$- D_{db}(\mathbf{k})d(\mathbf{k}, r)\overline{v}_d(\mathbf{k}, r)D_{ac}(\mathbf{p})d(\mathbf{p}, s)u_c(\mathbf{p}, s)$$

$$- C_{db}(\mathbf{k})d^\dagger(\mathbf{k}, r)\overline{u}_d(\mathbf{k}, r)C_{ac}(\mathbf{p})d^\dagger(\mathbf{p}, s)v_c(\mathbf{p}, s)$$

$$\left. + C_{db}(\mathbf{k})d^\dagger(\mathbf{k}, r)\overline{u}_d(\mathbf{k}, r)D_{ac}(\mathbf{p})d(\mathbf{p}, s)u_c(\mathbf{p}, s) \right).$$

Now consider the following two terms in the above expansion:

$$-C_{ac}(\mathbf{p})d^{\dagger}(\mathbf{p},s)v_c(\mathbf{p},s)C_{db}(\mathbf{k})d^{\dagger}(\mathbf{k},r)\overline{u}_d(\mathbf{k},r)$$

$$- C_{db}(\mathbf{k})d^{\dagger}(\mathbf{k},r)\overline{u}_d(\mathbf{k},r)C_{ac}(\mathbf{p})d^{\dagger}(\mathbf{p},s)v_c(\mathbf{p},s)$$

$$= d^{\dagger}(\mathbf{p},s)d^{\dagger}(\mathbf{k},r)\Big[-C_{ac}(\mathbf{p})C_{db}(\mathbf{k}) + C_{db}(\mathbf{k})C_{ac}(\mathbf{p})\Big]v_c(\mathbf{p},s)\overline{u}_d(\mathbf{k},r)$$

$$= -d^{\dagger}(\mathbf{p},s)d^{\dagger}(\mathbf{k},r)\Big[C_{ac}(\mathbf{p}),C_{db}(\mathbf{k})\Big]v_c(\mathbf{p},s)\overline{u}_d(\mathbf{k},r)$$

$$= 0,$$

where we made use of Eq. (5.73). A similar calculation yields:

$$-D_{ac}(\mathbf{p})d(\mathbf{p},s)u_c(\mathbf{p},s)D_{db}(\mathbf{k})d(\mathbf{k},r)\overline{v}_d(\mathbf{k},r)$$

$$- D_{db}(\mathbf{k})d(\mathbf{k},r)\overline{v}_d(\mathbf{k},r)D_{ac}(\mathbf{p})d(\mathbf{p},s)u_c(\mathbf{p},s) = 0.$$

Hence:

$$\{Q_a,\overline{Q}_b\} = \frac{m\lambda^2}{2}\sum_{s,r}\int d^3p \int d^3k$$

$$\left(\Big\{C_{ac}(\mathbf{p})d^{\dagger}(\mathbf{p},s),D_{db}(\mathbf{k})d(\mathbf{k},r)\Big\}v_c(\mathbf{p},s)\overline{v}_d(\mathbf{k},r)\right.$$

$$\left.+ \Big\{D_{ac}(\mathbf{p})d(\mathbf{p},s),C_{db}(\mathbf{k})d^{\dagger}(\mathbf{k},r)\Big\}u_c(\mathbf{p},s)\overline{u}_d(\mathbf{k},r)\right).$$

Now the two anticommutators in this expansion can be rewritten as follows

$$\{C\,d^{\dagger},D\,d\} = C\,d^{\dagger}\,D\,d + D\,d\,C\,d^{\dagger}$$

$$= C\,D\,d^{\dagger}\,d + D\,C\,d\,d^{\dagger}$$

$$= C\,D\,d^{\dagger}\,d + D\,C\,d\,d^{\dagger} + D\,C\,d^{\dagger}\,d - D\,C\,d^{\dagger}\,d$$

$$= [C,D]\,d^{\dagger}\,d + D\,C\,\{d,d^{\dagger}\}.$$

Thus we can rewrite the anticommutator of two supercharges in the form:

$$\{Q_a,\overline{Q}_b\} = \frac{m\lambda^2}{2}\sum_{s,r}\int d^3p \int d^3k \left(\Big[[C_{ac}(\mathbf{p}),D_{db}(\mathbf{k})]d^{\dagger}(\mathbf{p},s)d(\mathbf{k},r)\right.$$

$$+ D_{db}(\mathbf{k})C_{ac}(\mathbf{p})\{d(\mathbf{k},r),d^{\dagger}(\mathbf{p},s)\}\Big]v_c(\mathbf{p},s)\overline{v}_d(\mathbf{k},r)$$

$$+ \Big[[C_{db}(\mathbf{k}),D_{ac}(\mathbf{p})]d^{\dagger}(\mathbf{k},r)d(\mathbf{p},s)$$

$$\left.+ D_{ac}(\mathbf{p})C_{db}(\mathbf{k})\{d(\mathbf{p},s),d^{\dagger}(\mathbf{k},r)\}\Big]u_c(\mathbf{p},s)\overline{u}_d(\mathbf{k},r)\right)$$

$$\underset{(5.72)}{\overset{(5.46)}{=}} \frac{m\lambda^2}{2}\sum_{s,r}\int d^3p\,d^3k\Big\{\Big([\delta_{ac}\delta_{db}-\gamma^5_{ac}\gamma^5_{db}]\delta(\mathbf{p}-\mathbf{k})d^\dagger(\mathbf{p},s)d(\mathbf{k},r)$$

$$+D_{db}(\mathbf{k})C_{ac}(\mathbf{p})\delta_{rs}\delta(\mathbf{p}-\mathbf{k})\Big)v_c(\mathbf{p},s)\overline{v}_d(\mathbf{k},r)$$

$$+\Big([\delta_{db}\delta_{ac}-\gamma^5_{db}\gamma^5_{ac}]\delta(\mathbf{p}-\mathbf{k})d^\dagger(\mathbf{k},r)d(\mathbf{p},s)$$

$$+D_{ac}(\mathbf{p})C_{db}(\mathbf{k})\delta_{rs}\delta(\mathbf{p}-\mathbf{k})\Big)u_c(\mathbf{p},s)\overline{u}_d(\mathbf{k},r)\Big\}$$

$$= \frac{m\lambda^2}{2}\sum_{s,r}\int d^3p\,\Big\{\Big([\delta_{ac}\delta_{db}-\gamma^5_{ac}\gamma^5_{db}]d^\dagger(\mathbf{p},s)d(\mathbf{p},r)$$

$$+D_{db}(\mathbf{p})C_{ac}(\mathbf{p})\delta_{rs}\Big)v_c(\mathbf{p},s)\overline{v}_d(\mathbf{p},r)$$

$$+\Big([\delta_{db}\delta_{ac}-\gamma^5_{db}\gamma^5_{ac}]d^\dagger(\mathbf{p},r)d(\mathbf{p},s)$$

$$+D_{ac}(\mathbf{p})C_{db}(\mathbf{p})\delta_{rs}\Big)u_c(\mathbf{p},s)\overline{u}_d(\mathbf{p},r)\Big\}$$

$$= \frac{m\lambda^2}{2}\sum_{s,r}\int d^3p\,\Big\{d^\dagger(\mathbf{p},r)d(\mathbf{p},s)\big[\delta_{ac}\delta_{db}-\gamma^5_{ac}\gamma^5_{db}\big]$$

$$\times\big[v_c(\mathbf{p},r)\overline{v}_d(\mathbf{p},s)+u_c(\mathbf{p},s)\overline{u}_d(\mathbf{p},r)\big]$$

$$+\delta_{rs}\Big(D_{db}(\mathbf{p})C_{ac}(\mathbf{p})v_c(\mathbf{p},r)\overline{v}_d(\mathbf{p},s)$$

$$+D_{ac}(\mathbf{p})C_{db}(\mathbf{p})u_c(\mathbf{p},s)\overline{u}_d(\mathbf{p},r)\Big)\Big\}$$

$$= \frac{m\lambda^2}{2}\sum_{s,r}\int d^3p\,\Big\{d^\dagger(\mathbf{p},r)d(\mathbf{p},s)\big[\delta_{ac}\delta_{db}-\gamma^5_{ac}\gamma^5_{db}\big]$$

$$\times\big[v_c(\mathbf{p},r)\overline{v}_d(\mathbf{p},s)+u_c(\mathbf{p},s)\overline{u}_d(\mathbf{p},r)\big]\Big\}$$

$$+\frac{m\lambda^2}{2}\int d^3p\,\Big[-D_{db}(\mathbf{p})C_{ac}(\mathbf{p})\big(\Lambda_-\big)_{cd}+D_{ac}(\mathbf{p})C_{db}(\mathbf{p})\big(\Lambda_+\big)_{cd}\Big]$$

$$= I_1+I_2.$$

In the last step we made use of Eqs. (5.62) and (5.63). We now consider the second integral I_2 and use Eqs. (5.50), (5.58) and (5.59). Then we obtain:

$$I_2 = \frac{m\lambda^2}{2}\int d^3p\,\Big[D_{ac}(\mathbf{p})C_{db}(\mathbf{p})\big(\Lambda_+\big)_{cd}-D_{db}(\mathbf{p})C_{ac}(\mathbf{p})\big(\Lambda_-\big)_{cd}\Big]$$

$$= \frac{m\lambda^2}{2}\int d^3p\,\Big[\Big(a^\dagger(\mathbf{p})\delta_{ac}-i\gamma^5_{ac}b^\dagger(\mathbf{p})\Big)\Big(a(\mathbf{p})\delta_{db}-i\gamma^5_{db}b(\mathbf{p})\Big)\Big(\frac{\not{p}+m}{2m}\Big)_{cd}$$

$$-\Big(a^\dagger(\mathbf{p})\delta_{db}-i\gamma^5_{db}b^\dagger(\mathbf{p})\Big)\Big(a(\mathbf{p})\delta_{ac}-i\gamma^5_{ac}b(\mathbf{p})\Big)\Big(\frac{-\not{p}+m}{2m}\Big)_{cd}\Big]$$

$$
= \frac{m\lambda^2}{2} \int d^3p \left[a^\dagger(\mathbf{p})a(\mathbf{p}) \left(\frac{\not{p}+m}{2m} \right)_{ab} - ia^\dagger(\mathbf{p})b(\mathbf{p})\gamma^5_{db} \left(\frac{\not{p}+m}{2m} \right)_{ad} \right.
$$

$$
- ib^\dagger(\mathbf{p})a(\mathbf{p})\gamma^5_{ac} \left(\frac{\not{p}+m}{2m} \right)_{cb} - b^\dagger(\mathbf{p})b(\mathbf{p})\gamma^5_{ac}\gamma^5_{db} \left(\frac{\not{p}+m}{2m} \right)_{cd}
$$

$$
- a^\dagger(\mathbf{p})a(\mathbf{p}) \left(\frac{-\not{p}+m}{2m} \right)_{ab} + ia^\dagger(\mathbf{p})b(\mathbf{p})\gamma^5_{ac} \left(\frac{-\not{p}+m}{2m} \right)_{cb}
$$

$$
\left. + ib^\dagger(\mathbf{p})a(\mathbf{p})\gamma^5_{db} \left(-\frac{\not{p}+m}{2m} \right)_{ad} + b^\dagger(\mathbf{p})b(\mathbf{p})\gamma^5_{db}\gamma^5_{ac} \left(\frac{-\not{p}+m}{2m} \right)_{cd} \right]
$$

$$
= \frac{m\lambda^2}{2} \int d^3p \left[a^\dagger(\mathbf{p})a(\mathbf{p}) \left(\frac{\not{p}}{m} \right)_{ab} - ia^\dagger(\mathbf{p})b(\mathbf{p}) \left(\frac{\not{p}+m}{2m}\gamma^5 \right)_{ab} \right.
$$

$$
- ib^\dagger(\mathbf{p})a(\mathbf{p}) \left(\gamma^5\frac{\not{p}+m}{2m} \right)_{ab} - b^\dagger(\mathbf{p})b(\mathbf{p}) \left(\gamma^5\frac{\not{p}+m}{2m}\gamma^5 \right)_{ab}
$$

$$
+ ia^\dagger(\mathbf{p})b(\mathbf{p}) \left(\gamma^5\frac{-\not{p}+m}{2m} \right)_{ab} + ib^\dagger(\mathbf{p})a(\mathbf{p}) \left(-\frac{\not{p}+m}{2m}\gamma^5 \right)_{ab}
$$

$$
\left. + b^\dagger(\mathbf{p})b(\mathbf{p}) \left(\gamma^5\frac{-\not{p}+m}{2m}\gamma^5 \right)_{ab} \right]
$$

$$
= \frac{m\lambda^2}{2} \int d^3p \left[a^\dagger(\mathbf{p})a(\mathbf{p}) \left(\frac{\not{p}}{m} \right)_{ab} + b^\dagger(\mathbf{p})b(\mathbf{p}) \left(\frac{\not{p}}{m} \right)_{ab} \right]
$$

$$
= \frac{\lambda^2}{2} \int d^3p\,\not{p}_{ab} \left[a^\dagger(\mathbf{p})a(\mathbf{p}) + b^\dagger(\mathbf{p})b(\mathbf{p}) \right],
$$

taking into account Eqs. (1.163a), (1.163b) and (1.164). Hence the anticommutator of two supercharges becomes:

$$
\left\{ Q_a, \overline{Q}_b \right\} = \frac{m\lambda^2}{2} \sum_{s,r} \int d^3p \left\{ d^\dagger(\mathbf{p}, r)d(\mathbf{p}, s) \left[\delta_{ac}\delta_{db} - \gamma^5_{ac}\gamma^5_{db} \right] \right.
$$

$$
\times \left[v_c(\mathbf{p}, r)\overline{v}_d(\mathbf{p}, s) + u_c(\mathbf{p}, s)\overline{u}_d(\mathbf{p}, r) \right] \Big\}
$$

$$
+ \frac{\lambda^2}{2} \int d^3p\,\not{p}_{ab} \left[a^\dagger(\mathbf{p})a(\mathbf{p}) + b^\dagger(\mathbf{p})b(\mathbf{p}) \right]. \tag{5.74}
$$

We now demonstrate in another lengthy calculation that the first integral I_1 can be rewritten in a simpler form. To this end we start by introducing sixteen linearly independent 4×4-matrices (cf. Eq. (3.15)):

$$
\mathbb{1}_{4\times4}, \gamma^\mu, \gamma^\mu\gamma^5, \sigma^{4\mu\nu}, \gamma^5.
$$

Any 4×4-matrix can be written as a linear combination of these matrices. In particular, we can write

$$
M_{cd}(\mathbf{p}, r, s) := v_c(\mathbf{p}, r) \otimes \overline{v}_d(\mathbf{p}, s) + u_c(\mathbf{p}, s) \otimes \overline{u}_d(\mathbf{p}, r)
$$

$$
= \left(m_0\mathbb{1}_{4\times4} + m_\mu\gamma^\mu + m_{\mu\nu}\sigma^{4\mu\nu} + n_\mu\gamma^\mu\gamma^5 + n_5\gamma^5 \right)_{cd}. \tag{5.75}
$$

Then we find:

$$\left(\delta_{ac}\delta_{db} - \gamma_{ac}^5\gamma_{db}^5\right) M_{cd}(\mathbf{p}, r, s)$$

$$= \left(\delta_{ac}\delta_{db} - \gamma_{ac}^5\gamma_{db}^5\right)\left(v_c(\mathbf{p}, r) \otimes \bar{v}_d(\mathbf{p}, s) + u_c(\mathbf{p}, s) \otimes \bar{u}_d(\mathbf{p}, r)\right)$$

$$= \left(\delta_{ac}\delta_{db} - \gamma_{ac}^5\gamma_{db}^5\right)\left[m_0\mathbb{1}_{4\times4} + m_\mu\gamma^\mu + m_{\mu\nu}\sigma^{4\mu\nu} + n_\mu\gamma^\mu\gamma^5 + n_5\gamma^5\right]_{cd}$$

$$= \left(m_0\mathbb{1}_{4\times4} + m_\mu\gamma^\mu + m_{\mu\nu}\sigma^{4\mu\nu} + n_\mu\gamma^\mu\gamma^5 + n_5\gamma^5\right)_{ab}$$

$$\quad - \left(\gamma^5\left[m_0\mathbb{1}_{4\times4} + m_\mu\gamma^\mu + m_{\mu\nu}\sigma^{4\mu\nu} + n_\mu\gamma^\mu\gamma^5 + n_5\gamma^5\right]\gamma^5\right)_{ab}$$

$$\overset{(*)}{=} m_0\delta_{ab} + m_\mu\gamma_{ab}^\mu + m_{\mu\nu}\left(\sigma^{4\mu\nu}\right)_{ab} + n_\mu\left(\gamma^\mu\gamma^5\right)_{ab}$$

$$\quad + n_5\gamma_{ab}^5 - m_0\delta_{ab} + m_\mu\gamma_{ab}^\mu - m_{\mu\nu}\left(\sigma^{4\mu\nu}\right)_{ab} + n_\mu\left(\gamma^\mu\gamma^5\right)_{ab} - n_5\gamma_{ab}^5$$

$$= 2\left(m_\mu\gamma^\mu + n_\mu\gamma^\mu\gamma^5\right)_{ab}.$$

In step $(*)$ we made use of Eqs. (1.163a), (1.163b) and (1.164).

Proposition 5.10:

The coefficients m_μ and n_μ are given by the following trace relations:

$$m_\mu = \frac{1}{4} \text{Tr}\left[\gamma_\mu M\right], \tag{5.76}$$

$$n_\mu = -\frac{1}{4} \text{Tr}\left[\gamma_\mu\gamma^5 M\right]. \tag{5.77}$$

Proof: These relations can be shown as follows.

$$\frac{1}{4} \text{Tr}\left[\gamma_\mu M\right] \overset{(5.75)}{=} \frac{1}{4} \text{Tr}\left[m_0\gamma_\mu + m_\rho\gamma_\mu\gamma^\rho + m_{\rho\sigma}\gamma_\mu\sigma^{4\rho\sigma}\right.$$

$$\left. + n_\rho\gamma_\mu\gamma^\rho\gamma^5 + n_5\gamma_\mu\gamma^5\right]$$

$$= \frac{m_0}{4} \text{Tr}\left[\gamma_\mu\right] + \frac{m_\rho}{4} \text{Tr}\left[\gamma_\mu\gamma^\rho\right] + \frac{m_{\rho\sigma}}{4} \text{Tr}\left[\gamma_\mu\sigma^{4\rho\sigma}\right]$$

$$\quad + \frac{n_\rho}{4} \text{Tr}\left[\gamma_\mu\gamma^\rho\gamma^5\right] + \frac{n_5}{4} \text{Tr}\left[\gamma_\mu\gamma^5\right]$$

$$= \frac{1}{4}m_\rho 4\delta_\mu{}^\rho = m_\mu.$$

Here we made use of the fact that

$$\text{Tr}\left[\gamma_\mu\right] = \text{Tr}\left[\gamma^5\gamma_\mu\right] = \text{Tr}\left[\gamma_\mu\gamma^\rho\gamma^5\right] = 0.$$

In a similar way we get:

$$-\frac{1}{4}\,\mathrm{Tr}\,\left[\gamma_\mu\gamma^5 M\right] \stackrel{(5.75)}{=} -\frac{1}{4}\,\mathrm{Tr}\,\left[m_0\gamma_\mu\gamma^5 + m_\rho\gamma_\mu\gamma^5\gamma^\rho + m_{\rho\sigma}\gamma_\mu\gamma^5\sigma^{4\rho\sigma}\right.$$
$$\left. +n_\rho\gamma_\mu\gamma^5\gamma^\rho\gamma^5 + n_5\gamma_\mu\gamma^5\gamma^5\right]$$
$$= \frac{1}{4}n_\rho 4\delta_\mu{}^\rho = n_\mu.$$

This completes the proof of Proposition 5.10.

Then Eq. (5.74) becomes

$$\left\{Q_a,\overline{Q}_b\right\} = \frac{m\lambda^2}{2}\sum_{r,s}\int d^3 p\,\left\{d^\dagger(\mathbf{p},r)d(\mathbf{p},s)2\left(m_\mu\gamma^\mu + n_\mu\gamma^\mu\gamma^5\right)_{ab}\right.$$
$$+\frac{\lambda^2}{2}\int d^3 p\,\slashed{p}_{ab}\left[a^\dagger(\mathbf{p})a(\mathbf{p}) + b^\dagger(\mathbf{p})b(\mathbf{p})\right]$$
$$= \frac{m\lambda^2}{4}\sum_{r,s}\int d^3 p\,\left\{d^\dagger(\mathbf{p},r)d(\mathbf{p},s)\left(\mathrm{Tr}\,\left[\gamma_\mu M(\mathbf{p},r,s)\right]\gamma^\mu\right.\right.$$
$$\left.- \mathrm{Tr}\,\left[\gamma_\mu\gamma^5 M(\mathbf{p},r,s)\right]\gamma^\mu\gamma^5\right)_{ab}$$
$$+\frac{\lambda^2}{2}\int d^3 p\,\slashed{p}_{ab}\left[a^\dagger(\mathbf{p})a(\mathbf{p}) + b^\dagger(\mathbf{p})b(\mathbf{p})\right]. \tag{5.78}$$

In deriving this expression, we used Eqs. (5.76) and (5.77). Evaluating the traces we find:

$$\mathrm{Tr}\,\left[\gamma_\mu M(\mathbf{p},r,s)\right] = (\gamma_\mu)_{ab} M_{ba}(\mathbf{p},r,s)$$
$$\stackrel{(5.75)}{=} (\gamma_\mu)_{ab}\left(v_b(\mathbf{p},r)\overline{v}_a(\mathbf{p},s) + u_b(\mathbf{p},s)\overline{u}_a(\mathbf{p},r)\right)$$
$$= \overline{v}(\mathbf{p},s)\gamma_\mu v(\mathbf{p},r) + \overline{u}(\mathbf{p},r)\gamma_\mu u(\mathbf{p},s). \tag{5.79}$$

In a similar way one gets

$$\mathrm{Tr}\,\left[\gamma_\mu\gamma^5 M(\mathbf{p},r,s)\right] = \overline{v}(\mathbf{p},s)\gamma_\mu\gamma^5 v(\mathbf{p},r) + \overline{u}(\mathbf{p},r)\gamma_\mu\gamma^5 u(\mathbf{p},s). \tag{5.80}$$

Using Eq. (5.61) we have

$$\overline{u}(\mathbf{p},r)\gamma_\mu u(\mathbf{p},s) = \overline{u}(\mathbf{p},r)\gamma_\mu\Lambda_+(\mathbf{p})u(\mathbf{p},s)$$
$$\stackrel{(5.58)}{=} \overline{u}(\mathbf{p},r)\gamma_\mu\left(\frac{\slashed{p}+m}{2m}\right)u(\mathbf{p},s)$$

$$= \frac{1}{2m}\overline{u}(\mathbf{p},r)\gamma_\mu\left(p_\rho\gamma^\rho + m\right)u(\mathbf{p},s)$$

$$= \frac{1}{2m}\overline{u}(\mathbf{p},r)\left(p_\rho\gamma_\mu\gamma^\rho + m\gamma_\mu\right)u(\mathbf{p},s)$$

$$= \frac{1}{2m}\overline{u}(\mathbf{p},r)\left(p_\rho 2\delta_\mu{}^\rho - p_\rho\gamma^\rho\gamma_\mu + m\gamma_\mu\right)u(\mathbf{p},s)$$

$$= \frac{1}{2m}\overline{u}(\mathbf{p},r)\left(2p_\mu - \slashed{p}\gamma_\mu + m\gamma_\mu\right)u(\mathbf{p},s)$$

$$= \overline{u}(\mathbf{p},r)\frac{p_\mu}{m}u(\mathbf{p},s) + \overline{u}(\mathbf{p},r)\Lambda_-(\mathbf{p})\gamma_\mu u(\mathbf{p},s)$$

$$= \frac{p_\mu}{m}\overline{u}(\mathbf{p},r)u(\mathbf{p},s).$$

In the last step we made use of the projection property $\overline{u}(\mathbf{p},r)\Lambda_-(\mathbf{p}) = 0$. Since the momentum space solutions $u(\mathbf{p},r)$ satisfy the orthogonality relation (5.31), we get:

$$\overline{u}(\mathbf{p},r)\gamma_\mu u(\mathbf{p},s) = \frac{p_\mu}{m}\delta_{rs}. \tag{5.81}$$

A corresponding relation holds for $\overline{v}(\mathbf{p},s)\gamma_\mu v(\mathbf{p},r)$, i.e.

$$\overline{v}(\mathbf{p},s)\gamma_\mu v(\mathbf{p},r) = \frac{p_\mu}{m}\delta_{rs}. \tag{5.82}$$

Inserting Eqs. (5.79) to (5.82) into the anticommutator (5.78), we obtain:

$$\begin{aligned}
\{Q_a,\overline{Q}_b\} &= \frac{\lambda^2}{2}\int d^3p\,\slashed{p}_{ab}\left[a^\dagger(\mathbf{p})a(\mathbf{p}) + b^\dagger(\mathbf{p})b(\mathbf{p})\right] \\
&\quad + \frac{m\lambda^2}{4}\sum_{r,s}\int d^3p\,d^\dagger(\mathbf{p},r)d(\mathbf{p},s) \\
&\quad \times \left\{\left[\overline{v}(\mathbf{p},s)\gamma_\mu v(\mathbf{p},r) + \overline{u}(\mathbf{p},r)\gamma_\mu u(\mathbf{p},s)\right]\gamma^\mu_{ab}\right. \\
&\quad \left. - \left[\overline{v}(\mathbf{p},s)\gamma_\mu\gamma^5 v(\mathbf{p},r) + \overline{u}(\mathbf{p},r)\gamma_\mu\gamma^5 u(\mathbf{p},s)\right]\left(\gamma^\mu\gamma^5\right)_{ab}\right\} \\
&= \frac{\lambda^2}{2}\int d^3p\,\slashed{p}_{ab}\left[a^\dagger(\mathbf{p})a(\mathbf{p}) + b^\dagger(\mathbf{p})b(\mathbf{p})\right] \\
&\quad + \frac{m\lambda^2}{4}\sum_{r,s}\int d^3p\,d^\dagger(\mathbf{p},r)d(\mathbf{p},s)\left\{\left[\frac{p_\mu}{m}\delta_{rs} + \frac{p_\mu}{m}\delta_{rs}\right]\gamma^\mu_{ab}\right. \\
&\quad \left. - \left[\overline{v}(\mathbf{p},s)\gamma_\mu\gamma^5 v(\mathbf{p},r) + \overline{u}(\mathbf{p},r)\gamma_\mu\gamma^5 u(\mathbf{p},s)\right]\left(\gamma^\mu\gamma^5\right)_{ab}\right\} \\
&= \frac{\lambda^2}{2}\int d^3p\,\slashed{p}_{ab}\left[a^\dagger(\mathbf{p})a(\mathbf{p}) + b^\dagger(\mathbf{p})b(\mathbf{p}) + \sum_r d^\dagger(\mathbf{p},r)d(\mathbf{p},r)\right]
\end{aligned}$$

$$-\frac{m\lambda^2}{4}\sum_{r,s}\int d^3p\, d^\dagger(\mathbf{p},r)d(\mathbf{p},s)\Big\{\overline{v}(\mathbf{p},s)\gamma_\mu\gamma^5 v(\mathbf{p},r)$$

$$+\overline{u}(\mathbf{p},r)\gamma_\mu\gamma^5 u(\mathbf{p},s)\Big](\gamma^\mu\gamma^5)_{ab}. \tag{5.83}$$

We now convince ourselves that the contribution of the term proportional to $(\gamma^\mu\gamma^5)_{ab}$ vanishes. This can be shown to follow from symmetry properties of both sides. Multiplying Eq. (5.83) from the right by the charge conjugation matrix C_{db} as in the calculations of the proof of Eq. (3.27), and using $C=-C^\top$, then

$$\big\{Q_a,Q_d\big\}$$

$$=\frac{\lambda^2}{2}\int d^3p\, p_\mu(\gamma^\mu C^\top)_{ad}\Big[a^\dagger(\mathbf{p})a(\mathbf{p})+b^\dagger(\mathbf{p})b(\mathbf{p})+\sum_r d^\dagger(\mathbf{p},r)d(\mathbf{p},r)\Big]$$

$$-\frac{m\lambda^2}{4}\sum_{r,s}\int d^3p\, d^\dagger(\mathbf{p},r)d(\mathbf{p},s)\Big\{\overline{v}(\mathbf{p},s)\gamma_\mu\gamma^5 v(\mathbf{p},r)$$

$$+\overline{u}(\mathbf{p},r)\gamma_\mu\gamma^5 u(\mathbf{p},s)\Big](\gamma^\mu\gamma^5 C^\top)_{ad}. \tag{5.84}$$

Now, according to Eqs. (3.16) and (3.19b), the first part proportional to $(\gamma^\mu C)$, is symmetric in its indices a and d, whereas the second part is antisymmetric in a and d. The left hand side of Eq. (5.84) is symmetric in a and d, so the coefficient of the $\gamma^\mu\gamma^5 C$-term must vanish.

Thus we conclude that the anticommutator of two supercharges Q_a and $\overline{Q}_b$ in the Wess–Zumino model is given by

$$\boxed{\big\{Q_a,\overline{Q}_b\big\}=\frac{\lambda^2}{2}\int d^3p\, \slashed{p}_{ab}\Big[a^\dagger(\mathbf{p})a(\mathbf{p})+b^\dagger(\mathbf{p})b(\mathbf{p})+\sum_r d^\dagger(\mathbf{p},r)d(\mathbf{p},r)\Big].}$$
$$\tag{5.85}$$

In Sec. 5.6 we will see that the energy-momentum operator is given by

$$P^\mu=\int d^3p\, p^\mu\Big[a^\dagger(\mathbf{p})a(\mathbf{p})+b^\dagger(\mathbf{p})b(\mathbf{p})+\sum_r d^\dagger(\mathbf{p},r)d(\mathbf{p},r)\Big]. \tag{5.86}$$

Thus if we choose $\lambda=2$ we obtain

$$\big\{Q_a,\overline{Q}_b\big\}=2P_\mu(\gamma^\mu)_{ab}. \tag{5.87}$$

This result shows that the Super–Poincaré algebra can indeed be represented in terms of commutators and anticommutators of linear operators in Fock space. We now demonstrate that the fourth of relations (4.1) is also satisfied.

Proposition 5.11:

Given the momentum operator of Eq. (5.86) and the spinor charge of Eq. (5.49), then the commutator of these operators vanishes, *i.e.*

$$\left[P^\mu, Q_a\right] = 0. \tag{5.88}$$

Proof: Using the momentum space expansions of the operators P^μ and Q_a, *i.e.* Eqs. (5.49) and (5.86), we get

$$\left[P^\mu, Q_a\right] = i(2m)^{1/2}\left[\int d^3p\, p^\mu \left[a^\dagger(\mathbf{p})a(\mathbf{p}) + b^\dagger(\mathbf{p})b(\mathbf{p}) + \sum_r d^\dagger(\mathbf{p}, r)d(\mathbf{p}, r)\right],\right.$$

$$\left.\sum_s \int d^3k \left\{C(\mathbf{k})d^\dagger(\mathbf{k}, s)v(\mathbf{k}, s) - D(\mathbf{k})d(\mathbf{k}, s)u(\mathbf{k}, s)\right\}_a\right]$$

$$= i(2m)^{1/2}\int d^3p\, p^\mu \int d^3k \sum_s \left[a^\dagger(\mathbf{p})a(\mathbf{p}) + b^\dagger(\mathbf{p})b(\mathbf{p})\right.$$

$$+ \sum_r d^\dagger(\mathbf{p}, r)d(\mathbf{p}, r), C_{ab}(\mathbf{k})d^\dagger(\mathbf{k}, s)v_b(\mathbf{k}, s)$$

$$\left. - D_{ab}(\mathbf{k})d(\mathbf{k}, s)u_b(\mathbf{k}, s)\right]$$

$$= i(2m)^{1/2}\int d^3p\, p^\mu \int d^3k \sum_s$$

$$\times \left\{\left[a^\dagger(\mathbf{p})a(\mathbf{p}), C_{ab}(\mathbf{k})d^\dagger(\mathbf{k}, s)\right]v_b(\mathbf{k}, s)\right.$$

$$- \left[a^\dagger(\mathbf{p})a(\mathbf{p}), D_{ab}(\mathbf{k})d(\mathbf{k}, s)\right]u_b(\mathbf{k}, s)$$

$$+ \left[b^\dagger(\mathbf{p})b(\mathbf{p}), C_{ab}(\mathbf{k})d^\dagger(\mathbf{k}, s)\right]v_b(\mathbf{k}, s)$$

$$- \left[b^\dagger(\mathbf{p})b(\mathbf{p}), D_{ab}(\mathbf{k})d(\mathbf{k}, s)\right]u_b(\mathbf{k}, s)$$

$$+ \sum_r \left[d^\dagger(\mathbf{p}, r)d(\mathbf{p}, r), C_{ab}(\mathbf{k})d^\dagger(\mathbf{k}, s)\right]v_b(\mathbf{k}, s)$$

$$\left. - \sum_r \left[d^\dagger(\mathbf{p}, r)d(\mathbf{p}, r), D_{ab}(\mathbf{k})d(\mathbf{k}, s)\right]u_b(\mathbf{k}, s)\right\}.$$

We work out the various commutators separately:

$$\left[a^\dagger(\mathbf{p})a(\mathbf{p}), C_{ab}(\mathbf{k})d^\dagger(\mathbf{k}, s)\right] = a^\dagger(\mathbf{p})\left[a(\mathbf{p}), C_{ab}(\mathbf{k})d^\dagger(\mathbf{k}, s)\right]$$

$$+ \left[a^\dagger(\mathbf{p}), C_{ab}(\mathbf{k})d^\dagger(\mathbf{k}, s)\right]a(\mathbf{p})$$

$$= a^\dagger(\mathbf{p})\Big[a(\mathbf{p}), C_{ab}(\mathbf{k})\Big]d^\dagger(\mathbf{k}, s)$$

$$+ a^\dagger(\mathbf{p})C_{ab}(\mathbf{k})\Big[a(\mathbf{p}), d^\dagger(\mathbf{k}, s)\Big]$$

$$+ C_{ab}(\mathbf{k})\Big[a^\dagger(\mathbf{p}), d^\dagger(\mathbf{k}, s)\Big]a(\mathbf{p})$$

$$+ \Big[a^\dagger(\mathbf{p}), C_{ab}(\mathbf{k})\Big]d^\dagger(\mathbf{k}, s)a(\mathbf{p})$$

$$= a^\dagger(\mathbf{p})\Big[a(\mathbf{p}), a(\mathbf{k})\delta_{ab} - i\gamma_{ab}^5 b(\mathbf{k})\Big]d^\dagger(\mathbf{k}, s)$$

$$+ \Big[a^\dagger(\mathbf{p}), a(\mathbf{k})\delta_{ab} - i\gamma_{ab}^5 b(\mathbf{k})\Big]d^\dagger(\mathbf{k}, s)a(\mathbf{p}).$$

Here we made use of Eq. (5.50) and took into account that the operators $a(\mathbf{p})$ and $a^\dagger(\mathbf{p})$ commute with the fermionic creation operator $d^\dagger(\mathbf{k})$. Furthermore, the operators $a(\mathbf{p})$ and $a^\dagger(\mathbf{p})$ satisfy

$$\Big[a(\mathbf{p}), a(\mathbf{k})\Big] = \Big[a^\dagger(\mathbf{p}), a^\dagger(\mathbf{k})\Big] = 0; \quad \Big[a(\mathbf{p}), a^\dagger(\mathbf{k})\Big] = \delta(\mathbf{p} - \mathbf{k}),$$

as well as

$$\Big[a(\mathbf{p}), b(\mathbf{k})\Big] = \Big[a^\dagger(\mathbf{p}), b(\mathbf{k})\Big] = \Big[a(\mathbf{p}), b^\dagger(\mathbf{k})\Big] = \Big[a^\dagger(\mathbf{p}), b^\dagger(\mathbf{k})\Big] = 0,$$

so that the the final expression for the first commutator becomes:

$$\Big[a^\dagger(\mathbf{p})a(\mathbf{p}), C_{ab}(\mathbf{k})d^\dagger(\mathbf{k}, s)\Big]v_b(\mathbf{k}, s) = -\delta(\mathbf{p} - \mathbf{k})d^\dagger(\mathbf{k}, s)a(\mathbf{p})v_a(\mathbf{k}, s).$$

In a similar way one can work out the other commutators which give:

$$-\Big[a^\dagger(\mathbf{p})a(\mathbf{p}), D_{ab}(\mathbf{k})d(\mathbf{k}, s)\Big]u_b(\mathbf{k}, s) = -\delta(\mathbf{p} - \mathbf{k})a^\dagger(\mathbf{p})d(\mathbf{k}, s)u_a(\mathbf{k}, s),$$

$$\Big[b^\dagger(\mathbf{p})b(\mathbf{p}), C_{ab}(\mathbf{k})d^\dagger(\mathbf{k}, s)\Big]v_b(\mathbf{k}, s) = i\gamma_{ab}^5\delta(\mathbf{p} - \mathbf{k})d^\dagger(\mathbf{k}, s)b(\mathbf{p})v_b(\mathbf{k}, s),$$

$$-\Big[b^\dagger(\mathbf{p})b(\mathbf{p}), D_{ab}(\mathbf{k})d(\mathbf{k}, s)\Big]u_b(\mathbf{k}, s) = i\gamma_{ab}^5\delta(\mathbf{p} - \mathbf{k})b^\dagger(\mathbf{p})d(\mathbf{k}, s)u_b(\mathbf{k}, s),$$

and

$$\sum_r\Big[d^\dagger(\mathbf{p}, r)d(\mathbf{p}, r), C_{ab}(\mathbf{k})d^\dagger(\mathbf{k}, s)\Big]v_b(\mathbf{k}, s)$$

$$= \delta(\mathbf{p} - \mathbf{k})d^\dagger(\mathbf{p}, s)\Big\{a(\mathbf{k})v_a(\mathbf{k}, s) - i\gamma_{ab}^5 b(\mathbf{k})v_b(\mathbf{k}, s)\Big\},$$

$$\sum_r\Big[d^\dagger(\mathbf{p}, r)d(\mathbf{p}, r), D_{ab}(\mathbf{k})d(\mathbf{k}, s)\Big]u_b(\mathbf{k}, s)$$

$$= \delta(\mathbf{p} - \mathbf{k})d(\mathbf{p}, s)\Big\{a^\dagger(\mathbf{k})u_a(\mathbf{k}, s) - i\gamma_{ab}^5 b^\dagger(\mathbf{k})u_b(\mathbf{k}, s)\Big\}.$$

These relations can easily be surmised by inserting the defining expression (5.50) of the operators C and D and observing that in each case only one type of operators does not commute with the others.

Now, inserting these commutators in our expression for the commutator of the operators P_μ and Q_a, we have:

$$\left[P^\mu, Q_a\right] = i(2m)^{1/2} \int d^3p\, p^\mu \int d^3k \sum_s$$

$$\times \left\{ -a(\mathbf{p})d^\dagger(\mathbf{k}, s)v_a(\mathbf{k}, s) - a^\dagger(\mathbf{p})d(\mathbf{k}, s)u_a(\mathbf{k}, s) \right.$$
$$+ i\gamma^5_{ab}b(\mathbf{p})d^\dagger(\mathbf{k}, s)v_b(\mathbf{k}, s) + i\gamma^5_{ab}b^\dagger(\mathbf{p})d(\mathbf{k}, s)u_b(\mathbf{k}, s)$$
$$+ a(\mathbf{k})d^\dagger(\mathbf{p}, s)v_a(\mathbf{k}, s) - i\gamma^5_{ab}b(\mathbf{k})d^\dagger(\mathbf{p}, s)v_b(\mathbf{k}, s)$$
$$\left. + a^\dagger(\mathbf{k})d(\mathbf{p}, s)u_a(\mathbf{k}, s) - i\gamma^5_{ab}b^\dagger(\mathbf{k})d(\mathbf{p}, s)u_b(\mathbf{k}, s) \right\}\delta(\mathbf{p} - \mathbf{k}).$$

Integration over d^3k yields:

$$\left[P^\mu, Q_a\right] = i(2m)^{1/2} \int d^3p\, p^\mu \sum_s$$

$$\times \left\{ -a(\mathbf{p})d^\dagger(\mathbf{p}, s)v_a(\mathbf{p}, s) - a^\dagger(\mathbf{p})d(\mathbf{p}, s)u_a(\mathbf{p}, s) \right.$$
$$+ i\gamma^5_{ab}b(\mathbf{p})d^\dagger(\mathbf{p}, s)v_b(\mathbf{p}, s) + i\gamma^5_{ab}b^\dagger(\mathbf{p})d(\mathbf{p}, s)u_b(\mathbf{p}, s)$$
$$+ a(\mathbf{p})d^\dagger(\mathbf{p}, s)v_a(\mathbf{p}, s) - i\gamma^5_{ab}b(\mathbf{p})d^\dagger(\mathbf{p}, s)v_b(\mathbf{p}, s)$$
$$\left. + a^\dagger(\mathbf{p})d(\mathbf{p}, s)u_a(\mathbf{p}, s) - i\gamma^5_{ab}b^\dagger(\mathbf{p})d(\mathbf{p}, s)u_b(\mathbf{p}, s) \right\} = 0.$$

This completes the proof of Proposition 5.11.

Proposition 5.11 shows that within the Wess–Zumino model the Fock-space representations of the energy-momentum operator P_μ and the supercharge Q_a are commuting operators, consistent with the requirement of the super-Poincaré algebra.

5.6   The Energy-Momentum Operator of the Wess–Zumino Model

The goal of this section is the derivation of the energy-momentum operator of the Wess–Zumino model as given in Eq. (5.86). In particular we will see that in the Wess–Zumino model the Hamilton-operator and the three-momentum operator are normal ordered, *i.e.*

$$H =: H :, \qquad P_i =: P_i : . \qquad (5.89)$$

Here : : denotes *normal ordering*, *i.e.* all creation operators stand to the left of all annihilation operators.[2] This feature can be observed in the expansion (5.86) of the energy-momentum four vector P_μ. This property is a reflection of the fact that supersymmetric theories possess the same number of bosonic and fermionic degrees of freedom for a given mass. Furthermore, the first of Eqs. (5.89) implies stringent consequences for the energy of the vacuum state as will be discussed in Chapter 9.

We recall first the following aspects of the Wess–Zumino model. The Lagrangian density of the free massive Wess–Zumino model is given by Eq. (5.10), where $A(x)$ is a real scalar field, $B(x)$ is a real pseudoscalar field, and $\psi(x)$ is a spin-1/2 Majorana field. We saw earlier that these fields have the following plane-wave expansions:

$$A(x) = \frac{1}{(2\pi)^{3/2}} \int \frac{d^3p}{(2\omega_p)^{1/2}} \left[a(\mathbf{p})e^{-ipx} + a^\dagger(\mathbf{p})e^{ipx} \right] \qquad (5.90)$$

with

$$\omega_p = p_0 = \left(\mathbf{p}^2 + m^2 \right)^{1/2}, \qquad (5.91)$$

and bosonic creation and annihilation operators satisfying

$$\left. \begin{aligned}
\left[a(\mathbf{p}), a(\mathbf{k}) \right] &= 0, \\
\left[a^\dagger(\mathbf{p}), a^\dagger(\mathbf{k}) \right] &= 0, \\
\left[a(\mathbf{p}), a^\dagger(\mathbf{k}) \right] &= \delta(\mathbf{p} - \mathbf{k}).
\end{aligned} \right\} \qquad (5.92)$$

The pseudoscalar field $B(x)$ has the plane-wave expansion:

$$B(x) = \frac{1}{(2\pi)^{3/2}} \int \frac{d^3p}{(2\omega_p)^{1/2}} \left[b(\mathbf{p})e^{-ipx} + b^\dagger(\mathbf{p})e^{ipx} \right] \qquad (5.93)$$

with

$$\left. \begin{aligned}
\left[b(\mathbf{p}), b(\mathbf{k}) \right] &= 0, \\
\left[b^\dagger(\mathbf{p}), b^\dagger(\mathbf{k}) \right] &= 0, \\
\left[b(\mathbf{p}), b^\dagger(\mathbf{k}) \right] &= \delta(\mathbf{p} - \mathbf{k}).
\end{aligned} \right\} \qquad (5.94)$$

---

[2]For an introduction to this subject see *e.g.* J.D. Bjorken and S.D. Drell [18], Chap. 12 or C. Itzykson and J.-B. Zuber [60], Chap. 3.

Finally, the Majorana spinor field $\psi(x)$ has the expansion

$$\psi(x) = \frac{1}{(2\pi)^{3/2}} \sum_s \int d^3p \left(\frac{m}{\omega_p}\right)^{1/2} \left[d(\mathbf{p}, s)u(\mathbf{p}, s)e^{-ipx} + d^\dagger(\mathbf{p}, s)v(\mathbf{p}, s)e^{ipx} \right],$$

$$(5.95)$$

where the fermionic creation and annihilation operators satisfy the anticommutation relations

$$\left.\begin{array}{rcl} \{d(\mathbf{p}, s), d(\mathbf{k}, r)\} & = & 0, \\ \{d^\dagger(\mathbf{p}, s), d^\dagger(\mathbf{k}, r)\} & = & 0, \\ \{d(\mathbf{p}, s), d^\dagger(\mathbf{k}, r)\} & = & \delta_{rs}\delta(\mathbf{p} - \mathbf{k}). \end{array}\right\} \qquad (5.96)$$

The energy-momentum operator is defined by

$$P_\mu = \int d^3x T_{0\mu}, \quad P_0 = H, \qquad (5.97)$$

where $T_{\mu\nu}$ is the canonical energy-momentum tensor, defined by:

$$T_{\nu\mu} = -\eta_{\nu\mu}\mathcal{L} + \sum_i \frac{\partial \mathcal{L}}{\partial\left(\frac{\partial \Phi_i}{\partial x_\nu}\right)} \frac{\partial \Phi_i}{\partial x^\mu}. \qquad (5.98)$$

Here, $\mathcal{L}$ is the Langrangian density Eq. (5.10), and the sum is to be taken over the various fields of the theory. Substituting Eq. (5.98) into Eq. (5.97) we obtain

$$P_\mu = \int d^3x T_{0\mu} = \int d^3x \left\{ \sum_i \frac{\partial \mathcal{L}}{\partial\left(\frac{\partial \Phi_i}{\partial x_0}\right)} \frac{\partial \Phi_i}{\partial x^\mu} - \eta_{0\mu}\mathcal{L} \right\}. \qquad (5.99)$$

For the Wess–Zumino model, the energy-momentum operator (5.99) becomes

$$P_\mu = \int d^3x \left\{ \frac{\partial \mathcal{L}}{\partial\left(\frac{\partial A(x)}{\partial x_0}\right)} \frac{\partial A(x)}{\partial x^\mu} + \frac{\partial \mathcal{L}}{\partial\left(\frac{\partial B(x)}{\partial x_0}\right)} \frac{\partial B(x)}{\partial x^\mu} \right.$$

$$\left. + \frac{\partial \mathcal{L}}{\partial\left(\frac{\partial \psi(x)}{\partial x_0}\right)} \frac{\partial \psi(x)}{\partial x^\mu} - \eta_{0\mu}\mathcal{L} \right\}. \qquad (5.100)$$

Calculating the various terms appearing in Eq. (5.100) and taking into account thereby the Lagrangian density (5.10) as well as the plane-wave expansions (5.90), (5.93) and (5.95) we find:

$$\frac{\partial \mathcal{L}}{\partial\left(\frac{\partial A(x)}{\partial x_0}\right)} = \frac{\partial A(x)}{\partial x^0} = \frac{i}{(2\pi)^{3/2}} \int \frac{d^3p}{(2/\omega_p)^{1/2}} \left[-a(\mathbf{p})e^{-ipx} + a^\dagger(\mathbf{p})e^{ipx} \right],$$

$$(5.101)$$

$$\frac{\partial A(x)}{\partial x^\mu} = \frac{i}{(2\pi)^{3/2}} \int d^3p\, p_\mu (2\omega_p)^{-1/2} \left[-a(\mathbf{p})e^{-ipx} + a^\dagger(\mathbf{p})e^{ipx} \right], \quad (5.102)$$

$$\frac{\partial \mathcal{L}}{\partial \left(\frac{\partial B(x)}{\partial x_0} \right)} = \frac{\partial B(x)}{\partial x^0} = \frac{i}{(2\pi)^{3/2}} \int \frac{d^3p}{(2/\omega_p)^{1/2}} \left[-b(\mathbf{p})e^{-ipx} + b^\dagger(\mathbf{p})e^{ipx} \right],$$

$$(5.103)$$

$$\frac{\partial B(x)}{\partial x^\mu} = \frac{i}{(2\pi)^{3/2}} \int d^3p\, p_\mu (2\omega_p)^{-1/2} \left[-b(\mathbf{p})e^{-ipx} + b^\dagger(\mathbf{p})e^{ipx} \right], \quad (5.104)$$

$$\frac{\partial \mathcal{L}}{\partial \left(\frac{\partial \psi(x)}{\partial x_0} \right)} = \frac{i}{2}\overline{\psi}\gamma^0 = \frac{i}{2}\frac{1}{(2\pi)^{3/2}} \sum_s \int d^3p \left(\frac{m}{\omega_p} \right)^{1/2}$$

$$\times \left\{ d^\dagger(\mathbf{p},s)\overline{u}(\mathbf{p},s)\gamma^0 e^{ipx} + d(\mathbf{p},s)\overline{v}(\mathbf{p},s)\gamma^0 e^{-ipx} \right\}, \quad (5.105)$$

$$\frac{\partial \psi(x)}{\partial x^\mu} = \frac{i}{(2\pi)^{3/2}} \sum_s \int d^3p \left(\frac{m}{\omega_p} \right)^{1/2} p_\mu$$

$$\times \left\{ -d(\mathbf{p},s)u(\mathbf{p},s)e^{-ipx} + d^\dagger(\mathbf{p},s)v(\mathbf{p},s)e^{ipx} \right\}. \quad (5.106)$$

5.6.1 The Hamilton Operator

We now calculate the Hamiltonian of the Wess–Zumino model. This operator is given by:

$$H = P_0 = \int d^3x \left\{ \frac{\partial \mathcal{L}}{\partial \left(\frac{\partial A(x)}{\partial x_0} \right)} \frac{\partial A(x)}{\partial x^0} + \frac{\partial \mathcal{L}}{\partial \left(\frac{\partial B(x)}{\partial x_0} \right)} \frac{\partial B(x)}{\partial x^0} \right.$$

$$\left. + \frac{\partial \mathcal{L}}{\partial \left(\frac{\partial \psi(x)}{\partial x_0} \right)} \frac{\partial \psi(x)}{\partial x^0} - \eta_{00}\mathcal{L} \right\}$$

$$\equiv P_0^A + P_0^B + P_0^\psi - \eta_{00} \int \mathcal{L} d^3x. \quad (5.107)$$

Here

$$P_0^A = \int d^3x \frac{\partial \mathcal{L}}{\partial \left(\frac{\partial A(x)}{\partial x_0} \right)} \frac{\partial A(x)}{\partial x^0}$$

$$\overset{(5.101)}{\underset{(5.102)}{=}} -\frac{1}{(2\pi)^3} \int d^3x \int d^3p \int d^3k \frac{1}{2}(\omega_p\omega_k)^{1/2}$$

$$\times \left[-a(\mathbf{p})e^{-ipx} + a^\dagger(\mathbf{p})e^{ipx} \right]\left[-a(\mathbf{k})e^{-ikx} + a^\dagger(\mathbf{k})e^{ikx} \right]$$

$$
= -\frac{1}{(2\pi)^3} \int d^3x \int d^3p \int d^3k \frac{1}{2} (\omega_p\omega_k)^{1/2}
$$
$$
\times \Big[a(\mathbf{p})a(\mathbf{k})e^{-i(p+k)x} + a^\dagger(\mathbf{p})a^\dagger(\mathbf{k})e^{i(p+k)x}
$$
$$
-a(\mathbf{p})a^\dagger(\mathbf{k})e^{-i(p-k)x} - a^\dagger(\mathbf{p})a(\mathbf{k})e^{i(p-k)x} \Big]
$$
$$
= -\int d^3p \int d^3k \frac{1}{2} (\omega_p\omega_k)^{1/2} \Big[a(\mathbf{p})a(\mathbf{k})e^{-2i\omega_p t}\delta(\mathbf{p}+\mathbf{k})
$$
$$
+a^\dagger(\mathbf{p})a^\dagger(\mathbf{k})e^{2i\omega_p t}\delta(\mathbf{p}+\mathbf{k}) - a(\mathbf{p})a^\dagger(\mathbf{k})\delta(\mathbf{p}-\mathbf{k}) - a^\dagger(\mathbf{p})a(\mathbf{k})\delta(\mathbf{p}-\mathbf{k}) \Big].
$$

Thus we obtain the following contribution of the $A(x)$-field to the Hamiltonian:

$$
P_0^A = -\frac{1}{2} \int d^3p\, \omega_p \Big[a(\mathbf{p})a(-\mathbf{p})e^{-2i\omega_p t} - a(\mathbf{p})a^\dagger(\mathbf{p})
$$
$$
- a^\dagger(\mathbf{p})a(\mathbf{p}) + a^\dagger(\mathbf{p})a^\dagger(-\mathbf{p})e^{2i\omega_p t} \Big]. \tag{5.108}
$$

A similar calculation leads to the following contribution of the $B(x)$ field to the Hamiltonian:

$$
P_0^B = -\frac{1}{2} \int d^3p\, \omega_p \Big[b(\mathbf{p})b(-\mathbf{p})e^{-2i\omega_p t} - b(\mathbf{p})b^\dagger(\mathbf{p})
$$
$$
- b^\dagger(\mathbf{p})b(\mathbf{p}) + b^\dagger(\mathbf{p})b^\dagger(-\mathbf{p})e^{2i\omega_p t} \Big]. \tag{5.109}
$$

In order to calculate the contribution of the Majorana spinor field $\psi(x)$ to the energy, *i.e.* to evaluate P_0^ψ, we require a number of intermediate results.

Proposition 5.12:

The following relations can be shown to hold:

$$
\bar{u}(\mathbf{p},s)\gamma^0 u(\mathbf{p},r) = \frac{p_0}{m}\,\delta_{rs}, \tag{5.110a}
$$
$$
\bar{v}(\mathbf{p},s)\gamma^0 v(\mathbf{p},r) = \frac{p_0}{m}\,\delta_{rs}, \tag{5.110b}
$$
$$
\bar{u}(\mathbf{p},s)\gamma^0 v(-\mathbf{p},r) = 0, \tag{5.110c}
$$
$$
\bar{v}(\mathbf{p},s)\gamma^0 u(-\mathbf{p},r) = 0. \tag{5.110d}
$$

Proof: We start with the proof of Eq. (5.110a):

$$
\bar{u}(\mathbf{p},s)\gamma^0 u(\mathbf{p},r) \overset{(5.61)}{=} \bar{u}(\mathbf{p},s)\gamma^0\Lambda_+(p)u(\mathbf{p},r)
$$
$$
\overset{(5.58)}{=} \bar{u}(\mathbf{p},s)\gamma^0\Big(\frac{\slashed{p}+m}{2m}\Big)u(\mathbf{p},r)
$$
$$
= \bar{u}(\mathbf{p},s)\gamma^0\frac{1}{2m}\Big(p_0\gamma^0 - \mathbf{p}\cdot\boldsymbol{\gamma} + m\Big)u(\mathbf{p},r)
$$

$$= \ \bar{u}(\mathbf{p},s)\frac{1}{2m}\Big(p_0\gamma^0 + \mathbf{p}\cdot\boldsymbol{\gamma} + m\Big)\gamma^0 u(\mathbf{p},r)$$

$$= \ \bar{u}(\mathbf{p},s)\frac{1}{2m}\Big(2p_0\gamma^0 - \not{p} + m\Big)\gamma^0 u(\mathbf{p},r)$$

$$= \ \frac{p_0}{m}\bar{u}(\mathbf{p},s)u(\mathbf{p},r) + \bar{u}(\mathbf{p},s)\Lambda_-(p)\gamma^0 u(\mathbf{p},r) \overset{(5.31)}{=} \frac{p_0}{m}\delta_{rs},$$

since $\bar{u}(\mathbf{p},s)\Lambda_-(p) = 0$. The proof of Eq. (5.110b) proceeds as follows:

$$\bar{v}(\mathbf{p},s)\gamma^0 v(\mathbf{p},r) \overset{(5.61)}{=} \bar{v}(\mathbf{p},s)\gamma^0\Lambda_-(p)v(\mathbf{p},r)$$

$$\overset{(5.59)}{=} \bar{v}(\mathbf{p},s)\gamma^0\Big(\frac{-\not{p}+m}{2m}\Big)v(\mathbf{p},r)$$

$$= \ \bar{v}(\mathbf{p},s)\gamma^0\frac{1}{2m}\Big(-p_0\gamma^0 + \mathbf{p}\cdot\boldsymbol{\gamma} + m\Big)v(\mathbf{p},r)$$

$$= \ \bar{v}(\mathbf{p},s)\frac{1}{2m}\Big(-2p_0\gamma^0 + p_0\gamma^0 - \mathbf{p}\cdot\boldsymbol{\gamma} + m\Big)\gamma^0 v(\mathbf{p},r)$$

$$= \ -\frac{p_0}{m}\bar{v}(\mathbf{p},s)v(\mathbf{p},r) + \bar{v}(\mathbf{p},s)\Big(\frac{\not{p}+m}{2m}\Big)\gamma^0 v(\mathbf{p},r)$$

$$\overset{(5.31)}{=} \frac{p_0}{m}\delta_{rs} + \bar{v}(\mathbf{p},s)\Lambda_+(p)\gamma^0 v(\mathbf{p},r) \ = \ \frac{p_0}{m}\delta_{rs}.$$

Equation (5.110c) is shown as follows:

$$\bar{u}(\mathbf{p},s)\gamma^0 v(-\mathbf{p},r) \overset{(5.61)}{=} \bar{u}(\mathbf{p},s)\gamma^0\Lambda_-(p_0,-\mathbf{p})v(-\mathbf{p},r)$$

$$= \ \bar{u}(\mathbf{p},s)\gamma^0\frac{1}{2m}\Big(-p_0\gamma^0 - \mathbf{p}\cdot\boldsymbol{\gamma} + m\Big)v(-\mathbf{p},r)$$

$$= \ \bar{u}(\mathbf{p},s)\frac{1}{2m}\Big(-p_0\gamma^0 + \mathbf{p}\cdot\boldsymbol{\gamma} + m\Big)\gamma^0 v(-\mathbf{p},r)$$

$$= \ \bar{u}(\mathbf{p},s)\Lambda_-(p)\gamma^0 v(-\mathbf{p},r) \overset{(5.66)}{=} 0.$$

Finally, Eq. (5.110d),

$$\bar{v}(\mathbf{p},s)\gamma^0 u(-\mathbf{p},r) \overset{(5.61)}{=} \bar{v}(\mathbf{p},s)\gamma^0\Lambda_+(p_0,-\mathbf{p})u(-\mathbf{p},r)$$

$$= \ \bar{v}(\mathbf{p},s)\gamma^0\frac{1}{2m}\Big(p_0\gamma^0 + \mathbf{p}\cdot\boldsymbol{\gamma} + m\Big)u(-\mathbf{p},r)$$

$$= \ \bar{v}(\mathbf{p},s)\frac{1}{2m}\Big(p_0\gamma^0 - \mathbf{p}\cdot\boldsymbol{\gamma} + m\Big)\gamma^0 u(-\mathbf{p},r)$$

$$= \ \bar{v}(\mathbf{p},s)\Lambda_+(p)\gamma^0 u(-\mathbf{p},r) \overset{(5.67)}{=} 0.$$

This completes the proof of Proposition 5.12.

Now we can work out the contribution of the Majorana spinor field $\psi(x)$ to the Hamilton operator of the Wess–Zumino model. We have:

$$P_0^\psi = \int d^3x \frac{\partial \mathcal{L}}{\partial\left(\frac{\partial\psi(x)}{\partial x_0}\right)} \frac{\partial\psi(x)}{\partial x^0} = \frac{i}{2} \int d^3x \overline{\psi}(x)\gamma^0 \partial_0 \psi(x)$$

$$= -\frac{1}{2(2\pi)^3} \int d^3x \sum_{s,r} \int d^3p \int d^3k \left(\frac{m}{\omega_k}\right)^{1/2} \left(\frac{m}{\omega_p}\right)^{1/2} \omega_k$$

$$\times \left\{ d^\dagger(\mathbf{p},s)\overline{u}(\mathbf{p},s)\gamma^0 e^{ipx} + d(\mathbf{p},s)\overline{v}(\mathbf{p},s)\gamma^0 e^{-ipx} \right\}$$

$$\times \left\{ -d(\mathbf{k},r)u(\mathbf{k},r)e^{-ikx} + d^\dagger(\mathbf{k},r)v(\mathbf{k},r)e^{ikx} \right\}$$

$$= -\frac{1}{2(2\pi)^3} \int d^3x \sum_{s,r} \int d^3p \int d^3k \left(\frac{m}{\omega_p}\right)^{1/2} (m\omega_k)^{1/2}$$

$$\times \Big\{ -d^\dagger(\mathbf{p},s)d(\mathbf{k},r)\overline{u}(\mathbf{p},s)\gamma^0 u(\mathbf{k},r)e^{i(p-k)x}$$

$$+ d(\mathbf{p},s)d^\dagger(\mathbf{k},r)\overline{v}(\mathbf{p},s)\gamma^0 v(\mathbf{k},r)e^{-i(p-k)x}$$

$$+ d^\dagger(\mathbf{p},s)d^\dagger(\mathbf{k},r)\overline{u}(\mathbf{p},s)\gamma^0 v(\mathbf{k},r)e^{i(p+k)x}$$

$$- d(\mathbf{p},s)d(\mathbf{k},r)\overline{v}(\mathbf{p},s)\gamma^0 u(\mathbf{k},r)e^{-i(p+k)x} \Big\}$$

$$= -\frac{1}{2}\sum_{s,r} \int d^3p \int d^3k \left(\frac{m}{\omega_p}\right)^{1/2} (m\omega_k)^{1/2}$$

$$\times \Big\{ -d^\dagger(\mathbf{p},s)d(\mathbf{k},r)\overline{u}(\mathbf{p},s)\gamma^0 u(\mathbf{k},r)\delta(\mathbf{p}-\mathbf{k})$$

$$+ d(\mathbf{p},s)d^\dagger(\mathbf{k},r)\overline{v}(\mathbf{p},s)\gamma^0 v(\mathbf{k},r)\delta(\mathbf{p}-\mathbf{k})$$

$$+ d^\dagger(\mathbf{p},s)d^\dagger(\mathbf{k},r)\overline{u}(\mathbf{p},s)\gamma^0 v(\mathbf{k},r)e^{2i\omega_p t}\delta(\mathbf{p}+\mathbf{k})$$

$$- d(\mathbf{p},s)d(\mathbf{k},r)\overline{v}(\mathbf{p},s)\gamma^0 u(\mathbf{k},r)e^{-2i\omega_p t}\delta(\mathbf{p}+\mathbf{k}) \Big\}$$

$$= -\frac{1}{2}\sum_{s,r} \int d^3p\, m\Big\{ -d^\dagger(\mathbf{p},s)d(\mathbf{p},r)\overline{u}(\mathbf{p},s)\gamma^0 u(\mathbf{p},r)$$

$$+ d(\mathbf{p},s)d^\dagger(\mathbf{p},r)\overline{v}(\mathbf{p},s)\gamma^0 v(\mathbf{p},r)$$

$$+ d^\dagger(\mathbf{p},s)d^\dagger(-\mathbf{p},r)\overline{u}(\mathbf{p},s)\gamma^0 v(-\mathbf{p},r)e^{2i\omega_p t}$$

$$- d(\mathbf{p},s)d(-\mathbf{p},r)\overline{v}(\mathbf{p},s)\gamma^0 u(-\mathbf{p},r)e^{-2i\omega_p t} \Big\}$$

$$\overset{(*)}{=} -\frac{1}{2}\sum_{r} \int d^3p\, p_0 \Big\{ -d^\dagger(\mathbf{p},r)d(\mathbf{p},r) + d(\mathbf{p},r)d^\dagger(\mathbf{p},r) \Big\}$$

$$= \sum_{r} \int d^3p\, p_0 \Big\{ d^\dagger(\mathbf{p},r)d(\mathbf{p},r) - \frac{1}{2}\delta(0) \Big\}.$$

In step $(*)$ we used relations (5.110a) to (5.110d) to work out the various terms of this contribution. Thus, in summary we have

$$P_0^{\psi} = \sum_r \int d^3 p \, p_0 \left\{ d^{\dagger}(\mathbf{p}, r) d(\mathbf{p}, r) - \frac{1}{2} \delta(\mathbf{0}) \right\}. \tag{5.111}$$

As can be seen from Eq. (5.107) in order to complete the calculation of the Hamiltonian, the remaining term to be determined is the Lagrangian. Thus:

$$
\begin{aligned}
L &= \int d^3 x \, \mathcal{L}_{\text{free}} \\
&= \int d^3 x \Big\{ \frac{1}{2} \big(\partial_\mu A(x) \big) \big(\partial^\mu A(x) \big) - \frac{1}{2} m^2 A^2(x) \\
&\quad + \frac{1}{2} \big(\partial_\mu B(x) \big) \big(\partial^\mu B(x) \big) - \frac{1}{2} m^2 B^2(x) \\
&\quad + \frac{1}{2} \overline{\psi}(x) (i \partial\!\!\!/ - m) \psi(x) \Big\}.
\end{aligned}
$$

The contribution of the spinor field $\psi(x)$ vanishes due to the equation of motion, *i.e.* Eq. (5.5) in the case of $g = 0$. Inserting the Fourier expansions of the scalar field $A(x)$ and the pseudoscalar field $B(x)$ yields:

$$
\begin{aligned}
L = \int d^3 x \Big\{ &\frac{1}{2} \Big[\frac{i}{(2\pi)^{3/2}} \int d^3 p \frac{p_\mu}{(2\omega_p)^{1/2}} \Big(-a(\mathbf{p}) e^{-ipx} + a^{\dagger}(\mathbf{p}) e^{ipx} \Big) \\
&\times \frac{i}{(2\pi)^{3/2}} \int d^3 k \frac{k^\mu}{(2\omega_k)^{1/2}} \Big(-a(\mathbf{k}) e^{-ikx} + a^{\dagger}(\mathbf{k}) e^{ikx} \Big) \Big] \\
&- \frac{m^2}{2(2\pi)^3} \int \frac{d^3 p}{(2\omega_p)^{1/2}} \Big(a(\mathbf{p}) e^{-ipx} + a^{\dagger}(\mathbf{p}) e^{ipx} \Big) \\
&\times \int \frac{d^3 k}{(2\omega_k)^{1/2}} \Big(a(\mathbf{k}) e^{-ikx} + a^{\dagger}(\mathbf{k}) e^{ikx} \Big) \\
&- \frac{1}{2(2\pi)^3} \Big[\int d^3 p \frac{p_\mu}{(2\omega_p)^{1/2}} \Big(-b(\mathbf{p}) e^{-ipx} + b^{\dagger}(\mathbf{p}) e^{ipx} \Big) \\
&\times \int d^3 k \frac{k^\mu}{(2\omega_k)^{1/2}} \Big(-b(\mathbf{k}) e^{-ikx} + b^{\dagger}(\mathbf{k}) e^{ikx} \Big) \Big] \\
&- \frac{m^2}{2(2\pi)^3} \int \frac{d^3 p}{(2\omega_p)^{1/2}} \Big(b(\mathbf{p}) e^{-ipx} + b^{\dagger}(\mathbf{p}) e^{ipx} \Big) \\
&\times \int \frac{d^3 k}{(2\omega_k)^{1/2}} \Big(b(\mathbf{k}) e^{-ikx} + b^{\dagger}(\mathbf{k}) e^{ikx} \Big) \Big\}
\end{aligned}
$$

$$
= \int d^3x \left\{ -\frac{1}{2(2\pi)^3} \int d^3p \int d^3k \, \frac{p_\mu k^\mu}{2(\omega_p \omega_k)^{1/2}} \left[a(\mathbf{p})a(\mathbf{k})e^{-i(p+k)x} \right. \right.
$$

$$
\left. + a^\dagger(\mathbf{p})a^\dagger(\mathbf{k})e^{i(p+k)x} - a(\mathbf{p})a^\dagger(\mathbf{k})e^{-i(p-k)x} - a^\dagger(\mathbf{p})a(\mathbf{k})e^{i(p-k)x} \right]
$$

$$
- \frac{m^2}{2(2\pi)^3} \int d^3p \int \frac{d^3k}{2(\omega_p \omega_k)^{1/2}} \left[a(\mathbf{p})a(\mathbf{k})e^{-i(p+k)x} \right.
$$

$$
\left. + a^\dagger(\mathbf{p})a^\dagger(\mathbf{k})e^{i(p+k)x} + a(\mathbf{p})a^\dagger(\mathbf{k})e^{-i(p-k)x} + a^\dagger(\mathbf{p})a(\mathbf{k})e^{i(p-k)x} \right]
$$

$$
- \frac{1}{2(2\pi)^3} \int d^3p \int d^3k \, \frac{p_\mu k^\mu}{2(\omega_p \omega_k)^{1/2}} \left[b(\mathbf{p})b(\mathbf{k})e^{-i(p+k)x} \right.
$$

$$
\left. + b^\dagger(\mathbf{p})b^\dagger(\mathbf{k})e^{i(p+k)x} - b(\mathbf{p})b^\dagger(\mathbf{k})e^{-i(p-k)x} - b^\dagger(\mathbf{p})b(\mathbf{k})e^{i(p-k)x} \right]
$$

$$
- \frac{m^2}{2(2\pi)^3} \int d^3p \int \frac{d^3k}{2(\omega_p \omega_k)^{1/2}} \left[b(\mathbf{p})b(\mathbf{k})e^{-i(p+k)x} \right.
$$

$$
\left. \left. + b^\dagger(\mathbf{p})b^\dagger(\mathbf{k})e^{i(p+k)x} + b(\mathbf{p})b^\dagger(\mathbf{k})e^{-i(p-k)x} + b^\dagger(\mathbf{p})b(\mathbf{k})e^{i(p-k)x} \right] \right\}
$$

$$
= -\frac{1}{2} \int d^3p \int d^3k \, \frac{p_\mu k^\mu}{2(\omega_p \omega_k)^{1/2}} \left[a(\mathbf{p})a(\mathbf{k})\delta(\mathbf{p}+\mathbf{k})e^{-2i\omega_p t} \right.
$$

$$
\left. + a^\dagger(\mathbf{p})a^\dagger(\mathbf{k})\delta(\mathbf{p}+\mathbf{k})e^{2i\omega_p t} - a(\mathbf{p})a^\dagger(\mathbf{k})\delta(\mathbf{p}-\mathbf{k}) - a^\dagger(\mathbf{p})a(\mathbf{k})\delta(\mathbf{p}-\mathbf{k}) \right]
$$

$$
- \frac{m^2}{2} \int d^3p \int \frac{d^3k}{2(\omega_p \omega_k)^{1/2}}
$$

$$
\times \left[a(\mathbf{p})a(\mathbf{k})\delta(\mathbf{p}+\mathbf{k})e^{-2i\omega_p t} + a^\dagger(\mathbf{p})a^\dagger(\mathbf{k})\delta(\mathbf{p}+\mathbf{k})e^{2i\omega_p t} \right.
$$

$$
\left. + a(\mathbf{p})a^\dagger(\mathbf{k})\delta(\mathbf{p}-\mathbf{k}) + a^\dagger(\mathbf{p})a(\mathbf{k})\delta(\mathbf{p}-\mathbf{k}) \right]
$$

$$
- \frac{1}{2} \int d^3p \int d^3k \, \frac{p_\mu k^\mu}{2(\omega_p \omega_k)^{1/2}}
$$

$$
\times \left[b(\mathbf{p})b(\mathbf{k})\delta(\mathbf{p}+\mathbf{k})e^{-2i\omega_p t} + b^\dagger(\mathbf{p})b^\dagger(\mathbf{k})\delta(\mathbf{p}+\mathbf{k})e^{2i\omega_p t} \right.
$$

$$
\left. - b(\mathbf{p})b^\dagger(\mathbf{k})\delta(\mathbf{p}-\mathbf{k}) - b^\dagger(\mathbf{p})b(\mathbf{k})\delta(\mathbf{p}-\mathbf{k}) \right]
$$

$$
- \frac{m^2}{2} \int d^3p \int \frac{d^3k}{2(\omega_p \omega_k)^{1/2}}
$$

$$
\times \left[b(\mathbf{p})b(\mathbf{k})\delta(\mathbf{p}+\mathbf{k})e^{-2i\omega_p t} + b^\dagger(\mathbf{p})b^\dagger(\mathbf{k})\delta(\mathbf{p}+\mathbf{k})e^{2i\omega_p t} \right.
$$

$$
\left. + b(\mathbf{p})b^\dagger(\mathbf{k})\delta(\mathbf{p}-\mathbf{k}) + b^\dagger(\mathbf{p})b(\mathbf{k})\delta(\mathbf{p}-\mathbf{k}) \right].
$$

Hence the Lagrangian becomes:

$$L = -\frac{1}{2} \int \frac{d^3p}{2\omega_p} (p_0^2 + \mathbf{p}^2 + m^2) \left[a(\mathbf{p})a(-\mathbf{p})e^{-2i\omega_p t} + a^\dagger(\mathbf{p})a^\dagger(-\mathbf{p})e^{2i\omega_p t} \right]$$
$$- \frac{1}{2} \int \frac{d^3p}{2\omega_p} (p_0^2 + \mathbf{p}^2 + m^2) \left[b(\mathbf{p})b(-\mathbf{p})e^{-2i\omega_p t} + b^\dagger(\mathbf{p})b^\dagger(-\mathbf{p})e^{2i\omega_p t} \right].$$
$$(5.112)$$

All other terms vanish as a result of the relation $p^2 = m^2$. Collecting now all contributions to the Hamilton operator, we obtain the following expression:

$$H = P_0^A + P_0^B + P_0^\psi - \eta_{00}L$$
$$= \int d^3p \frac{\omega_p}{2} \left[a(\mathbf{p})a^\dagger(\mathbf{p}) + a^\dagger(\mathbf{p})a(\mathbf{p}) - a(\mathbf{p})a(-\mathbf{p})e^{-2i\omega_p t} \right.$$
$$\left. - a^\dagger(\mathbf{p})a^\dagger(-\mathbf{p})e^{2i\omega_p t} \right]$$
$$+ \int d^3p \frac{\omega_p}{2} \left[b(\mathbf{p})b^\dagger(\mathbf{p}) + b^\dagger(\mathbf{p})b(\mathbf{p}) - b(\mathbf{p})b(-\mathbf{p})e^{-2i\omega_p t} \right.$$
$$\left. - b^\dagger(\mathbf{p})b^\dagger(-\mathbf{p})e^{2i\omega_p t} \right]$$
$$+ \sum_r \int d^3p\, \omega_p \left\{ d^\dagger(\mathbf{p},r)d(\mathbf{p},r) - \frac{1}{2}\delta(0) \right\}$$
$$+ \frac{1}{2} \int \frac{d^3p}{2\omega_p} (p_0^2 + \mathbf{p}^2 + m^2) \left[a(\mathbf{p})a(-\mathbf{p})e^{-2i\omega_p t} + a^\dagger(\mathbf{p})a^\dagger(-\mathbf{p})e^{2i\omega_p t} \right]$$
$$+ \frac{1}{2} \int \frac{d^3p}{2\omega_p} (p_0^2 + \mathbf{p}^2 + m^2) \left[b(\mathbf{p})b(-\mathbf{p})e^{-2i\omega_p t} + b^\dagger(\mathbf{p})b^\dagger(-\mathbf{p})e^{2i\omega_p t} \right]$$
$$= \int d^3p\, \omega_p \left\{ a^\dagger(\mathbf{p})a(\mathbf{p}) + \frac{1}{2}\delta(0) + b^\dagger(\mathbf{p})b(\mathbf{p}) + \frac{1}{2}\delta(0) \right.$$
$$\left. + \sum_r \left(d^\dagger(\mathbf{p},r)d(\mathbf{p},r) - \frac{1}{2}\delta(0) \right) \right\}$$
$$+ \frac{1}{2} \int d^3p \left[-\omega_p + \frac{1}{2\omega_p}(\omega_p^2 + \mathbf{p}^2 + m^2) \right]$$
$$\left[a(\mathbf{p})a(-\mathbf{p})e^{-2i\omega_p t} + a^\dagger(\mathbf{p})a^\dagger(-\mathbf{p})e^{2i\omega_p t} \right]$$
$$+ \frac{1}{2} \int d^3p \left[-\omega_p + \frac{1}{2\omega_p}(\omega_p^2 + \mathbf{p}^2 + m^2) \right]$$
$$\left[b(\mathbf{p})b(-\mathbf{p})e^{-2i\omega_p t} + b^\dagger(\mathbf{p})b^\dagger(-\mathbf{p})e^{2i\omega_p t} \right].$$

Now

$$-\omega_p + \frac{1}{2\omega_p}(\omega_p^2 + \mathbf{p}^2 + m^2) = \frac{1}{2\omega_p}(-\omega_p^2 + \mathbf{p}^2 + m^2) = \frac{1}{2\omega_p}(-p^2 + m^2) = 0.$$

Hence, the Hamilton operator is

$$H = \int d^3p\, \omega_p \Big\{ a^\dagger(\mathbf{p})a(\mathbf{p}) + \frac{1}{2}\delta(\mathbf{0}) + b^\dagger(\mathbf{p})b(\mathbf{p}) + \frac{1}{2}\delta(\mathbf{0})$$
$$+ \sum_{r=1,2} \Big(d^\dagger(\mathbf{p},r)d(\mathbf{p},r) - \frac{1}{2}\delta(\mathbf{0}) \Big) \Big\}.$$

Our result is therefore

$$H = \int d^3p\, \omega_p \Big\{ a^\dagger(\mathbf{p})a(\mathbf{p}) + b^\dagger(\mathbf{p})b(\mathbf{p}) + \sum_{r=1,2} d^\dagger(\mathbf{p},r)d(\mathbf{p},r) \Big\}. \qquad (5.113)$$

From Eq. (5.113) we see that the Hamilton operator is normal ordered, *i.e.* all terms have annihilation operators to the right of the creation operators. As can be seen from the derivation of that expression, this is due to the fact that the zero-point energies of the bosonic fields cancel exactly the zero-point-energies of the fermionic degrees of freedom. The crucial point is, that in the Wess–Zumino model we have an equal number of bosonic and fermionic degrees of freedom such that the cancellation of the zero-point energies can occur.

5.6.2   The Three-Momentum P_i

The three-momentum operator $P_i, i = 1, 2, 3$, of the Wess–Zumino model is defined by:

$$P_i = \int d^3x \left\{ \frac{\partial \mathcal{L}}{\partial \left(\frac{\partial A(x)}{\partial x_0} \right)} \frac{\partial A(x)}{\partial x^i} + \frac{\partial \mathcal{L}}{\partial \left(\frac{\partial B(x)}{\partial x_0} \right)} \frac{\partial B(x)}{\partial x^i} + \frac{\partial \mathcal{L}}{\partial \left(\frac{\partial \psi(x)}{\partial x_0} \right)} \frac{\partial \psi(x)}{\partial x^i} \right\}.$$

$$(5.114)$$

Evaluating this expression we have

$$P_i = \int d^3x \left\{ \frac{\partial A(x)}{\partial x_0} \frac{\partial A(x)}{\partial x^i} + \frac{\partial B(x)}{\partial x_0} \frac{\partial B(x)}{\partial x^i} + \frac{i}{2} \bar{\psi}\gamma^0 \frac{\partial \psi(x)}{\partial x^i} \right\}$$
$$= \int d^3x \left\{ \frac{i}{(2\pi)^{3/2}} \int d^3p \Big(\frac{\omega_p}{2} \Big)^{1/2} \Big[-a(\mathbf{p})e^{-ipx} + a^\dagger(\mathbf{p})e^{ipx} \Big] \right.$$
$$\left. \times \frac{-i}{(2\pi)^{3/2}} \int \frac{d^3k\, k_i}{(2\omega_k)^{1/2}} \Big[a(\mathbf{k})e^{-ikx} - a^\dagger(\mathbf{k})e^{ikx} \Big] \right\}$$

$$+ \int d^3x \left\{ \frac{i}{(2\pi)^{3/2}} \int d^3p \left(\frac{\omega_p}{2}\right)^{1/2} \left[-b(\mathbf{p})e^{-ipx} + b^\dagger(\mathbf{p})e^{ipx}\right] \right.$$

$$\times \frac{-i}{(2\pi)^{3/2}} \int \frac{d^3k\,k_i}{(2\omega_k)^{1/2}} \left[b(\mathbf{k})e^{-ikx} - b^\dagger(\mathbf{k})e^{ikx}\right] \right\}$$

$$+ \int d^3x \left\{ \frac{i}{2(2\pi)^{3/2}} \sum_r \int d^3p \left(\frac{m}{\omega_p}\right)^{1/2} \right.$$

$$\left[d^\dagger(\mathbf{p},r)\bar{u}(\mathbf{p},r)\gamma^0 e^{ipx} + d(\mathbf{p},r)\bar{v}(\mathbf{p},r)\gamma^0 e^{-ipx}\right]$$

$$\times \frac{-i}{2(2\pi)^{3/2}} \sum_s \int d^3k \left(\frac{m}{\omega_k}\right)^{1/2} k_i$$

$$\times \left. \left[d(\mathbf{k},s)u(\mathbf{k},s)e^{-ikx} - d^\dagger(\mathbf{k},s)v(\mathbf{k},s)e^{+ikx}\right] \right\}.$$

Here we used the momentum space expansions (5.90), (5.93) and (5.95) for the scalar field $A(x)$, the pseudoscalar field $B(x)$ and the Majorana field $\psi(x)$. Working out the products yields:

$$P_i = \frac{1}{(2\pi)^3} \int d^3x \int d^3p \int d^3k \left(\frac{\omega_p}{2}\right)^{1/2} \frac{k_i}{(2\omega_k)^{1/2}}$$

$$\left[-a(\mathbf{p})a(\mathbf{k})e^{-i(p+k)x} + a(\mathbf{p})a^\dagger(\mathbf{k})e^{-i(p-k)x} \right.$$

$$+ a^\dagger(\mathbf{p})a(\mathbf{k})e^{i(p-k)x} - a^\dagger(\mathbf{p})a^\dagger(\mathbf{k})e^{i(p+k)x}$$

$$- b(\mathbf{p})b(\mathbf{k})e^{-i(p+k)x} + b(\mathbf{p})b^\dagger(\mathbf{k})e^{-i(p-k)x}$$

$$\left. + b^\dagger(\mathbf{p})b(\mathbf{k})e^{i(p-k)x} - b^\dagger(\mathbf{p})b^\dagger(\mathbf{k})e^{i(p+k)x}\right]$$

$$+ \frac{1}{2(2\pi)^3} \int d^3x \sum_{r,s} \int d^3p \int d^3k \frac{m}{(\omega_p\omega_k)^{1/2}} k_i$$

$$\times \left[d^\dagger(\mathbf{p},r)d(\mathbf{k},s)\bar{u}(\mathbf{p},r)\gamma^0 u(\mathbf{k},s)e^{i(p-k)x}\right.$$

$$- d(\mathbf{p},r)d^\dagger(\mathbf{k},s)\bar{v}(\mathbf{p},r)\gamma^0 v(\mathbf{k},s)e^{-i(p-k)x}$$

$$+ d^\dagger(\mathbf{p},r)d^\dagger(\mathbf{k},s)\bar{u}(\mathbf{p},r)\gamma^0 v(\mathbf{k},s)e^{i(p+k)x}$$

$$\left. + d(\mathbf{p},r)d(\mathbf{k},s)\bar{v}(\mathbf{p},r)\gamma^0 u(\mathbf{k},s)e^{-i(p+k)x}\right].$$

Integration over x yields

$$
\begin{aligned}
P_i = \int d^3p \int d^3k \left(\frac{\omega_p}{2}\right)^{1/2} \frac{k_i}{(2\omega_k)^{1/2}}
\end{aligned}
$$

$$
\times \left[-a(\mathbf{p})a(\mathbf{k})\delta(\mathbf{p}+\mathbf{k})e^{-2i\omega_p t} + a(\mathbf{p})a^\dagger(\mathbf{k})\delta(\mathbf{p}-\mathbf{k}) \right.
$$

$$
+ a^\dagger(\mathbf{p})a(\mathbf{k})\delta(\mathbf{p}-\mathbf{k}) - a^\dagger(\mathbf{p})a^\dagger(\mathbf{k})\delta(\mathbf{p}+\mathbf{k})e^{2i\omega_p t}
$$

$$
- b(\mathbf{p})b(\mathbf{k})\delta(\mathbf{p}+\mathbf{k})e^{-2i\omega_p t} + b(\mathbf{p})b^\dagger(\mathbf{k})\delta(\mathbf{p}-\mathbf{k})
$$

$$
\left. + b^\dagger(\mathbf{p})b(\mathbf{k})\delta(\mathbf{p}-\mathbf{k}) - b^\dagger(\mathbf{p})b^\dagger(\mathbf{k})\delta(\mathbf{p}+\mathbf{k})e^{2i\omega_p t} \right]
$$

$$
+ \frac{1}{2}\sum_{r,s} \int d^3p \int d^3k \frac{m}{(\omega_p\omega_k)^{1/2}} k_i
$$

$$
\times \left[d^\dagger(\mathbf{p},r)d(\mathbf{k},s)\bar{u}(\mathbf{p},r)\gamma^0 u(\mathbf{k},s)\delta(\mathbf{p}-\mathbf{k}) \right.
$$

$$
- d(\mathbf{p},r)d^\dagger(\mathbf{k},s)\bar{v}(\mathbf{p},r)\gamma^0 v(\mathbf{k},s)\delta(\mathbf{p}-\mathbf{k})
$$

$$
- d^\dagger(\mathbf{p},r)d^\dagger(\mathbf{k},s)\bar{u}(\mathbf{p},r)\gamma^0 v(\mathbf{k},s)\delta(\mathbf{p}+\mathbf{k})e^{2i\omega_p t}
$$

$$
\left. + d(\mathbf{p},r)d(\mathbf{k},s)\bar{v}(\mathbf{p},r)\gamma^0 u(\mathbf{k},s)\delta(\mathbf{p}+\mathbf{k})e^{-2i\omega_p t} \right]
$$

$$
= \frac{1}{2}\int d^3p\, p_i \left[a(\mathbf{p})a^\dagger(\mathbf{p}) + a^\dagger(\mathbf{p})a(\mathbf{p}) + b(\mathbf{p})b^\dagger(\mathbf{p}) + b^\dagger(\mathbf{p})b(\mathbf{p}) \right.
$$

$$
+ a(\mathbf{p})a(-\mathbf{p})e^{-2i\omega_p t} + a^\dagger(\mathbf{p})a^\dagger(-\mathbf{p})e^{2i\omega_p t}
$$

$$
\left. + b(\mathbf{p})b(-\mathbf{p})e^{-2i\omega_p t} + b^\dagger(\mathbf{p})b^\dagger(-\mathbf{p})e^{2i\omega_p t} \right]
$$

$$
+ \frac{1}{2}\sum_{r,s} \int d^3p \frac{m}{\omega_p} p_i \left[d^\dagger(\mathbf{p},r)d(\mathbf{p},s)\bar{u}(\mathbf{p},s)\gamma^0 u(\mathbf{p},s) \right.
$$

$$
\left. - d(\mathbf{p},r)d^\dagger(\mathbf{p},s)\bar{v}(\mathbf{p},r)\gamma^0 v(\mathbf{p},s) \right]
$$

$$
- \frac{1}{2}\sum_{r,s} \int d^3p \frac{m}{\omega_p} p_i \left[-d^\dagger(\mathbf{p},r)d^\dagger(-\mathbf{p},s)\bar{u}(\mathbf{p},r)\gamma^0 v(-\mathbf{p},s)e^{2i\omega_p t} \right.
$$

$$
\left. + d(\mathbf{p},r)d^\dagger(-\mathbf{p},s)\bar{v}(\mathbf{p},r)\gamma^0 u(-\mathbf{p},s)e^{-2i\omega_p t} \right].
$$

The time-dependent terms in this momentum-space expansion all vanish due to

$$
\int_{-\infty}^{+\infty} p_i a(\mathbf{p})a(-\mathbf{p})e^{-2i\omega_p t} d^3p \overset{(*)}{=} -\int_{-\infty}^{+\infty} p_i a(-\mathbf{p})a(\mathbf{p})e^{-2i\omega_p t} d^3p
$$

$$
\overset{(5.92)}{=} -\int_{-\infty}^{+\infty} p_i a(\mathbf{p})a(-\mathbf{p})e^{-2i\omega_p t} d^3p
$$

where in step $(*)$ we replace $\mathbf{p}$ by $-\mathbf{p}$. Hence this integral vanishes,

$$\int_{-\infty}^{+\infty} p_i a(\mathbf{p}) a(-\mathbf{p}) e^{-2i\omega_p t} d^3 p = 0.$$

Furthermore, using Eqs. (5.110a) to (5.110d) we obtain:

$$P_i = \frac{1}{2} \int d^3 p\, p_i \Big[a(\mathbf{p}) a^\dagger(\mathbf{p}) + a^\dagger(\mathbf{p}) a(\mathbf{p}) + b(\mathbf{p}) b^\dagger(\mathbf{p}) + b^\dagger(\mathbf{p}) b(\mathbf{p}) \Big]$$
$$+ \frac{1}{2} \sum_r \int d^3 p\, p_i \Big[d^\dagger(\mathbf{p}, r) d(\mathbf{p}, r) - d(\mathbf{p}, r) d^\dagger(\mathbf{p}, r) \Big].$$

Using Eqs.(5.92), (5.94) and (5.96) we obtain:

$$P_i = \int d^3 p\, p_i \Big[a(\mathbf{p}) a^\dagger(\mathbf{p}) + \frac{1}{2}\delta(\mathbf{0}) + b(\mathbf{p}) b^\dagger(\mathbf{p}) + \frac{1}{2}\delta(\mathbf{0})$$
$$+ \sum_r \Big(d^\dagger(\mathbf{p}, r) d(\mathbf{p}, r) - \frac{1}{2}\delta(\mathbf{0}) \Big) \Big],$$

i.e. the three-momentum of the Wess–Zumino model has the form:

$$P_i = \int d^3 p\, p_i \Big[a(\mathbf{p}) a^\dagger(\mathbf{p})) + b(\mathbf{p}) b^\dagger(\mathbf{p}) + \sum_r d^\dagger(\mathbf{p}, r) d(\mathbf{p}, r) \Big]. \qquad (5.115)$$

As in the case of the Hamilton operator, the three-momentum operator of the Wess–Zumino model has the interesting property to be normal-ordered, *i.e.*

$$P_i =: P_i : . \qquad (5.116)$$

Combining Eqs. (5.113) and (5.115) we obtain the four-momentum operator

$$P_\mu = \int d^3 p\, p_\mu \Big[a(\mathbf{p}) a^\dagger(\mathbf{p})) + b(\mathbf{p}) b^\dagger(\mathbf{p}) + \sum_r d^\dagger(\mathbf{p}, r) d(\mathbf{p}, r) \Big]. \qquad (5.117)$$

This is the expression (5.86) referred to earlier.

5.7 Infinitesimal Supersymmetry Transformations

We now demonstrate that the spinor charges Q_a of the Wess–Zumino model, given by Eq. (5.49) with $\lambda = 2$ as in Eq. (5.87), generate the supersymmetry transformations (5.11a) to (5.11d) of the local fields $A(x)$, $B(x)$ and $\psi(x)$ of the massive free Wess–Zumino model.

We recall that in classical mechanics[3] one writes the generating function F of an infinitesimal canonical transformation

$$F = F_{\mathrm{id}} + \epsilon\, G,$$

where F_{id} is the generating function of the identity transformation in phase space, G is the generator of the infinitesimal transformation, and ϵ is the appropriate infinitesimal parameter. It is then shown that an arbitrary function u of the canonical variables undergoes an infinitesimal change δu under the transformation which is given by

$$\delta u = \epsilon\,[u, G].$$

In classical mechanics the bracket denotes the *Poisson bracket*.

We now demonstrate that in the present case of the Wess–Zumino model the corresponding relations for the fields are given by the operator relations

$$\delta A(x) = -i\big[\bar\epsilon\, Q, A(x)\big] = -i\bar\epsilon_a\big[Q_a, A(x)\big], \qquad (5.118\mathrm{a})$$

$$\delta B(x) = -i\big[\bar\epsilon\, Q, B(x)\big] = -i\bar\epsilon_a\big[Q_a, B(x)\big], \qquad (5.118\mathrm{b})$$

$$\delta \psi(x) = -i\big[\bar\epsilon\, Q, \psi(x)\big] = -i\bar\epsilon_a\{Q_a, \psi(x)\}. \qquad (5.118\mathrm{c})$$

Here, $\bar\epsilon$ is an infinitesimal, x-independet Grassmann parameter and as usual, $\bar\epsilon$ denotes the Dirac adjoint, *i.e.* $\bar\epsilon = \epsilon^\dagger \gamma^0$. The variations

$$\delta A(x), \quad \delta B(x) \quad \text{and} \quad \delta \psi(x)$$

are given by relations (5.11a) to (5.11d). We verify relations (5.118a) to (5.118c) by inserting into the commutators the Fourier expansions obtained earlier for the relevant quantities.

Proof of Eq.(5.118a): Using Eqs. (5.47) and (5.49) we get for the commutator:

$$-i\big[\bar\epsilon Q, A(x)\big] = -i\bar\epsilon_a\big[Q_a, A(x)\big]$$

$$= \sqrt{2m}\,\bar\epsilon_a \sum_s \int d^3p\,(2\pi)^{-3/2} \int d^3k\,(2\omega_k)^{-1/2}$$

$$\times \Big[C_{ab}(\mathbf{p}) d^\dagger(\mathbf{p}, s) v_b(\mathbf{p}, s) - D_{ab}(\mathbf{p}) d(\mathbf{p}, s) u_b(\mathbf{p}, s),$$

$$a(\mathbf{k}) e^{-ikx} + a^\dagger(\mathbf{k}) e^{ikx} \Big]$$

---

[3]See *e.g.* H. Goldstein [49], Chap. 8 or H.J.W. Müller-Kirsten [75].

$$= \bar{\epsilon}_a \sum_s (2\pi)^{-3/2} \int d^3p \int d^3k \Big(\frac{m}{\omega_k}\Big)^{1/2}$$

$$\Big\{ \Big[C_{ab}(\mathbf{p}), a(\mathbf{k})\Big] d^\dagger(\mathbf{p},s) v_b(\mathbf{p},s) e^{-ikx}$$

$$+ \Big[C_{ab}(\mathbf{p}), a^\dagger(\mathbf{k})\Big] d^\dagger(\mathbf{p},s) v_b(\mathbf{p},s) e^{ikx}$$

$$- \Big[D_{ab}(\mathbf{p}), a(\mathbf{k})\Big] d(\mathbf{p},s) u_b(\mathbf{p},s) e^{-ikx}$$

$$- \Big[D_{ab}(\mathbf{p}), a^\dagger(\mathbf{k})\Big] d(\mathbf{p},s) u_b(\mathbf{p},s) e^{-ikx} \Big\}.$$

Now, using Eqs. (5.50) and (5.92) we evaluate each of the commutators separately:

$$\Big[C_{ab}(\mathbf{p}), a(\mathbf{k})\Big] = \Big[a(\mathbf{p})\delta_{ab} - i\gamma^5_{ab} b(\mathbf{p}), a(\mathbf{k})\Big]$$

$$= \Big[a(\mathbf{p}), a(\mathbf{k})\Big]\delta_{ab} - i\gamma^5_{ab}\Big[b(\mathbf{p}), a(\mathbf{k})\Big] = 0, \qquad (5.119a)$$

$$\Big[C_{ab}(\mathbf{p}), a^\dagger(\mathbf{k})\Big] = \Big[a(\mathbf{p})\delta_{ab} - i\gamma^5_{ab} b(\mathbf{p}), a^\dagger(\mathbf{k})\Big]$$

$$= \Big[a(\mathbf{p}), a^\dagger(\mathbf{k})\Big]\delta_{ab} - i\gamma^5_{ab}\Big[b(\mathbf{p}), a^\dagger(\mathbf{k})\Big] = \delta_{ab}\delta(\mathbf{p}-\mathbf{k}),$$

$$(5.119b)$$

$$\Big[D_{ab}(\mathbf{p}), a(\mathbf{k})\Big] = \Big[a^\dagger(\mathbf{p})\delta_{ab} - i\gamma^5_{ab} b^\dagger(\mathbf{p}), a(\mathbf{k})\Big]$$

$$= \Big[a^\dagger(\mathbf{p}), a(\mathbf{k})\Big]\delta_{ab} - i\gamma^5_{ab}\Big[b^\dagger(\mathbf{p}), a(\mathbf{k})\Big] = -\delta_{ab}\delta(\mathbf{p}-\mathbf{k}),$$

$$(5.119c)$$

$$\Big[D_{ab}(\mathbf{p}), a^\dagger(\mathbf{k})\Big] = \Big[a^\dagger(\mathbf{p})\delta_{ab} - i\gamma^5_{ab} b^\dagger(\mathbf{p}), a^\dagger(\mathbf{k})\Big]$$

$$= \Big[a^\dagger(\mathbf{p}), a^\dagger(\mathbf{k})\Big]\delta_{ab} - i\gamma^5_{ab}\Big[b^\dagger(\mathbf{p}), a^\dagger(\mathbf{k})\Big] = 0. \qquad (5.119d)$$

With these results we find:

$$-i\bar{\epsilon}_a\Big[Q_a, A(x)\Big] = \bar{\epsilon}_a \sum_s (2\pi)^{-3/2} \int d^3p \int d^3k \Big(\frac{m}{\omega_k}\Big)^{1/2}$$

$$\times \Big\{ \delta_{ab}\delta(\mathbf{p}-\mathbf{k}) d^\dagger(\mathbf{p},s) v_b(\mathbf{p},s) e^{ikx}$$

$$+ \delta_{ab}\delta(\mathbf{p}-\mathbf{k}) d(\mathbf{p},s) u_b(\mathbf{p},s) e^{-ikx} \Big\}$$

$$= \bar{\epsilon}_a \sum_s (2\pi)^{-3/2} \int d^3p \Big(\frac{m}{\omega_p}\Big)^{1/2}$$

$$\times \Big\{ d^\dagger(\mathbf{p},s) v_a(\mathbf{p},s) e^{ipx} + d(\mathbf{p},s) u_a(\mathbf{p},s) e^{-ipx} \Big\}$$

$$\overset{(5.95)}{=} \bar{\epsilon}_a \psi_a(x) \overset{(5.11a)}{=} \delta A(x).$$

This is the result we wanted to derive.

Proof of Eq. (5.118b): Proceeding as before

$$-i\big[\bar{\epsilon}Q, B(x)\big] = -i\bar{\epsilon}_a\big[Q_a, B(x)\big]$$

$$= \sqrt{2m}\,\bar{\epsilon}_a \sum_s \int d^3p\,(2\pi)^{-3/2} \int d^3k\,(2\omega_k)^{-1/2}$$

$$\times \Big[C_{ab}(\mathbf{p})d^\dagger(\mathbf{p}, s)v_b(\mathbf{p}, s) - D_{ab}(\mathbf{p})d(\mathbf{p}, s)u_b(\mathbf{p}, s),$$

$$b(\mathbf{k})e^{-ikx} + b^\dagger(\mathbf{k})e^{ikx}\Big]$$

$$= \bar{\epsilon}_a \sum_s (2\pi)^{-3/2} \int d^3p \int d^3k \Big(\frac{m}{\omega_k}\Big)^{1/2}$$

$$\times \Big\{\big[C_{ab}(\mathbf{p}), b(\mathbf{k})\big]d^\dagger(\mathbf{p}, s)v_b(\mathbf{p}, s)e^{-ikx}$$

$$+ \big[C_{ab}(\mathbf{p}), b^\dagger(\mathbf{k})\big]d^\dagger(\mathbf{p}, s)v_b(\mathbf{p}, s)e^{ikx}$$

$$- \big[D_{ab}(\mathbf{p}), b(\mathbf{k})\big]d(\mathbf{p}, s)u_b(\mathbf{p}, s)e^{-ikx}$$

$$- \big[D_{ab}(\mathbf{p}), b^\dagger(\mathbf{k})\big]d(\mathbf{p}, s)u_b(\mathbf{p}, s)e^{+ikx}\Big\}.$$

$$= \bar{\epsilon}_a \sum_s (2\pi)^{-3/2} \int d^3p \int d^3k \Big(\frac{m}{\omega_k}\Big)^{1/2}$$

$$\times \Big\{-i\gamma^5_{ab}\delta(\mathbf{p} - \mathbf{k})d^\dagger(\mathbf{p}, s)v_b(\mathbf{p}, s)e^{ikx}$$

$$- i\gamma^5_{ab}\delta(\mathbf{p} - \mathbf{k})d(\mathbf{p}, s)u_b(\mathbf{p}, s)e^{ikx}\Big\},$$

where we used the following commutation relations which may be derived in a similar way as Eqs. (5.119a) to (5.119d):

$$\big[C_{ab}(\mathbf{p}), b(\mathbf{k})\big] = \big[D_{ab}(\mathbf{p}), b^\dagger(\mathbf{k})\big] = 0,$$

$$\big[C_{ab}(\mathbf{p}), b^\dagger(\mathbf{k})\big] = -\big[D_{ab}(\mathbf{p}), b(\mathbf{k})\big] = -i\gamma^5_{ab}\delta(\mathbf{p} - \mathbf{k}).$$

Hence we find:

$$-i\big[\bar{\epsilon}Q, B(x)\big] = -i\bar{\epsilon}_a\gamma^5_{ab}(2\pi)^{-3/2} \sum_s \int d^3p \Big(\frac{m}{\omega_p}\Big)^{1/2}$$

$$\times \Big\{d(\mathbf{p}, s)u_b(\mathbf{p}, s)e^{-ipx} + d^\dagger(\mathbf{p}, s)v_b(\mathbf{p}, s)e^{ipx}\Big\}$$

$$\overset{(5.95)}{=} -i\bar{\epsilon}\gamma^5\psi(x) \overset{(5.11b)}{=} \delta B(x).$$

This completes the proof of Eq. (5.118b). Before we can demonstrate the validity of Eq. (5.118c), we need some more intermediate results.

Proposition 5.13:

The following relations hold:

$$\sum_r u_a(\mathbf{p},r)v_b(\mathbf{p},r) = \left(\frac{\slashed{p}+m}{2m}C\right)_{ab}, \tag{5.120a}$$

$$\sum_r v_a(\mathbf{p},r)u_b(\mathbf{p},r) = \left(\frac{\slashed{p}-m}{2m}C\right)_{ab}, \tag{5.120b}$$

where C is the charge conjugation matrix.

Proof: Consider:

$$u_a(\mathbf{p},r)v_b(\mathbf{p},r) \overset{(5.42)}{=} v_a^c(\mathbf{p},r)v_b(\mathbf{p},r) \overset{(5.40)}{=} C_{am}\overline{v}_m^\top(\mathbf{p},r)v_b(\mathbf{p},r)$$

$$= C_{am}v_b(\mathbf{p},r)\overline{v}_m^\top(\mathbf{p},r) \overset{(5.63)}{=} C_{am}\left(\Lambda_-\right)_{bm}$$

$$= \left(\Lambda_-C^\top\right)_{ba} \overset{(1.191)}{=} -\left(\Lambda_-C\right)_{ba}$$

$$= -\left(\Lambda_-C\right)_{ab}^\top \overset{(*)}{=} \left(\Lambda_+C\right)_{ab} \overset{(5.58)}{=} \left(\frac{\slashed{p}+m}{2m}C\right)_{ab}$$

In step $(*)$ we make use of the following relation that holds between the projection operators:

$$\Lambda_+C = \frac{\slashed{p}+m}{2m}C = \frac{CC^{-1}\slashed{p}C+mC}{2m} = \frac{-C\slashed{p}^\top+mC}{2m} = C\frac{-\slashed{p}^\top+m}{2m}.$$

Therefore

$$\left(\Lambda_+C\right)^\top = \frac{-\slashed{p}+m}{2m}C^\top = \Lambda_-C^\top = -\Lambda_-C.$$

Hence

$$\left(\Lambda_-C\right)^\top = -\Lambda_+C.$$

In an analogous way one shows Eq. (1.135b), using $C^\top = -C$:

$$v_a(\mathbf{p},r)u_b(\mathbf{p},r) \overset{(5.41)}{=} u_a^c(\mathbf{p},r)u_b(\mathbf{p},r) \overset{(5.39)}{=} C_{am}\overline{u}_m^\top(\mathbf{p},r)u_b(\mathbf{p},r)$$

$$= C_{am}u_b(\mathbf{p},r)\overline{u}_m^\top(\mathbf{p},r) \overset{(5.62)}{=} C_{am}\left(\Lambda_+(\mathbf{p})\right)_{bm}$$

$$= \left(\Lambda_+C^\top\right)_{ba} \overset{(1.191)}{=} -\left(\Lambda_+C\right)_{ba}$$

$$= -\left(\Lambda_+C\right)_{ab}^\top = -\left(\Lambda_-C^\top\right)_{ab} \overset{(5.58)}{=} \left(\frac{\slashed{p}-m}{2m}C\right)_{ab}.$$

This completes the proof of Proposition 5.13.

With the results of Proposition 5.13 at hand we can finally show Eq. (5.118c).

Proof of Eq. (5.118c): We have to evaluate the commutator

$$-i\Big[\bar{\epsilon}Q, \psi_a(x)\Big] = -i\Big(\bar{\epsilon}_b Q_b \psi_a(x) - \psi_a(x)\bar{\epsilon}_b Q_b\Big) = -i\bar{\epsilon}_a\Big\{Q_b, \psi_a(x)\Big\}$$

$$= \frac{(2m)^{1/2}}{(2\pi)^{3/2}} \bar{\epsilon}_b \sum_{r,s} \int d^3p \int d^3k \Big(\frac{m}{\omega_k}\Big)^{1/2}$$

$$\times \Big\{C_{bc}(\mathbf{p})d^\dagger(\mathbf{p},s)v_c(\mathbf{p},s) - D_{bc}(\mathbf{p})d(\mathbf{p},s)u_c(\mathbf{p},s),$$

$$+ d(\mathbf{k},r)u_a(\mathbf{k},r)e^{-ikx} + d^\dagger(\mathbf{k},r)v_a(\mathbf{k},r)e^{ikx}\Big\}.$$

Here we substitute the momentum space expansions Eqs. (5.49) and (5.95). Thus we find

$$-i\Big[\bar{\epsilon}Q, \psi_a(x)\Big]$$

$$= \frac{(2m)^{1/2}}{(2\pi)^{3/2}} \bar{\epsilon}_b \sum_{r,s} \int d^3p \int d^3k \Big(\frac{m}{\omega_k}\Big)^{1/2}$$

$$\times \Big[\big\{d^\dagger(\mathbf{p},s), d(\mathbf{k},r)\big\}C_{bd}(\mathbf{p})v_d(\mathbf{p},s)u_a(\mathbf{k},r)e^{-ikx}$$

$$+ \big\{d^\dagger(\mathbf{p},s), d^\dagger(\mathbf{k},r)\big\}C_{bd}(\mathbf{p})v_d(\mathbf{p},s)v_a(\mathbf{k},r)e^{ikx}$$

$$- \big\{d(\mathbf{p},s), d(\mathbf{k},r)\big\}D_{bd}(\mathbf{p})u_d(\mathbf{p},s)u_a(\mathbf{k},r)e^{-ikx}$$

$$- \big\{d(\mathbf{p},s), d^\dagger(\mathbf{k},r)\big\}D_{bd}(\mathbf{p})u_d(\mathbf{p},s)v_a(\mathbf{k},r)e^{-ikx}\Big]$$

$$\overset{(5.96)}{=} \frac{(2m)^{1/2}}{(2\pi)^{3/2}} \bar{\epsilon}_b \sum_{r} \int d^3p \Big(\frac{m}{\omega_p}\Big)^{1/2}\Big[C_{bd}(\mathbf{p})v_d(\mathbf{p},r)u_a(\mathbf{p},r)e^{-ipx}$$

$$- D_{bd}(\mathbf{p})u_d(\mathbf{p},r)v_a(\mathbf{p},r)e^{ipx}\Big]$$

$$\overset{(5.120a)}{\underset{(5.120b)}{=}} \frac{(2m)^{1/2}}{(2\pi)^{3/2}} \bar{\epsilon}_b \int d^3p \Big(\frac{m}{\omega_p}\Big)^{1/2}\Big[C_{bd}(\mathbf{p})\Big(\frac{\not{p}-m}{2m}C\Big)_{da}e^{-ipx}$$

$$- D_{bd}(\mathbf{p})\Big(\frac{\not{p}+m}{2m}C\Big)_{da}e^{ipx}\Big]$$

$$= \frac{\bar{\epsilon}_b}{(2\pi)^{3/2}} \int d^3p \frac{1}{(2\omega_p)^{1/2}}\Big[\big[a(\mathbf{p})\delta_{bd} - i\gamma^5_{bd}b(\mathbf{p})\big]\big[(\not{p}-m)C\big]_{da}e^{-ipx}$$

$$- \big(a^\dagger(\mathbf{p})\delta_{bd} - i\gamma^5_{bd}b^\dagger(\mathbf{p})\big)\big((\not{p}+m)C\big)_{da}e^{ipx}\Big]$$

$$= \bar{\epsilon}_b \,(2\pi)^{-3/2} \int d^3p (2\omega_p)^{-1/2}$$

$$\times \Big[a(\mathbf{p})(\not p - m)Ce^{-ipx} - ib(\mathbf{p})\gamma^5(\not p - m)Ce^{-ipx}$$

$$- a^\dagger(\mathbf{p})(\not p + m)Ce^{ipx} + ib^\dagger(\mathbf{p})\gamma^5(\not p + m)Ce^{ipx} \Big]_{ba}.$$

We now use the following equations:

$$\big(\not p - m\big)e^{-ipx} = \big(i\not\partial - m\big)e^{-ipx}, \quad \big(\not p + m\big)e^{ipx} = -\big(i\not\partial - m\big)e^{ipx}.$$

Then we get:

$$
\begin{aligned}
-i\Big[\bar{\epsilon}Q, \psi_a(x)\Big] \;=\;& \bar{\epsilon}_b \,(2\pi)^{-3/2} \int d^3p (2\omega_p)^{-1/2} \\
& \times \Big[a(\mathbf{p})(i\not\partial - m)Ce^{-ipx} + a^\dagger(\mathbf{p})(i\not\partial - m)Ce^{ipx} \\
& \quad -ib(\mathbf{p})\gamma^5(i\not\partial - m)Ce^{-ipx} - ib^\dagger(\mathbf{p})\gamma^5(i\not\partial - m)Ce^{ipx} \Big]_{ba} \\
=\;& \bar{\epsilon}_b \Big[(i\not\partial - m)C(2\pi)^{-3/2} \int d^3p (2\omega_p)^{-1/2} \\
& \times \Big(a(\mathbf{p})e^{-ipx} + a^\dagger(\mathbf{p})e^{ipx} \Big) \\
& \quad -i\gamma^5(i\not\partial - m)C(2\pi)^{-3/2} \int d^3p (2\omega_p)^{-1/2} \\
& \times \Big(b(\mathbf{p})e^{-ipx} + b^\dagger(\mathbf{p})e^{ipx} \Big) \Big]_{ba} \\
\overset{(5.90)}{\underset{(5.93)}{=}}\;& \bar{\epsilon}_b \Big[(i\not\partial - m)CA(x) - i\gamma^5(i\not\partial - m)CB(x) \Big]_{ba} \\
\overset{(1.189)}{=}\;& \bar{\epsilon}_b \Big[C\big(-i\not\partial^\top - m\big)A(x) - i\gamma^5 C\big(-i\not\partial^\top - m\big)B(x) \Big]_{ba} \\
\overset{(1.196)}{=}\;& \bar{\epsilon}_b \Big[C\big(-i\not\partial^\top - m\big)A(x) - iC\gamma^{5\top}\big(-i\not\partial^\top - m\big)B(x) \Big]_{ba} \\
=\;& \bar{\epsilon}_b C_{bd} \Big[\big(A(x) - i\gamma^{5\top}B(x) \big)\big(-i\overleftarrow{\not\partial}^\top - m\big) \Big]_{da} \\
=\;& \Big[\epsilon^\top \big(A(x) - i\gamma^{5\top}B(x) \big)^\top (-1)\big(i\overleftarrow{\not\partial} + m\big)^\top \Big]_a \\
=\;& \Big(-(i\not\partial + m)\big(A(x) - i\gamma^5 B(x)\big)\epsilon \Big)_a \overset{(5.11c)}{=} \delta\,\psi_a(x).
\end{aligned}
$$

These results demonstrate that the spinor charge Q_a defined by Eq. (5.49) gives the correct supersymmetry transformation of the scalar field $A(x)$, the pseudoscalar field $B(x)$, and the Majorana spinor field $\psi(x)$ in the Wess–Zumino model.

Chapter 6

Superspace Formalism and Superfields

6.1 Superspace

In order to be able to construct supersymmetric models, one wants to have a formalism in which supersymmetry is inherently manifest like Lorentz invariance in electrodynamics. Such a formalism is the superfield formalism introduced by A. Salam and J. Strathdee.[1] Consider, for instance, a theory formulated in three-dimensional Euclidean space. In general one cannot expect such a theory to be invariant under Lorentz transformations or to be invariant under the transformations of the Poincaré group. In order to obtain such a relativistically invariant theory one is forced to extend Euclidean 3-space to a flat pseudo-Riemannian 4-space, the Minkowski space. Introducing time as the additional coordinate, it is possible to formulate the theory in a relativistically invariant form. Analogously it is not possible to have a manifestly supersymmetric theory in Minkowski space. To obtain a formalism which achieves this we have to extend Minkowski space to *superspace*. Elements of superspace are so-called *supercoordinates* which consist of the usual four Minkowski spacetime coordinates and four constant (*i.e.* x_μ-independent) anticommuting Grassmann numbers.[2] If we formulate the theory in terms of the two-component Weyl spinor formalism, the latter are

$$\{\theta_A\}_{A=1,2} \quad \text{and} \quad \{\overline{\theta}_{\dot{B}}\}_{\dot{B}=\dot{1},\dot{2}}$$

---

[1] See A. Salam and J. Strathdee [99] and S. Ferrara, J. Wess and B. Zumino [40].

[2] See also the discussion in Chap. 31.3 of R. Penrose [87]. A comprehensive presentation of supermanifolds can be found in the monograph by B. DeWitt [26].

two-component Weyl spinors which transform under the self-representation of $SL(2,\mathbb{C})$ and the complex conjugate self-representation of $SL(2,\mathbb{C})$ respectively, and are considered to be independent. On the other hand, if we work with four-dimensional Majorana spinors, the additional parameters are constant, anticommuting Grassmann numbers

$$\{\epsilon_a\}_{a=1,2,3,4}$$

with ϵ_a satisfying the Majorana condition (1.203). The resulting "*superspace*" therefore has eight dimensions.

With the help of the anticommuting Grassmann parameters we can transform the graded Lie algebra (involving both commutators and anticommutators) into a regular Lie algebra (which involves only commutators) by writing the elements of the spinor sector of the algebra

$$\theta^A Q_A, \quad \overline{\theta}_{\dot A}\overline{Q}^{\dot A}.$$

The anticommutation relations of the two-component Weyl spinors are given by

$$\{\theta_A, \theta_B\} = 0, \quad \{\overline{\theta}_{\dot A}, \overline{\theta}_{\dot B}\} = 0, \quad \{\theta_A, \overline{\theta}_{\dot B}\} = 0, \tag{6.1}$$

and an element of superspace is given by the supercoordinate $(x_\mu, \theta_A, \overline{\theta}_{\dot A})$. We now reformulate the supersymmetry algebra (4.1) entirely in terms of commutators. As is evident from Eq. (3.29) the relevant part of the Super–Poincaré algebra in the Weyl formalism has the anticommutation relations

$$
\begin{aligned}
\{Q_A, Q_B\} &= 0, \\
\{\overline{Q}_{\dot A}, \overline{Q}_{\dot B}\} &= 0, \\
\{Q_A, \overline{Q}_{\dot B}\} &= 2\sigma^\mu_{A\dot B} P_\mu.
\end{aligned}
\tag{6.2}
$$

Proposition 6.1:

The anticommutation relations (6.2) can be rewritten as the commutators:

$$
\left.
\begin{aligned}
[\theta^A Q_A, \overline{\theta}_{\dot B}\overline{Q}^{\dot B}] &= 2\theta^A \sigma^\mu_{A\dot B}\overline{\theta}^{\dot B} P_\mu, \\
[\theta^A Q_A, \theta^B Q_B] &= 0, \\
[\overline{\theta}_{\dot A}\overline{Q}^{\dot A}, \overline{\theta}_{\dot B}\overline{Q}^{\dot B}] &= 0,
\end{aligned}
\right\}
\tag{6.3}
$$

where θ^A and $\overline{\theta}_{\dot A}$ are anticommuting Grassmann numbers which also anticommute with the spinor charges Q_A and $\overline{Q}_{\dot A}$. In Eqs. (6.3) we use the summation convention for two-component Weyl spinors as in Eqs. (1.92a), (1.92b) and (1.93).

Proof: We start with the anticommutator

$$\{Q_A, \overline{Q}_{\dot{B}}\} = 2\sigma^\mu_{A\dot{B}} P_\mu,$$

$$\text{i.e.} \quad Q_A\overline{Q}_{\dot{B}} + \overline{Q}_{\dot{B}}Q_A = 2\sigma^\mu_{A\dot{B}} P_\mu.$$

Multiplying this equation from the left by θ^A and from the right by $\overline{\theta}^{\dot{B}}$, we obtain

$$\theta^A Q_A \overline{Q}_{\dot{B}} \overline{\theta}^{\dot{B}} + \theta^A \overline{Q}_{\dot{B}} Q_A \overline{\theta}^{\dot{B}} = 2\theta^A \sigma^\mu_{A\dot{B}} \overline{\theta}^{\dot{B}} P_\mu.$$

We now assume that $Q_A, \overline{Q}_{\dot{A}}$ are linear operators with the appropriate transformation properties under $SL(2, \mathbb{C})$. Then

$$Q_A \sim \frac{\partial}{\partial\theta^A}, \qquad \overline{Q}_{\dot{A}} \sim \frac{\partial}{\partial\overline{\theta}^{\dot{A}}},$$

and so

$$(Q_A\overline{\theta}_{\dot{B}}) = 0, \qquad (\overline{Q}_{\dot{B}}\theta_A) = 0.$$

The left hand side of the above equation can then be written (with quantities in brackets representing c-numbers):

$$-\theta^A Q_A \overline{\theta}^{\dot{B}} \overline{Q}_{\dot{B}} + \theta^A Q_A(\overline{Q}_{\dot{B}}\overline{\theta}^{\dot{B}}) - \overline{Q}_{\dot{B}}\theta^A Q_A \overline{\theta}^{\dot{B}} + (\overline{Q}_{\dot{B}}\theta^A)Q_A\overline{\theta}^{\dot{B}}$$

$$= -\theta^A Q_A \overline{\theta}^{\dot{B}} \overline{Q}_{\dot{B}} + \theta^A Q_A(\overline{Q}_{\dot{B}}\overline{\theta}^{\dot{B}}) + \overline{Q}_{\dot{B}}\theta^A \overline{\theta}^{\dot{B}} Q_A$$

$$\quad - \overline{Q}_{\dot{B}}\theta^A(Q_A\overline{\theta}^{\dot{B}}) + (\overline{Q}_{\dot{B}}\theta^A)Q_A\overline{\theta}^{\dot{B}}$$

$$= -\theta^A Q_A \overline{\theta}^{\dot{B}} \overline{Q}_{\dot{B}} + \theta^A Q_A(\overline{Q}_{\dot{B}}\overline{\theta}^{\dot{B}}) - \overline{Q}_{\dot{B}}\overline{\theta}^{\dot{B}}\theta^A Q_A$$

$$= -\theta^A Q_A \overline{\theta}^{\dot{B}} \overline{Q}_{\dot{B}} + \theta^A Q_A(\overline{Q}_{\dot{B}}\overline{\theta}^{\dot{B}}) + \overline{\theta}^{\dot{B}}\overline{Q}_{\dot{B}}\theta^A Q_A - (\overline{Q}_{\dot{B}}\overline{\theta}^{\dot{B}})\theta^A Q_A$$

$$= +\theta^A Q_A \overline{\theta}_{\dot{B}} \overline{Q}^{\dot{B}} - \overline{\theta}_{\dot{B}} \overline{Q}^{\dot{B}} \theta^A Q_A,$$

since (using Eqs. (1.85) and (1.86))

$$\overline{\theta}^{\dot{B}}\overline{Q}_{\dot{B}} = -\overline{\theta}_{\dot{A}}\epsilon^{\dot{A}\dot{B}}\epsilon_{\dot{B}\dot{C}}\overline{Q}^{\dot{C}}$$

$$= -\overline{\theta}_{\dot{A}}\delta^{\dot{A}}_{\dot{C}}\overline{Q}^{\dot{C}} = -\overline{\theta}_{\dot{A}}\overline{Q}^{\dot{A}} = -(\overline{\theta}\,\overline{Q}).$$

Hence returning to the original equation, we obtain

$$(\theta Q)(\overline{\theta}\,\overline{Q}) - (\overline{\theta}\,\overline{Q})(\theta Q) = 2(\theta\sigma^\mu\overline{\theta})P_\mu,$$

and so the first of Eqs. (6.6). The remaining two relations can be shown to hold in a similar way.

For later calculations we need some more commutators which are readily verified. We state them here without proof.

Proposition 6.2:

The following relations can be shown to hold:

$$[P_\mu, (\theta Q)] = 0, \qquad [P_\mu, (\overline{\theta}\,\overline{Q})] = 0,$$
$$[(\theta Q), (\theta \sigma_\mu \overline{\theta})] = 0, \qquad [(\overline{\theta}\,\overline{Q}), (\theta \sigma_\mu \overline{\theta})] = 0. \tag{6.4}$$

We have seen earlier that one way to find irreducible representations of the supersymmetry algebra is to proceed as in the case of the Poincaré algebra and to find Casimir operators and then to construct the appropriate Fock space. The superspace formalism is an alternative method to find irreducible representations of the supersymmetry algebra, and is particularly useful in performing calculations. In order to be able to write down explicit representations of the supersymmetry algebra in terms of linear operators acting on functions in superspace, it is necessary to have at our disposal a calculus for differentiating with respect to Grassmann numbers. We therefore introduce this first.

6.2   Grassmann Differentiation

We consider a set of N discrete numbers $\{a_1, a_2, \ldots, a_N\}$ obeying

$$\{a_i, a_j\} = 0, \quad \forall \, i, j = 1, 2, \ldots, N.$$

We can construct a differential calculus for these Grassmann variables, and similarly — as we shall consider later — an integral calculus. However, the numbers $a_i, i = 1, 2, \ldots, N$, are discrete objects; for this reason the derivative is defined formally as

$$\frac{\partial a_i}{\partial a_j} := \delta_{ij}. \tag{6.5a}$$

This is simply a definition of the symbol on the left hand side and is not to be looked at as the ratio of two infinitesimal increments.

To obtain a product rule we must take into account the anticommutative character of the variables a_i. The product rule is then given by the relation:

$$\frac{\partial}{\partial a_p}(a_{i_1} a_{i_2} \ldots a_{i_r}) \;=\; \delta_{p i_1} a_{i_2} \ldots a_{i_r} - \delta_{p i_2} a_{i_1} a_{i_3} \ldots a_{i_r} + \cdots$$

$$(-1)^{r-1} \delta_{p i_r} a_{i_1} a_{i_2} \ldots a_{i_{r-1}}. \tag{6.5b}$$

Equation (6.5b) implies that $\partial/\partial a_i$ is an *"antiderivative"*. We also have the following relation

$$\left\{\frac{\partial}{\partial a_p}, a_r\right\} = \frac{\partial}{\partial a_p} a_r + a_r \frac{\partial}{\partial a_p}$$

$$= \frac{\partial a_r}{\partial a_p} - a_r \frac{\partial}{\partial a_p} + a_r \frac{\partial}{\partial a_p} \quad \text{(with Eq. (6.5b))}$$

$$= \frac{\partial a_r}{\partial a_p} = \delta_{rp} \quad \text{(with Eq. (6.5a))}.$$

Hence

$$\left\{\frac{\partial}{\partial a_p}, a_r\right\} = \delta_{pr}. \tag{6.5c}$$

Given any function $f(a_1, a_2, a_3, \ldots, a_N)$, we have

$$\frac{\partial}{\partial a_i}\frac{\partial}{\partial a_j} f(a_1, a_2, a_3, \ldots, a_N) = -\frac{\partial}{\partial a_j}\frac{\partial}{\partial a_i} f(a_1, a_2, a_3, \ldots, a_N),$$

so that

$$\left\{\frac{\partial}{\partial a_i}, \frac{\partial}{\partial a_j}\right\} f(a_1, a_2, a_3, \ldots, a_N) = 0. \tag{6.5d}$$

As an example we consider

$$f(a_1, a_2, a_3, \ldots, a_N) = a_r a_p.$$

Then

$$\frac{\partial}{\partial a_i}\left(\frac{\partial}{\partial a_j} a_r a_p\right) = \frac{\partial}{\partial a_i}(\delta_{jr} a_p - a_r \delta_{jp}) \quad \text{(with Eqs. (6.5a), (6.5b))}$$

$$= \delta_{jr}\delta_{ip} - \delta_{ir}\delta_{jp} \quad \text{(with Eq. (6.5a))}.$$

On the other hand

$$\frac{\partial}{\partial a_j}\left(\frac{\partial}{\partial a_i} a_r a_p\right) = \frac{\partial}{\partial a_j}(\delta_{ir} a_p - a_r \delta_{ip}) = \delta_{ir}\delta_{jp} - \delta_{jr}\delta_{ip} = -\frac{\partial}{\partial a_i}\left(\frac{\partial}{\partial a_j} a_r a_p\right).$$

Hence

$$\left\{\frac{\partial}{\partial a_i}, \frac{\partial}{\partial a_j}\right\} a_r a_p = 0.$$

We now consider the two-component Weyl spinors θ^A and $\overline{\theta}^{\dot{A}}$, and define

$$\partial_A := \frac{\partial}{\partial \theta^A}, \qquad \partial^A; = \frac{\partial}{\partial \theta_A},$$

$$\overline{\partial}_{\dot{A}} := \frac{\partial}{\partial \overline{\theta}^{\dot{A}}}, \qquad \overline{\partial}^{\dot{A}} := \frac{\partial}{\partial \overline{\theta}_{\dot{A}}}. \tag{6.5e}$$

According to Eq. (6.5a) differentiation with respect to θ is defined as

$$
\begin{aligned}
\partial_A \theta^B &= \frac{\partial}{\partial \theta^A} \theta^B = \delta_A^{\ B}, \\
\partial^A \theta_B &= \frac{\partial}{\partial \theta_A} \theta_B = \delta^A_{\ B}, \\
\partial_A \theta_B &= \frac{\partial}{\partial \theta^A}(\epsilon_{BC}\theta^C) = \epsilon_{BC}\delta^C_{\ A} = \epsilon_{BA}, \\
\overline{\partial}_{\dot{A}} \overline{\theta}^{\dot{B}} &= \frac{\partial}{\partial \overline{\theta}^{\dot{A}}} \overline{\theta}^{\dot{B}} = \delta_{\dot{A}}^{\ \dot{B}}, \\
\overline{\partial}^{\dot{A}} \overline{\theta}_{\dot{B}} &= \frac{\partial}{\partial \overline{\theta}_{\dot{A}}} \overline{\theta}_{\dot{B}} = \delta^{\dot{A}}_{\ \dot{B}}.
\end{aligned}
\tag{6.5f}
$$

We can use the metric tensor to raise or lower indices of derivatives.

Proposition 6.3:

Undotted indices of derivatives can be raised or lowered with the formulas:

$$
\epsilon^{AB}\partial_B = -\partial^A, \qquad \partial_A = -\epsilon_{AB}\partial^B.
\tag{6.5g}
$$

Proof: We have:

$$
\epsilon^{AB}\partial_B \theta^C = \epsilon^{AB} \frac{\partial}{\partial \theta^B} \theta^C = \epsilon^{AB}\delta_B^{\ C} = \epsilon^{AC},
$$

and

$$
\begin{aligned}
-\partial^A \theta^C &= -\partial^A \epsilon^{CD}\theta_D = -\epsilon^{CD} \frac{\partial}{\partial \theta_A} \theta_D \\
&= -\epsilon^{CD}\delta_D^{\ A} = -\epsilon^{CA} = \epsilon^{AC} = \epsilon^{AB}\partial_B \theta^C.
\end{aligned}
$$

This implies for the operators

$$
\epsilon^{AB}\partial_B = -\partial^A,
$$

and the other relation follows similarly. We conclude therefore that:

$$
\partial^A \theta^C = -\epsilon^{AC}, \qquad \text{and similarly} \qquad \partial_A \theta_C = -\epsilon_{AC}.
\tag{6.5h}
$$

Equation (6.5d) implies for our special case:

$$
\{\partial_A, \partial_B\} = 0.
\tag{6.5i}
$$

Analogous formulas hold for $\overline{\partial}$. Thus

$$\overline{\partial}^{\dot{A}} = \overline{\partial}_{\dot{B}}\epsilon^{\dot{B}\dot{A}} \quad \text{or} \quad \left.\begin{array}{r} \overline{\partial}^{\dot{A}} = -\epsilon^{\dot{A}\dot{B}}\overline{\partial}_{\dot{B}}, \\ \text{and} \quad \overline{\partial}_{\dot{A}} = -\epsilon_{\dot{A}\dot{B}}\overline{\partial}^{\dot{B}} \end{array}\right\}, \tag{6.5j}$$

implying

$$\overline{\partial}^{\dot{A}}\overline{\theta}^{\dot{B}} = -\epsilon^{\dot{A}\dot{B}}, \quad \text{and similarly} \quad \overline{\partial}_{\dot{A}}\overline{\theta}_{\dot{B}} = -\epsilon_{\dot{A}\dot{B}}, \tag{6.5k}$$

since (using Eq. (1.86))

$$\overline{\partial}_{\dot{A}}\overline{\theta}_{\dot{B}} = \overline{\partial}_{\dot{A}}\epsilon_{\dot{B}\dot{C}}\overline{\theta}^{\dot{C}} = -\epsilon_{\dot{B}\dot{C}}\delta_{\dot{A}}^{\dot{C}} = \epsilon_{\dot{B}\dot{A}} = -\epsilon_{\dot{A}\dot{B}}, \tag{6.5l}$$

and as a consequence of Eq. (6.5d):

$$\{\overline{\partial}_{\dot{A}}, \overline{\partial}_{\dot{B}}\} = 0.$$

In addition, since θ and $\overline{\theta}$ are considered to be independent, we demand

$$\{\overline{\partial}_{\dot{A}}, \theta^B\} = 0, \qquad \{\partial_A, \overline{\theta}^{\dot{B}}\} = 0, \tag{6.5m}$$

and hence

$$\overline{\partial}_{\dot{A}}\theta^B = 0, \tag{6.5n}$$

$$\partial_A\overline{\theta}^{\dot{B}} = 0. \tag{6.5o}$$

Proposition 6.4:

We have

$$\partial_A\theta^2 = 2\theta_A, \qquad \overline{\partial}_{\dot{A}}\overline{\theta}^2 = -2\overline{\theta}_{\dot{A}}. \tag{6.5p}$$

Proof: Consider

$$\begin{aligned} \partial_A\theta^2 &= \partial_A(\theta^B\theta_B) = (\partial_A\theta^B)\theta_B - \theta^B(\partial_A\theta_B) \\ &= \delta_A^B\theta_B - \theta^B[-\epsilon_{AD}\partial^D\theta_B] = \theta_A + \theta^B\epsilon_{AB} \overset{(1.84)}{=} 2\theta_A. \end{aligned}$$

Similarly:

$$\overline{\partial}_{\dot{A}}(\overline{\theta}^2) = \overline{\partial}_{\dot{A}}(\overline{\theta}_{\dot{B}}\overline{\theta}^{\dot{B}}) = -2\overline{\theta}_{\dot{A}}.$$

Proposition 6.5:

The following relations hold:

$$\epsilon^{AB}\partial_A\partial_B(\theta\theta) = \ 4 \ = \partial^B\partial_B(\theta\theta),$$
$$\epsilon_{\dot{A}\dot{B}}\overline{\partial}^{\dot{A}}\overline{\partial}^{\dot{B}}(\overline{\theta}\,\overline{\theta}) = \ 4 \ = \overline{\partial}_{\dot{B}}\overline{\partial}^{\dot{B}}(\overline{\theta}\,\overline{\theta}). \tag{6.5q}$$

Proof: Consider:

$$\epsilon^{AB}\partial_A\partial_B(\theta\theta) \ \overset{(6.5p)}{=} \ 2\epsilon^{AB}\partial_A\theta_B = 2\epsilon^{AB}(-\epsilon_{AD}\partial^D\theta_B)$$
$$= \ 2\epsilon^{AB}(-\epsilon_{AD}\delta^D_B) = 2\epsilon^{AB}(-\epsilon_{AB})$$
$$= \ 2\epsilon^{AB}\epsilon_{BA} = 2\delta^A_A = 4.$$

Similarly

$$\epsilon_{\dot{A}\dot{B}}\overline{\partial}^{\dot{A}}\overline{\partial}^{\dot{B}}\overline{\theta}^2 \ = \ 2\epsilon_{\dot{A}\dot{B}}\overline{\partial}^{\dot{A}}\overline{\theta}^{\dot{B}} = -2\epsilon_{\dot{A}\dot{B}}\epsilon^{\dot{A}\dot{B}}$$
$$\overset{(6.5k)}{=} \ 2\epsilon_{\dot{A}\dot{B}}\epsilon^{\dot{B}\dot{A}} = 2\delta^{\dot{B}}_{\dot{A}} = 4.$$

6.3   Supersymmetry Transformations in the Weyl Formalism

6.3.1   Finite Supersymmetry Transformations

In Chapter 5 we discussed infinitesimal variations of the fields appearing in the Wess–Zumino model. Such changes of the fields correspond to infinitesimal supersymmetry transformations. The consideration can be extended to finite supersymmetry transformations in a natural way in the context of the superspace formalism. Following A. Salam and J. Strathdee [99] we consider the action of the supersymmetry group on the space of left cosets with respect to the subgroup of homogeneous Lorentz transformations (*i.e.* there are no terms in $M_{\mu\nu}$ in the arguments of the exponentials of Eq. (6.6) below; we consider only the subgroup of transformations with the homogeneous Lorentz transformations factorized out). This space is a homogeneous space on which the factor group acts transitively, and as described above, it is an eight-dimensional space which is parametrized in terms of Minkowski space-time coordinates x_μ and Weyl spinors $\theta_A, \overline{\theta}_{\dot{A}}, A = 1, 2, \dot{A} = \dot{1}, \dot{2}$. It is this space which is called *superspace*. Since we factorize out the homogeneous Lorentz transformations, we construct and define the following operators:

$$\left. \begin{array}{ll} L(x_\mu, \theta_A, \overline{\theta}^{\dot{A}}) & := \exp(-ix_\mu P^\mu + i\theta Q + i\overline{\theta}\,\overline{Q}), \\ L_1(x_\mu, \theta_A, \overline{\theta}^{\dot{A}}) & := \exp(-ix_\mu P^\mu + i\theta Q) \cdot \exp(i\overline{\theta}\,\overline{Q}), \\ L_2(x_\mu, \theta_A, \overline{\theta}^{\dot{A}}) & := \exp(-ix_\mu P^\mu + i\overline{\theta}\,\overline{Q}) \cdot \exp(i\theta Q). \end{array} \right\} \tag{6.6}$$

Here $L(x, \theta, \overline{\theta})$ is a unitary operator. The operators L_1 and L_2 are also unitary and are related to $L(x, \theta, \overline{\theta})$ as in Eq. (6.9) below. The quantities $L(x, \theta, \overline{\theta}), L_1(x, \theta, \overline{\theta})$ and $L_2(x, \theta, \overline{\theta})$ are operators which describe three different but equivalent actions of the supersymmetry group on functions defined on superspace. Hence Eq. (6.6) leads to three different but equivalent realizations of one and the same supersymmetry transformation, and hence, as we shall see, to three different definitions of superfields. In Eq. (6.6) the quantities P_μ, Q_A and $\overline{Q}^{\dot{A}}$ denote Hermitian operators which act on functions in superspace. These operators correspond to the basic elements of the Super–Poincaré algebra. It should be noted that we use the same symbols for these operators and their corresponding Lie algebra elements. It should also be noted from Eq. (6.6) that θ and $\overline{\theta}$ have dimension $+1/2$ in length since $Q, \overline{Q}$ have dimension $-1/2$ as follows *e.g.* from Eq. (6.2).

A general (Lorentz scalar or pseudoscalar) *superfield* Φ is an operator-valued function defined on superspace which is to be understood in terms of its power series expansion in θ and $\overline{\theta}$. Since θ and $\overline{\theta}$ are anti-commuting Grassmann numbers, this power series is finite, *i.e.*

$$
\begin{aligned}
\Phi(x, \theta, \overline{\theta}) = \ & f(x) + \theta^A \phi_A(x) + \overline{\theta}_{\dot{A}} \overline{\chi}^{\dot{A}}(x) + (\theta\theta) m(x) + (\overline{\theta}\,\overline{\theta}) n(x) \\
& + (\theta \sigma^\mu \overline{\theta}) V_\mu(x) + (\theta\theta) \overline{\theta}_{\dot{A}} \overline{\lambda}^{\dot{A}}(x) + (\overline{\theta}\,\overline{\theta}) \theta^A \psi_A(x) \\
& + (\theta\theta)(\overline{\theta}\,\overline{\theta}) d(x).
\end{aligned} \tag{6.7}
$$

In constructing this expression one recalls, of course, that

$$
(\theta\theta) \equiv \theta^A \theta_A, \qquad (\overline{\theta}\,\overline{\theta}) \equiv \overline{\theta}_{\dot{A}} \overline{\theta}^{\dot{A}}
$$

are Lorentz scalars (*cf.* Eqs. (1.92a) and (1.92b)), $\theta \sigma^\mu \overline{\theta}$ is a Lorentz vector (*cf.* Eq. (1.117)) and so on. One can easily convince oneself that the sum of terms of Eq. (6.7) exhausts all possibilities of nonvanishing combinations of powers of θ and $\overline{\theta}$; *e.g.* with Eq. (1.134) the term $\overline{\theta}\overline{\sigma}^\mu\theta$ which does not appear in Eq. (6.7) can be reduced to $-\theta\sigma^\mu\overline{\theta}$. All higher powers of Grassmann numbers $\theta, \overline{\theta}$ vanish as explained earlier. The quantities $f(x), \phi_A(x), \overline{\chi}^{\dot{A}}(x)$ *etc.* are called *component fields*. The power series expansion (6.7) will be discussed in detail later.

The three different operators (6.6) lead to three different types of superfields $\Phi(x, \theta, \overline{\theta}), \Phi_1(x, \theta, \overline{\theta}), \Phi_2(x, \theta, \overline{\theta})$ respectively. These are related by the following identity:

$$
\Phi(x_\mu, \theta, \overline{\theta}) = \Phi_1(x_\mu + i\theta\sigma_\mu\overline{\theta}, \theta, \overline{\theta}) = \Phi_2(x_\mu - i\theta\sigma_\mu\overline{\theta}, \theta, \overline{\theta}). \tag{6.8}
$$

Before we demonstrate the validity of this relation, we show that the following connection holds between the operators (6.6):

$$L(x, \theta, \overline{\theta}) = L_1(x + i\theta\sigma\overline{\theta}\theta, \overline{\theta}) = L_2(x - i\theta\sigma\overline{\theta}, \theta, \overline{\theta}). \qquad (6.9)$$

In order to prove Eq. (6.9) we need the *Baker–Campbell–Hausdorff formula*[3]

$$\exp(A) \cdot \exp(B) = \exp\left(A + B + \tfrac{1}{2}[A, B] + \tfrac{1}{12}[A, [A, B]] - \tfrac{1}{12}[B, [B, A]] + \cdots\right). \qquad (6.10)$$

Hence

$$
\begin{aligned}
L_1(x + i\theta\sigma_\mu\overline{\theta}, \theta, \overline{\theta}) \overset{(6.6)}{=}\ & \exp\{-i(x_\mu + i\theta\sigma_\mu\overline{\theta})P^\mu + i\theta Q\}\exp\{i\overline{\theta}\,\overline{Q}\} \\
=\ & \exp\{-i(x_\mu + i\theta\sigma_\mu\overline{\theta})P^\mu + i\theta Q + i\overline{\theta}\,\overline{Q} \\
& + \tfrac{1}{2}[-ix_\mu P^\mu + \theta\sigma_\mu\overline{\theta}P^\mu + i\theta Q, i\overline{\theta}\,\overline{Q}] + 0\} \\
=\ & \exp\{-i(x_\mu + i\theta\sigma_\mu\overline{\theta})P^\mu + i\theta Q + i\overline{\theta}\,\overline{Q} \\
& + \tfrac{1}{2}x_\mu[P^\mu, \overline{\theta}\,\overline{Q}] - \tfrac{1}{2}[\theta Q, \overline{\theta}\,\overline{Q}] + \tfrac{i}{2}\theta\sigma_\mu\overline{\theta}[P^\mu, \overline{\theta}\,\overline{Q}]\},
\end{aligned}
$$

and with Eqs. (6.3) and (6.4) this becomes

$$
\begin{aligned}
L_1(x + i\theta\sigma_\mu\overline{\theta}, \theta, \overline{\theta}) =\ & \exp\{-i(x_\mu + i\theta\sigma_\mu\overline{\theta})P^\mu + i\theta Q + i\overline{\theta}\,\overline{Q} - \theta\sigma_\mu\overline{\theta}P^\mu\} \\
=\ & \exp\{-ix_\mu P^\mu + i\theta Q + i\overline{\theta}\,\overline{Q}\} \overset{(6.6)}{=} L(x, \theta, \overline{\theta}).
\end{aligned}
$$

It should be noted that all higher order commutators, *e.g.* $[A, [A, B]]$, in the expression of the Baker–Campbell–Hausdorff formula vanish, because the commutators close into P_μ *e.g.* formally (using Eqs. (6.3) and (6.4)):

$$
\begin{aligned}
[P + \theta Q, [P + \theta Q, \overline{\theta}\,\overline{Q}]] =\ & [P + \theta Q, [P, \overline{\theta}\,\overline{Q}] + [\theta Q, \overline{\theta}\,\overline{Q}]] \\
=\ & [P + \theta Q, P] = [P, P] + [\theta Q, P] = 0.
\end{aligned}
$$

Consider in the same way

$$
\begin{aligned}
L_2(x - i\theta\sigma\overline{\theta}, \theta, \overline{\theta}) =\ & \exp\{-i(x_\mu - i\theta\sigma_\mu\overline{\theta})P^\mu + i\overline{\theta}\,\overline{Q}\}\exp\{i\theta Q\} \\
\overset{(6.10)}{=}\ & \exp\{-i(x_\mu - i\theta\sigma_\mu\overline{\theta})P^\mu + i\overline{\theta}\,\overline{Q} + i\theta Q \\
& + \tfrac{1}{2}[-ix_\mu P^\mu - \theta\sigma_\mu\overline{\theta}P^\mu + i\overline{\theta}\,\overline{Q}, i\theta Q]\} \\
=\ & \exp\{-ix_\mu P^\mu - \theta\sigma_\mu\overline{\theta}P^\mu + i\overline{\theta}\,\overline{Q} + i\theta Q \\
& + \tfrac{1}{2}x_\mu[P^\mu, \theta Q] - \tfrac{1}{2}[\overline{\theta}\,\overline{Q}, \theta Q] - \tfrac{i}{2}\theta\sigma_\mu\overline{\theta}[P^\mu, \theta Q]\} \\
=\ & \exp\{-ix_\mu P^\mu - \theta\sigma_\mu\overline{\theta}P^\mu + i\overline{\theta}\,\overline{Q} + i\theta Q + \theta\sigma_\mu\overline{\theta}P^\mu\} \\
=\ & \exp\{-ix_\mu P^\mu + i\overline{\theta}\,\overline{Q} + i\theta Q\} \overset{(6.6)}{=} L(x, \theta, \overline{\theta}).
\end{aligned}
$$

---

[3]See *e.g.* W. Miller [72], p. 161.

We can now demonstrate that the relation (6.8) holds. For a given field configuration

$$\Phi_0 \equiv \Phi(x_0, \theta_0, \overline{\theta}_0)$$

we have

$$
\begin{aligned}
\Phi(x, \theta, \overline{\theta}) &:= L(x, \theta, \overline{\theta}) \Phi_0 L^{-1}(x, \theta, \overline{\theta}) \\
&\overset{(6.9)}{=} L_1(x + i\theta\sigma\overline{\theta}, \theta, \overline{\theta}) \Phi_0 L_1^{-1}(x + i\theta\sigma\overline{\theta}, \theta, \overline{\theta}) \\
&=: \Phi_1(x + i\theta\sigma\overline{\theta}, \theta, \overline{\theta}),
\end{aligned}
\tag{6.11}
$$

and

$$
\begin{aligned}
\Phi(x, \theta, \overline{\theta}) &:= L(x, \theta, \overline{\theta}) \Phi_0 L^{-1}(x, \theta, \overline{\theta}) \\
&= L_2(x - i\theta\sigma\overline{\theta}, \theta, \overline{\theta}) \Phi_0 L_2^{-1}(x - i\theta\sigma\overline{\theta}, \theta, \overline{\theta}) \\
&=: \Phi_2(x - i\theta\sigma\overline{\theta}, \theta, \overline{\theta}).
\end{aligned}
\tag{6.12}
$$

Proposition 6.6:

Under a *finite supersymmetry transformation* denoted by T_α the superfields Φ, Φ_1 and Φ_2 undergo the following transformations:

$$T_\alpha \Phi(x, \theta, \overline{\theta}) = \Phi(x + i\theta\sigma\overline{\alpha} - i\alpha\sigma\overline{\theta}, \theta + \alpha, \overline{\theta} + \overline{\alpha}), \tag{6.13}$$
$$T_\alpha \Phi_1(x, \theta, \overline{\theta}) = \Phi_1(x + 2i\theta\sigma\overline{\alpha} + i\alpha\sigma\overline{\alpha}, \theta + \alpha, \overline{\theta} + \overline{\alpha}), \tag{6.14}$$
$$T_\alpha \Phi_2(x, \theta, \overline{\theta}) = \Phi_2(x - 2i\alpha\sigma\overline{\theta} - i\alpha\sigma\overline{\alpha}, \theta + \alpha, \overline{\theta} + \overline{\alpha}). \tag{6.15}$$

Proof: Considering the left hand side of Eq. (6.13), we have:

$$
\begin{aligned}
T_\alpha \Phi(x, \theta, \overline{\theta}) &= L(0, \alpha, \overline{\alpha}) \Phi(x, \theta, \overline{\theta}) L^{-1}(0, \alpha, \overline{\alpha}) \\
&= L(0, \alpha, \overline{\alpha})[L(x, \theta, \overline{\theta}) \Phi_0 L^{-1}(x, \theta, \overline{\theta})] L^{-1}(0, \alpha, \overline{\alpha}) \\
&= [L(0, \alpha, \overline{\alpha}) L(x, \theta, \overline{\theta})] \Phi_0 [L(0, \alpha, \overline{\alpha}) L(x, \theta, \overline{\theta})]^{-1} \quad (6.16)
\end{aligned}
$$

Now, using Eq. (6.6) and then Eq. (6.10), we obtain:

$$
\begin{aligned}
L(0, \alpha, \overline{\alpha}) L(x, \theta, \overline{\theta}) &= \exp\{i\alpha Q + i\overline{\alpha}\overline{Q}\} \exp\{-ix_\mu P^\mu + i\theta Q + i\overline{\theta}\,\overline{Q}\} \\
&= \exp\{-ix_\mu P^\mu + i(\theta + \alpha)Q + i(\overline{\theta} + \overline{\alpha})\overline{Q} \\
&\quad + \frac{1}{2}[i\alpha Q + i\overline{\alpha}\overline{Q}, -ix_\mu P^\mu + i\theta Q + i\overline{\theta}\,\overline{Q}]\},
\end{aligned}
$$

and then

$$
\begin{aligned}
L(0,\alpha,\overline{\alpha})L(x,\theta,\overline{\theta}) &= \exp\{-ix_\mu P^\mu + i(\theta+\alpha)Q + i(\overline{\theta}+\overline{\alpha})\overline{Q} \\
&\quad +\frac{1}{2}x_\mu[\alpha Q, P^\mu] - \frac{1}{2}[\alpha Q, \theta Q] - \frac{1}{2}[\alpha Q, \overline{\theta}\,\overline{Q}] \\
&\quad +\frac{1}{2}x_\mu[\overline{\alpha}\overline{Q}, P^\mu] - \frac{1}{2}[\overline{\alpha}\overline{Q}, \theta Q] - \frac{1}{2}[\overline{\alpha}\overline{Q}, \overline{\theta Q}]\} \\
&= \exp\{-ix_\mu P^\mu + i(\theta+\alpha)Q + i(\overline{\theta}+\overline{\alpha})\overline{Q} \\
&\quad -\alpha\sigma_\mu\overline{\theta}P^\mu + \theta\sigma_\mu\overline{\alpha}P^\mu\},
\end{aligned}
$$

and with Eqs. (6.3) and (6.4) this becomes

$$
\begin{aligned}
L(0,\alpha,\overline{\alpha})L(x,\theta,\overline{\theta}) &= \exp\{-i[x_\mu - i\alpha\sigma_\mu\overline{\theta} + i\theta\sigma_\mu\overline{\alpha}]P^\mu \\
&\quad +i(\theta+\alpha)Q + i(\overline{\theta}+\overline{\alpha})\overline{Q}\} \\
&\overset{(6.6)}{=} L(x - i\alpha\sigma\overline{\theta} + i\theta\sigma\overline{\alpha}, \theta+\alpha, \overline{\theta}+\overline{\alpha}). \quad (6.17)
\end{aligned}
$$

Inserting Eq. (6.17) into Eq. (6.16), we obtain

$$
\begin{aligned}
T_\alpha\Phi(x,\theta,\overline{\theta}) &= L(x - i\alpha\sigma\overline{\theta} + i\theta\sigma\overline{\alpha}, \theta+\alpha, \overline{\theta}+\overline{\alpha})\Phi_0 \\
&\quad \times L^{-1}(x - i\alpha\sigma\overline{\theta} + i\theta\sigma\overline{\alpha}, \theta+\alpha, \overline{\theta}+\overline{\alpha}) \\
&= \Phi(x - i\alpha\sigma\overline{\theta} + i\theta\sigma\overline{\alpha}, \theta+\alpha, \overline{\theta}+\overline{\alpha}).
\end{aligned}
$$

This proves Eq. (6.13). In order to prove Eq. (6.14), we consider:

$$
\begin{aligned}
T_\alpha\Phi_1(x,\theta,\overline{\theta}) &= L_1(i\alpha\sigma\overline{\alpha},\alpha,\overline{\alpha})\Phi_1(x,\theta,\overline{\theta})L_1^{-1}(i\alpha\sigma\overline{\alpha},\alpha,\overline{\alpha}) \\
&= L_1(i\alpha\sigma\overline{\alpha},\alpha,\overline{\alpha})\big[L_1(x,\theta,\overline{\theta})\Phi_0 L_1^{-1}(x,\theta,\overline{\theta})\big]L_1^{-1}(i\alpha\sigma\overline{\alpha},\alpha,\overline{\alpha}) \\
&= \big[L_1(i\alpha\sigma\overline{\alpha},\alpha,\overline{\alpha})L_1(x,\theta,\overline{\theta})\big]\Phi_0\big[L_1(i\alpha\sigma\overline{\alpha},\alpha,\overline{\alpha})L_1(x,\theta,\overline{\theta})\big]^{-1}.
\end{aligned}
$$

Using Eq. (6.6) we have

$$
\begin{aligned}
L_1(i\alpha\sigma\overline{\alpha},\alpha,\overline{\alpha})L_1(x,\theta,\overline{\theta}) &= \exp\{-i(i\alpha\sigma_\mu\overline{\alpha})P^\mu + i\alpha Q\}\exp\{i\overline{\alpha}\overline{Q}\} \\
&\quad \times \exp\{-ix_\mu P^\mu + i\theta Q\}\exp\{i\overline{\theta}\,\overline{Q}\}.
\end{aligned}
$$

In the next step we use the *Baker–Campbell–Hausdorff formula* in the form of the following expression:

$$
\begin{aligned}
\exp\{A\}\exp\{B\} &= \exp\left\{A + B + \tfrac{1}{2}[A,B]\right\} \\
&= \exp\left\{A + B - \tfrac{1}{2}[A,B] + [A,B]\right\} \\
&= \exp\left\{B + A + \tfrac{1}{2}[B,A] + [A,B]\right\} \\
&= \exp\{B + [A,B]\}\exp\{A\}. \quad (6.18)
\end{aligned}
$$

Setting

$$e^A = e^{i\overline{\alpha}\,\overline{Q}} \quad \text{and} \quad e^B = e^{-ix_\mu P^\mu + i\theta Q},$$

we obtain with Eq. (6.18):

$$
\begin{aligned}
&\exp\{a\sigma_\mu\overline{\alpha}P^\mu + ia Q\}\exp\{i\overline{\alpha}\,\overline{Q}\}\exp\{-ix_\mu P^\mu + i\theta Q\}\exp\{i\overline{\theta}\,\overline{Q}\} \\
\overset{(6.18)}{=}\ &\exp\{a\sigma_\mu\overline{\alpha}P^\mu + ia Q\}\exp\{-ix_\mu P^\mu + i\theta Q + [i\overline{\alpha}\,\overline{Q}, -ix_\mu P^\mu + i\theta Q]\} \\
&\times \exp\{i\overline{\alpha}\,\overline{Q}\}\exp\{i\overline{\theta}\,\overline{Q}\},
\end{aligned}
$$

and with Eqs. (6.5b), (6.4), and in the last step with Eq. (6.10), this becomes

$$
\begin{aligned}
&\exp\{a\sigma_\mu\overline{\alpha}P^\mu + ia Q\}\exp\{i\overline{\alpha}\,\overline{Q}\}\exp\{-ix_\mu P^\mu + i\theta Q\}\exp\{i\overline{\theta}\,\overline{Q}\} \\
=\ &\exp\{a\sigma_\mu\overline{\alpha}P^\mu + ia Q\}\exp\{-ix_\mu P^\mu + i\theta Q - [\overline{\alpha}\,\overline{Q}, \theta Q]\}\exp\{i(\overline{\theta} + \overline{\alpha})\overline{Q}\} \\
=\ &\exp\{a\sigma_\mu\overline{\alpha}P^\mu + ia Q\}\exp\{-ix_\mu P^\mu + i\theta Q + 2\theta\sigma_\mu\overline{\alpha}P^\mu\}\exp\{i(\overline{\theta} + \overline{\alpha})\overline{Q}\} \\
=\ &\exp\{a\sigma_\mu\overline{\alpha}P^\mu + ia Q\}\exp\{-i(x_\mu + 2i\theta\sigma_\mu\overline{\alpha})P^\mu + i\theta Q\}\exp\{i(\overline{\theta} + \overline{\alpha})\overline{Q}\} \\
=\ &\exp\{-i(x_\mu + 2i\theta\sigma_\mu\overline{\alpha} + ia\sigma_\mu\overline{\alpha})P^\mu + i(\theta + \alpha)Q\}\exp\{i(\overline{\theta} + \overline{\alpha})\overline{Q}\}.
\end{aligned}
$$

In the last step we took into account that the commutator of the two exponents vanishes. Hence finally:

$$L_1(ia\sigma\overline{\alpha}, \alpha, \overline{\alpha})L_1(x, \theta, \overline{\theta}) = L_1(x + 2i\theta\sigma\overline{\alpha} + ia\sigma\overline{\alpha}, \theta + \alpha, \overline{\theta} + \overline{\alpha}), \quad (6.19)$$

and we have:

$$
\begin{aligned}
T_\alpha\Phi_1(x, \theta, \overline{\theta}) &= [L_1(ia\sigma\overline{\alpha}, \alpha, \overline{\alpha})L_1(x, \theta, \overline{\theta})]\Phi_0[L_1(ia\sigma\overline{\alpha}, \alpha, \overline{\alpha})L_1(x, \theta, \overline{\theta})]^{-1} \\
&= L_1(x + 2i\theta\sigma\overline{\alpha} + ia\sigma\overline{\alpha}, \theta + \alpha, \overline{\theta} + \overline{\alpha})\Phi_0 \\
&\quad L_1^{-1}(x + 2i\theta\sigma\overline{\alpha} + ia\sigma\overline{\alpha}, \theta + \alpha, \overline{\theta} + \overline{\alpha}) \\
&= \Phi(x_\mu + 2i\theta\sigma_\mu\overline{\alpha} + ia\sigma_\mu\overline{\alpha}, \theta + \alpha, \overline{\theta} + \overline{\alpha}).
\end{aligned}
$$

In Eqs. (6.13) to (6.15) the symbol T_α refers to a particular supersymmetry transformation. Thus T_α acting on superfields Φ corresponds to the application of the operator $L(0, \alpha, \overline{\alpha})$, whereas T_α acting on a superfield of type 1, *i.e.* Φ_1, requires the operator $L_1(ia\sigma\overline{\alpha}, \alpha, \overline{\alpha})$. This is due to the fact that according to Eq. (6.9) the operators $L(0, \alpha, \overline{\alpha})$ and $L_1(ia\sigma\overline{\alpha}, \alpha, \overline{\alpha})$ generate the same supersymmetry transformation in different representations. Correspondingly T_α acting on superfields of type 2, *i.e.* Φ_2, requires operators

$L_2(-i\alpha\sigma\overline{\alpha}, \alpha, \overline{\alpha})$. Consider:

$$
\begin{aligned}
T_\alpha \Phi_2(x, \theta, \overline{\theta}) &= L_2(-i\alpha\sigma\overline{\alpha}, \alpha, \overline{\alpha}) \Phi_2(x, \theta, \overline{\theta}) L_2^{-1}(-i\alpha\sigma\overline{\alpha}, \alpha, \overline{\alpha}) \\
&= L_2(-i\alpha\sigma\overline{\alpha}, \alpha, \overline{\alpha}) [L_2(x, \theta, \overline{\theta}) \Phi_0 L_2^{-1}(x, \theta, \overline{\theta})] L_2^{-1}(-i\alpha\sigma\overline{\alpha}, \alpha, \overline{\alpha}) \\
&= [L_2(-i\alpha\sigma\overline{\alpha}, \alpha, \overline{\alpha}) L_2(x, \theta, \overline{\theta})] \\
&\quad \times \Phi_0 [L_2(-i\alpha\sigma\overline{\alpha}, \alpha, \overline{\alpha}) L_2(x, \theta, \overline{\theta})]^{-1} \\
&= L_2(x_\mu - 2i\alpha\sigma_\mu\overline{\theta} - i\alpha\sigma_\mu\overline{\alpha}, \theta + \alpha, \overline{\theta} + \overline{\alpha}) \Phi_0 \\
&\quad \times L_2^{-1}(x_\mu - 2i\alpha\sigma_\mu\overline{\theta} - i\alpha\sigma_\mu\overline{\alpha}, \theta + \alpha, \overline{\theta} + \overline{\alpha}) \quad \text{(see below)} \\
&= \Phi_2(x - 2i\alpha\sigma\overline{\theta} - i\alpha\sigma\overline{\alpha}, \theta + \alpha, \overline{\theta} + \overline{\alpha}),
\end{aligned}
$$

where, using Eq. (6.6),

$$
\begin{aligned}
&L_2(-i\alpha\sigma\overline{\alpha}, \alpha, \overline{\alpha}) L_2(x, \theta, \overline{\theta}) \\
&= \exp\{-i(-i\alpha\sigma_\mu\overline{\alpha})P^\mu + i\overline{\alpha}\,\overline{Q}\} \exp\{i\alpha Q\} \exp\{-ix_\mu P^\mu + i\overline{\theta}\,\overline{Q}\} \exp\{i\theta Q\},
\end{aligned}
$$

and with Eq. (6.18)

$$
\begin{aligned}
&L_2(-i\alpha\sigma\overline{\alpha}, \alpha, \overline{\alpha}) L_2(x, \theta, \overline{\theta}) \\
&= \exp\{-\alpha\sigma_\mu\overline{\alpha}P^\mu + i\overline{\alpha}\,\overline{Q}\} \exp\{-ix_\mu P^\mu + i\overline{\theta}\,\overline{Q} + [i\alpha Q, i\overline{\theta}\,\overline{Q}]\} \\
&\quad \times \exp\{i(\theta + \alpha)Q\} \\
&= \exp\{-\alpha\sigma_\mu\overline{\alpha}P^\mu + i\overline{\alpha}\,\overline{Q}\} \exp\{-ix_\mu P^\mu + i\overline{\theta}\,\overline{Q} - 2\alpha\sigma_\mu\overline{\theta}P^\mu\} \\
&\quad \times \exp\{i(\theta + \alpha)Q\} \\
&= \exp\{-\alpha\sigma_\mu\overline{\alpha}P^\mu + i\overline{\alpha}\,\overline{Q}\} \exp\{-i(x_\mu - 2i\alpha\sigma_\mu\overline{\theta})P^\mu + i\overline{\theta}\,\overline{Q}\} \\
&\quad \times \exp\{i(\theta + \alpha)Q\} \\
&= \exp\{-i(x_\mu - 2i\alpha\sigma_\mu\overline{\theta} - i\alpha\sigma_\mu\overline{\alpha})P^\mu + i(\overline{\theta} + \overline{\alpha})\overline{Q}\} \exp\{i(\theta + \alpha)Q\} \\
&= L_2(x_\mu - 2i\alpha\sigma_\mu\overline{\theta} - i\alpha\sigma_\mu\overline{\alpha}, \theta + \alpha, \overline{\theta} + \overline{\alpha}).
\end{aligned}
$$

This completes the proof of the last of relations (6.13) to (6.15).

6.3.2   Infinitesimal Supersymmetry Transformations and Differential Operator Representations of the Generators

From Eqs. (6.13) to (6.15) we can derive the transformation properties of the three types of superfields in infinitesimal form, *i.e.* we now consider T_α of Eqs. (6.13) to (6.15) for infinitesimal α. Starting with the superfield Φ we have:

$$
\begin{aligned}
\delta_s \Phi \overset{(6.13)}{=}\ & T_\alpha \Phi(x, \theta, \overline{\theta}) - \Phi(x, \theta, \overline{\theta}) \\
=\ & \Phi(x + i\theta\sigma\overline{\alpha} - i\alpha\sigma\overline{\theta}, \theta + \alpha, \overline{\theta} + \overline{\alpha}) - \Phi(x, \theta, \overline{\theta}).
\end{aligned}
$$

We now derive the differential operator representation of the generators $Q, \overline{Q}$ in much the same way as the differential operator representation of the momentum operator P is obtained. Thus, considering the translation of the variable x of some function $f(x)$ we have

$$
\begin{aligned}
\delta_t f(x) &= f(x + a) - f(x) = a \frac{df}{dx} + \cdots \\
&= e^{iPa} f(x) e^{-iPa} - f(x) \\
&= ia(Pf - fP) + \cdots \\
&= ia(Pf)_c + \cdots,
\end{aligned}
$$

where $(Pf)_c$ is a c-number. Equating the expression in the last line and that in the first line yields the operator representation $P = -id/dx$. Proceeding analogously in the present case we obtain (with the appropriate Taylor expansions)

$$
\begin{aligned}
\delta_s \Phi &= \Phi(x, \theta, \overline{\theta}) + i(\theta \sigma^\mu \alpha - \alpha \sigma^\mu \overline{\theta}) \partial_\mu \Phi(x, \theta, \overline{\theta}) \\
&\quad + \alpha \frac{\partial}{\partial \theta} \Phi(x, \theta, \overline{\theta}) + \overline{\alpha} \frac{\partial}{\partial \overline{\theta}} \Phi(x, \theta, \overline{\theta}) + \cdots - \Phi(x, \theta, \overline{\theta}) \\
&= \left\{ \alpha \frac{\partial}{\partial \theta} + \overline{\alpha} \frac{\partial}{\partial \overline{\theta}} + i(\theta \sigma^\mu \overline{\alpha} - \alpha \sigma^\mu \overline{\theta}) \partial_\mu + \cdots \right\} \Phi(x, \theta, \overline{\theta}). \quad (6.20)
\end{aligned}
$$

But we have also (using Eq. (6.6))

$$
\begin{aligned}
\delta_s \Phi &= L(0, \alpha, \overline{\alpha}) \Phi(x, \theta, \overline{\theta}) L^{-1}(0, \alpha, \overline{\alpha}) - \Phi(x, \theta, \overline{\theta}) \\
&= \exp\{i\alpha Q + i\overline{\alpha}\overline{Q}\} \Phi(x, \theta, \overline{\theta}) \exp\{-i\alpha Q - i\overline{\alpha}\overline{Q}\} - \Phi(x, \theta, \overline{\theta}) \\
&= (1 + i\alpha Q + i\overline{\alpha}\overline{Q} + \cdots) \Phi(x, \theta, \overline{\theta})(1 - i\alpha Q - i\overline{\alpha}\overline{Q} + \cdots) - \Phi(x, \theta, \overline{\theta}) \\
&= i\alpha Q \Phi(x, \theta, \overline{\theta}) + i\overline{\alpha}\overline{Q} \Phi(x, \theta, \overline{\theta}) - i\Phi(x, \theta, \overline{\theta}) \alpha Q - i\Phi(x, \theta, \overline{\theta}) \overline{\alpha}\overline{Q} + \cdots \\
&= i[\alpha Q, \Phi(x, \theta, \overline{\theta})] + i[\overline{\alpha}\overline{Q}, \Phi(x, \theta, \overline{\theta})] + \cdots, \quad (6.21)
\end{aligned}
$$

where the derivatives with respect to Grassmann numbers are to be understood in the sense of the definition given in Sec. 6.2. Considering any function $F(x, \theta, \overline{\theta})$ on superspace, we have:

$$
\begin{aligned}
i[\alpha Q&, \Phi(x, \theta, \overline{\theta})] F(x, \theta, \overline{\theta}) \\
&= i\{\alpha Q \Phi(x, \theta, \overline{\theta}) - \Phi(x, \theta, \overline{\theta}) \alpha Q\} F(x, \theta, \overline{\theta}) \\
&= i\alpha Q(\Phi(x, \theta, \overline{\theta}) F(x, \theta, \overline{\theta})) - \Phi(x, \theta, \overline{\theta})(i\alpha Q F(x, \theta, \overline{\theta})) \\
&= i\alpha [Q \Phi(x, \theta, \overline{\theta})] F(x, \theta, \overline{\theta}) + i\Phi(x, \theta, \overline{\theta}) \alpha [Q F(x, \theta, \overline{\theta})] \\
&\quad - i\Phi(x, \theta, \overline{\theta}) \alpha [Q F(x, \theta, \overline{\theta})] \\
&= i\alpha [Q \Phi(x, \theta, \overline{\theta})] F(x, \theta, \overline{\theta}). \quad (6.22)
\end{aligned}
$$

Hence in the linear approximation we can rewrite Eq. (6.21) as

$$\begin{aligned}
\delta_s\Phi &= i\alpha Q\Phi(x,\theta,\overline{\theta}) + i\overline{\alpha}\overline{Q}\Phi(x,\theta,\overline{\theta}) \\
&= [i\alpha^A Q_A + i\overline{\alpha}_{\dot{A}}\overline{Q}^{\dot{A}}]\Phi(x,\theta,\overline{\theta}).
\end{aligned} \tag{6.23}$$

Comparing Eqs. (6.20) and (6.23) we obtain the following differential operator representation of the group generators:

$$Q_A = -i(\partial_A - i\sigma^\mu_{A\dot{B}}\overline{\theta}^{\dot{B}}\partial_\mu), \qquad \partial_A \equiv \frac{\partial}{\partial\theta^A}, \tag{6.24}$$

$$\overline{Q}^{\dot{A}} = -i(\overline{\partial}^{\dot{A}} - i\theta^A\sigma^\mu_{A\dot{B}}\epsilon^{\dot{B}\dot{A}}\partial_\mu)$$

$$= -i(\overline{\partial}^{\dot{A}} - i(\overline{\sigma}^\mu\theta)^{\dot{A}}\partial_\mu), \qquad \overline{\partial}^{\dot{A}} \equiv \frac{\partial}{\partial\overline{\theta}_{\dot{A}}}, \tag{6.25}$$

since, using Eqs. (1.106a) and (1.83),

$$\begin{aligned}
(\overline{\sigma}^\mu\theta)^{\dot{A}} &= \overline{\sigma}^{\mu\dot{A}A}\theta_A = \epsilon^{AB}\epsilon^{\dot{A}\dot{B}}\sigma^\mu_{B\dot{B}}\theta_A \\
&= \epsilon^{BA}\theta_A\sigma^\mu_{B\dot{B}}\epsilon^{\dot{B}\dot{A}} = \theta^B\sigma^\mu_{B\dot{B}}\epsilon^{\dot{B}\dot{A}}.
\end{aligned}$$

We can proceed in a similar way for type 1 superfields Φ_1. Then, using Eq. (6.14),

$$\begin{aligned}
\delta_s\Phi_1 &= T_\alpha\Phi_1(x,\theta,\overline{\theta}) - \Phi(x,\theta,\overline{\theta}) \\
&= \Phi_1(x + 2i\theta\sigma\overline{\alpha} + i\alpha\sigma\overline{\alpha}, \theta + \alpha, \overline{\theta} + \overline{\alpha}) - \Phi_1(x,\theta,\overline{\theta}) \\
&= 2i\theta\sigma_\mu\overline{\alpha}\partial^\mu\Phi_1(x,\theta,\overline{\theta}) + \alpha^A\frac{\partial}{\partial\theta^A}\Phi_1(x,\theta,\overline{\theta}) + \overline{\alpha}_{\dot{A}}\frac{\partial}{\partial\overline{\theta}_{\dot{A}}}\Phi_1(x,\theta,\overline{\theta}) + \cdots \\
&= \alpha^A\frac{\partial}{\partial\theta^A}\Phi_1(x,\theta,\overline{\theta}) + \overline{\alpha}_{\dot{A}}\left(\frac{\partial}{\partial\overline{\theta}_{\dot{A}}} - 2i\theta^A\sigma^\mu_{A\dot{B}}\epsilon^{\dot{B}\dot{A}}\partial_\mu\right)\Phi_1(x,\theta,\overline{\theta}).
\end{aligned} \tag{6.26}$$

On the other hand we also have

$$\begin{aligned}
\delta_s\Phi_1 &= L_1(i\alpha\sigma\overline{\alpha}, \alpha, \overline{\alpha})\Phi_1(x,\theta,\overline{\theta})L_1^{-1}(i\alpha\sigma\overline{\alpha}, \alpha, \overline{\alpha}) - \Phi_1(x,\theta,\overline{\theta}) \\
&\overset{(6.6)}{=} \exp\{\alpha\sigma_\mu\overline{\alpha}P^\mu + i\alpha Q^{(1)}\}\exp\{i\overline{\alpha}\overline{Q}^{(1)}\}\Phi_1(x,\theta,\overline{\theta}) \\
&\quad \times \exp\{-\alpha\sigma_\mu\overline{\alpha}P^\mu + i\alpha Q^{(1)}\}\exp\{-i\overline{\alpha}\overline{Q}^{(1)}\} - \Phi_1(x,\theta,\overline{\theta}) \\
&\overset{(6.10)}{\underset{(6.4)}{=}} (1 + i\alpha Q^{(1)} + \cdots)(1 + i\overline{\alpha}\overline{Q}^{(1)})\Phi_1(x,\theta,\overline{\theta}) \\
&\quad \times (1 - i\alpha Q^{(1)} + \cdots)(1 - i\overline{\alpha}\overline{Q}^{(1)}) - \Phi_1(x,\theta,\overline{\theta}) \\
&= i[\alpha Q^{(1)}, \Phi_1(x,\theta,\overline{\theta})] + i[\overline{\alpha}\overline{Q}^{(1)}, \Phi_1(x,\theta,\overline{\theta})] + \cdots \\
&= i\alpha^A(Q_A^{(1)}\Phi_1(x,\theta,\overline{\theta})) + i\overline{\alpha}_{\dot{A}}(\overline{Q}^{(1)\dot{A}}\Phi_1(x,\theta,\overline{\theta})) + \cdots.
\end{aligned} \tag{6.27}$$

Comparing Eqs. (6.26) and (6.27) we obtain the following representations of the group generators:

$$Q_A^{(1)} = -i\frac{\partial}{\partial\theta^A} \equiv -i\partial_A, \tag{6.28}$$

$$\overline{Q}^{(1)\dot{A}} = -i\left(\frac{\partial}{\partial\overline{\theta}_{\dot{A}}} - 2i\theta^A\sigma_{A\dot{B}}^{\mu}\epsilon^{\dot{B}\dot{A}}\partial_\mu\right). \tag{6.29}$$

For convenience we add the following relation which is obtained with the help of Eq. (6.5j):

$$\overline{Q}_{\dot{A}}^{(1)} = \epsilon_{\dot{A}\dot{B}}\overline{Q}^{(1)\dot{B}} = -i(-\overline{\partial}_{\dot{A}} + 2i\theta^A\sigma_{A\dot{A}}^{\mu}\partial_\mu).$$

Finally, for superfields of type 2, i.e. Φ_2, the variation under infinitesimal supersymmetry transformations is given by

$$
\begin{aligned}
\delta_s\Phi_2 &= T_\alpha\Phi_2(x,\theta,\overline{\theta}) - \Phi_2(x,\theta,\overline{\theta})\\
&\stackrel{(6.15)}{=} \Phi_2(x - 2i\alpha\sigma\overline{\theta} - i\alpha\sigma\overline{\alpha}, \theta+\alpha, \overline{\theta}+\overline{\alpha}) - \Phi_2(x,\theta,\overline{\theta})\\
&= \Phi_2(x,\theta,\overline{\theta}) - 2i\alpha\sigma^\mu\overline{\theta}\partial_\mu\Phi_2(x,\theta,\overline{\theta}) + \alpha\frac{\partial}{\partial\theta}\Phi_2(x,\theta,\overline{\theta})\\
&\quad + \overline{\alpha}\frac{\partial}{\partial\overline{\theta}}\Phi_2(x,\theta,\overline{\theta}) + \cdots - \Phi_2(x,\theta,\overline{\theta})\\
&= \alpha^A\left[\frac{\partial}{\partial\theta^A} - 2i\sigma_{A\dot{B}}^{\mu}\overline{\theta}^{\dot{B}}\partial_\mu\right]\Phi_2(x,\theta,\overline{\theta}) + \overline{\alpha}_{\dot{A}}\frac{\partial}{\partial\overline{\theta}_{\dot{A}}}\Phi_2(x,\theta,\overline{\theta}) + \cdots.
\end{aligned}
\tag{6.30}
$$

On the other hand:

$$\delta_s\Phi_2(x,\theta,\overline{\theta}) = L_2(-i\alpha\sigma\overline{\alpha},\alpha,\overline{\alpha})\Phi_2(x,\theta,\overline{\theta})L_2^{-1}(-i\alpha\sigma\overline{\alpha},\alpha,\overline{\alpha}) - \Phi_2(x,\theta,\overline{\theta}),$$

and using Eq. (6.6) this becomes

$$
\begin{aligned}
\delta_s\Phi_2(x,\theta,\overline{\theta}) &\stackrel{(6.6)}{=} \exp\left\{-\alpha\sigma_\mu\overline{\alpha}P^\mu + i\overline{\alpha}\overline{Q}^{(2)}\right\}\exp\left\{i\alpha Q^{(2)}\right\}\Phi_2(x,\theta,\overline{\theta})\\
&\quad \times \exp\left\{\alpha\sigma_\mu\overline{\alpha}P^\mu - i\overline{\alpha}\overline{Q}^{(2)}\right\}\exp\left\{-i\alpha Q^{(2)}\right\} - \Phi_2(x,\theta,\overline{\theta})\\
&= (1 + i\overline{\alpha}\overline{Q}^{(2)} + \cdots)(1 + i\alpha Q^{(2)} + \cdots)\Phi_2(x,\theta,\overline{\theta})\\
&\quad \times (1 - i\overline{\alpha}\overline{Q}^{(2)} + \cdots)(1 - i\alpha Q^{(2)} + \cdots) - \Phi_2(x,\theta,\overline{\theta})\\
&= i\left[\overline{\alpha}\overline{Q}^{(2)}, \Phi_2(x,\theta,\overline{\theta})\right] + i\left[\alpha Q^{(2)}, \Phi_2(x,\theta,\overline{\theta})\right] + \cdots\\
&= i\overline{\alpha}_{\dot{A}}(\overline{Q}^{(2)\dot{A}}\Phi_2(x,\theta,\overline{\theta})) + i(Q_A^{(2)}\Phi_2(x,\theta,\overline{\theta})) + \cdots.
\end{aligned}
$$

Comparing this expression with that of Eq. (6.30), we obtain the following representation of the generators:

$$Q_A^{(2)} = -i\left(\frac{\partial}{\partial\theta^A} - 2i\sigma_{A\dot{B}}^\mu \overline{\theta}^{\dot{B}} \partial_\mu\right), \tag{6.31}$$

$$\overline{Q}^{(2)\dot{A}} = -i\frac{\partial}{\partial\overline{\theta}_{\dot{A}}} \equiv -i\overline{\partial}^{\dot{A}}, \qquad \overline{Q}_{\dot{A}}^{(2)} = i\overline{\partial}_{\dot{A}}. \tag{6.32}$$

Equations (6.24), (6.25), (6.28), and (6.29), (6.31), (6.32), are three different representations of the spinor charges of the Super–Poincaré algebra as differential operators acting in superspace. For reasons of consistency these operators must obey the same anticommutation relations as the spinor charges of the algebra. It should be noted that the construction of the differential operator representations of the spinor charges was possible only as a result of the spinor extension of spacetime. We now verify the following proposition.

Proposition 6.7:

Given $Q_A, \overline{Q}_{\dot{A}}$ as in Eqs. (6.24), (6.25), we have:

$$\{Q_A, Q_B\} = \{\overline{Q}_{\dot{A}}, \overline{Q}_{\dot{B}}\} = 0, \tag{6.33}$$

$$\{Q_A, \overline{Q}_{\dot{B}}\} = -2i\sigma_{A\dot{B}}^\mu \partial_\mu = 2\sigma_{A\dot{B}}^\mu P_\mu. \tag{6.34}$$

Proof: Consider:

$$-\{Q_A, Q_B\} = \{\partial_A - i\sigma_{A\dot{B}}^\mu \overline{\theta}^{\dot{B}} \partial_\mu, \partial_B - i\sigma_{B\dot{C}}^\nu \overline{\theta}^{\dot{C}} \partial_\nu\}$$

$$= \{\partial_A, \partial_B\} - i\{\partial_A, \sigma_{B\dot{C}}^\nu \overline{\theta}^{\dot{C}} \partial_\nu\} - i\{\sigma_{A\dot{B}}^\mu \overline{\theta}^{\dot{B}} \partial_\mu, \partial_B\}$$

$$-\{\sigma_{A\dot{B}}^\mu \overline{\theta}^{\dot{B}} \partial_\mu, \sigma_{B\dot{C}}^\nu \overline{\theta}^{\dot{C}} \partial_\nu\} = 0,$$

since (see Eq. (6.5i))

$$\{\partial_A, \partial_B\} = 0, \qquad \partial_A \overline{\theta}^{\dot{C}} = 0, \qquad \{\overline{\theta}^{\dot{B}}, \overline{\theta}^{\dot{C}}\} = 0.$$

This establishes the first of relations (6.33). Before we proceed we lower the index of $\overline{Q}^{\dot{A}}$. Thus, using Eq. (1.86) we have (using Eq. (6.5j)):

$$\overline{Q}_{\dot{A}} = \epsilon_{\dot{A}\dot{B}} \overline{Q}^{\dot{B}} = \epsilon_{\dot{A}\dot{B}} \frac{1}{i}\{\overline{\partial}^{\dot{B}} - i\theta^A \sigma_{A\dot{C}}^\mu \epsilon^{\dot{C}\dot{B}} \partial_\mu\}$$

$$= \frac{1}{i}\{-\overline{\partial}_{\dot{A}} - i\theta^A \sigma_{A\dot{C}}^\mu \epsilon^{\dot{C}\dot{B}} \epsilon_{\dot{A}\dot{B}} \partial_\mu\} = \frac{1}{i}\{-\overline{\partial}_{\dot{A}} + i\theta^A \sigma_{A\dot{C}}^\mu \delta_{\dot{A}}^{\dot{C}} \partial_\mu\}.$$

Hence

$$\overline{Q}_{\dot{A}} = i(\overline{\partial}_{\dot{A}} - i\theta^A \sigma^\mu_{A\dot{A}} \partial_\mu). \tag{6.35}$$

We now have

$$
\begin{aligned}
-\{\overline{Q}_{\dot{A}}, \overline{Q}_{\dot{B}}\} &= \{-\overline{\partial}_{\dot{A}} + i\theta^A \sigma^\mu_{A\dot{A}} \partial_\mu, -\overline{\partial}_{\dot{B}} + i\theta^B \sigma^\nu_{B\dot{B}} \partial_\nu\} \\
&= \{\overline{\partial}_{\dot{A}}, \overline{\partial}_{\dot{B}}\} - i\{\overline{\partial}_{\dot{A}}, \theta^B \sigma^\nu_{B\dot{B}} \partial_\nu\} - i\{\theta^A \sigma^\mu_{A\dot{A}} \partial_\mu, \overline{\partial}_{\dot{B}}\} \\
&\quad - \{\theta^A \sigma^\mu_{A\dot{A}} \partial_\mu, \theta^B \sigma^\nu_{B\dot{B}} \partial_\nu\} \\
&= 0,
\end{aligned}
$$

using Eq. (6.5n). Finally consider

$$
\begin{aligned}
-\{Q_A, \overline{Q}_{\dot{B}}\} &= \{\partial_A - i\sigma^\mu_{A\dot{C}} \overline{\theta}^{\dot{C}} \partial_\mu, -\overline{\partial}_{\dot{B}} + i\theta^C \sigma^\nu_{C\dot{B}} \partial_\nu\} \\
&= -\{\partial_A, \overline{\partial}_{\dot{B}}\} + i\{\partial_A, \theta^C \sigma^\nu_{C\dot{B}} \partial_\nu\} + i\{\sigma^\mu_{A\dot{C}} \overline{\theta}^{\dot{C}} \partial_\mu, \overline{\partial}_{\dot{B}}\} \\
&\quad + \{\sigma^\mu_{A\dot{C}} \overline{\theta}^{\dot{C}} \partial_\mu, \theta^C \sigma^\nu_{C\dot{B}} \partial_\nu\} \\
&= i\left(\frac{\partial}{\partial\theta^A} \theta^C\right) \sigma^\nu_{C\dot{B}} \partial_\nu + i\left(\frac{\partial}{\partial\overline{\theta}^{\dot{B}}} \sigma^\mu_{A\dot{C}} \overline{\theta}^{\dot{C}}\right) \partial_\mu \\
&= i\delta^C_A \sigma^\nu_{C\dot{B}} \partial_\nu + i\sigma^\mu_{A\dot{B}} \partial_\mu = i\sigma^\mu_{A\dot{B}} \partial_\mu + i\sigma^\mu_{A\dot{B}} \\
&= 2i\sigma^\mu_{A\dot{B}} \partial_\mu = -2\sigma^\mu_{A\dot{B}} P_\mu. \tag{6.36}
\end{aligned}
$$

Similarly we can establish the following result.

Proposition 6.8:

The four differential operators defined by Eqs. (6.28), (6.29) and (6.31), (6.32), satisfy the following relations:

$$\{Q^{(1)}_A, Q^{(1)}_B\} = \{\overline{Q}^{(1)}_{\dot{A}}, \overline{Q}^{(1)}_{\dot{B}}\} = 0, \tag{6.37}$$

$$\{Q^{(1)}_A, \overline{Q}^{(1)}_{\dot{B}}\} = -2i\sigma^\mu_{A\dot{B}} \partial_\mu = 2\sigma^\mu_{A\dot{B}} P_\mu, \tag{6.38}$$

and

$$\{Q^{(2)}_A, Q^{(2)}_B\} = \{\overline{Q}^{(2)}_{\dot{A}}, \overline{Q}^{(2)}_{\dot{B}}\} = 0, \tag{6.39}$$

$$\{Q^{(2)}_A, \overline{Q}^{(2)}_{\dot{B}}\} = -2i\sigma^\mu_{A\dot{B}} \partial_\mu = 2\sigma^\mu_{A\dot{B}} P_\mu. \tag{6.40}$$

Proof: Equations (6.37) are readily verified. Consider (using Eqs. (6.28) and (6.29))

$$
\begin{aligned}
\{Q_A^{(1)}, \overline{Q}_{\dot{A}}^{(1)}\} &= -\{\partial_A, -\overline{\partial}_{\dot{A}} + 2i\theta^B \sigma_{B\dot{A}}^{\mu}\partial_\mu\} \\
&= \{\partial_A, \overline{\partial}_{\dot{A}}\} - 2i\{\partial_A, \theta^B \sigma_{B\dot{A}}^{\mu}\partial_\mu\} \\
&= -2i\frac{\partial}{\partial\theta^A}(\theta^B \sigma_{B\dot{A}}^{\mu}\partial_\mu) = -2i\delta_A^B \sigma_{B\dot{A}}^{\mu}\partial_\mu \\
&= -2i\sigma_{A\dot{A}}^{\mu}\partial_\mu = 2\sigma_{A\dot{A}}^{\mu}P_\mu.
\end{aligned}
$$

Equations (6.39) are also readily seen to hold. Consider therefore Eq. (6.40). We have (using Eqs. (6.31), (6.32))

$$
\begin{aligned}
\{Q_A^{(2)}, \overline{Q}_{\dot{B}}^{(2)}\} &= -\{\partial_A - 2i\sigma_{A\dot{C}}^{\mu}\overline{\theta}^{\dot{C}}\partial_\mu, -\overline{\partial}_{\dot{B}}\} = \{\partial_A, \overline{\partial}_{\dot{B}}\} - 2i\{\sigma_{A\dot{C}}^{\mu}\overline{\theta}^{\dot{C}}\partial_\mu, \overline{\partial}_{\dot{B}}\} \\
&= -2i\overline{\partial}_{\dot{B}}(\sigma_{A\dot{C}}^{\mu}\overline{\theta}^{\dot{C}}\partial_\mu) = -2i\sigma_{A\dot{C}}^{\mu}\delta_{\dot{B}}^{\dot{C}}\partial_\mu \\
&= -2i\sigma_{A\dot{B}}^{\mu}\partial_\mu = 2\sigma_{A\dot{B}}^{\mu}P_\mu.
\end{aligned}
$$

6.4   Consistency with the Majorana Formalism

We can verify the consistency of the anticommutators above with those of the Majorana formulation, *i.e.* Eq. (3.27). In order to see this, we need in addition to Eqs. (6.24), (6.25) and (6.35) the derivative representation of Q^A. This may be obtained from Eq. (6.24). Thus:

$$
\begin{aligned}
Q^A &= \epsilon^{AB}Q_B = -i\epsilon^{AB}\left(\partial_B - i\sigma_{B\dot{B}}^{\mu}\overline{\theta}^{\dot{B}}\partial_\mu\right) \\
&\overset{(6.5g)}{=} i\left(\partial^A + i\epsilon^{AB}\sigma_{B\dot{B}}^{\mu}\overline{\theta}^{\dot{B}}\partial_\mu\right) \\
&= i\{\partial^A + i\epsilon^{AB}(\epsilon_{BD}\epsilon_{\dot{B}\dot{D}}\overline{\sigma}^{\mu\dot{D}D})\overline{\theta}^{\dot{B}}\partial_\mu\} \\
&\overset{(1.106a)}{=} i(\partial^A - i\delta_D^A\epsilon_{\dot{D}\dot{B}}\overline{\theta}^{\dot{B}}\overline{\sigma}^{\mu\dot{D}D}\partial_\mu) \\
&\overset{(1.86)}{=} i(\partial^A - i\overline{\theta}_{\dot{D}}\overline{\sigma}^{\mu\dot{D}A}\partial_\mu).
\end{aligned}
$$

This result together with Eqs. (6.24), (6.25) and (6.35) allows us to write:

$$
\begin{pmatrix} Q_A \\ \overline{Q}^{\dot{A}} \end{pmatrix} = -i\left[\begin{pmatrix} \frac{\partial}{\partial\theta^A} \\ \frac{\partial}{\partial\overline{\theta}_{\dot{A}}} \end{pmatrix} - i\begin{pmatrix} 0 & \sigma_{A\dot{B}}^{\mu} \\ \overline{\sigma}^{\mu\dot{A}B} & 0 \end{pmatrix}\begin{pmatrix} \theta_B \\ \overline{\theta}^{\dot{B}} \end{pmatrix}\partial_\mu\right],
$$

and

$$(Q^A, \overline{Q}_{\dot{A}}) = -i\left[\left(-\frac{\partial}{\partial\theta_A}, -\frac{\partial}{\partial\overline{\theta}^{\dot{A}}}\right)\right.$$
$$\left. + i(\theta^B, \overline{\theta}_{\dot{B}})\begin{pmatrix} 0 & \sigma^\mu_{B\dot{A}} \\ \overline{\sigma}^{\mu\dot{B}A} & 0 \end{pmatrix}\partial_\mu\right].$$

The Majorana character of these spinors is evident since the same Q appears in both Weyl spinors. Defining as usual

$$(\theta)_a = \begin{pmatrix} \theta_B \\ \overline{\theta}^{\dot{B}} \end{pmatrix} \qquad \text{so that} \qquad (\overline{\theta})_a = (\theta^B, \overline{\theta}_{\dot{B}}),$$

we have

$$\left(\frac{\partial}{\partial\theta}\right)_a = \left(\frac{\partial}{\partial\theta_A}, \frac{\partial}{\partial\overline{\theta}^{\dot{A}}}\right), \qquad \left(\frac{\partial}{\partial\overline{\theta}}\right)_a = \begin{pmatrix} \frac{\partial}{\partial\theta^A} \\ \frac{\partial}{\partial\overline{\theta}_{\dot{A}}} \end{pmatrix}.$$

Hence we can rewrite the above relations in the form

$$(Q) = -i\left(\frac{\partial}{\partial\overline{\theta}} - i\gamma^\mu\theta\partial_\mu\right),$$
$$(\overline{Q}) = -i\left(-\frac{\partial}{\partial\theta} + i\overline{\theta}\gamma^\mu\partial_\mu\right),$$

and the anticommutator (3.27), *i.e.*

$$\{Q_a, \overline{Q}_b\} = 2\gamma^\mu_{ab}P_\mu,$$

is readily verified. Together with this anticommutator we also have

$$\{Q_a, Q_b\} = -2(\gamma^\mu C)_{ab}P_\mu,$$

and

$$\{\overline{Q}_a, \overline{Q}_b\} = 2(C^{-1}\gamma^\mu)_{ab}P_\mu.$$

The first of these relations has been obtained before (see Eq. (3.22), where $a = -2$ as stated in the Corollary following Eq. (3.27)). We can verify the last relation by considering (since for Majorana spinors $\psi = \psi^C = C\overline{\psi}^\top, \psi^\top = \overline{\psi}C^\top = \overline{\psi}C^{-1}$)

$$\begin{aligned} \{Q_a, Q_b\} &= \{\overline{Q}_{a'}, \overline{Q}_{b'}\}C^\top_{a'a}C^\top_{b'b} \\ &= 2C_{aa'}(C^{-1}\gamma^\mu)_{a'b'}C^\top_{b'b}P_\mu \\ &= 2(\gamma^\mu C^\top)_{ab}P_\mu \\ &= -2(\gamma^\mu C)_{ab}P_\mu, \qquad \text{since} \quad C^\top = -C. \end{aligned}$$

We can verify that the derivative representations of $Q_a, \overline{Q}_a$ indeed satisfy the above anticommutation relations among themselves. Thus

$$\{Q_a, Q_b\} = -2\frac{\partial}{\partial\overline{\theta}_a}[-i(\gamma^\mu\theta)_b\partial_\mu].$$

Using Eq. (3.28b) this becomes

$$
\begin{aligned}
\{Q_a, Q_b\} &= -2(\gamma^\mu)_{bc}P_\mu\frac{\partial}{\partial\overline{\theta}_a}\theta_c = -2(\gamma^\mu)_{bc}P_\mu C_{ca} \\
&= -2(\gamma^\mu C)_{ba}P_\mu = -2(\gamma^\mu C)_{ab}P_\mu,
\end{aligned}
$$

since γ^μ is symmetric. Correspondingly

$$
\begin{aligned}
\{\overline{Q}_a, \overline{Q}_b\} &= -2\left[-\frac{\partial}{\partial\theta_a}i(\overline{\theta}\gamma^\mu)_b\partial_\mu\right] = -2\frac{\partial}{\partial\theta_a}(\overline{\theta}\gamma^\mu)_b P_\mu \\
&\overset{(3.28c)}{=} -2(C\gamma^\mu)_{ab}P_\mu \\
&= 2(C^{-1}\gamma^\mu)_{ab}P_\mu, \qquad \text{since} \quad C^\top = -C.
\end{aligned}
$$

6.5   Covariant Derivatives

Covariant derivatives are derivatives which are useful in the construction of manifestly supersymmetric Lagrangians. We can define three types of covariant derivatives corresponding to the three types of superfields introduced previously.

We define as covariant derivatives for superfields $\Phi(x, \theta, \overline{\theta})$ the operators

$$
\begin{aligned}
D_A &:= \partial_A + i\sigma^\mu_{A\dot{B}}\overline{\theta}^{\dot{B}}\partial_\mu, \\
D^A &:= \epsilon^{AB}D_B = -\partial^A - i\overline{\theta}_{\dot{C}}\overline{\sigma}^{\mu\dot{C}A}\partial_\mu;
\end{aligned}
\tag{6.41}
$$

and

$$
\begin{aligned}
\overline{D}_{\dot{A}} &:= -\overline{\partial}_{\dot{A}} - i\theta^B\sigma^\mu_{B\dot{A}}\partial_\mu, \\
\overline{D}^{\dot{A}} &:= \epsilon^{\dot{A}\dot{B}}\overline{D}_{\dot{B}} = \overline{\partial}^{\dot{A}} + i\overline{\sigma}^{\mu\dot{A}C}\theta_C\partial_\mu.
\end{aligned}
\tag{6.42}
$$

Proposition 6.9:

The operators D_A and $\overline{D}_{\dot{A}}$ are covariant derivatives, *i.e.* D_A and $\overline{D}_{\dot{A}}$ are invariant under supersymmetry transformations in the sense that

$$[D_A, \delta_s] = 0, \tag{6.43a}$$

$$[\overline{D}_{\dot{A}}, \delta_s] = 0. \tag{6.43b}$$

Proof: Consider, using Eq. (6.23),

$$
\begin{aligned}
[D_A, \delta_s]\Phi(x, \theta, \overline{\theta}) &= [D_A, i\alpha Q + i\overline{\alpha}\overline{Q}]\Phi(x, \theta, \overline{\theta}) \\
&= i[D_A, \alpha^B Q_B]\Phi(x, \theta, \overline{\theta}) + i[D_A, \overline{\alpha}_{\dot{B}}\overline{Q}^{\dot{B}}]\Phi(x, \theta, \overline{\theta}) \\
&= i(D_A \alpha^B Q_B - \alpha^B Q_B D_A)\Phi(x, \theta, \overline{\theta}) \\
&\quad + i(D_A \overline{\alpha}_{\dot{B}}\overline{Q}^{\dot{B}} - \overline{\alpha}_{\dot{B}}\overline{Q}^{\dot{B}} D_A)\Phi(x, \theta, \overline{\theta}) \\
&= -i\alpha^B(D_A Q_B + Q_B D_A)\Phi(x, \theta, \overline{\theta}) \\
&\quad - i\overline{\alpha}_{\dot{B}}(D_A \overline{Q}^{\dot{B}} + \overline{Q}^{\dot{B}} D_A)\Phi(x, \theta, \overline{\theta}) \\
&= -i\alpha^B\{D_A, Q_B\}\Phi(x, \theta, \overline{\theta}) - i\overline{\alpha}_{\dot{B}}\{D_A, \overline{Q}^{\dot{B}}\}\Phi(x, \theta, \overline{\theta}).
\end{aligned}
$$

Thus in order to prove Eq. (6.43a) we have to demonstrate that

$$\{D_A, Q_B\} = 0, \qquad \{D_A, \overline{Q}^{\dot{B}}\} = 0. \tag{6.44}$$

Consider:

$$
\begin{aligned}
\{D_A, Q_B\} &= -i\{\partial_A + i\sigma^\mu_{A\dot{B}}\overline{\theta}^{\dot{B}}\partial_\mu, \partial_B - i\sigma^\nu_{B\dot{C}}\overline{\theta}^{\dot{C}}\partial_\nu\} \\
&= -i\{\partial_A, \partial_B\} - \{\partial_A, \sigma^\nu_{B\dot{C}}\overline{\theta}^{\dot{C}}\partial_\nu\} \\
&\quad + \{\sigma^\mu_{A\dot{B}}\overline{\theta}^{\dot{B}}\partial_\mu, \partial_B\} - i\{\sigma^\mu_{A\dot{B}}\overline{\theta}^{\dot{B}}\partial_\mu, \sigma^\nu_{B\dot{C}}\overline{\theta}^{\dot{C}}\partial_\nu\} \\
&= 0, \tag{6.45}
\end{aligned}
$$

since

$$\{\partial_A, \partial_B\} = \{\partial_A, \overline{\theta}^{\dot{B}}\} = \{\overline{\theta}^{\dot{B}}, \overline{\theta}^{\dot{C}}\} = 0.$$

Now consider:

$$
\begin{aligned}
i\{D_A, \overline{Q}^{\dot{B}}\} &= \{\partial_A + i\sigma^\mu_{A\dot{C}}\overline{\theta}^{\dot{C}}\partial_\mu, \overline{\partial}^{\dot{B}} - i\theta^C \sigma^\nu_{C\dot{D}}\epsilon^{\dot{D}\dot{B}}\partial_\nu\} \\
&= \{\partial_A, \overline{\partial}^{\dot{B}}\} - i\{\partial_A, \theta^C \sigma^\nu_{C\dot{D}}\epsilon^{\dot{D}\dot{B}}\partial_\nu\} + i\{\sigma^\mu_{A\dot{C}}\overline{\theta}^{\dot{C}}\partial_\mu, \overline{\partial}^{\dot{B}}\} \\
&\quad + \{\sigma^\mu_{A\dot{C}}\overline{\theta}^{\dot{C}}\partial_\mu, \theta^C \sigma^\nu_{C\dot{D}}\epsilon^{\dot{D}\dot{B}}\partial_\nu\} \\
&= -i\sigma^\nu_{A\dot{D}}\epsilon^{\dot{D}\dot{B}}\partial_\nu + i\sigma^\mu_{A\dot{C}}(\overline{\partial}^{\dot{B}}\overline{\theta}^{\dot{C}})\partial_\mu.
\end{aligned}
$$

Using Eq. (6.5j) we have:

$$\overline{\partial}^{\dot{B}}\overline{\theta}^{\dot{C}} = -\epsilon^{\dot{B}\dot{D}}\overline{\partial}_{\dot{D}}\overline{\theta}^{\dot{C}} = -\epsilon^{\dot{B}\dot{D}}\delta^{\dot{C}}_{\dot{D}} = -\epsilon^{\dot{B}\dot{C}}. \tag{6.46}$$

Hence

$$
\begin{aligned}
i\{D_A, \overline{Q}^{\dot{B}}\} &= -i\sigma^\nu_{A\dot{D}}\epsilon^{\dot{D}\dot{B}}\partial_\nu + i\sigma^\mu_{A\dot{C}}(-\epsilon^{\dot{B}\dot{C}})\partial_\mu \\
&= -i\sigma^\nu_{A\dot{D}}\epsilon^{\dot{D}\dot{B}}\partial_\nu + i\sigma^\nu_{A\dot{C}}\epsilon^{\dot{C}\dot{B}}\partial_\nu \\
&= 0.
\end{aligned}
\tag{6.47}
$$

With Eqs. (6.45) and (6.47) we therefore obtain Eq. (6.43a). Now,

$$
\begin{aligned}
[\overline{D}_{\dot{A}}, \delta_s]\Phi(x,\theta,\overline{\theta}) &= [\overline{D}_{\dot{A}}, i\alpha^B Q_B + i\overline{\alpha}_{\dot{B}}\overline{Q}^{\dot{B}}]\Phi(x,\theta,\overline{\theta}) \\
&= ([\overline{D}_{\dot{A}}, i\alpha^B Q_B] + [\overline{D}_{\dot{A}}, i\overline{\alpha}_{\dot{B}}\overline{Q}^{\dot{B}}])\Phi(x,\theta,\overline{\theta}) \\
&= (-i\alpha^B\{\overline{D}_{\dot{A}}, Q_B\} - i\overline{\alpha}_{\dot{B}}\{\overline{D}_{\dot{A}}, \overline{Q}^{\dot{B}}\})\Phi(x,\theta,\overline{\theta}).
\end{aligned}
$$

Hence in order to demonstrate Eq. (6.43b) we have to show that

$$
\{\overline{D}_{\dot{A}}, Q_B\} = \{\overline{D}_{\dot{A}}, \overline{Q}^{\dot{B}}\} = 0.
$$

Consider

$$
\begin{aligned}
i\{\overline{D}_{\dot{A}}, Q_B\} &= \{-\overline{\partial}_{\dot{A}} - i\theta^B\sigma^\mu_{B\dot{A}}\partial_\mu, \partial_B - i\sigma^\mu_{B\dot{C}}\overline{\theta}^{\dot{C}}\partial_\mu\} \\
&= -\{\overline{\partial}_{\dot{A}}, \partial_B\} + i\{\overline{\partial}_{\dot{A}}, \sigma^\mu_{B\dot{C}}\overline{\theta}^{\dot{C}}\partial_\mu\} \\
&\quad -i\{\theta^C\sigma^\mu_{C\dot{A}}\partial_\mu, \partial_B\} - \{\theta^C\sigma^\mu_{C\dot{A}}\partial_\mu, \sigma^\mu_{B\dot{C}}\overline{\theta}^{\dot{C}}\partial_\mu\} \\
&= -i\sigma^\mu_{B\dot{C}}(\overline{\partial}_{\dot{A}}\overline{\theta}^{\dot{C}})\partial_\mu - i\partial_B\theta^C\sigma^\mu_{C\dot{A}}\partial_\mu \\
&= i\sigma^\mu_{B\dot{C}}\delta^{\dot{C}}_{\dot{A}}\partial_\mu - i\delta^C_B\sigma^\mu_{C\dot{A}}\partial_\mu \\
&= i\sigma^\mu_{B\dot{A}}\partial_\mu - i\sigma^\mu_{B\dot{A}}\partial_\mu = 0,
\end{aligned}
$$

and

$$
\begin{aligned}
i\{\overline{D}_{\dot{A}}, \overline{Q}^{\dot{B}}\} &= \{-\overline{\partial}_{\dot{A}} - i\theta^B\sigma^\mu_{B\dot{A}}\partial_\mu, \overline{\partial}^{\dot{B}} - i\theta^C\sigma^\mu_{C\dot{D}}\epsilon^{\dot{D}\dot{B}}\partial_\mu\} \\
&= -\{\overline{\partial}_{\dot{A}}, \overline{\partial}^{\dot{B}}\} + i\{\overline{\partial}_{\dot{A}}, \theta^C\sigma^\mu_{C\dot{D}}\epsilon^{\dot{D}\dot{B}}\partial_\mu\} - i\{\theta^B\sigma^\mu_{B\dot{A}}\partial_\mu, \overline{\partial}^{\dot{B}}\} \\
&\quad -\{\theta^B\sigma^\mu_{B\dot{A}}\partial_\mu, \theta^C\sigma^\mu_{C\dot{D}}\epsilon^{\dot{D}\dot{B}}\partial_\mu\} = 0,
\end{aligned}
$$

since

$$
\{\overline{\partial}_{\dot{A}}, \overline{\partial}^{\dot{B}}\} = 0, \quad \overline{\partial}_{\dot{A}}\theta^C = 0, \quad \text{and} \quad \{\theta^B, \theta^C\} = 0.
$$

Hence we obtain the result

$$
[\overline{D}_{\dot{A}}, \delta_s]\Phi(x,\theta,\overline{\theta}) = 0.
$$

This completes the proof of Proposition 6.9.

In a similar way we can find covariant derivatives for superfields of types 1 and 2.

Proposition 6.10:

The operators

$$D_A^{(1)} := \partial_A + 2i\sigma^\mu_{A\dot{B}}\overline{\theta}^{\dot{B}}\partial_\mu, \tag{6.48}$$

$$\overline{D}_{\dot{A}}^{(1)} := -\overline{\partial}_{\dot{A}}, \tag{6.49}$$

$$D_A^{(2)} := \partial_A, \tag{6.50}$$

$$\overline{D}_{\dot{A}}^{(2)} := -\overline{\partial}_{\dot{A}} - 2i\theta^B\sigma^\mu_{B\dot{A}}\partial_\mu, \tag{6.51}$$

are *covariant derivatives, i.e.*

$$[D_A^{(1)}, \delta_s]\Phi_1(x, \theta, \overline{\theta}) = 0,$$

$$[\overline{D}_{\dot{A}}^{(1)}, \delta_s]\Phi_1(x, \theta, \overline{\theta}) = 0, \tag{6.52}$$

$$[D_A^{(2)}, \delta_s]\Phi_2(x, \theta, \overline{\theta}) = 0,$$

$$[\overline{D}_{\dot{A}}^{(2)}, \delta_s]\Phi_2(x, \theta, \overline{\theta}) = 0. \tag{6.53}$$

Proof: Consider (using Eq. (6.27))

$$\begin{aligned}
[D_A^{(1)}, \delta_s]\Phi_1(x, \theta, \overline{\theta}) &= [D_A^{(1)}, i\alpha Q^{(1)} + i\overline{\alpha}\overline{Q}^{(1)}]\Phi_1(x, \theta, \overline{\theta}) \\
&= -i\alpha^B\{D_A^{(1)}, Q_B^{(1)}\}\Phi_1(x, \theta, \overline{\theta}) \\
&\quad -i\overline{\alpha}_{\dot{A}}\{D_A^{(1)}, \overline{Q}^{(1)\dot{A}}\}\Phi_1(x, \theta, \overline{\theta}).
\end{aligned}$$

But with Eqs. (6.48) and (6.28),

$$\{D_A^{(1)}, Q_B^{(1)}\} = -i\{\partial_A + 2i\sigma^\mu_{A\dot{B}}\overline{\theta}^{\dot{B}}\partial_\mu, \partial_B\} = -i\{\partial_A, \partial_B\} + \{\sigma^\mu_{A\dot{B}}\overline{\theta}^{\dot{B}}\partial_\mu, \partial_B\}$$

and with Eqs. (6.48) and (6.29):

$$\begin{aligned}
i\{D_A^{(1)}, \overline{Q}^{(1)\dot{A}}\} &= \{\partial_A + 2i\sigma^\mu_{A\dot{B}}\overline{\theta}^{\dot{B}}\partial_\mu, \overline{\partial}^{\dot{A}} - 2i\theta^C\sigma^\mu_{C\dot{B}}\epsilon^{\dot{B}\dot{A}}\partial_\mu\} \\
&= \{\partial_A, \overline{\partial}^{\dot{A}}\} - 2i\{\partial_A, \theta^C\sigma^\mu_{C\dot{B}}\epsilon^{\dot{B}\dot{A}}\partial_\mu\} \\
&\quad +2i\{\sigma^\mu_{A\dot{B}}\overline{\theta}^{\dot{B}}\partial_\mu, \overline{\partial}^{\dot{A}}\} + 4\{\sigma^\mu_{A\dot{B}}\overline{\theta}^{\dot{B}}\partial_\mu, \theta^C\sigma^\mu_{C\dot{B}}\epsilon^{\dot{B}\dot{A}}\partial_\mu\} \\
&= -2i(\partial_A\theta^C)\sigma^\mu_{C\dot{B}}\epsilon^{\dot{B}\dot{A}}\partial_\mu + 2i\sigma^\mu_{A\dot{B}}(\overline{\partial}^{\dot{A}}\overline{\theta}^{\dot{B}})\partial_\mu \\
&\overset{(6.47)}{=} -2i\delta_A^C\sigma^\mu_{C\dot{B}}\epsilon^{\dot{B}\dot{A}}\partial_\mu + 2i\sigma^\mu_{A\dot{B}}\epsilon^{\dot{B}\dot{A}}\partial_\mu \\
&= 0.
\end{aligned}$$

This demonstrates that $D_A^{(1)}$ is a covariant derivative for type 1 superfields. Now consider (using Eq. (6.27))

$$
\begin{aligned}
[\overline{D}_{\dot{A}}^{(1)}, \delta_s]\Phi_1(x,\theta,\overline{\theta}) &= [\overline{D}_{\dot{A}}^{(1)}, i\alpha Q^{(1)} + i\overline{\alpha}\,\overline{Q}^{(1)}]\Phi_1(x,\theta,\overline{\theta}) \\
&= -i\alpha^A\{\overline{D}_{\dot{A}}^{(1)}, Q_A^{(1)}\}\Phi_1(x,\theta,\overline{\theta}) \\
&\quad -i\overline{\alpha}_{\dot{B}}\{\overline{D}_{\dot{A}}^{(1)}, \overline{Q}^{(1)\dot{B}}\}\Phi_1(x,\theta,\overline{\theta}).
\end{aligned}
$$

Now, using Eqs. (6.28), (6.49),

$$
\{\overline{D}_{\dot{A}}^{(1)}, Q_A^{(1)}\} = -i\{-\overline{\partial}_{\dot{A}}, \partial_A\} = 0,
$$

and, using Eqs. (6.49) and (6.29),

$$
\{\overline{D}_{\dot{A}}^{(1)}, \overline{Q}^{(1)\dot{B}}\} = -i\{-\overline{\partial}_{\dot{A}}, \overline{\partial}^{\dot{B}} - 2i\theta^A\sigma^\mu_{A\dot{C}}\epsilon^{\dot{C}\dot{B}}\partial_\mu\} = 0.
$$

Hence

$$
[\overline{D}_{\dot{A}}^{(1)}, \delta_s]\Phi_1(x,\theta,\overline{\theta}) = 0,
$$

and Eq. (6.52) is verified.

In order to verify Eq. (6.53) we consider

$$
\begin{aligned}
[D_A^{(2)}, \delta_s]\Phi_2(x,\theta,\overline{\theta}) &= [D_A^{(2)}, i\alpha Q^{(2)} + i\overline{\alpha}\,\overline{Q}^{(2)}]\Phi_2(x,\theta,\overline{\theta}) \\
&= -i\alpha^B\{D_A^{(2)}, Q_B^{(2)}\}\Phi_2(x,\theta,\overline{\theta}) \\
&\quad -i\overline{\alpha}_{\dot{B}}\{D_A^{(2)}, \overline{Q}^{(2)\dot{B}}\}\Phi_2(x,\theta,\overline{\theta}).
\end{aligned}
$$

Again we have to evaluate anticommutators. Thus, using Eqs. (6.50) and (6.31), we have

$$
\begin{aligned}
\{D_A^{(2)}, Q_B^{(2)}\} &= -i\{\partial_A, \partial_B - 2i\sigma^\mu_{B\dot{C}}\overline{\theta}^{\dot{C}}\partial_\mu\} \\
&= -i\{\partial_A, \partial_B\} - 2\{\partial_A, \sigma^\mu_{B\dot{C}}\overline{\theta}^{\dot{C}}\partial_\mu\} = 0.
\end{aligned}
$$

Analogously, using Eqs. (6.50) and (6.32), we have

$$
\{D_A^{(2)}, \overline{Q}^{(2)\dot{B}}\} = \{\partial_A, \overline{\partial}^{\dot{B}}\} = 0.
$$

Hence

$$
[\overline{D}_{\dot{A}}^{(2)}, \delta_s]\Phi_2(x,\theta,\overline{\theta}) = 0.
$$

Finally consider:

$$[\overline{D}_{\dot{A}}^{(2)}, \delta_s]\Phi_2(x,\theta,\overline{\theta}) = [\overline{D}_{\dot{A}}^{(2)}, i\alpha Q^{(2)} + i\overline{\alpha}\overline{Q}^{(2)}]\Phi_2(x,\theta,\overline{\theta})$$

$$= -i\alpha^A\{\overline{D}_{\dot{A}}^{(2)}, Q_A^{(2)}\}\Phi_2(x,\theta,\overline{\theta})$$

$$-i\overline{\alpha}_{\dot{B}}\{\overline{D}_{\dot{A}}^{(2)}, \overline{Q}^{(2)\dot{B}}\}\Phi_2(x,\theta,\overline{\theta}).$$

Using Eqs. (6.31) and (6.51) we have

$$i\{\overline{D}_{\dot{A}}^{(2)}, Q_A^{(2)}\} = \{-\overline{\partial}_{\dot{A}} - 2i\theta^B\sigma_{B\dot{A}}^\mu\partial_\mu, \partial_A - 2i\sigma_{A\dot{C}}^\mu\overline{\theta}^{\dot{C}}\partial_\mu\}$$

$$= -\{\overline{\partial}_{\dot{A}}, \partial_A\} + 2i\{\overline{\partial}_{\dot{A}}, \sigma_{A\dot{C}}^\mu\overline{\theta}^{\dot{C}}\partial_\mu\}$$

$$-2i\{\theta^B\sigma_{B\dot{A}}^\mu\partial_\mu, \partial_A\} - 4\{\theta^B\sigma_{B\dot{A}}^\mu\partial_\mu, \sigma_{A\dot{C}}^\nu\overline{\theta}^{\dot{C}}\partial_\nu\}$$

$$= 2i\sigma_{A\dot{C}}^\mu(\overline{\partial}_{\dot{A}}\overline{\theta}^{\dot{C}})\partial_\mu - 2i(\partial_A\theta^B)\sigma_{B\dot{A}}^\mu\partial_\mu$$

$$= 2i\sigma_{A\dot{A}}^\mu\partial_\mu - 2i\sigma_{A\dot{A}}^\mu\partial_\mu = 0.$$

The covariant derivatives have important properties which we summarize in the following proposition.

Proposition 6.11:

The covariant derivatives obey the following algebra:

(a)

$$\left.\begin{array}{rcl}
\{D_A, D_B\} &=& \{\overline{D}_{\dot{A}}, \overline{D}_{\dot{B}}\} = 0, \\
D^3 = \overline{D}^3 &=& 0, \\
\{D_A, \overline{D}_{\dot{B}}\} &=& -2i\sigma_{A\dot{B}}^\mu\partial_\mu = 2\sigma_{A\dot{B}}^\mu P_\mu, \\
\{D^A, \overline{D}^{\dot{B}}\} &=& -2i\overline{\sigma}^{\mu\dot{B}A}\partial_\mu,
\end{array}\right\} \quad (6.54)$$

(b)

$$\left.\begin{array}{rcl}
\{D_A^{(1)}, D_B^{(1)}\} &=& \{\overline{D}_{\dot{A}}^{(1)}, \overline{D}_{\dot{B}}^{(1)}\} = 0, \\
\{D_A^{(1)}, \overline{D}_{\dot{B}}^{(1)}\} &=& -2i\sigma_{A\dot{B}}^\mu\partial_\mu = 2\sigma_{A\dot{B}}^\mu P_\mu,
\end{array}\right\} \quad (6.55)$$

(c)

$$\left.\begin{array}{rcl}
\{D_A^{(2)}, D_B^{(2)}\} &=& \{\overline{D}_{\dot{A}}^{(2)}, \overline{D}_{\dot{B}}^{(2)}\} = 0, \\
\{D_A^{(2)}, \overline{D}_{\dot{B}}^{(2)}\} &=& -2i\sigma_{A\dot{B}}^\mu\partial_\mu = 2\sigma_{A\dot{B}}^\mu P_\mu.
\end{array}\right\} \quad (6.56)$$

Proof:

(a)

$$\begin{aligned}
\{D_A, D_B\} &= \{\partial_A + i\sigma^\mu_{A\dot{B}}\overline{\theta}^{\dot{B}}\partial_\mu, \partial_B + i\sigma^\nu_{B\dot{C}}\overline{\theta}^{\dot{C}}\partial_\nu\} \\
&= \{\partial_A, \partial_B\} + i\{\partial_A, \sigma^\nu_{B\dot{C}}\overline{\theta}^{\dot{C}}\partial_\nu\} + i\{\sigma^\mu_{A\dot{B}}\overline{\theta}^{\dot{B}}\partial_\mu, \partial_B\} \\
&\quad -\{\sigma^\mu_{A\dot{B}}\overline{\theta}^{\dot{B}}\partial_\mu, \sigma^\nu_{B\dot{C}}\overline{\theta}^{\dot{C}}\partial_\nu\} = 0.
\end{aligned}$$

Also,

$$\begin{aligned}
\{\overline{D}_{\dot{A}}, \overline{D}_{\dot{B}}\} &= \{-\overline{\partial}_{\dot{A}} - i\theta^B\sigma^\mu_{B\dot{A}}\partial_\mu, -\overline{\partial}_{\dot{B}} - i\theta^C\sigma^\nu_{C\dot{B}}\partial_\nu\} \\
&= \{\overline{\partial}_{\dot{A}}, \overline{\partial}_{\dot{B}}\} + i\{\overline{\partial}_{\dot{A}}, \theta^C\}\sigma^\nu_{C\dot{B}}\partial_\nu + i\{\theta^B, \overline{\partial}_{\dot{B}}\}\sigma^\mu_{B\dot{A}}\partial_\mu \\
&\quad -\{\theta^B, \theta^C\}\sigma^\mu_{B\dot{A}}\partial_\mu\sigma^\nu_{C\dot{B}}\partial_\nu = 0,
\end{aligned}$$

and

$$\begin{aligned}
\{D_A, \overline{D}_{\dot{B}}\} &= \{\partial_A + i\sigma^\mu_{A\dot{C}}\overline{\theta}^{\dot{C}}\partial_\mu, -\overline{\partial}_{\dot{B}} - i\theta^B\sigma^\nu_{B\dot{B}}\partial_\nu\} \\
&= -\{\partial_A, \overline{\partial}_{\dot{B}}\} - i\{\partial_A, \theta^B\sigma^\nu_{B\dot{B}}\partial_\nu\} - i\{\sigma^\mu_{A\dot{C}}\overline{\theta}^{\dot{C}}\partial_\mu, \overline{\partial}_{\dot{B}}\} \\
&\quad + \sigma^\mu_{A\dot{C}}\partial_\mu\{\overline{\theta}^{\dot{C}}, \theta^B\}\sigma^\nu_{B\dot{B}}\partial_\nu \\
&= -i(\partial_A\theta^B)\sigma^\nu_{B\dot{B}}\partial_\nu - i\sigma^\mu_{A\dot{C}}(\overline{\partial}_{\dot{B}}\overline{\theta}^{\dot{C}})\partial_\mu \\
&= -i\sigma^\nu_{A\dot{B}}\partial_\nu - i\sigma^\mu_{A\dot{B}}\partial_\mu \\
&= -2i\sigma^\nu_{A\dot{B}}\partial_\nu,
\end{aligned}$$

and

$$\begin{aligned}
\{D^A, \overline{D}^{\dot{B}}\} &= \epsilon^{AC}\epsilon^{\dot{B}\dot{E}}(D_C\overline{D}_{\dot{E}} + \overline{D}_{\dot{E}}D_C) = \epsilon^{AC}\epsilon^{\dot{B}\dot{E}}(-2i\sigma^\mu_{C\dot{E}}\partial_\mu) \\
&= -2i\overline{\sigma}^{\mu\dot{B}A}\partial_\mu.
\end{aligned}$$

(b)

$$\begin{aligned}
\{D^{(1)}_A, D^{(1)}_B\} &= \{\partial_A + 2i\sigma^\mu_{A\dot{B}}\overline{\theta}^{\dot{B}}\partial_\mu, \partial_B + 2i\sigma^\nu_{B\dot{C}}\overline{\theta}^{\dot{C}}\partial_\nu\} \\
&= \{\partial_A, \partial_B\} + 2i\sigma^\nu_{B\dot{C}}\partial_\nu\{\partial_A, \overline{\theta}^{\dot{C}}\} + 2i\sigma^\mu_{A\dot{B}}\partial_\mu\{\overline{\theta}^{\dot{B}}, \partial_B\} \\
&\quad -4\sigma^\mu_{A\dot{B}}\partial_\mu\sigma^\nu_{B\dot{C}}\partial_\nu\{\overline{\theta}^{\dot{B}}, \overline{\theta}^{\dot{C}}\} = 0.
\end{aligned}$$

Also

$$\{\overline{D}^{(1)}_{\dot{A}}, \overline{D}^{(1)}_{\dot{B}}\} = \{\overline{\partial}_{\dot{A}}, \overline{\partial}_{\dot{B}}\} = 0,$$

and

$$\begin{aligned}
\{D_A^{(1)}, \overline{D}_{\dot{B}}^{(1)}\} &= \{\partial_A + 2i\sigma^\mu_{A\dot{C}}\overline{\theta}^{\dot{C}}\partial_\mu, -\overline{\partial}_{\dot{B}}\} \\
&= -\{\partial_A, \overline{\partial}_{\dot{B}}\} - 2i\sigma^\mu_{A\dot{C}}\partial_\mu\{\overline{\theta}^{\dot{C}}, \overline{\partial}_{\dot{B}}\} \\
&= -2i\sigma^\mu_{A\dot{C}}\partial_\mu\delta^{\dot{C}}_{\dot{B}} = -2i\sigma^\mu_{A\dot{B}}\partial_\mu.
\end{aligned}$$

(c)

$$\{D_A^{(2)}, D_B^{(2)}\} = \{\partial_A, \partial_B\} = 0.$$

Also:

$$\begin{aligned}
\{\overline{D}_{\dot{A}}^{(2)}, \overline{D}_{\dot{B}}^{(2)}\} &= \{-\overline{\partial}_{\dot{A}} - 2i\theta^B\sigma^\mu_{B\dot{A}}\partial_\mu, -\overline{\partial}_{\dot{B}} - 2i\theta^C\sigma^\nu_{C\dot{B}}\partial_\nu\} \\
&= \{\overline{\partial}_{\dot{A}}, \overline{\partial}_{\dot{B}}\} + 2i\{\overline{\partial}_{\dot{A}}, \theta^C\}\sigma^\nu_{C\dot{B}}\partial_\nu \\
&\quad + 2i\{\theta^B, \overline{\partial}_{\dot{B}}\}\sigma^\mu_{B\dot{A}}\partial_\mu - 4\{\theta^B, \theta^C\}\sigma^\mu_{B\dot{A}}\partial_\mu\sigma^\nu_{C\dot{B}}\partial_\nu \\
&= 0,
\end{aligned}$$

and

$$\begin{aligned}
\{D_A^{(2)}, \overline{D}_{\dot{A}}^{(2)}\} &= \{\partial_A, -\overline{\partial}_{\dot{A}} - 2i\theta^B\sigma^\mu_{B\dot{A}}\partial_\mu\} \\
&= -\{\partial_A, \overline{\partial}_{\dot{A}}\} - 2i\{\partial_A, \theta^B\}\sigma^\mu_{B\dot{A}}\partial_\mu \\
&= -2i\delta^B_A\sigma^\mu_{B\dot{A}}\partial_\mu = -2i\sigma^\mu_{A\dot{A}}\partial_\mu.
\end{aligned}$$

This completes the proof of Proposition 6.11

6.6 Projection Operators

In later sections we will be concerned with superfields which satisfy certain constraints. These special types of fields can also be obtained from the most general form of a superfield by the application of *projection operators*. It is convenient to define these projection operators here after the introduction of the covariant derivatives. We begin by proving a set of relations which are of considerable use in calculations.

Proposition 6.12:

The following relations hold:

$$[D_A, \overline{D}^2] = -4i\sigma^\mu_{A\dot{A}}\overline{D}^{\dot{A}}\partial_\mu, \tag{6.57}$$

$$[D^A, \overline{D}^2] = 4i\overline{D}_{\dot{C}}\overline{\sigma}^{\mu\dot{C}A}\partial_\mu, \tag{6.58}$$

$$[\overline{D}_{\dot{A}}, D^2] = 4iD^A\sigma^\mu_{A\dot{A}}\partial_\mu, \tag{6.59}$$

$$[\overline{D}^{\dot{A}}, D^2] = -4i\overline{\sigma}^{\mu\dot{A}A}D_A\partial_\mu, \tag{6.60}$$

$$\overline{D}\overline{\sigma}^\mu D = -D\sigma^\mu\overline{D} - 4i\partial^\mu, \tag{6.61}$$

$$[D^2, \overline{D}^2] = -8i(D\sigma^\mu\overline{D})\partial_\mu + 16\,\Box, \tag{6.62}$$

$$[\overline{D}^2, D^2] = -8i(\overline{D}\overline{\sigma}^\mu D)\partial_\mu + 16\,\Box, \tag{6.63}$$

$$\sigma^\mu_{A\dot{A}}\sigma^\nu_{B\dot{B}} + \sigma^\nu_{A\dot{A}}\sigma^\mu_{B\dot{B}} = \eta^{\mu\nu}\epsilon_{AB}\epsilon_{\dot{A}\dot{B}} + 4(\sigma^{\mu\rho}\epsilon^\top)_{AB}(\epsilon\overline{\sigma}^\nu_\rho)_{\dot{A}\dot{B}}, \tag{6.64}$$

$$D\sigma^{\mu\nu}\epsilon^\top D = 0, \quad \overline{D}\epsilon\overline{\sigma}^{\mu\nu}\overline{D} = 0, \tag{6.65}$$

$$D^A\overline{D}^2 D_A = \overline{D}_{\dot{A}}D^2\overline{D}^{\dot{A}}. \tag{6.66}$$

Proof:

(a) We have Eq. (6.54), *i.e.*

$$\{D_A, \overline{D}_{\dot{A}}\} = -2i\sigma^\mu_{A\dot{A}}\partial_\mu.$$

Multiplying this equation from the right by $\overline{D}^{\dot{A}}$, we obtain

$$D_A\overline{D}^2 + \overline{D}_{\dot{A}}D_A\overline{D}^{\dot{A}} = -2i\sigma^\mu_{A\dot{A}}\overline{D}^{\dot{A}}\partial_\mu. \tag{6.67}$$

Alternatively, multiplying Eq. (6.54) by $\epsilon^{\dot{A}\dot{B}}$ we obtain

$$(D_A\overline{D}_{\dot{A}} + \overline{D}_{\dot{A}}D_A)\epsilon^{\dot{A}\dot{B}} = -2i\sigma^\mu_{A\dot{A}}\partial_\mu\epsilon^{\dot{A}\dot{B}}.$$

Using Eq. (1.85), *i.e.*

$$\overline{D}^{\dot{B}} = -\overline{D}_{\dot{A}}\epsilon^{\dot{A}\dot{B}},$$

this becomes

$$-(D_A\overline{D}^{\dot{B}} + \overline{D}^{\dot{B}}D_A) = -2i\sigma^\mu_{A\dot{A}}\partial_\mu\epsilon^{\dot{A}\dot{B}}.$$

For $\dot{B} = \dot{A}$ this equation yields

$$D_A\overline{D}^{\dot{A}} = -\overline{D}^{\dot{A}}D_A + 2i\sigma^\mu_{A\dot{B}}\partial_\mu\epsilon^{\dot{D}\dot{A}}.$$

Inserting this into Eq. (6.67), we obtain

$$D_A\overline{D}^2 + \overline{D}_{\dot{A}}(-\overline{D}^{\dot{A}}D_A + 2i\sigma^\mu_{A\dot{D}}\partial_\mu\epsilon^{\dot{D}\dot{A}}) = -2i\sigma^\mu_{A\dot{A}}\overline{D}^{\dot{A}}\partial_\mu,$$

i.e.

$$[D_A, \overline{D}^2] = -2i(\sigma^\mu_{A\dot{A}}\overline{D}^{\dot{A}} + \overline{D}_{\dot{A}}\sigma^\mu_{A\dot{D}}\epsilon^{\dot{D}\dot{A}})\partial_\mu \stackrel{(1.85)}{=} -4i\sigma^\mu_{A\dot{A}}\overline{D}^{\dot{A}}\partial_\mu.$$

This is what had to be shown.

(b) Equation (6.58) follows by multiplying Eq. (6.57) by ϵ^{BA} and using Eqs. (1.83) and (1.106a).

(c) We proceed as under (a) with the relation

$$\{D_A, \overline{D}_{\dot{A}}\} = -2i\sigma^\mu_{A\dot{A}}\partial_\mu.$$

Hence

$$D^2\overline{D}_{\dot{A}} + D^A\overline{D}_{\dot{A}}D_A = -2iD^A\sigma^\mu_{A\dot{A}}\partial_\mu. \tag{6.68}$$

Also

$$\epsilon^{AB}(D_B\overline{D}_{\dot{A}} + \overline{D}_{\dot{A}}D_B) = \epsilon^{AB}(-2i\sigma^\mu_{B\dot{A}}\partial_\mu).$$

Using Eq. (1.83) this gives:

$$D^A\overline{D}_{\dot{A}} + \overline{D}_{\dot{A}}D^A = -2i\epsilon^{AB}\sigma^\mu_{B\dot{A}}\partial_\mu.$$

Inserting this result into Eq. (6.68) we obtain

$$\begin{aligned}
[D^2, \overline{D}_{\dot{A}}] &= 2i\epsilon^{AB}\sigma^\mu_{B\dot{A}}\partial_\mu D_A - 2iD^A\sigma^\mu_{A\dot{A}}\partial_\mu \\
&= -4iD^A\sigma^\mu_{A\dot{A}}\partial_\mu,
\end{aligned}$$

with the help of Eq. (1.83).

(d) The result (6.60) follows by multiplying Eq. (6.59) by $\epsilon^{\dot{B}\dot{A}}$ and using Eqs. (1.85) and (1.106a).

(e) Consider

$$\begin{aligned}
\overline{D}_{\dot{A}}\overline{\sigma}^{\mu\dot{A}A}D_A &\stackrel{(6.54)}{=} (-2i\sigma^\rho_{A\dot{A}}\partial_\rho - D_A\overline{D}_{\dot{A}})\overline{\sigma}^{\mu\dot{A}A} \\
&\stackrel{(1.106a)}{=} -D_A\overline{D}_{\dot{A}}\epsilon^{AB}\epsilon^{\dot{A}\dot{B}}\sigma^\mu_{B\dot{B}} - 2i\sigma^\rho_{A\dot{A}}\overline{\sigma}^{\mu\dot{A}A}\partial_\rho \\
&\stackrel{(1.83)}{\underset{(1.84)}{=}} -(D\sigma^\mu\overline{D}) - 2i\text{Tr}(\sigma^\rho\overline{\sigma}^\mu)\partial_\rho \\
&\stackrel{(1.109a)}{=} -(D\sigma^\mu\overline{D}) - 4i\partial^\mu.
\end{aligned}$$

(f) Consider

$$
\begin{aligned}
[D^2, \overline{D}^2] \;\;&=\;\; D^A[D_A, \overline{D}^2] + [D^A, \overline{D}^2]D_A \\
&\overset{(1.83)}{=}\; D^A[D_A, \overline{D}^2] + \epsilon^{AB}[D_B, \overline{D}^2]\epsilon_{AC}D^C \\
&=\; D^A[D_A, \overline{D}^2] - [D_A, \overline{D}^2]D^A \\
&\overset{(6.57)}{=}\; D^A[-4i\sigma^\mu_{A\dot{A}}\overline{D}^{\dot{A}}\partial_\mu] + [4i\sigma^\mu_{A\dot{A}}\overline{D}^{\dot{A}}\partial_\mu]D^A \\
&=\; 4i\sigma^\mu_{A\dot{A}}\partial_\mu[\overline{D}^{\dot{A}}D^A - D^A\overline{D}^{\dot{A}}] \\
&\overset{(6.54)}{=}\; 4i\sigma^\mu_{A\dot{A}}\partial_\mu[-2D^A\overline{D}^{\dot{A}} - 2i\overline{\sigma}^{\rho A\dot{A}}\partial_\rho] \\
&=\; -8i(D\sigma^\mu\overline{D})\partial_\mu + 8(\sigma^\mu_{A\dot{A}}\overline{\sigma}^{\rho\dot{A}A})\partial_\mu\partial_\rho \\
&\overset{(1.109a)}{=}\; -8i(D\sigma^\mu\overline{D})\partial_\mu + 16\,\square.
\end{aligned}
$$

(g) The result (6.63) follows immediately from Eqs. (6.61) and (6.62).

(h) Consider

$$
\begin{aligned}
\eta^{\mu\nu}\epsilon_{AB}\epsilon_{\dot{A}\dot{B}} \;&\overset{(1.109a)}{=}\; \frac{1}{2}\mathrm{Tr}(\sigma^\mu\overline{\sigma}^\nu)\epsilon_{AB}\epsilon_{\dot{A}\dot{B}} \\
&=\; \frac{1}{2}\sigma^\mu_{D\dot{C}}\overline{\sigma}^{\nu\dot{C}D}\epsilon_{AB}\epsilon_{\dot{A}\dot{B}} \\
&\overset{(1.78a)}{=}\; \frac{1}{2}\sigma^\mu_{D\dot{E}}(\epsilon^{\dot{E}\dot{F}}\epsilon_{\dot{F}\dot{C}})\overline{\sigma}^{\nu\dot{C}K}(\epsilon_{KL}\epsilon^{LD})\epsilon_{AB}\epsilon_{\dot{A}\dot{B}} \\
&\overset{(1.106b)}{=}\; -\frac{1}{2}\sigma^\mu_{D\dot{E}}\epsilon^{\dot{E}\dot{F}}\sigma^\nu_{L\dot{F}}\epsilon^{LD}\epsilon_{AB}\epsilon_{\dot{A}\dot{B}}.
\end{aligned}
$$

Now from the explicit form of the ϵ-matrices we deduce:

$$
\begin{aligned}
\epsilon^{AB}\epsilon_{CD} &= \delta^A_D\delta^B_C - \delta^A_C\delta^B_D, \\
\epsilon^{\dot{A}\dot{B}}\epsilon_{\dot{C}\dot{D}} &= \delta^{\dot{A}}_{\dot{D}}\delta^{\dot{B}}_{\dot{C}} - \delta^{\dot{A}}_{\dot{C}}\delta^{\dot{B}}_{\dot{D}}.
\end{aligned}
$$

Hence

$$
\begin{aligned}
\eta^{\mu\nu}\epsilon_{AB}\epsilon_{\dot{A}\dot{B}} &= -\frac{1}{2}\sigma^\mu_{D\dot{E}}\sigma^\nu_{L\dot{F}}(\delta^{\dot{E}}_{\dot{B}}\delta^{\dot{F}}_{\dot{A}} - \delta^{\dot{E}}_{\dot{A}}\delta^{\dot{F}}_{\dot{B}})(\delta^L_B\delta^D_A - \delta^L_A\delta^D_B) \\
&= -\frac{1}{2}(\sigma^\mu_{A\dot{B}}\sigma^\nu_{B\dot{A}} - \sigma^\mu_{B\dot{B}}\sigma^\nu_{A\dot{A}} - \sigma^\mu_{A\dot{A}}\sigma^\nu_{B\dot{B}} + \sigma^\mu_{B\dot{A}}\sigma^\nu_{A\dot{B}}).
\end{aligned}
$$

On the other hand:

$$
\begin{aligned}
4(\sigma^{\mu\rho}\epsilon^\top)_{AB}(\epsilon\overline{\sigma}^\nu_\rho)_{\dot{A}\dot{B}} \;&\overset{\substack{(1.138a)\\(1.138b)}}{=}\; -\frac{1}{4}[(\sigma^\mu_{A\dot{C}}\overline{\sigma}^{\rho\dot{C}C} - \sigma^{\ \rho}_{A\dot{C}}\overline{\sigma}^{\mu\dot{C}C})\epsilon^\top_{CB}] \\
&\qquad\qquad \times[\epsilon_{\dot{A}\dot{B}}(\overline{\sigma}^{\dot{E}D}_\rho\sigma^\nu_{D\dot{B}} - \overline{\sigma}^{\nu\dot{E}D}\sigma_{\rho D\dot{B}})] \\
&=\; -\frac{1}{4}[T_1 + T_2 + T_3 + T_4],
\end{aligned}
$$

where:

$$
\begin{aligned}
T_1 &= \sigma^\mu_{A\dot{C}}\overline{\sigma}^{\rho\dot{C}C}\epsilon^\top_{CB}\epsilon_{\dot{A}\dot{E}}\overline{\sigma}^{\dot{E}F}_\rho \delta^D_F \sigma^\nu_{D\dot{B}} \\
&= \sigma^\mu_{A\dot{C}}\overline{\sigma}^{\rho\dot{C}C}\epsilon^\top_{CB}\epsilon_{\dot{A}\dot{E}}\overline{\sigma}^{\dot{E}F}_\rho \epsilon_{FK}\epsilon^{KD}\sigma^\nu_{D\dot{B}} \\
&\overset{(1.106a)}{=} \sigma^\mu_{A\dot{C}}\overline{\sigma}^{\rho\dot{C}C}\epsilon^\top_{CB}(-\sigma_{\rho K\dot{A}})\epsilon^{KD}\sigma^\nu_{D\dot{B}} \\
&\overset{(1.110)}{=} \sigma^\mu_{A\dot{C}}\epsilon^\top_{CB}\epsilon^{DK}\sigma^\nu_{D\dot{B}}2\delta^C_K\delta^{\dot{C}}_{\dot{A}} \\
&= -2\sigma^\mu_{A\dot{A}}\epsilon_{BC}\epsilon^{CD}\sigma^\nu_{D\dot{B}} = -2\sigma^\mu_{A\dot{A}}\delta^D_B\sigma^\nu_{D\dot{B}} = -2\sigma^\mu_{A\dot{A}}\sigma^\nu_{B\dot{B}};
\end{aligned}
$$

$$
\begin{aligned}
T_2 &= -\sigma^\rho_{A\dot{C}}\overline{\sigma}^{\mu\dot{C}C}\epsilon^\top_{CB}\epsilon_{\dot{A}\dot{E}}\overline{\sigma}^{\dot{E}D}_\rho\sigma^\nu_{D\dot{B}} \overset{(1.110)}{=} -2\delta^D_A\delta^{\dot{E}}_{\dot{C}}\overline{\sigma}^{\mu\dot{C}C}\epsilon^\top_{CB}\epsilon_{\dot{A}\dot{E}}\sigma^\nu_{D\dot{B}} \\
&= -2\overline{\sigma}^{\mu\dot{C}C}\epsilon^\top_{CB}\epsilon_{\dot{A}\dot{C}}\sigma^\nu_{A\dot{B}} = -2\sigma^\mu_{B\dot{A}}\sigma^\nu_{A\dot{B}};
\end{aligned}
$$

$$
\begin{aligned}
T_3 &= -\sigma^\mu_{A\dot{G}}\overline{\sigma}^{\rho\dot{G}C}\epsilon^\top_{CB}\epsilon_{\dot{A}\dot{E}}\overline{\sigma}^{\nu\dot{E}D}\sigma_{\rho D\dot{B}} \\
&\overset{(1.110)}{=} -\sigma^\mu_{A\dot{G}}\epsilon^\top_{CB}\epsilon_{\dot{A}\dot{E}}\overline{\sigma}^{\nu\dot{E}D}\left(2\delta^C_D\delta^{\dot{G}}_{\dot{B}}\right) \\
&= -2\sigma^\mu_{A\dot{B}}\epsilon^\top_{CB}\epsilon_{\dot{A}\dot{E}}\overline{\sigma}^{\nu\dot{E}C} \overset{(1.106b)}{=} -2\sigma^\mu_{A\dot{B}}\sigma^\nu_{B\dot{A}};
\end{aligned}
$$

$$
\begin{aligned}
T_4 &= \sigma^\rho_{A\dot{G}}\overline{\sigma}^{\mu\dot{G}C}\epsilon^\top_{CB}\epsilon_{\dot{A}\dot{E}}\overline{\sigma}^{\nu\dot{E}D}\sigma_{\rho D\dot{B}} \\
&\overset{(1.106b)}{=} \sigma^\rho_{A\dot{G}}\overline{\sigma}^{\mu\dot{G}C}\epsilon^\top_{CB}\epsilon_{\dot{A}\dot{E}}\overline{\sigma}^{\nu\dot{E}D}\epsilon_{DK}\epsilon_{\dot{B}\dot{K}}\overline{\sigma}^{\dot{K}K}_\rho \\
&\overset{(1.110)}{=} \overline{\sigma}^{\mu\dot{G}C}\epsilon^\top_{CB}\epsilon_{\dot{A}\dot{E}}\overline{\sigma}^{\nu\dot{E}D}\epsilon_{DK}\epsilon_{\dot{B}\dot{K}}2\delta^K_A\delta^{\dot{K}}_{\dot{G}} \\
&= 2\overline{\sigma}^{\mu\dot{K}C}\epsilon^\top_{CB}\epsilon_{\dot{A}\dot{E}}\overline{\sigma}^{\nu\dot{E}D}\epsilon_{DA}\epsilon_{\dot{B}\dot{K}} \overset{(1.106b)}{=} -2\sigma^\mu_{B\dot{B}}\sigma^\nu_{A\dot{A}}.
\end{aligned}
$$

Hence

$$
4(\sigma^{\mu\rho}\epsilon^\top)_{AB}(\epsilon\overline{\sigma}^\nu_\rho)_{\dot{A}\dot{B}} = \frac{1}{2}[\sigma^\mu_{A\dot{A}}\sigma^\nu_{B\dot{B}} + \sigma^\mu_{B\dot{A}}\sigma^\nu_{A\dot{B}} + \sigma^\mu_{A\dot{B}}\sigma^\nu_{B\dot{A}} + \sigma^\mu_{B\dot{B}}\sigma^\nu_{A\dot{A}}],
$$

and so

$$
\eta^{\mu\nu}\epsilon_{AB}\epsilon_{\dot{A}\dot{B}} + 4(\sigma^{\mu\rho}\epsilon^\top)_{AB}(\epsilon\overline{\sigma}^\nu_\rho)_{\dot{A}\dot{B}} = \sigma^\mu_{A\dot{A}}\sigma^\nu_{B\dot{B}} + \sigma^\mu_{B\dot{B}}\sigma^\nu_{A\dot{A}},
$$

as had to be shown.

(i) Consider

$$
\begin{aligned}
D\sigma^{\mu\rho}\epsilon^\top D &= D^C(\sigma^{\mu\rho}\epsilon^\top)_{CE}D^E \\
&\overset{(1.149)}{=} \frac{1}{2}D^C(\sigma^{\mu\rho}\epsilon^\top)_{CE}D^E + \frac{1}{2}D^C(\sigma^{\mu\rho}\epsilon^\top)_{EC}D^E \\
&\overset{(6.54)}{=} \frac{1}{2}D^C(\sigma^{\mu\rho}\epsilon^\top)_{CE}D^E - \frac{1}{2}D^E(\sigma^{\mu\rho}\epsilon^\top)_{EC}D^C = 0.
\end{aligned}
$$

Analogously

$$\overline{D}(\epsilon\bar{\sigma}^{\mu\rho})\overline{D} \overset{(1.149)}{=} \frac{1}{2}\overline{D}^{\dot{A}}(\epsilon\bar{\sigma}^{\mu\rho})_{\dot{A}\dot{C}}\overline{D}^{\dot{C}} + \frac{1}{2}\overline{D}^{\dot{A}}(\epsilon\bar{\sigma}^{\mu\rho})_{\dot{C}\dot{A}}\overline{D}^{\dot{C}}$$

$$\overset{(6.54)}{=} \frac{1}{2}\overline{D}^{\dot{A}}(\epsilon\bar{\sigma}^{\mu\rho})_{\dot{A}\dot{C}}\overline{D}^{\dot{C}} - \frac{1}{2}\overline{D}^{\dot{C}}(\epsilon\bar{\sigma}^{\mu\rho})_{\dot{C}\dot{A}}\overline{D}^{\dot{A}} = 0.$$

(j) From Eqs. (6.57) and (6.59) we obtain:

$$D^A(D_A\overline{D}^2 - \overline{D}^2 D_A) = -4iD^A\sigma^\mu_{A\dot{A}}\overline{D}^{\dot{A}}\partial_\mu,$$

$$(\overline{D}_{\dot{A}}D^2 - D^2\overline{D}_{\dot{A}})\overline{D}^{\dot{A}} = 4iD^A\sigma^\mu_{A\dot{A}}\overline{D}^{\dot{A}}\partial_\mu.$$

Adding we obtain

$$D^A\overline{D}^2 D_A = \overline{D}_{\dot{A}}D^2\overline{D}^{\dot{A}}.$$

This completes the proof of the relations (6.57) to (6.66).

We now define a set of operators and then verify that they are projection operators, *i.e.* that each is idempotent and their sum is the identity operator.

Definition: We define the operators π_+, π_-, π_T by the relations

$$\pi_+ := -\frac{1}{16\Box}\overline{D}^2 D^2, \tag{6.69}$$

$$\pi_- := -\frac{1}{16\Box}D^2\overline{D}^2, \tag{6.70}$$

$$\pi_T := \frac{1}{8\Box}\overline{D}_{\dot{A}}D^2\overline{D}^{\dot{A}} = \frac{1}{8\Box}D^A\overline{D}^2 D_A, \tag{6.71}$$

$$\pi_0 := \pi_+ + \pi_-. \tag{6.72}$$

Proposition 6.13:

The operators (6.69), (6.70), and (6.71) are projection operators.

Proof: Consider Eq. (6.69). The square of this operator is

$$\pi_+\pi_+ = \frac{1}{16\Box}\overline{D}^2 D^2 \frac{1}{16\Box}\overline{D}^2 D^2 = \left(\frac{1}{16\Box}\right)^2 \overline{D}^2 D^2\overline{D}^2 D^2$$

$$\overset{(6.62)}{=} \left(\frac{1}{16\Box}\right)^2 \overline{D}^2 D^2(D^2\overline{D}^2 + 8iD\sigma^\mu\overline{D}\partial_\mu - 16\Box)$$

$$\overset{(6.54)}{=} \left(\frac{1}{16\Box}\right)^2 \overline{D}^2 D^2(-16\Box) = \pi_+,$$

since by Eq. (6.54) $D^3 = D^4 = 0$. Similarly

$$\pi_-\pi_- = \left(\frac{1}{16\Box}\right)^2 D^2 \overline{D}^2 D^2 \overline{D}^2$$

$$\overset{(6.62)}{=} \left(\frac{1}{16\Box}\right)^2 D^2 \overline{D}^2 (\overline{D}^2 D^2 + 8i\overline{D}\overline{\sigma}^\mu D \partial_\mu - 16\,\Box) = \pi_-,$$

where we used Eq. (6.54). We also have

$$\pi_+\pi_- = \pi_-\pi_+ = 0,$$

since $D^4 = 0 = \overline{D}^4$, and similarly

$$\pi_\pm \pi_T = \pi_T \pi_\pm = 0.$$

Next consider

$$\pi_T\pi_T = \left(\frac{1}{8\Box}\right)^2 \overline{D}_{\dot{A}} D^2 \overline{D}^{\dot{A}} \overline{D}_{\dot{B}} D^2 \overline{D}^{\dot{B}}$$

$$\overset{(6.59)}{\underset{(6.60)}{=}} \left(\frac{1}{8\Box}\right)^2 \overline{D}_{\dot{A}} (\overline{D}^{\dot{A}} D^2 + 4i\overline{\sigma}^{\mu \dot{A} A} D_A \partial_\mu)$$

$$\times (D^2 \overline{D}_{\dot{B}} + 4i D^E \sigma^\rho_{E\dot{B}} \partial_\rho) \overline{D}^{\dot{B}}$$

$$= \left(\frac{1}{8\Box}\right)^2 \overline{D}_{\dot{A}} (4i\overline{\sigma}^{\mu \dot{A} A} D_A \partial_\mu)(4i D^E \sigma^\rho_{E\dot{B}} \partial_\rho) \overline{D}^{\dot{B}},$$

since $D^3 = D^4 = 0$. Continuing we have

$$\pi_T\pi_T = -16\left(\frac{1}{8\Box}\right)^2 \left[\overline{D}_{\dot{A}} \overline{\sigma}^{\mu \dot{A} A} D_A D^E \sigma^\rho_{E\dot{B}} \overline{D}^{\dot{B}}\right] \partial_\mu \partial_\rho$$

$$\overset{(1.106a)}{=} -\left(\frac{1}{2\Box}\right)^2 \left[\overline{D}_{\dot{A}} \epsilon^{AC} \epsilon^{\dot{A}\dot{C}} \sigma^\mu_{C\dot{C}} D_A D^E \sigma^\rho_{E\dot{B}} \overline{D}^{\dot{B}}\right] \partial_\mu \partial_\rho$$

$$\overset{(1.83)}{=} -\left(\frac{1}{2\Box}\right)^2 \left[\overline{D}^{\dot{C}} \sigma^\mu_{C\dot{C}} D^C D^E \sigma^\rho_{E\dot{B}} \overline{D}^{\dot{B}}\right.$$

$$\left. + \overline{D}^{\dot{C}} \sigma^\rho_{C\dot{C}} D^C D^E \sigma^\mu_{E\dot{B}} \overline{D}^{\dot{B}}\right] \partial_\mu \partial_\rho$$

$$\overset{(6.64)}{=} -\frac{1}{2}\left(\frac{1}{2\Box}\right)^2 \left[\eta^{\mu\rho} \epsilon_{CE} \epsilon_{\dot{C}\dot{B}} \overline{D}^{\dot{C}} D^C D^E \overline{D}^{\dot{B}}\right.$$

$$\left. + 4(\sigma^{\mu\rho} \epsilon^\top)_{CE} (\epsilon \overline{\sigma}^\nu_\rho)_{\dot{C}\dot{B}} \overline{D}^{\dot{C}} D^C D^E \overline{D}^{\dot{B}}\right] \partial_\mu \partial_\rho$$

$$\overset{(1.86)}{=} -\frac{1}{2}\left(\frac{1}{2\Box}\right)^2\left[-\eta^{\mu\rho}\overline{D}_{\dot{B}}D^2\overline{D}^{\dot{B}}\right.$$

$$\left.+4\overline{D}^{\dot{C}}(D\sigma^{\mu\rho}\epsilon^\top D)(\epsilon\bar\sigma_\rho^{\;\nu})_{\dot{C}\dot{B}}\overline{D}^{\dot{B}}\right]\partial_\mu\partial_\rho$$

$$\overset{(6.65)}{=} \frac{1}{8\Box}\overline{D}_{\dot{B}}D^2\overline{D}^{\dot{B}} = \pi_T.$$

Finally we consider the sum of the three operators, $i.e.$

$$\pi_+ + \pi_- + \pi_T = \frac{1}{16\Box}[2\overline{D}_{\dot{A}}D^2\overline{D}^{\dot{A}} - \overline{D}^2D^2 - D^2\overline{D}^2]$$

$$\overset{(6.66)}{=} \frac{1}{16\Box}[\overline{D}_{\dot{A}}D^2\overline{D}^{\dot{A}} + D^A\overline{D}^2D_A - \overline{D}^2D^2 - D^2\overline{D}^2]$$

$$= \frac{1}{16\Box}[(\overline{D}_{\dot{A}}D^2 - D^2\overline{D}_{\dot{A}})\overline{D}^{\dot{A}} + (D^A\overline{D}^2 - \overline{D}^2D^A)D_A]$$

$$\overset{(6.58)}{\underset{(6.59)}{=}} \frac{1}{16\Box}[4iD^A\sigma^\mu_{A\dot{A}}\partial_\mu\overline{D}^{\dot{A}} + 4i\overline{D}_{\dot{C}}\bar\sigma^{\mu\dot{C}A}\partial_\mu D_A]$$

$$= \frac{1}{16\Box}4i(D\sigma^\mu\overline{D} + \overline{D}\bar\sigma^\mu D)\partial_\mu$$

$$= \frac{1}{16\Box}4i(-4i\partial^\mu)\partial_\mu = \mathbb{1}.$$

Hence the sum of the operators is the identity operator.

6.7   Constraints

From the definition of the projection operators we obtain the following constraints:

$$\overline{D}\pi_+ = 0, \qquad (\text{since } \overline{D}^3 = 0), \tag{6.73}$$

$$D\pi_- = 0, \qquad (\text{since } D^3 = 0). \tag{6.74}$$

Superfields

$$\Phi_\mp = \pi_\pm\Phi \tag{6.75}$$

satisfying these constraints are called respectively *left-handed* and *right-handed chiral superfields* in analogy to left-handed and right-handed fermionic fields

$$\psi_{L,R} = d_\mp\psi, \qquad d_\mp = \frac{1}{2}(1 \mp \gamma_5),$$

constructed from the fermionic field ψ, since $\psi_{L,R}$ obey the constraints

$$d_\pm\psi_{L,R} = 0.$$

6.8 Transformations of Component Fields

We introduced superfields as operator-valued functions defined on super-space, which are to be understood in terms of their power series expansions in the Grassmann variables θ and $\overline{\theta}$ (see Eq. (6.7)), *i.e.*

$$\Phi(x,\theta,\overline{\theta}) = f(x) + \theta\phi(x) + \overline{\theta}\overline{\chi}(x) + (\theta\theta)m(x) + (\overline{\theta}\,\overline{\theta})n(x) + (\theta\sigma^{\mu}\overline{\theta})V_{\mu}(x)$$
$$+ (\theta\theta)\overline{\theta}\,\overline{\lambda}(x) + (\overline{\theta}\,\overline{\theta})(\theta\psi(x)) + (\theta\theta)(\overline{\theta}\,\overline{\theta})d(x). \tag{6.76}$$

The quantities

$$f(x),\ \phi(x),\ \overline{\chi}(x),\ m(x),\ n(x),\ V_{\mu}(x),\ \overline{\lambda}(x),\ \psi(x),\ d(x)$$

are called *component fields*. Their geometric character is determined by their transformation properties under the Lorentz group and the additional requirement that $\Phi(x,\theta,\overline{\theta})$ be a Lorentz scalar or pseudoscalar. From these conditions we deduce that:

$$
\begin{aligned}
f(x), m(x), n(x) &\quad \text{are complex scalar/pseudo-scalar fields,} \\
\psi(x), \phi(x) &\quad \text{are left-handed Weyl spinor fields,} \\
\overline{\chi}(x), \overline{\lambda}(x) &\quad \text{are right-handed Weyl spinor fields,} \\
V_{\mu}(x) &\quad \text{is a Lorentz four-vector field,} \\
d(x) &\quad \text{is a scalar field.}
\end{aligned}
\tag{6.77}
$$

Thus a superfield is a short way to denote a *finite multiplet of fields*.

In an expansion such as that of Eq. (6.76) in the 2×2 Weyl formulation, the function $f(x)$ need not have a well-defined parity. We have seen before that the Dirac equation is invariant under parity transformations whereas the individual Weyl equations are not. The latter is reflected in the expansion (6.76). Thus $f(x)$ may be the sum of scalar and pseudoscalar contributions, and correspondingly $m(x)$ may be the sum of pseudoscalar and scalar contributions which are such that if we rewrite the superfield in the 4×4 Dirac formulation it will have a well-defined property under parity transformations, *i.e.* as a scalar or pseudoscalar.

Our next task is the computation of the transformation laws of the component fields with respect to supersymmetry transformations. This will enable us to obtain the Weyl representation of the transformations (5.11a) to (5.11d) of the Wess–Zumino model.

The transformation law for superfields is defined as

$$\delta_s \Phi(x,\theta,\overline{\theta}) = \delta_s' f(x) + \theta^A \delta_s' \phi_A(x) + \overline{\theta}_{\dot{A}} \delta_s' \overline{\chi}^{\dot{A}}(x) + (\theta\theta)\delta_s' m(x)$$
$$+ (\overline{\theta}\,\overline{\theta})\delta_s' n(x) + (\theta\sigma^\mu\overline{\theta})\delta_s' V_\mu(x) + (\theta\theta)\overline{\theta}_{\dot{A}}\delta_s' \overline{\lambda}^{\dot{A}}(x)$$
$$+ (\overline{\theta}\,\overline{\theta})\theta^A \delta_s' \psi_A(x) + (\theta\theta)(\overline{\theta}\,\overline{\theta})\delta_s' d(x). \qquad (6.78)$$

On the other hand, according to Eq. (6.20), we have the infinitesimal super-symmetry variation

$$\delta_s\Phi(x,\theta,\overline{\theta}) = \left[\alpha^A \partial_A + \overline{\alpha}_{\dot{A}}\overline{\partial}^{\dot{A}} + i\theta\sigma^\mu\overline{\alpha}\partial_\mu - i\alpha\sigma^\mu\overline{\theta}\partial_\mu\right]\Phi(x,\theta,\overline{\theta})$$
$$= \left[\alpha^A \partial_A + \overline{\alpha}_{\dot{A}}\overline{\partial}^{\dot{A}} + i\theta\sigma^\mu\overline{\alpha}\partial_\mu - i\alpha\sigma^\mu\overline{\theta}\partial_\mu\right] \times \left\{ f(x) + \theta^B \phi_B(x) \right.$$
$$+ \overline{\theta}_{\dot{B}}\overline{\chi}^{\dot{B}}(x) + (\theta\theta)m(x) + (\overline{\theta}\,\overline{\theta})n(x) + \theta\sigma^\nu\overline{\theta}V_\nu(x)$$
$$\left. + (\theta\theta)\overline{\theta}_{\dot{B}}\overline{\lambda}^{\dot{B}}(x) + (\overline{\theta}\,\overline{\theta})\theta^B \psi_B(x) + (\theta\theta)(\overline{\theta}\,\overline{\theta})d(x) \right\},$$

where we used Eq. (6.76). Multiplying out this implies

$$\delta_s\Phi(x,\theta,\overline{\theta}) = \alpha^A \phi_A(x) + 2\alpha^A\theta_A m(x) + \alpha^A\sigma^\nu_{A\dot{B}}\overline{\theta}^{\dot{B}}V_\nu(x) + 2\alpha^A\theta_A\overline{\theta}_{\dot{B}}\overline{\lambda}^{\dot{B}}(x)$$
$$+ (\overline{\theta}\,\overline{\theta})\alpha^A\psi_A(x) + 2\alpha^A\theta_A(\overline{\theta}\,\overline{\theta})d(x) + \overline{\alpha}_{\dot{A}}\overline{\chi}^{\dot{A}}(x) + 2\overline{\alpha}_{\dot{A}}\overline{\theta}^{\dot{A}}n(x)$$
$$- \overline{\alpha}_{\dot{A}}\theta^A\sigma^\nu_{A\dot{B}}\overline{\partial}^{\dot{A}}\overline{\theta}^{\dot{B}}V_\nu(x) + (\theta\theta)\overline{\alpha}_{\dot{A}}\overline{\lambda}^{\dot{A}}(x) + 2\overline{\alpha}_{\dot{A}}\overline{\theta}^{\dot{A}}\theta^B \psi_B(x)$$
$$+ 2(\theta\theta)\overline{\alpha}_{\dot{A}}\overline{\theta}^{\dot{A}}d(x) + i\Big[(\theta\sigma^\mu\overline{\alpha})\partial_\mu f(x) + (\theta\sigma^\mu\overline{\alpha})\theta^A \partial_\mu\phi_A(x)$$
$$+ (\theta\sigma^\mu\overline{\alpha})\overline{\theta}_{\dot{A}}\partial_\mu\overline{\chi}^{\dot{A}}(x) + (\theta\sigma^\mu\overline{\alpha})(\overline{\theta}\,\overline{\theta})\partial_\mu n(x)$$
$$+ (\theta\sigma^\mu\overline{\alpha})(\theta\sigma^\nu\overline{\theta})\partial_\mu V_\nu(x) + (\theta\sigma^\mu\overline{\alpha})(\overline{\theta}\,\overline{\theta})\theta^A \partial_\mu\psi_A(x)\Big]$$
$$- i\Big[(\alpha\sigma^\mu\overline{\theta})\partial_\mu f(x) + (\alpha\sigma^\mu\overline{\theta})\overline{\theta}_{\dot{A}}\partial_\mu\overline{\chi}^{\dot{A}}(x) + (\alpha\sigma^\mu\overline{\theta})\theta^B \partial_\mu\phi_B(x)$$
$$+ (\alpha\sigma^\mu\overline{\theta})(\theta\theta)\partial_\mu m(x) + (\alpha\sigma^\mu\overline{\theta})(\theta\sigma^\nu\overline{\theta})\partial_\mu V_\nu(x)$$
$$+ (\alpha\sigma^\mu\overline{\theta})(\theta\theta)\overline{\theta}_{\dot{A}}\partial_\mu\overline{\lambda}^{\dot{A}}(x)\Big],$$

where we used the fact that the third power of any Grassmann number in the Weyl representation vanishes. Hence (using Eqs. (1.97) and (6.5k))

$$\delta_s\Phi(x,\theta,\overline{\theta}) = \alpha^A \phi_A(x) + \overline{\alpha}_{\dot{A}}\overline{\chi}^{\dot{A}}(x) + \theta^A\{2\alpha_A m(x) + i\sigma^\mu_{A\dot{B}}\overline{\alpha}^{\dot{B}}f(x)$$
$$+ \overline{\alpha}_{\dot{A}}\sigma^\nu_{A\dot{B}}\overline{\partial}^{\dot{A}}\overline{\theta}^{\dot{B}}V_\nu(x)\} + \overline{\theta}_{\dot{A}}\{2\overline{\alpha}^{\dot{A}}n(x) + i(\alpha\sigma^\mu)_{\dot{B}}\epsilon^{\dot{B}\dot{A}}\partial_\mu f(x)$$
$$- (\alpha\sigma^\mu)_{\dot{B}}\epsilon^{\dot{B}\dot{A}}V_\mu(x)\} + (\theta\theta)\overline{\alpha}_{\dot{A}}\overline{\lambda}^{\dot{A}}(x) + \underbrace{i(\theta\sigma^\mu\overline{\alpha})\theta^A \partial_\mu\phi_A(x)}_{=:T_1}$$

$$+(\overline{\theta}\,\overline{\theta})\alpha^A\psi_A(x)\underbrace{-i(\alpha\sigma^\mu\overline{\theta})\overline{\theta}_{\dot A}\partial_\mu\overline{\chi}^{\dot A}(x)}_{=:T_2}+(\theta\theta)\overline{\theta}_{\dot A}2\overline{\alpha}^{\dot A}d(x)$$

$$+\underbrace{i(\theta\sigma^\mu\overline{\alpha})(\theta\sigma^\nu\overline{\theta})\partial_\mu V_\nu(x)}_{=:T_3}\underbrace{-i(\alpha\sigma^\mu\overline{\theta})(\theta\theta)\partial_\mu m(x)}_{=:T_4}$$

$$+(\overline{\theta}\,\overline{\theta})\theta^A 2\alpha_A d(x)+i(\overline{\theta}\,\overline{\theta})\theta^A\sigma^\mu_{A\dot B}\overline{\alpha}^{\dot B}\partial_\mu n(x)$$

$$\underbrace{-i(\alpha\sigma^\mu\overline{\theta})(\theta\sigma^\nu\overline{\theta})\partial_\mu V_\nu(x)}_{=:T_5}+\underbrace{i(\theta\sigma^\mu\overline{\alpha})(\overline{\theta}\,\overline{\theta})\theta^A\partial_\mu\psi_A(x)}_{=:T_6}$$

$$\underbrace{-i(\alpha\sigma^\mu\overline{\theta})(\theta\theta)\overline{\theta}_{\dot A}\partial_\mu\overline{\lambda}^{\dot A}(x)}_{=:T_7}+2\alpha^A\theta_A\overline{\theta}_{\dot B}\overline{\lambda}^{\dot B}(x)$$

$$+2\overline{\alpha}^{\dot A}\overline{\theta}_{\dot A}\theta^B\psi_B(x)+i(\theta\sigma^\mu\overline{\alpha})\overline{\theta}_{\dot A}\partial_\mu\overline{\chi}^{\dot A}(x)$$

$$-i(\alpha\sigma^\mu\overline{\theta})\theta^A\partial_\mu\phi_A(x). \tag{6.79}$$

Now, since $\{\overline{\alpha}^{\dot A},\theta^B\}=0$, we have (the first equality)

$$
\begin{aligned}
T_1 &= i\theta^A\sigma^\mu_{A\dot A}\overline{\alpha}^{\dot A}\theta^B\partial_\mu\phi_B(x)=-i\theta^A\theta^B\sigma^\mu_{A\dot A}\overline{\alpha}^{\dot A}\partial_\mu\phi_B(x)\\
&\overset{(1.100a)}{=}\frac{1}{2}i\epsilon^{AB}(\theta\theta)\sigma^\mu_{A\dot A}\overline{\alpha}^{\dot A}\partial_\mu\phi_B(x)\\
&= (\theta\theta)\left[-\frac{i}{2}\partial_\mu\phi^A(x)\sigma^\mu_{A\dot A}\overline{\alpha}^{\dot A}\right]=(\theta\theta)\left[-\frac{i}{2}\partial_\mu\phi(x)\sigma^\mu\overline{\alpha}\right], \tag{6.80}
\end{aligned}
$$

where in the second last step we used $\{\overline{\alpha}^{\dot A},\phi_B\}=0$. We also have

$$
\begin{aligned}
T_2 &= -i(\alpha\sigma^\mu\overline{\theta})\overline{\theta}_{\dot A}\partial_\mu\overline{\chi}^{\dot A}(x)=-i\alpha^A\sigma^\mu_{A\dot B}\overline{\theta}^{\dot B}\overline{\theta}_{\dot A}\partial_\mu\overline{\chi}^{\dot A}(x)\\
&\overset{(1.85)}{=}-i\alpha^A\sigma^\mu_{A\dot B}\epsilon^{\dot B\dot C}\overline{\theta}_{\dot C}\overline{\theta}_{\dot A}\partial_\mu\overline{\chi}^{\dot A}(x)\overset{(1.100d)}{=}\frac{i}{2}\alpha^A\sigma^\mu_{A\dot B}\epsilon^{\dot B\dot C}\epsilon_{\dot C\dot A}(\overline{\theta}\,\overline{\theta})\partial_\mu\overline{\chi}^{\dot A}(x)\\
&= (\overline{\theta}\,\overline{\theta})\left\{\frac{i}{2}\alpha^A\sigma^\mu_{A\dot A}\partial_\mu\overline{\chi}^{\dot A}(x)\right\}=(\overline{\theta}\,\overline{\theta})\left\{\frac{i}{2}\alpha\sigma^\mu\partial_\mu\overline{\chi}(x)\right\}, \tag{6.81}
\end{aligned}
$$

where we used $\{\alpha^A,\overline{\theta}_{\dot A}\}=0$. Using Eqs. (1.97) and (1.137), we have

$$T_3=i(\theta\sigma^\mu\overline{\alpha})(\theta\sigma^\nu\overline{\theta})\partial_\mu V_\nu(x)=\frac{i}{2}(\theta\theta)(\overline{\theta}\overline{\alpha})\partial^\mu V_\mu(x). \tag{6.82}$$

Also:

$$
\begin{aligned}
T_4 &= -i(\alpha\sigma^\mu\overline{\theta})(\theta\theta)\partial_\mu m(x)=-i(\theta\theta)(\alpha^A\sigma^\mu_{A\dot B}\overline{\theta}^{\dot B})\partial_\mu m(x)\\
&\overset{(1.85)}{=}-i(\theta\theta)(\alpha^A\sigma^\mu_{A\dot C}\epsilon^{\dot C\dot B}\overline{\theta}_{\dot B})\partial_\mu m(x)=i(\theta\theta)\overline{\theta}_{\dot A}(\alpha\sigma^\mu\epsilon)^{\dot A}\partial_\mu m(x). \tag{6.83}
\end{aligned}
$$

Again from Eq. (1.137) we obtain:

$$T_5 = -i(\alpha\sigma^\mu\overline{\theta})(\theta\sigma^\nu\overline{\theta})\partial_\mu V_\nu(x) = -\frac{i}{2}(\overline{\theta}\,\overline{\theta})(\theta\alpha)\partial^\mu V_\mu(x). \qquad (6.84)$$

Furthermore:

$$
\begin{aligned}
T_6 &= i(\theta\sigma^\mu\overline{\alpha})(\overline{\theta}\,\overline{\theta})\theta^A\partial_\mu\psi_A(x) = i\theta^B\sigma^\mu_{B\dot{C}}\overline{\alpha}^{\dot{C}}(\overline{\theta}\,\overline{\theta})\theta^A\partial_\mu\psi_A(x) \\
&= -i\theta^B\theta^A(\overline{\theta}\,\overline{\theta})\sigma^\mu_{B\dot{C}}\overline{\alpha}^{\dot{C}}\partial_\mu\psi_A(x) = \frac{i}{2}(\theta\theta)(\overline{\theta}\,\overline{\theta})\epsilon^{BA}\sigma^\mu_{B\dot{C}}\overline{\alpha}^{\dot{C}}\partial_\mu\psi_A(x) \\
&\overset{(1.100a)}{\underset{(*)}{=}} \frac{i}{2}(\theta\theta)(\overline{\theta}\,\overline{\theta})\partial_\mu\psi_A(x)\epsilon^{AB}\sigma^\mu_{B\dot{C}}\overline{\alpha}^{\dot{C}} \\
&= \frac{i}{2}(\theta\theta)(\overline{\theta}\,\overline{\theta})\partial_\mu\psi^A(x)\sigma^\mu_{A\dot{C}}\overline{\alpha}^{\dot{C}} = (\theta\theta)(\overline{\theta}\,\overline{\theta})\left\{\frac{i}{2}(\partial_\mu\psi(x)\sigma^\mu\overline{\alpha})\right\}, \quad (6.85)
\end{aligned}
$$

in step (*) we used $\epsilon^{BA} = -\epsilon^{AB}$, $\{\overline{\alpha}, \psi\} = 0$, and finally

$$
\begin{aligned}
T_7 &= i(\alpha\sigma^\mu\overline{\theta})(\theta\theta)\overline{\theta}_{\dot{A}}\partial_\mu\overline{\lambda}^{\dot{A}}(x) = i(\theta\theta)\alpha^A\sigma^\mu_{A\dot{B}}\overline{\theta}^{\dot{B}}\overline{\theta}_{\dot{A}}\partial_\mu\overline{\lambda}^{\dot{A}}(x) \\
&= i(\theta\theta)\alpha^A\sigma^\mu_{A\dot{B}}\epsilon^{\dot{B}\dot{C}}\overline{\theta}_{\dot{C}}\overline{\theta}_{\dot{A}}\partial_\mu\overline{\lambda}^{\dot{A}}(x) = -\frac{i}{2}(\theta\theta)\alpha^A\sigma^\mu_{A\dot{B}}\epsilon^{\dot{B}\dot{C}}\epsilon_{\dot{C}\dot{A}}(\overline{\theta}\,\overline{\theta})\partial_\mu\overline{\lambda}^{\dot{A}}(x) \\
&= -\frac{i}{2}(\theta\theta)(\overline{\theta}\,\overline{\theta})\alpha^A\sigma^\mu_{A\dot{A}}\partial_\mu\overline{\lambda}^{\dot{A}}(x) = (\theta\theta)(\overline{\theta}\,\overline{\theta})\left\{-\frac{i}{2}\alpha\sigma^\mu\partial_\mu\overline{\lambda}(x)\right\}. \quad (6.86)
\end{aligned}
$$

Substituting expressions (6.80) to (6.86) into Eq. (6.79) we obtain

$$
\begin{aligned}
\delta_s\Phi(x,\theta,\overline{\theta}) &= \alpha^A\phi_A(x) + \overline{\alpha}_{\dot{A}}\overline{\chi}^{\dot{A}}(x) + \theta^A\{2\alpha_A m(x) + i(\sigma^\mu\overline{\alpha})_A\partial_\mu f(x) \\
&\quad + (\sigma^\nu\overline{\alpha})_A V_\nu(x)\} + \overline{\theta}_{\dot{A}}\{2\overline{\alpha}^{\dot{A}}n(x) + i(\alpha\sigma^\mu\epsilon)^{\dot{A}}\partial_\mu f(x) \\
&\quad - (\alpha\sigma^\nu\epsilon)^{\dot{A}}V_\nu(x)\} + (\theta\theta)\left\{\overline{\alpha}\overline{\lambda}(x) - \frac{i}{2}\partial_\mu\phi(x)\sigma^\mu\overline{\alpha}\right\} \\
&\quad + (\overline{\theta}\,\overline{\theta})\left\{\alpha\psi(x) + \frac{i}{2}\alpha\sigma^\mu\partial_\mu\overline{\chi}(x)\right\} + (\theta\theta)\overline{\theta}_{\dot{A}}\left\{2\overline{\alpha}^{\dot{A}}d(x)\right. \\
&\quad \left. + \frac{i}{2}\overline{\alpha}^{\dot{A}}\partial^\mu V_\mu(x) + i(\alpha\sigma^\mu\epsilon)^{\dot{A}}\partial_\mu m(x)\right\} + (\overline{\theta}\,\overline{\theta})\theta^A\left\{2\alpha_A d(x)\right. \\
&\quad \left. + i(\sigma^\mu\overline{\alpha})_A\partial_\mu n(x) - \frac{i}{2}\alpha_A\partial^\mu V_\mu(x)\right\} \\
&\quad + (\theta\theta)(\overline{\theta}\,\overline{\theta})\frac{i}{2}[\partial_\mu\psi(x)\sigma^\mu\overline{\alpha} - \alpha\sigma^\mu\partial_\mu\overline{\lambda}(x)] + 2(\alpha\theta)\overline{\theta}\,\overline{\lambda}(x) \\
&\quad + 2(\overline{\alpha}\overline{\theta})\theta\psi(x) + i(\theta\sigma^\mu\overline{\alpha})\overline{\theta}\partial_\mu\overline{\chi}(x) - i(\alpha\sigma^\mu\overline{\theta})\theta\partial_\mu\phi(x).
\end{aligned}
$$

$$(6.87)$$

We now *Fierz transform* the last four terms of this Eq. (6.87). Thus

$$2(\alpha\theta)(\theta\overline{\lambda}(x)) \quad = \quad 2(\alpha\theta)\overline{\theta}_{\dot{A}}\overline{\lambda}^{\dot{A}}(x) \overset{(1.97)}{=} 2(\theta\alpha)\overline{\theta}_{\dot{A}}\overline{\lambda}^{\dot{A}}(x)$$
$$\overset{(1.133)}{=} \quad (\theta\sigma^\mu\overline{\theta})(\alpha\sigma_\mu\overline{\lambda}(x)). \tag{6.88}$$

In order to *Fierz transform* the term

$$(\overline{\alpha}\,\overline{\theta})(\theta\psi(x)),$$

we require a formula analogous to Eq. (1.133). Let $\overline{\phi}, \overline{\psi}$ and χ be two-component Weyl spinors. Then

$$
\begin{aligned}
(\overline{\phi}\,\overline{\psi})\chi^B \quad &= \quad (\overline{\phi}_{\dot{A}}\overline{\psi}^{\dot{A}})\chi^B = \overline{\phi}_{\dot{A}}\epsilon^{\dot{A}\dot{C}}\overline{\psi}_{\dot{C}}\epsilon^{BD}\chi_D \\
&= \quad \overline{\phi}_{\dot{A}}\epsilon^{\dot{A}\dot{C}}\delta_{\dot{C}}^{\dot{D}}\overline{\psi}_{\dot{D}}\epsilon^{BD}\delta_D^F\chi_F = \overline{\phi}_{\dot{A}}\epsilon^{\dot{A}\dot{C}}\epsilon^{BD}\overline{\psi}_{\dot{D}}\delta_{\dot{C}}^{\dot{D}}\delta_D^F\chi_F \\
&\overset{(1.110)}{=} \quad \frac{1}{2}\overline{\phi}_{\dot{A}}\epsilon^{\dot{A}\dot{C}}\epsilon^{BD}\overline{\psi}_{\dot{D}}\sigma_{D\dot{C}}^{\mu}\overline{\sigma}_\mu^{\dot{D}F}\chi_F = \frac{1}{2}\overline{\psi}_{\dot{D}}\overline{\sigma}^{\dot{D}F}\chi_F\overline{\phi}_{\dot{A}}\epsilon^{\dot{A}\dot{C}}\epsilon^{BD}\sigma_{D\dot{C}}^{\mu} \\
&\overset{(1.106a)}{=} \quad \frac{1}{2}(\overline{\psi}\overline{\sigma}_\mu\chi)\overline{\phi}_{\dot{A}}\overline{\sigma}^{\mu\dot{A}B} = -\frac{1}{2}(\chi\sigma_\mu\overline{\psi})(\overline{\phi}\overline{\sigma}^\mu)^B,
\end{aligned}
\tag{6.89}
$$

where we used Eq. (1.134) in the last step. Hence (using Eqs. (6.89) and (1.134))

$$2(\overline{\alpha}\,\overline{\theta})(\theta\psi(x)) = -(\theta\sigma^\mu\overline{\theta})(\overline{\alpha}\,\overline{\sigma}_\mu\psi(x)) = (\theta\sigma^\mu\overline{\theta})(\psi(x)\sigma_\mu\overline{\alpha}). \tag{6.90}$$

The third term we Fierz transform is

$$
\begin{aligned}
-i(\alpha\sigma^\mu\overline{\theta})(\theta\partial_\mu\phi(x)) &= -i\alpha^A\sigma_{A\dot{B}}^\mu\overline{\theta}^{\dot{B}}\theta^C\partial_\mu\phi_C(x) \\
&= \quad -i\alpha^A\sigma_{A\dot{B}}^\mu\theta^C\partial_\mu\phi_C(x)\overline{\theta}^{\dot{B}} = -i\alpha^A\sigma_{A\dot{B}}^\mu\epsilon^{\dot{B}\dot{C}}\theta^C\partial_\mu\phi_C(x)\overline{\theta}_{\dot{C}} \\
&\overset{(1.133)}{=} \quad -\frac{i}{2}\alpha^A\sigma_{A\dot{B}}^\mu\epsilon^{\dot{B}\dot{C}}(\theta\sigma_\rho\overline{\theta})(\partial_\mu\phi(x)\sigma^\rho)_{\dot{C}} \\
&= \quad -\frac{i}{2}(\theta\sigma_\rho\overline{\theta})\alpha^A\sigma_{A\dot{B}}^\mu\epsilon^{\dot{B}\dot{C}}\partial_\mu\phi^B(x)\sigma_{B\dot{C}}^\rho \\
&= \quad -\frac{i}{2}(\theta\sigma_\rho\overline{\theta})\alpha^A\partial_\mu\phi^B(x)\sigma_{A\dot{B}}^\mu\epsilon^{\dot{B}\dot{C}}\sigma_{B\dot{C}}^\rho \\
&\overset{(1.100a)}{=} \quad \frac{i}{4}(\theta\sigma_\rho\overline{\theta})(\alpha\partial_\mu\phi(x))\epsilon^{AB}\sigma_{A\dot{B}}^\mu\epsilon^{\dot{B}\dot{C}}\sigma_{B\dot{C}}^\rho \\
&= \quad -\frac{i}{4}(\theta\sigma_\rho\overline{\theta})(\alpha\partial_\mu\phi(x))\epsilon^{BA}\sigma_{A\dot{B}}^\mu\epsilon^{\dot{B}\dot{C}}\sigma_{B\dot{C}}^\rho \\
&\overset{(1.107)}{=} \quad \frac{i}{4}(\theta\sigma_\rho\overline{\theta})(\alpha\partial_\mu\phi(x))(\overline{\sigma}^\mu)^{\dot{C}B}(\sigma^\rho)_{B\dot{C}} \\
&= \quad \frac{i}{4}(\theta\sigma_\rho\overline{\theta})(\alpha\partial_\mu\phi(x))\mathrm{Tr}[\overline{\sigma}^\mu\sigma^\rho] \overset{(1.109a)}{=} \frac{i}{2}(\theta\sigma_\rho\overline{\theta})(\alpha\partial_\mu\phi(x))\eta^{\mu\rho} \\
&= \quad \frac{i}{2}(\theta\sigma^\mu\overline{\theta})(\alpha\partial_\mu\phi(x)). \tag{6.91}
\end{aligned}
$$

The last contribution we want to Fierz transform is

$$i(\theta\sigma^\mu\overline{\alpha})(\overline{\theta}\partial_\mu\overline{\chi}(x)) = i\theta^A\sigma^\mu_{A\dot{B}}\overline{\alpha}^{\dot{B}}(\overline{\theta}\partial_\mu\overline{\chi}(x))$$

$$\overset{(1.98)}{=} -i\sigma^\mu_{A\dot{B}}\overline{\alpha}^{\dot{B}}(\partial_\mu\overline{\chi}(x)\overline{\theta})\theta^A \overset{(6.89)}{=} \frac{i}{2}\sigma^\mu_{A\dot{B}}\overline{\alpha}^{\dot{B}}(\theta\sigma_\nu\overline{\theta})(\partial_\mu\overline{\chi}(x)\overline{\sigma}^\nu)^A$$

$$= -\frac{i}{2}(\theta\sigma_\nu\overline{\theta})\partial_\mu\overline{\chi}_{\dot{A}}(x)\overline{\sigma}^{\nu\dot{A}A}\sigma^\mu_{A\dot{B}}\overline{\alpha}^{\dot{B}}$$

$$= -\frac{i}{2}(\theta\sigma_\nu\overline{\theta})\partial_\mu\overline{\chi}_{\dot{A}}(x)\overline{\alpha}_{\dot{C}}\overline{\sigma}^{\nu\dot{A}A}\sigma^\mu_{A\dot{B}}\epsilon^{\dot{B}\dot{C}}$$

$$\overset{(1.100d)}{=} \frac{i}{4}(\theta\sigma_\nu\overline{\theta})(\partial_\mu\overline{\chi}(x)\overline{\alpha})\epsilon_{\dot{A}\dot{C}}\overline{\sigma}^{\nu\dot{A}A}\sigma^\mu_{A\dot{B}}\epsilon^{\dot{B}\dot{C}}$$

$$= -\frac{i}{4}(\theta\sigma_\nu\overline{\theta})(\partial_\mu\overline{\chi}(x)\overline{\alpha})\overline{\sigma}^{\nu\dot{A}A}\sigma^\mu_{A\dot{B}}\delta^{\dot{B}}_{\dot{A}}$$

$$= -\frac{i}{4}(\theta\sigma_\nu\overline{\theta})(\partial_\mu\overline{\chi}(x)\overline{\alpha})\text{Tr}[\overline{\sigma}^\nu\sigma^\mu] \overset{(1.109a)}{=} -\frac{i}{2}(\theta\sigma_\nu\overline{\theta})(\partial_\mu\overline{\chi}(x)\overline{\alpha})\eta^{\nu\mu}$$

$$= -\frac{i}{2}(\theta\sigma_\mu\overline{\theta})(\partial^\mu\overline{\chi}(x)\overline{\alpha}). \tag{6.92}$$

This result could have been guessed from the *Fierz formula* (1.101). Hence the final expression for $\delta_s\Phi$ is given by

$$\begin{aligned}
\delta_s\Phi(x,\theta,\overline{\theta}) = \ & \alpha\phi(x) + \overline{\alpha}\,\overline{\chi}(x) \\
& +\theta\Big\{2\alpha m(x) + i(\sigma^\mu\overline{\alpha})\partial_\mu f(x) + (\sigma^\mu\overline{\alpha})V_\mu(x)\Big\} \\
& +\overline{\theta}\Big\{2\overline{\alpha}n(x) + i(\alpha\sigma^\mu\epsilon)\partial_\mu f(x) - (\alpha\sigma^\mu\epsilon)V_\mu(x)\Big\} \\
& +(\theta\theta)\Big\{\overline{\alpha}\,\overline{\lambda}(x) - \frac{i}{2}\partial_\mu\phi(x)\sigma^\mu\overline{\alpha}\Big\} \\
& +(\overline{\theta}\,\overline{\theta})\Big\{\alpha\psi(x) + \frac{i}{2}\alpha\sigma^\mu\partial_\mu\overline{\chi}(x)\Big\} \\
& +(\theta\sigma^\mu\overline{\theta})\Big\{\alpha\sigma_\mu\overline{\lambda}(x) + \psi(x)\sigma_\mu\overline{\alpha} + \frac{i}{2}\alpha\partial_\mu\phi(x) - \frac{i}{2}\partial_\mu\overline{\chi}(x)\overline{\alpha}\Big\} \\
& +(\theta\theta)\overline{\theta}\Big\{2\overline{\alpha}d(x) + \frac{i}{2}\overline{\alpha}\partial^\mu V_\mu(x) + i(\alpha\sigma^\mu\epsilon)\partial_\mu m(x)\Big\} \\
& +(\overline{\theta}\,\overline{\theta})\theta\Big\{2\alpha d(x) - \frac{i}{2}\alpha\partial^\mu V_\mu(x) + i(\sigma^\mu\overline{\alpha})\partial_\mu n(x)\Big\} \\
& +(\theta\theta)(\overline{\theta}\,\overline{\theta})\frac{i}{2}\Big\{\partial_\mu\psi(x)\sigma^\mu\overline{\alpha} + \alpha\sigma^\mu\partial_\mu\overline{\lambda}(x)\Big\}. \tag{6.93}
\end{aligned}$$

Comparing the coefficients of the same powers of θ in Eqs. (6.78) and (6.93), we obtain the transformation properties of the component fields. Thus

$$\delta_s' f(x) = \alpha\phi(x) + \overline{\alpha}\,\overline{\chi}(x),$$

$$\delta_s' \phi_A(x) = 2\alpha_A m(x) + (\sigma^\mu\overline{\alpha})_A\{i\partial_\mu f(x) + V_\mu(x)\},$$

$$\delta_s' \overline{\chi}^{\dot{A}}(x) = 2\overline{\alpha}^{\dot{A}} n(x) + (\alpha\sigma^\mu\epsilon)^{\dot{A}}\{i\partial_\mu f(x) - V_\mu(x)\},$$

$$\delta_s' m(x) = \overline{\alpha}\overline{\lambda}(x) - \frac{i}{2}\partial_\mu\phi(x)\sigma^\mu\overline{\alpha},$$

$$\delta_s' n(x) = \alpha\psi(x) + \frac{i}{2}\alpha\sigma^\mu\partial_\mu\overline{\chi}(x),$$

$$\delta_s' V_\mu(x) = \alpha\sigma_\mu\overline{\lambda}(x) + \psi(x)\sigma_\mu\overline{\alpha} + \frac{i}{2}\alpha\partial_\mu\phi(x) - \frac{i}{2}\partial_\mu\overline{\chi}(x)\overline{\alpha},$$

$$\delta_s' \overline{\lambda}^{\dot{A}}(x) = 2\overline{\alpha}^{\dot{A}} d(x) + \frac{i}{2}\overline{\alpha}^{\dot{A}}\partial^\mu V_\mu(x) + i(\alpha\sigma^\mu\epsilon)^{\dot{A}}\partial_\mu m(x),$$

$$\delta_s' \psi_A(x) = 2\alpha_A d(x) - \frac{i}{2}\alpha_A\partial^\mu V_\mu(x) + i(\sigma^\mu\overline{\alpha})_A\partial_\mu n(x),$$

$$\delta_s' d(x) = \frac{i}{2}\partial_\mu\psi(x)\sigma^\mu\overline{\alpha} + \frac{i}{2}\alpha\sigma^\mu\partial_\mu\overline{\lambda}(x)$$

$$\overset{(1.134)}{=} \frac{i}{2}\partial_\mu\psi(x)\sigma^\mu\overline{\alpha} - \frac{i}{2}\partial_\mu\overline{\lambda}(x)\overline{\sigma}^\mu\alpha. \qquad (6.94)$$

We make the very important observation that $\delta_s' d(x)$ is a total derivative (the significance of this observation will be seen later).

The set of Eqs. (6.94) gives the general transformation laws of the component fields under supersymmetry transformations. We shall see for special examples that linear combinations of superfields are again superfields because the generators Q and $\overline{Q}$ are linear differential operators. This means that superfields form linear representations of the supersymmetry algebra. In general the representations are highly reducible. However, one can eliminate the extra component fields by imposing *covariant constraints*. In this way superfields replace the problem of finding supersymmetry representations via the *Casimir invariants* by that of finding appropriate constraints. Such covariant constraints are, for example,

$$\overline{D}_{\dot{A}}\Phi(x,\theta,\overline{\theta}) = 0, \qquad (6.95)$$

$$D_A\Phi^\dagger(x,\theta,\overline{\theta}) = 0, \qquad (6.96)$$

$$\Phi(x,\theta,\overline{\theta}) = \Phi^\dagger(x,\theta,\overline{\theta}). \qquad (6.97)$$

Superfields satisfying the constraint (6.95) or (6.96) are called *chiral* or *scalar superfields*; superfields obeying the reality condition (6.97) are called *vector superfields*. The constraints (6.95), (6.96) can be expressed in terms of projection operators (6.74), (6.75).

Of course, a general superfield Φ can also be expressed in terms of four-component Grassmann variables θ_a which obey the Majorana condition, *i.e.* we can write

$$\Phi(x,\theta) = A(x) + \bar{\theta}\psi(x) + (\bar{\theta}\theta)F(x) + i\bar{\theta}\gamma^5\theta G(x) + \bar{\theta}\gamma^\mu\gamma^5\theta A_\mu(x)$$
$$+ (\bar{\theta}\theta)\bar{\theta}\chi(x) + (\bar{\theta}\theta)(\bar{\theta}\theta)D(x).$$

Now, of course, powers of θ higher than the fourth are zero, *i.e.*

$$\theta^n = 0, \quad \text{for} \quad n \geq 5.$$

Terms such as

$$\bar{\theta}\gamma^\mu\theta, \ \bar{\theta}\sigma^{\mu\nu}\theta,$$

do not appear in the expansion. They vanish on account of relations (1.209c), (1.209e) and (1.134), (1.151).

We have seen that the covariant derivatives D and $\overline{D}$ satisfy an algebra similar to that of the generators Q and $\overline{Q}$, and we have also seen that the latter act as Fock space annihilation and creation operators with respect to a Clifford vacuum. Equation (6.95) shows that $\overline{D}$ mimics this behaviour with respect to the field Φ. The somewhat ad hoc way of introducing the chirality constraints (6.95), (6.96) and the constraint (6.97) has been criticized [36] in view of their unclear connection with well defined particle representations, and a direct derivation of the fields from known irreducible particle representations has been given in the literature [36].

Chapter 7

Constrained Superfields and Supermultiplets

7.1 Chiral Superfields

Superfields which satisfy either of the constraints (6.95), (6.96), are called *chiral* or *scalar superfields*. A superfield $\Phi(x, \theta, \overline{\theta})$ which satisfies Eq. (6.95), *i.e.* $\overline{D}\Phi = 0$, is called a *left-handed chiral superfield*, and a superfield $\Phi(x, \theta, \overline{\theta})$ which satisfies Eq. (6.96), *i.e.* $D\Phi^\dagger = 0$, is called a *right-handed chiral superfield*. The origin of this terminology has been mentioned earlier, but will also become clear in the following.

We consider first left-handed chiral superfields. The constraint

$$\overline{D}_{\dot{A}}\Phi(x, \theta, \overline{\theta}) = 0, \qquad \overline{D}_{\dot{A}} = -\overline{\partial}_{\dot{A}} - i\theta^B \sigma^\mu_{B\dot{A}} \partial_\mu, \qquad (7.1)$$

(recall Eq. (6.42)) is easily solved in terms of new variables given by

$$\left. \begin{array}{rclcl} y^\mu & := & x^\mu + i\theta\sigma^\mu\overline{\theta}, & & \\ \theta'_A, & := & \theta_A, & \overline{\theta}'_{\dot{A}} := \overline{\theta}_{\dot{A}}. & \end{array} \right\} \qquad (7.2)$$

Proposition 7.1:

The new variables (7.2) satisfy the following conditions:

$$\begin{array}{lll} \text{(a)} & \overline{D}_{\dot{A}} y^\mu & = & 0, \\ \text{(b)} & \overline{D}_{\dot{A}} \theta_A & = & 0. \end{array} \qquad (7.3)$$

Proof: (a) We have

$$\overline{D}_{\dot{A}} y^\mu = \left\{ -\overline{\partial}_{\dot{A}} - i\theta^A \sigma^\nu_{A\dot{A}} \frac{\partial}{\partial x^\nu} \right\} (x^\mu + i\theta\sigma^\mu\overline{\theta}) = i\theta^B \sigma^\mu_{B\dot{A}} - i\theta^A \sigma^\nu_{A\dot{A}} \delta^\mu_\nu = 0,$$

since $\{\overline{\partial}_{\dot{A}}, \theta\} = 0$.

(b) This result is obvious.

Hence, in view of Eq. (7.3) any function of y^μ and θ satisfies the constraint (7.1). We now change the variables of the covariant derivatives.

Proposition 7.2:

In terms of the new set of variables (7.2), the covariant derivatives $D_A, \overline{D}_{\dot{A}}$ of Eqs. (6.41) and (6.42) become

$$D_A \quad \longrightarrow \quad D_A^{(1)} = \partial_A + 2i\sigma^\mu_{A\dot{A}} \overline{\theta}^{\dot{A}} \frac{\partial}{\partial y^\mu},$$

$$\overline{D}_{\dot{A}} \quad \longrightarrow \quad \overline{D}_{\dot{A}}^{(1)} = -\overline{\partial}_{\dot{A}}.$$

Verification: The covariant derivatives in the variables $x, \theta, \overline{\theta}$ are given by Eqs. (6.41) and (6.42), *i.e.*

$$D_A(x, \theta, \overline{\theta}) = \frac{\partial}{\partial \theta^A} + i\sigma^\mu_{A\dot{A}} \overline{\theta}^{\dot{A}} \frac{\partial}{\partial x^\mu},$$

$$\overline{D}_{\dot{A}}(x, \theta, \overline{\theta}) = -\frac{\partial}{\partial \overline{\theta}^{\dot{A}}} - i\theta^A \sigma^\mu_{A\dot{A}} \frac{\partial}{\partial x^\mu}.$$

Transforming these operators according to Eq. (7.2) we have

$$\frac{\partial}{\partial x^\mu} = \frac{\partial y^\nu}{\partial x^\mu} \frac{\partial}{\partial y^\nu} = \delta^\nu_\mu \frac{\partial}{\partial y^\nu} = \frac{\partial}{\partial y^\mu},$$

$$\frac{\partial}{\partial \theta^A} = \frac{\partial \theta'^B}{\partial \theta^A} \frac{\partial}{\partial \theta'^B} + \frac{\partial y^\mu}{\partial \theta^A} \frac{\partial}{\partial y^\mu} = \delta^B_A \frac{\partial}{\partial \theta'^B} + i\sigma^\mu_{A\dot{A}} \overline{\theta}'^{\dot{A}} \frac{\partial}{\partial y^\mu}$$

$$= \frac{\partial}{\partial \theta'^A} + i\sigma^\mu_{A\dot{A}} \overline{\theta}'^{\dot{A}} \frac{\partial}{\partial y^\mu} \qquad (\text{since} \quad \overline{\theta}' = \overline{\theta}),$$

$$\frac{\partial}{\partial \overline{\theta}^{\dot{A}}} = \frac{\partial \overline{\theta}'^{\dot{B}}}{\partial \overline{\theta}^{\dot{A}}} \frac{\partial}{\partial \overline{\theta}'^{\dot{B}}} + \frac{\partial y^\mu}{\partial \overline{\theta}^{\dot{A}}} \frac{\partial}{\partial y^\mu} = \frac{\partial}{\partial \overline{\theta}'^{\dot{A}}} - i\theta'^A \sigma^\mu_{A\dot{A}} \frac{\partial}{\partial y^\mu}.$$

Hence

$$D_A(x, \theta, \overline{\theta}) = \frac{\partial}{\partial \theta^A} + i\sigma^\mu_{A\dot{A}} \overline{\theta}^{\dot{A}} \frac{\partial}{\partial x^\mu}$$

$$= \frac{\partial}{\partial \theta'^A} + i\sigma^\mu_{A\dot{A}} \overline{\theta}'^{\dot{A}} \frac{\partial}{\partial y^\mu} + i\sigma^\mu_{A\dot{A}} \overline{\theta}'^{\dot{A}} \frac{\partial}{\partial y^\mu}$$

$$= \frac{\partial}{\partial \theta'^A} + 2i\sigma^\mu_{A\dot{A}} \overline{\theta}'^{\dot{A}} \frac{\partial}{\partial y^\mu} \overset{(6.48)}{=} D_A^{(1)}(y, \theta', \overline{\theta}'), \qquad (7.4\text{a})$$

and

$$
\begin{aligned}
\overline{D}_{\dot{A}}(x,\theta,\overline{\theta}) &= -\frac{\partial}{\partial\overline{\theta}^{\dot{A}}} - i\theta^A\sigma^\mu_{A\dot{A}}\frac{\partial}{\partial x^\mu} \\
&= -\frac{\partial}{\partial\overline{\theta}'^{\dot{A}}} + i\theta'^A\sigma^\mu_{A\dot{A}}\frac{\partial}{\partial y^\mu} - i\theta'^A\sigma^\mu_{A\dot{A}}\frac{\partial}{\partial y^\mu} = -\frac{\partial}{\partial\overline{\theta}'^{\dot{A}}} \\
&= \overline{D}_{\dot{A}}^{(1)}(y,\theta',\overline{\theta}').
\end{aligned} \tag{7.4b}
$$

In this representation the constraint (7.1) has the simple implication that

$$
\Phi(x,\theta,\overline{\theta}) = \Phi(y - i\theta\sigma\overline{\theta},\theta,\overline{\theta}) \equiv \Phi_1(y,\theta,\overline{\theta})
$$

(see Eq. (6.8)) does not possess any explicit dependence on $\overline{\theta}$, *i.e.*

$$
\overline{D}_{\dot{A}}^{(1)}\Phi_1(y,\theta,\overline{\theta}) = -\frac{\partial}{\partial\overline{\theta}^{\dot{A}}}\Phi_1(y,\theta,\overline{\theta}) = 0
$$

by Eq. (7.1), so that $\Phi_1(y,\theta,\overline{\theta}) \equiv \Phi_1(y,\theta)$, implying that Φ_1 has the power series expansion in θ:

$$
\Phi_1(y,\theta) = A(y) + \sqrt{2}\theta\psi(y) + (\theta\theta)F(y), \tag{7.5}
$$

and is independent of $\overline{\theta}$. From Eq. (7.5) we see that a superfield Φ_1 satisfying the constraint (7.1) depends on the two-component Weyl spinor $\psi(y)$ which, according to Eq. (1.60), is a left-handed Weyl spinor. This dependence of the superfield on the left-handed Weyl spinor $\psi(y)$ is the origin of the name *"left-handed chiral superfield"*.

In Eq. (7.5) the functions $A(y)$ and $F(y)$ describe complex scalar fields (A and F have no well defined parity, the change of parity being given by complex conjugation $A \to A^*, F \to F^*$ as explained earlier). From Eq. (7.5) we regain the original field $\Phi(x,\theta,\overline{\theta})$ satisfying the constraint (7.1) in the variables $x,\theta,\overline{\theta}$ by expanding the component fields in the following way. We have

$$
\begin{aligned}
\Phi_1(y,\theta) &= A(y) + \sqrt{2}\theta\psi(y) + (\theta\theta)F(y) \\
&= A(x + i\theta\sigma\overline{\theta}) + \sqrt{2}\theta\psi(x + i\theta\sigma\overline{\theta}) + (\theta\theta)F(x + i\theta\sigma\overline{\theta}) \\
&= A(x) + i\theta\sigma^\mu\overline{\theta}\partial_\mu A(x) - \frac{1}{2}(\theta\sigma_\mu\overline{\theta})(\theta\sigma_\nu\overline{\theta})\partial^\mu\partial^\nu A(x) \\
&\quad + \sqrt{2}\theta^A\psi_A(x) + \sqrt{2}i\theta^A(\theta\sigma^\mu\overline{\theta})\partial_\mu\psi_A(x) + (\theta\theta)F(x) \\
&= A(x) + i(\theta\sigma^\mu\overline{\theta})\partial_\mu A(x) - \frac{1}{4}(\theta\theta)(\overline{\theta}\,\overline{\theta})\Box A(x) + \sqrt{2}\theta^A\psi_A(x) \\
&\quad + \frac{i}{\sqrt{2}}(\theta\theta)\overline{\theta}_{\dot{A}}\partial_\mu\psi^A(x)\sigma^\mu_{A\dot{B}}\epsilon^{\dot{B}\dot{A}} + (\theta\theta)F(x),
\end{aligned} \tag{7.6}
$$

where we used Eqs. (1.137), (1.80) and (1.100a), and the term containing $\partial_\mu \psi_A$ was rewritten in the following way:

$$i\theta^A(\theta\sigma^\mu\overline{\theta})\partial_\mu\psi_A = i\theta^A(\theta^B\sigma^\mu_{B\dot{B}}\overline{\theta}^{\dot{B}})\partial_\mu\psi_A \overset{(1.100a)}{=} i\{-\frac{1}{2}\epsilon^{AB}(\theta\theta)\}\sigma^\mu_{B\dot{B}}\overline{\theta}^{\dot{B}}\partial_\mu\psi_A$$

$$= -\frac{i}{2}(\theta\theta)\overline{\theta}^{\dot{B}}\partial_\mu\psi_A\epsilon^{AB}\sigma^\mu_{B\dot{B}} = \frac{i}{2}(\theta\theta)\overline{\theta}_{\dot{C}}\partial_\mu\psi^B\sigma^\mu_{B\dot{B}}\epsilon^{\dot{B}\dot{C}}$$

$$= \frac{i}{2}(\theta\theta)\overline{\theta}_{\dot{A}}\partial_\mu\psi^A(x)\sigma^\mu_{A\dot{B}}\epsilon^{\dot{B}\dot{A}}.$$

As a consistency check we demonstrate (using Eq. (6.42)) the vanishing of

$$\overline{D}_{\dot{A}}\Phi(x,\theta,\overline{\theta}) = \{-\overline{\partial}_{\dot{A}} - i\theta^A\sigma^\mu_{A\dot{A}}\partial_\mu\}\Big\{A(x) + i(\theta\sigma^\mu\overline{\theta})\partial_\mu A(x) + \sqrt{2}\theta^A\psi_A(x)$$

$$-\frac{1}{4}(\theta\theta)(\overline{\theta}\,\overline{\theta})\Box A(x) + (\theta\theta)F(x) + \frac{i}{\sqrt{2}}(\theta\theta)\partial_\rho\psi(x)\sigma^\rho\overline{\theta}\Big\}$$

$$= i\theta^A\sigma^\rho_{A\dot{A}}\partial_\rho A(x) - \frac{1}{2}(\theta\theta)\overline{\theta}_{\dot{A}}\Box A(x) - \frac{i}{\sqrt{2}}(\theta\theta)\partial_\rho\psi^A(x)\sigma^\rho_{A\dot{A}}$$

$$-i\theta^A\sigma^\mu_{A\dot{A}}\partial_\mu A(x) + \theta^A\sigma^\mu_{A\dot{A}}(\theta\sigma^\rho\overline{\theta})\partial_\mu\partial_\rho A(x)$$

$$-i\sqrt{2}\theta^A\sigma^\mu_{A\dot{A}}\theta^B\partial_\mu\psi_B(x), \tag{7.7}$$

where we used $\theta^3 = 0$ and where

$$-\overline{\partial}_{\dot{A}}(\theta\theta)(\overline{\theta}\,\overline{\theta}) \overset{(6.5m)}{=} -(\theta\theta)\overline{\partial}_{\dot{A}}(\overline{\theta}_{\dot{B}}\overline{\theta}^{\dot{B}}) \overset{(6.5b)}{=} -(\theta\theta)\{(\overline{\partial}_{\dot{A}}\overline{\theta}_{\dot{B}})\overline{\theta}^{\dot{B}} - \overline{\theta}_{\dot{B}}(\overline{\partial}_{\dot{A}}\overline{\theta}^{\dot{B}})\}$$

$$\overset{(6.5k)}{=} -(\theta\theta)\{-\epsilon_{\dot{A}\dot{B}}\overline{\theta}^{\dot{B}} - \overline{\theta}_{\dot{B}}\delta_{\dot{A}}^{\dot{B}}\} = 2(\theta\theta)\overline{\theta}_{\dot{A}},$$

and therefore

$$-\overline{\partial}_{\dot{A}}\Big\{-\frac{1}{4}(\theta\theta)(\overline{\theta}\,\overline{\theta})\Box A(x)\Big\} = -\frac{1}{2}(\theta\theta)\overline{\theta}_{\dot{A}}\Box A(x).$$

Using Eq. (1.137) in the form

$$(\theta\sigma^\rho\overline{\theta})(\theta\sigma^\mu)_{\dot{A}} = \frac{1}{2}\eta^{\rho\mu}(\theta\theta)\overline{\theta}_{\dot{A}}, \tag{7.8}$$

and Eq. (1.100a), we may rewrite Eq. (7.7) as

$$\overline{D}_{\dot{A}}\Phi(x,\theta,\overline{\theta}) = -\frac{1}{2}(\theta\theta)\overline{\theta}_{\dot{A}}\Box A(x) - \frac{i}{\sqrt{2}}(\theta\theta)(\partial_\rho\psi(x)\sigma^\rho)_{\dot{A}}$$

$$+\frac{1}{2}(\theta\theta)\overline{\theta}_{\dot{A}}\Box A(x) + \frac{i}{\sqrt{2}}(\theta\theta)(\partial_\mu\psi(x)\sigma^\mu)_{\dot{A}} = 0.$$

This result states that the superfield (7.6) (with variables $x, \theta, \overline{\theta}$) is the most general solution of Eq. (7.1). It should be noted that the superfield (7.6) obtained in this way depends only on the fields $A(x), F(x)$ and $\psi(x)$, a dependence which is characteristic for a left-handed chiral superfield.

The superfield $\Phi^\dagger(x, \theta, \overline{\theta})$ satisfies the constraint

$$D_A \Phi^\dagger(x, \theta, \overline{\theta}) = 0, \tag{7.9}$$

and is called a *right-handed chiral superfield*. Again we make a shift of variables to solve the constraint equation (7.9) by introducing the variables

$$\left.\begin{aligned}
z^\mu &:= x^\mu - i\theta\sigma^\mu\overline{\theta}, \\
\theta'_A &:= \theta_A, \quad \overline{\theta}'_{\dot{A}} := \overline{\theta}_{\dot{A}}.
\end{aligned}\right\} \tag{7.10}$$

Proposition 7.3:

The new variables (7.10) satisfy the following conditions:

$$\text{(a)} \quad D_A z^\mu = 0, \quad \text{(b)} \quad D_A \overline{\theta}_{\dot{A}} = 0, \tag{7.11}$$

where (*cf.* Eq. (6.41))

$$D_A = \partial_A + i\sigma^\mu_{A\dot{A}}\overline{\theta}^{\dot{A}}\frac{\partial}{\partial x^\mu}.$$

Proof: (a) We have

$$D_A z^\mu = (\partial_A + i\sigma^\mu_{A\dot{A}}\overline{\theta}^{\dot{A}}\partial_\rho)(x^\mu - i\theta\sigma^\mu\overline{\theta}) = -i\sigma^\mu_{A\dot{A}}\overline{\theta}^{\dot{A}} + i\sigma^\rho_{A\dot{A}}\overline{\theta}^{\dot{A}}\frac{\partial x^\mu}{\partial x^\rho} = 0.$$

(b) This result is obvious. Hence, in view of Eq. (7.11) any function of z^μ and $\overline{\theta}$ satisfies the constraint (7.9). We now change the variables of the covariant derivatives.

Proposition 7.4:

In terms of the new set of variables (7.10) the covariant derivatives $D_A, \overline{D}_{\dot{A}}$ of Eqs. (6.41) and (6.42) become:

$$\begin{aligned}
D_A &\to D_A^{(2)} = \partial_A, \\
\overline{D}_{\dot{A}} &\to \overline{D}_{\dot{A}}^{(2)} = -\overline{\partial}_{\dot{A}} - 2i\theta^A\sigma^\mu_{A\dot{A}}\frac{\partial}{\partial z^\mu}, \\
D_A^{(2)} &\equiv D_A^{(2)}(z, \theta, \overline{\theta}), \\
\overline{D}_{\dot{A}}^{(2)} &\equiv \overline{D}_{\dot{A}}^{(2)}(z, \theta, \overline{\theta}).
\end{aligned} \tag{7.12}$$

Verification: The expressions (7.12) are easily verified in analogy to Eqs. (7.4a) and (7.4b). In view of Eq. (7.11) we can set:

$$\Phi^\dagger(z,\overline{\theta}) = A^*(z) + \sqrt{2}\,\overline{\theta}\,\overline{\psi}(z) + (\overline{\theta}\,\overline{\theta})F^*(z). \tag{7.13}$$

It should be observed that $\Phi^\dagger$ which is the Hermitian conjugate of the left-handed chiral superfield Φ is a type-2 superfield (*cf.* Eq. (6.12)). In this representation we have, of course,

$$D_A^{(2)}\Phi^\dagger(z,\overline{\theta}) = 0 = D_A^{(2)}\Phi_2(z,\overline{\theta}),$$

expressing the fact that $\Phi^\dagger(z,\overline{\theta})$ does not depend explicitly on θ. Expanding Eq. (7.13) we obtain

$$
\begin{aligned}
\Phi_2(z,\overline{\theta}) &= A^*(z) + \sqrt{2}\,\overline{\theta}\,\overline{\psi}(z) + (\overline{\theta}\,\overline{\theta})F^*(z) \\
&= A^*(x - i\theta\sigma\overline{\theta}) + \sqrt{2}\,\overline{\theta}\,\overline{\psi}(x - i\theta\sigma\overline{\theta}) + (\overline{\theta}\,\overline{\theta})F^*(x - i\theta\sigma\overline{\theta}) \\
&= A^*(x) - i\theta\sigma^\mu\overline{\theta}\partial_\mu A^*(x) - \frac{1}{2}(\theta\sigma^\mu\overline{\theta})(\theta\sigma^\nu\overline{\theta})\partial_\mu\partial_\nu A^*(x) \\
&\quad + \sqrt{2}\,\overline{\theta}\,\overline{\psi}(x) + i\sqrt{2}(\theta\sigma^\mu\overline{\theta})\overline{\theta}_{\dot{A}}\partial_\mu\overline{\psi}^{\dot{A}}(x) + (\overline{\theta}\,\overline{\theta})F^*(x).
\end{aligned}
$$

Rearranging the third and the fifth terms with the help of Eqs. (1.137) and (1.100d) we finally obtain

$$
\begin{aligned}
\Phi_2(z,\overline{\theta}) &= A^*(x) - i(\theta\sigma^\mu\overline{\theta})\partial_\mu A^*(x) - \frac{1}{4}(\theta\theta)(\overline{\theta}\,\overline{\theta})\Box A^*(x) + \sqrt{2}\,\overline{\theta}\,\overline{\psi}(x) \\
&\quad + (\overline{\theta}\,\overline{\theta})F^*(x) - \frac{i}{\sqrt{2}}(\overline{\theta}\,\overline{\theta})(\theta\sigma^\mu\partial_\mu\overline{\psi}(x)) \\
&= \Phi^\dagger(x,\theta,\overline{\theta}). \tag{7.14}
\end{aligned}
$$

The result is seen to be the Hermitian conjugate of that of Eq. (7.6). The sixth term of this expression follows from the fifth term of Eq. (7.6) upon using Eq. (1.135a).

As a consistency check we verify that

$$D_A\Phi^\dagger(x,\theta,\overline{\theta}) = 0, \tag{7.15}$$

where

$$D_A = \partial_A + i\sigma^\mu_{A\dot{A}}\overline{\theta}^{\dot{A}}\partial_\mu.$$

Consider:

$$
\begin{aligned}
D_A \Phi^\dagger(x,\theta,\overline{\theta}) &= \{\partial_A + i\sigma^\mu_{A\dot{A}}\overline{\theta}^{\dot{A}}\partial_\mu\}\Big\{ A^*(x) - i(\theta\sigma^\mu\overline{\theta})\partial_\mu A^*(x) \\
&\quad -\frac{1}{4}(\theta\theta)(\overline{\theta}\,\overline{\theta})\Box A^*(x) + \sqrt{2}\,\overline{\theta}\,\overline{\psi}(x) - \frac{i}{\sqrt{2}}(\overline{\theta}\,\overline{\theta})\theta\sigma^\mu\partial_\mu\overline{\psi}(x) \\
&\quad +(\overline{\theta}\,\overline{\theta})F^*(x)\Big\} \\
&= -i\sigma^\mu_{A\dot{A}}\overline{\theta}^{\dot{A}}\partial_\mu A^*(x) - \frac{1}{2}\theta_A(\overline{\theta}\,\overline{\theta})\Box A^*(x) \\
&\quad -\frac{i}{\sqrt{2}}(\overline{\theta}\,\overline{\theta})\sigma^\mu_{A\dot{B}}\partial_\mu\overline{\psi}^{\dot{B}}(x) + i\sigma^\mu_{A\dot{A}}\overline{\theta}^{\dot{A}}\partial_\mu A^*(x) \\
&\quad +\sigma^\mu_{A\dot{A}}\overline{\theta}^{\dot{A}}(\theta\sigma^\mu\overline{\theta})\partial_\mu\partial_\nu A^*(x) + i\sqrt{2}\sigma^\mu_{A\dot{A}}\overline{\theta}^{\dot{A}}\overline{\theta}_{\dot{B}}\partial^\mu\overline{\psi}^{\dot{B}}(x) \\
&= -\frac{1}{2}\theta_A(\overline{\theta}\,\overline{\theta})\Box A^*(x) - \frac{i}{\sqrt{2}}(\overline{\theta}\,\overline{\theta})\sigma^\mu_{A\dot{B}}\partial_\mu\overline{\psi}^{\dot{B}}(x) \\
&\quad +\frac{1}{2}\theta_A(\overline{\theta}\,\overline{\theta})\Box A^*(x) + \frac{i}{\sqrt{2}}(\overline{\theta}\,\overline{\theta})\sigma^\mu_{A\dot{B}}\partial_\mu\overline{\psi}^{\dot{B}}(x) = 0.
\end{aligned}
$$

Equation (7.15) states that the expression (7.14) is the most general solution of the constraint equation $D_A\Phi^\dagger = 0$. It should be observed that $\Phi^\dagger$ depends only on the complex scalar fields A^*, F^* and a right-handed chiral Weyl spinor field $\overline{\psi}(x)$. This is characteristic for right-handed chiral superfields.

Products of scalar superfields are again scalar superfields. In order to see this, let Φ_i and Φ_k be two scalar superfields with component expansions

$$
\begin{aligned}
\Phi_i &= A_i(y) + \sqrt{2}\theta\psi_i(y) + (\theta\theta)F_i(y), \\
\Phi_k &= A_k(y) + \sqrt{2}\theta\psi_k(y) + (\theta\theta)F_k(y).
\end{aligned}
$$

Then

$$
\begin{aligned}
\Phi_i\Phi_k &= \big[A_i(y) + \sqrt{2}\theta\psi_i(y) + (\theta\theta)F_i(y)\big]\big[A_k(y) + \sqrt{2}\theta\psi_k(y) + (\theta\theta)F_k(y)\big] \\
&= A_i(y)A_k(y) + 2(\theta\psi_i(y))(\theta\Psi_k(y)) + \sqrt{2}\theta A_i(y)\psi_k(y) \\
&\quad +(\theta\theta)A_i(y)F_k(y) + \sqrt{2}\theta\psi_i(y)A_k(y) + (\theta\theta)F_i(y)A_k(y) \\
&= A_i(y)A_k(y) + \sqrt{2}\theta[A_i(y)\psi_k(y) + \psi_i(y)A_k(y)] + (\theta\theta)[A_i(y)F_k(y) \\
&\quad +F_i(y)A_k(y) - \psi_i(y)\psi_k(y)]. \qquad\qquad (7.16)
\end{aligned}
$$

From the expansion (7.16) we see that $\Phi_i\Phi_k$ is again a function of y and θ, and therefore

$$
\overline{D}_{\dot{A}}(\Phi_i\Phi_k) = 0.
$$

Of course, this result can also be obtained by taking into account the fact that $\overline{D}_{\dot{A}}$ is a linear operator. Hence $\Phi_i\Phi_k$ is again a left-handed chiral superfield. Now consider the product of three left-handed chiral superfields, *i.e.*

$$
\begin{aligned}
\Phi_l\Phi_i\Phi_k = {}& A_l(y)A_i(y)A_k(y) + \sqrt{2}\theta\big[A_l(y)A_i(y)\psi_k(y) + A_l(y)\psi_i(y)A_k(y) \\
& + \psi_l(y)A_i(y)A_k(y)\big] + (\theta\theta)\big[A_l(y)A_i(y)F_k(y) + A_l(y)F_i(y)A_k(y) \\
& + F_l(y)A_i(y)A_k(y) - \psi_l(y)\psi_k(y)A_i(y) \\
& - \psi_l(y)\psi_i(y)A_k(y) - A_l(y)\psi_i(y)\psi_k(y)\big],
\end{aligned}
\tag{7.17}
$$

and again $\Phi_l\Phi_i\Phi_k$ is a left-handed chiral superfield satisfying the condition

$$
\overline{D}_{\dot{A}}(\Phi_l\Phi_i\Phi_k) = 0.
\tag{7.18}
$$

Similar results hold, of course, for the conjugate fields.

An important point in the construction of invariant actions (which will be considered later) is the transformation property of the highest component of a superfield. From the general expression (6.94) we obtain the transformation properties of the component fields of the left-handed chiral superfield (7.5) as (see below)

$$
\begin{aligned}
\delta_s' A(y) &= \sqrt{2}\alpha\psi(y), \\
\delta_s' \psi_A(y) &= \sqrt{2}\alpha_A F(y) + i\sqrt{2}\sigma^\mu_{A\dot{A}}\overline{\alpha}^{\dot{A}}\partial_\mu A(y), \\
\delta_s' F(y) &= -i\sqrt{2}\partial_\mu\psi(y)\sigma^\mu\overline{\alpha}.
\end{aligned}
\tag{7.19}
$$

These relations are obtained by first comparing Eq. (6.76) with Eq. (7.6), and making the following identifications:

$$
\begin{aligned}
f(x) &\longrightarrow A(x), \\
\phi(x) &\longrightarrow \sqrt{2}\psi(x), \\
\overline{\chi}(x) &\longrightarrow 0, \\
m(x) &\longrightarrow F(x), \\
n(x) &\longrightarrow 0, \\
V_\mu(x) &\longrightarrow i\partial_\mu A(x), \\
\overline{\lambda}^{\dot{A}}(x) &\longrightarrow -\frac{i}{\sqrt{2}}\partial_\mu\psi^A(x)\sigma^\mu_{A\dot{B}}\epsilon^{\dot{B}\dot{A}}, \\
\psi(x) &\longrightarrow 0, \\
d(x) &\longrightarrow -\frac{1}{4}\Box A(x).
\end{aligned}
$$

The appropriate substitutions in Eq. (6.94) then yield Eq. (7.19), now, of course, in conformity with the variable of Eq. (7.5), in terms of y. It is evident from Eq. (7.19) that the fields A, ψ, F constitute an irreducible representation of the supersymmetry algebra since their multiplet transforms under the supersymmetry transformation into itself. From Eq. (7.19) one can also see the significance of the auxiliary field F: The supersymmetry algebra is closed linearly in the off-shell case (*i.e.* in the presence of F), whereas one has a nonlinear representation of the supersymmetry algebra in the on-shell case.

An important aspect of the transformation properties of superfields is that the highest component field of Φ, *i.e.* F, transforms into a total spacetime derivative. A spacetime integral $\int d^4x$ of this quantity is thus invariant under supersymmetry transformations, because the supersymmetric variation of this component field (which is the total derivative) can be transformed into a surface integral which vanishes, provided the fields fall off sufficiently fast at infinity.

In order to be able to construct supersymmetric Lagrangians in terms of superfields we need the product of a right-handed chiral superfield and a left-handed chiral superfield. However, this product is neither chiral nor antichiral. Consider

$$\begin{aligned} \Phi_i^\dagger &= A_i^*(z) + \sqrt{2}\bar\theta\,\bar\psi_i(z) + (\bar\theta\,\bar\theta)F_i^*(z), \\ \Phi_j &= A_j(y) + \sqrt{2}\theta\psi_j(y) + (\theta\theta)F_j(y). \end{aligned}$$

Then

$$\begin{aligned} \Phi_i^\dagger\Phi_j &= \Big[A_i^*(z) + \sqrt{2}\bar\theta\,\bar\psi_i(z) + (\bar\theta\,\bar\theta)F_i^*(z)\Big]\Big[A_j(y) + \sqrt{2}\theta\psi_j(y) + (\theta\theta)F_j(y)\Big] \\ &= A_i^*(z)A_j(y) + \sqrt{2}\theta A_i^*(z)\psi_j(y) + \sqrt{2}\,\bar\theta\,\bar\psi_i(z)A_j(y) + (\theta\theta)A_i^*(z)F_j(y) \\ &\quad + (\bar\theta\,\bar\theta)F_i^*(z)A_j(y) + 2(\bar\theta\,\bar\psi_i(z))(\theta\psi_j(z)) + \sqrt{2}(\bar\theta\,\bar\theta)\theta\psi_j(y)F_i^*(z) \\ &\quad + \sqrt{2}(\theta\theta)\,\bar\theta\,\bar\psi_i(z)F_j(y) + (\theta\theta)(\bar\theta\,\bar\theta)F_i^*(z)F_j(y), \end{aligned}$$

and with $z = x - i\theta\sigma\bar\theta, y = x + i\theta\sigma\bar\theta$,

$$\begin{aligned} \Phi_i^\dagger\Phi_j &= A_i^*(x - i\theta\sigma\bar\theta)A_j(x + i\theta\sigma\bar\theta) + \sqrt{2}\theta A_i^*(x - i\theta\sigma\bar\theta)\psi_j(x + i\theta\sigma\bar\theta) \\ &\quad + \sqrt{2}\bar\theta\,\bar\psi_i(x - i\theta\sigma\bar\theta)A_j(x + i\theta\sigma\bar\theta) + (\theta\theta)A_i^*(x - i\theta\sigma\bar\theta)F_j(x + i\theta\sigma\bar\theta) \\ &\quad + (\bar\theta\,\bar\theta)F_i^*(x - i\theta\sigma\bar\theta)A_j(x + i\theta\sigma\bar\theta) + 2[\bar\theta\,\bar\psi_i(x - i\theta\sigma\bar\theta)][\theta\psi_j(x + i\theta\sigma\bar\theta)] \\ &\quad + \sqrt{2}(\bar\theta\,\bar\theta)\theta\psi_j(x + i\theta\sigma\bar\theta)F_i^*(x - i\theta\sigma\bar\theta) \\ &\quad + \sqrt{2}(\theta\theta)\bar\theta\psi_i^*(x - i\theta\sigma\bar\theta)F_j(x + i\theta\sigma\bar\theta) \\ &\quad + (\theta\theta)(\bar\theta\,\bar\theta)F_i^*(x - i\theta\sigma\bar\theta)F_j(x + i\theta\sigma\bar\theta). \end{aligned} \tag{7.20}$$

We consider various terms separately. Thus, using Eq. (1.137) in the first step,

$$
A_i^*(x - i\theta\sigma\overline{\theta})A_j(x + i\theta\sigma\overline{\theta})
$$

$$
= \left\{ A_i^*(x) - i\theta\sigma^\mu\overline{\theta}\partial_\mu A_i^*(x) - \frac{1}{4}(\theta\theta)(\overline{\theta}\,\overline{\theta})\Box A_i^*(x) \right\}
$$

$$
\times \left\{ A_j(x) - i\theta\sigma^\rho\overline{\theta}\partial_\rho A_j(x) - \frac{1}{4}(\theta\theta)(\overline{\theta}\overline{\theta})\Box A_j(x) \right\}
$$

$$
= A_i^*(x)A_j(x) + i\theta\sigma^\rho\overline{\theta}\partial_\rho A_j(x)A_i^*(x)
$$

$$
- \frac{1}{4}(\theta\theta)(\overline{\theta}\,\overline{\theta})\left\{ A_i^*(x)\Box A_j(x) + \Box A_i^*(x)A_j(x) \right\}
$$

$$
- i\theta\sigma^\mu\overline{\theta}\partial_\mu A_i^*(x)A_j(x) + (\theta\sigma^\mu\overline{\theta})(\theta\sigma^\rho\overline{\theta})\partial_\mu A_i^*(x)\partial_\rho A_j(x)
$$

$$
= A_i^*(x)A_j(x) + i(\theta\sigma^\rho\overline{\theta})\left[(\partial_\rho A_j(x))A_i^*(x) - (\partial_\rho A_i^*(x))A_j(x) \right]
$$

$$
+ (\theta\theta)(\overline{\theta}\,\overline{\theta})\left\{ -\frac{1}{4}A_i^*(x)\Box A_j(x) - \frac{1}{4}\Box A_i^*(x)A_j(x) \right.
$$

$$
\left. + \frac{1}{2}\partial_\mu A_i^*(x)\partial^\mu A_j(x) \right\}, \tag{7.21}
$$

where we used again Eq. (1.137). Next consider (with $(\theta\psi_j) = \theta^A\psi_{jA}$)

$$
\sqrt{2}\theta^A A_i^*(x - i\theta\sigma\overline{\theta})\psi_{jA}(x + i\theta\sigma\overline{\theta})
$$

$$
= \sqrt{2}\theta^A\left\{ (A_i^*(x) - i\theta\sigma^\mu\overline{\theta}\partial_\mu A_i^*(x))(\psi_{jA}(x) + i\theta\sigma^\nu\overline{\theta}\partial_\nu\psi_{jA}(x)) \right\}
$$

$$
= \sqrt{2}\theta^A\psi_{jA}(x)A_i^*(x) - i\sqrt{2}\theta^A\theta^B\sigma^\mu_{B\dot{A}}\overline{\theta}^{\dot{A}}\partial_\mu A_i^*(x)\psi_{jA}(x)
$$

$$
+ i\sqrt{2}\theta^A A_i^*(x)\theta^B\sigma^\mu_{B\dot{A}}\overline{\theta}^{\dot{A}}\partial_\mu\psi_{jA}(x)
$$

$$
= \sqrt{2}(\theta\psi_j(x))A_i^*(x) + \frac{i}{\sqrt{2}}(\theta\theta)\epsilon^{AB}\sigma^\mu_{B\dot{A}}\overline{\theta}^{\dot{A}}\partial_\mu A_i^*(x)\psi_{jA}(x)
$$

$$
- \frac{i}{\sqrt{2}}(\theta\theta)A_i^*(x)\epsilon^{AB}\sigma^\mu_{B\dot{C}}\overline{\theta}^{\dot{C}}\partial_\mu\psi_{jA}(x) \quad \text{(using Eq. (1.100a))}
$$

$$
= \sqrt{2}(\theta\psi_j(x))A_i^*(x) - \frac{i}{\sqrt{2}}(\theta\theta)\overline{\theta}^{\dot{A}}\psi_j^A(x)\sigma^\mu_{A\dot{A}}\partial_\mu A_i^*(x)
$$

$$
+ \frac{i}{\sqrt{2}}(\theta\theta)\overline{\theta}^{\dot{A}}A_i^*(x)\partial_\mu\psi_j^A(x)\sigma^\mu_{A\dot{A}} \quad \text{(using Eq. (1.83))}
$$

$$
= \sqrt{2}(\theta\psi_j(x))A_i^*(x) - \frac{i}{\sqrt{2}}(\theta\theta)\overline{\theta}^{\dot{A}}\sigma^\mu_{A\dot{A}}(\psi_j^A(x)\partial_\mu A_i^*(x)
$$

$$
- A_i^*(x)\partial_\mu\psi_j^A(x)). \tag{7.22}
$$

Now consider (the demonstration that the term underbraced T can be re-expressed as T' is demonstrated a step later)

$$U := \sqrt{2}\,\theta\,\overline{\psi}_i(x - i\theta\sigma\overline{\theta})A_j(x + i\theta\sigma\overline{\theta})$$

$$= \sqrt{2}\theta_{\dot{A}}\left(\overline{\psi}_i^{\dot{A}}(x) - i\theta\sigma^\mu\overline{\theta}\partial_\mu\overline{\psi}_i^{\dot{A}}(x)\right)\left(A_j(x) + i\theta\sigma^\mu\overline{\theta}\partial_\mu A_j(x)\right)$$

$$= \sqrt{2}(\overline{\theta}\,\overline{\psi}_i(x))A_j(x) + \underbrace{i\sqrt{2}\theta_{\dot{A}}\overline{\psi}_i^{\dot{A}}(x)\theta\sigma^\mu\overline{\theta}\partial_\mu A_j(x)}_{T}$$

$$- i\sqrt{2}\theta_{\dot{A}}(\theta\sigma^\mu\overline{\theta})\partial_\mu\overline{\psi}_i^{\dot{A}}(x)A_j(x).$$

Using Eqs. (1.98) and (1.85) this becomes

$$U = \sqrt{2}(\overline{\theta}\,\overline{\psi}_i(x))A_j(x) - i\sqrt{2}\psi_{i\dot{A}}(x)\overline{\theta}^{\dot{A}}\overline{\theta}^{\dot{B}}(\theta\sigma^\mu)_{\dot{B}}\partial_\mu A_j(x)$$

$$+ i\sqrt{2}\theta_{\dot{A}}\overline{\theta}_{\dot{B}}(\theta\sigma^\mu)_{\dot{C}}\epsilon^{\dot{C}\dot{B}}\partial_\mu\overline{\psi}_i^{\dot{A}}(x)A_j(x)$$

$$= \sqrt{2}\,(\overline{\theta}\,\overline{\psi}_i(x))A_j(x) - \underbrace{\frac{i}{\sqrt{2}}(\overline{\theta}\,\overline{\theta})(\theta\sigma^\mu\overline{\psi}_i(x))\partial_\mu A_j(x)}_{T'}$$

$$+ \frac{i}{\sqrt{2}}(\overline{\theta}\,\overline{\theta})\theta\sigma^\mu\partial_\mu\overline{\psi}_i(x)A_j(x) \qquad \text{(with Eq. (1.100c))}$$

$$= \sqrt{2}(\overline{\theta}\,\overline{\psi}_i(x))A_j(x)$$

$$+ (\overline{\theta}\,\overline{\theta})\theta^A\left[-\frac{i}{4}\sigma^\mu_{A\dot{A}}\left(\overline{\psi}_i^{\dot{A}}(x)\partial_\mu A_j(x) - \partial_\mu\overline{\psi}_i^{\dot{A}}(x)A_j(x)\right)\right]. \qquad (7.23)$$

In the expression U above the term underbraced T has been re-expressed in the following way as T':

$$\underbrace{i\sqrt{2}\theta_{\dot{A}}\overline{\psi}_i^{\dot{A}}(x)\theta^A\sigma^\mu_{A\dot{B}}\overline{\theta}^{\dot{B}}\partial_\mu A_j(x)}_{T} \overset{(1.98)}{=} i\sqrt{2}\psi_{i\dot{A}}(x)\overline{\theta}^{\dot{A}}\theta^A\sigma^\mu_{A\dot{B}}\overline{\theta}^{\dot{B}}\partial_\mu A_j(x)$$

$$= -i\sqrt{2}\psi_{i\dot{A}}(x)\overline{\theta}^{\dot{A}}\overline{\theta}^{\dot{B}}\theta^A\sigma^\mu_{A\dot{B}}\partial_\mu A_j(x)$$

$$\overset{(1.100c)}{=} -\frac{i}{\sqrt{2}}\psi_{i\dot{A}}(x)\epsilon^{\dot{A}\dot{B}}(\overline{\theta}\overline{\theta})\theta^A\sigma^\mu_{A\dot{B}}\partial_\mu A_j(x)$$

$$= \frac{i}{\sqrt{2}}\overline{\psi}_{i\dot{A}}(x)(\overline{\theta}\,\overline{\theta})\theta^A\sigma^\mu_{A\dot{B}}\epsilon^{\dot{B}\dot{A}}\partial_\mu A_j(x)$$

$$= \frac{-i}{\sqrt{2}}(\overline{\theta}\,\overline{\theta})\theta^A\sigma^\mu_{A\dot{B}}\epsilon^{\dot{B}\dot{A}}\overline{\psi}_{i\dot{A}}(x)\partial_\mu A_j(x)$$

$$= \underbrace{-\frac{i}{\sqrt{2}}(\overline{\theta}\,\overline{\theta})(\theta\sigma^\mu\overline{\psi}_i(x))\partial_\mu A_j(x)}_{T'}.$$

Also

$$(\theta\theta)A_i^*(x - i\theta\sigma\overline{\theta})F_j(x + i\theta\sigma\overline{\theta}) = (\theta\theta)A_i^*(x)F_j(x), \qquad (7.24)$$

$$(\overline{\theta}\,\overline{\theta})A_j(x + i\theta\sigma\overline{\theta})F_i^*(x - i\theta\sigma\overline{\theta}) = (\overline{\theta}\,\overline{\theta})A_j(x)F_i^*(x), \qquad (7.25)$$

and

$$2\overline{\theta}\,\overline{\psi}_i(x - i\theta\sigma\overline{\theta})\theta\psi_j(x + i\theta\sigma\overline{\theta}) = -2\overline{\theta}_{\dot{A}}\theta^A\overline{\psi}_i^{\dot{A}}(x - i\theta\sigma\overline{\theta})\psi_{jA}(x + i\theta\sigma\overline{\theta})$$

$$= -2\overline{\theta}_{\dot{A}}\theta^A\Big\{\overline{\psi}_i^{\dot{A}}(x)\psi_{jA}(x) + \underbrace{i\overline{\psi}_i^{\dot{A}}(x)\theta\sigma^\mu\overline{\theta}\partial_\mu\overline{\psi}_{jA}(x)}_{S_1}$$

$$\underbrace{-i\theta\sigma^\mu\overline{\theta}(\partial_\mu\overline{\psi}_i^{\dot{A}}(x))\psi_{jA}(x)}_{S_2}\Big\}$$

$$= 2(\overline{\theta}\,\overline{\psi}_i(x))(\theta\psi_j(x)) - \frac{i}{2}(\theta\theta)(\overline{\theta}\,\overline{\theta})\Big[\psi_j(x)\sigma^\mu\partial_\mu\overline{\psi}_i(x)$$

$$- \partial_\mu\psi_j(x)\sigma^\mu\overline{\psi}_i(x)\Big], \qquad (7.26)$$

since *e.g.* the term underbraced S_1 can be re-expressed as

$$\underbrace{-2i\overline{\theta}_{\dot{A}}\theta^A\overline{\psi}_i^{\dot{A}}(x)\theta\sigma^\mu\overline{\theta}\partial_\mu\overline{\psi}_{jA}(x)}_{S_1} = -2i\overline{\theta}_{\dot{A}}\overline{\theta}^{\dot{B}}\theta^A\theta^B\overline{\psi}_i^{\dot{A}}\sigma^\mu_{B\dot{B}}\partial_\mu\psi_{jA}$$

$$\overset{(1.100a)}{\underset{(1.100e)}{=}} -2i\big(\frac{1}{2}\delta_{\dot{A}}^{\dot{B}}(\overline{\theta}\,\overline{\theta})\big)\big(-\frac{1}{2}\epsilon^{AB}(\theta\theta)\big)$$

$$\times\overline{\psi}_i^{\dot{A}}\sigma^\mu_{B\dot{B}}\partial_\mu\psi_{jA}$$

$$= \frac{i}{2}(\overline{\theta}\,\overline{\theta})(\theta\theta)\overline{\psi}_i^{\dot{B}}\sigma^\mu_{B\dot{B}}\epsilon^{AB}\partial_\mu\psi_{jA}$$

$$= -\frac{i}{2}(\overline{\theta}\,\overline{\theta})(\theta\theta)(\partial_\mu\psi_{jA})\epsilon^{AB}\sigma^\mu_{B\dot{B}}\overline{\psi}_i^{\dot{B}}$$

$$= \frac{i}{2}(\overline{\theta}\,\overline{\theta})(\theta\theta)(\partial_\mu\epsilon^{BA}\psi_{jA})\sigma^\mu_{B\dot{B}}\overline{\psi}_i^{\dot{B}}$$

$$\overset{(1.72)}{=} \frac{i}{2}(\overline{\theta}\,\overline{\theta})(\theta\theta)(\partial_\mu\psi_j^B)\sigma^\mu_{B\dot{B}}\overline{\psi}_i^{\dot{B}}$$

$$= \frac{i}{2}(\theta\theta)(\overline{\theta}\,\overline{\theta})((\partial_\mu\psi_j)\sigma^\mu\overline{\psi}_i),$$

and S_2 is

$$\underbrace{2i\overline{\theta}_{\dot{A}}\theta^A(\theta\sigma^\mu\overline{\theta})(\partial_\mu\overline{\psi}_i^{\dot{A}}(x))\psi_{jA}(x)}_{S_2} = 2i\overline{\theta}_{\dot{A}}\theta^A\theta^B\sigma^\mu_{B\dot{B}}\overline{\theta}^{\dot{B}}(\partial_\mu\overline{\psi}_i^{\dot{A}}(x))\psi_{jA}(x)$$

$$= 2i(\overline{\theta}_{\dot{A}}\overline{\theta}^{\dot{B}})(\theta^A\theta^B)\sigma^\mu_{B\dot{B}}(\partial_\mu\overline{\psi}_i^{\dot{A}}(x))\psi_{jA}(x)$$

$$\underset{(1.100e)}{\overset{(1.100a)}{=}} -\frac{i}{2}\delta_{\dot{A}}^{\dot{B}}(\overline{\theta}\,\overline{\theta})\epsilon^{AB}(\theta\theta)\sigma_{B\dot{B}}^{\mu}(\partial_{\mu}\overline{\psi}_{i}^{\dot{A}}(x))\psi_{jA}(x)$$

$$= -\frac{i}{2}(\theta\theta)(\overline{\theta\theta})(\epsilon^{BA}\psi_{jA}(x))\delta_{\dot{A}}^{\dot{B}}\sigma_{B\dot{B}}^{\mu}(\partial_{\mu}\overline{\psi}_{i}^{\dot{A}}(x))$$

$$= -\frac{i}{2}(\theta\theta)(\overline{\theta}\,\theta)\psi_{j}^{B}(x)\sigma_{B\dot{B}}^{\mu}\partial_{\mu}\overline{\psi}_{i}^{\dot{B}}(x)$$

$$= -\frac{i}{2}(\theta\theta)(\overline{\theta}\,\theta)(\psi_{j}(x)\sigma^{\mu}\partial_{\mu}\overline{\psi}_{i}(x)).$$

Inserting Eqs. (7.21) to (7.26) into Eq. (7.20) we finally obtain

$$\begin{aligned}
\Phi_{i}^{\dagger}\Phi_{j}^{\dagger} &= A_{i}^{*}(x)A_{j}(x) + \sqrt{2}\theta\psi_{j}(x)A_{i}^{*}(x) + \sqrt{2}\,\overline{\theta}\,\overline{\psi}_{i}(x)A_{j}(x) \\
&\quad + (\theta\theta)A_{i}^{*}(x)F_{j}(x) + (\overline{\theta}\,\overline{\theta})F_{i}^{*}(x)A_{j}(x) + 2\overline{\theta}\,\overline{\psi}_{i}(x)\theta\psi_{j}(x) \\
&\quad + i(\theta\sigma^{\mu}\overline{\theta})[(\partial_{\mu}A_{j}(x))A_{i}^{*}(x) - (\partial_{\mu}A_{i}^{*}(x))A_{j}(x)] \\
&\quad - \sqrt{2}(\theta\theta)\overline{\theta}_{\dot{A}}\left\{\frac{i}{2}\sigma_{A\dot{B}}^{\mu}\epsilon^{\dot{B}\dot{A}}(\psi_{j}^{A}(x)\partial_{\mu}A_{i}^{*}(x) - A_{i}^{*}(x)\partial_{\mu}\psi_{j}^{A}(x)) \right. \\
&\quad \left. + \overline{\psi}_{i}^{\dot{A}}(x)F_{j}(x)\right\} + \sqrt{2}(\overline{\theta}\,\overline{\theta})\theta^{A}\left\{-\frac{i}{2}\sigma_{A\dot{A}}^{\mu}(\overline{\psi}_{i}^{\dot{A}}(x)\partial_{\mu}A_{j}(x) \right. \\
&\quad \left. - A_{j}(x)\partial_{\mu}\overline{\psi}_{i}^{\dot{A}}(x)) + \psi_{jA}(x)F_{i}^{*}(x)\right\} + (\theta\theta)(\overline{\theta}\,\overline{\theta}) \\
&\quad \times \left\{\frac{1}{2}\partial_{\mu}A_{i}^{*}(x)\partial^{\mu}A_{j}(x) - \frac{1}{4}A_{i}^{*}(x)\Box A_{j}(x) - \frac{1}{4}A_{j}(x)\Box A_{i}^{*}(x) \right. \\
&\quad \left. + \frac{i}{2}\partial_{\mu}\psi_{j}(x)\sigma^{\mu}\overline{\psi}_{i}(x) - \frac{i}{2}\psi_{j}(x)\sigma^{\mu}\partial_{\mu}\overline{\psi}_{i}(x) + F_{i}^{*}(x)F_{j}(x)\right\}. \quad (7.27a)
\end{aligned}$$

We will show later that for $i = j$ the $(\theta\theta)(\overline{\theta}\,\overline{\theta})$-component of this expression can be rewritten (*cf.* Eq. (8.28))

$$(\theta\theta)(\overline{\theta}\,\overline{\theta})\{-A_{i}^{*}(x)\Box A_{i}(x) + |F_{i}(x)|^{2} + i(\partial_{\mu}\overline{\psi}_{i}(x))\overline{\sigma}^{\mu}\psi_{i}(x)$$
$$+ \text{ total derivatives}\}. \quad (7.27b)$$

This expression will be required later at various points (in action integral (8.30); see also Eq. (9.74)).

In the above product the $(\theta\theta)(\overline{\theta}\,\overline{\theta})$-component transforms under supersymmetry transformations into a spacetime derivative (*i.e.* like $d(x)$ of Eq. (6.76) and $\delta_{s}'d(x)$ of Eq. (6.94)). We also observe that the product $\Phi_{i}^{\dagger}\Phi_{j}$ of antichiral and chiral superfields generates spin-1 component fields (*i.e.* the term proportional to $\theta\sigma^{\mu}\overline{\theta}$) in much the same way as the product of elementary Weyl spinors generates higher spin fields (see for example Eq. (1.152a)).

7.2   Vector Superfields, Generalized Gauge Transformations

Vector superfields satisfy the reality condition

$$V(x,\theta,\overline{\theta}) = V^{\dagger}(x,\theta,\overline{\theta}) \tag{7.28}$$

(see Eq. (6.97)). Like any other superfield the vector superfield V is defined in terms of its power series expansion in θ and $\overline{\theta}$, *i.e.*

$$V(x,\theta,\overline{\theta}) = C(x) + \theta\phi(x) + \overline{\theta}\,\overline{\chi}(x) + \theta\theta M(x) + \overline{\theta}\,\overline{\theta}N(x)\theta\sigma^{\mu}\overline{\theta}V_{\mu}(x)$$
$$+ (\theta\theta)\overline{\theta}\,\overline{\lambda}(x) + (\overline{\theta}\,\overline{\theta})\theta\psi(x) + (\theta\theta)(\overline{\theta}\,\overline{\theta})D(x).$$

The Hermitian conjugate is[1]

$$V^{\dagger}(x,\theta,\overline{\theta}) = C^{*}(x) + \overline{\theta}\,\overline{\phi}(x) + \theta\chi(x) + (\overline{\theta}\,\overline{\theta})M^{*}(x) + (\theta\theta)N^{*}(x)$$
$$+ (\theta\sigma^{\mu}\overline{\theta})V_{\mu}^{*}(x) + (\overline{\theta}\,\overline{\theta})\theta\lambda(x) + (\theta\theta)(\overline{\theta}\,\overline{\psi}(x)) + (\theta\theta)(\overline{\theta}\,\overline{\theta})D^{*}(x).$$

The reality condition (7.52) is satisfied if and only if

$$
\begin{aligned}
C(x) &= C^{*}(x) \longrightarrow C(x): \text{ a real scalar field,}\\
\phi(x) &= \chi(x),\\
M(x) &= N^{*}(x),\\
V_{\mu}(x) &= V_{\mu}^{*}(x) \longrightarrow V_{\mu}(x): \text{ a real vector field,}\\
\lambda(x) &= \psi(x),\\
D(x) &= D^{*}(x) \longrightarrow D(x): \text{ a real scalar field.}
\end{aligned}
$$

Hence a vector superfield obeying the constraint (7.28) has the general expansion

$$
\begin{aligned}
V(x,\theta,\overline{\theta}) = {}& C(x) + \theta\phi(x) + \overline{\theta}\,\overline{\phi}(x) + (\theta\theta)M(x) + (\overline{\theta}\,\overline{\theta})M^{*}(x)\\
& + (\theta\sigma_{\mu}\overline{\theta})V^{\mu}(x) + (\theta\theta)\overline{\theta}\,\overline{\lambda}(x) + (\overline{\theta}\,\overline{\theta})\theta\lambda(x)\\
& + (\theta\theta)(\overline{\theta}\,\overline{\theta})D(x),
\end{aligned}
\tag{7.29}
$$

where

$$
\begin{aligned}
C(x), V_{\mu}(x) \text{ and } D(x) & \quad \text{are real fields, and}\\
M(x), D(x), C(x) & \quad \text{are scalar fields,}\\
\lambda(x), \phi(x) & \quad \text{are spinor fields,}\\
V_{\mu}(x) & \quad \text{is a vector field.}
\end{aligned}
$$

---

[1]This can be verified with the help of Eqs. (1.135a) and (1.135b).

The vector field $V_\mu(x)$ lends its name to the entire multiplet $V(x, \theta, \overline{\theta})$. From $\delta'_s d(x)$ of Eq. (6.94) we know that under a supersymmetry transformation the $(\theta\theta)(\overline{\theta}\,\overline{\theta})$- component of $V(x, \theta, \overline{\theta})$ transforms into a spacetime derivative of $\lambda(x)$, *i.e.*

$$\delta'_s D(x) = \frac{i}{2}\{\partial_\mu \lambda(x)\sigma^\mu \overline{\alpha} - \partial_\mu \overline{\lambda}(x)\,\overline{\sigma}^\mu \alpha\}. \tag{7.30}$$

As explained earlier, this indicates that the $(\theta\theta)(\overline{\theta}\,\overline{\theta})$-component of the vector superfield is a candidate for a supersymmetric Lagrangian.

A particular example of a vector superfield is the product of a right-handed chiral superfield and a left-handed chiral superfield, $\Phi^\dagger \Phi$, as given by Eq. (7.27a), since in this case

$$(\Phi^\dagger \Phi)^\dagger = \Phi^\dagger (\Phi^\dagger)^\dagger = \Phi^\dagger \Phi, \tag{7.31}$$

and the reality condition (7.28) is satisfied.

Another important example of a vector superfield is the sum of a left-handed chiral superfield and a right-handed chiral superfield, since

$$(\Phi + \Phi^\dagger)^\dagger = \Phi^\dagger + \Phi = \Phi + \Phi^\dagger, \tag{7.32}$$

and again the reality condition (7.28) is satisfied. Expressing this sum in terms of component fields, we have (adding Eqs. (7.6) and (7.14))

$$\begin{aligned}
\Phi + \Phi^\dagger =\ & A(x) + A^*(x) + \sqrt{2}\theta\psi(x) + \sqrt{2}\,\overline{\theta}\,\overline{\psi}(x) + (\theta\theta)F(x) \\
& + (\overline{\theta}\,\overline{\theta})F^*(x) + i(\theta\sigma^\mu\overline{\theta})\partial_\mu[A(x) - A^*(x)] \\
& - \frac{i}{\sqrt{2}}(\theta\theta)\overline{\theta}\,\overline{\sigma}^\mu\partial_\mu\psi(x) - \frac{i}{\sqrt{2}}(\overline{\theta}\,\overline{\theta})\theta\sigma^\mu\partial_\mu\overline{\psi}(x) \\
& - \frac{1}{4}(\theta\theta)(\overline{\theta}\,\overline{\theta})\Box[A(x) + A^*(x)].
\end{aligned} \tag{7.33}$$

It is important to observe, that this combination of scalar superfields has the gradient $i\partial_\mu[A(x) - A^*(x)]$ as coefficient of $\theta\sigma^\mu\overline{\theta}$. We now consider a special choice of V which is such that certain components of V are invariant under the gauge transformations to be defined below. This is achieved by making in Eq. (7.29) the following replacements:

$$\lambda(x) \longrightarrow \lambda(x) - \frac{i}{2}\sigma^\mu\partial_\mu\overline{\phi}(x),$$

$$D(x) \longrightarrow D(x) - \frac{1}{4}\Box C(x).$$

Then

$$V(x,\theta,\overline{\theta}) = C(x) + \theta\phi(x) + \overline{\theta}\,\overline{\phi}(x) + (\theta\theta)M(x) + (\overline{\theta}\,\overline{\theta})M^*(x)$$
$$+ (\theta\sigma^\mu\overline{\theta})V_\mu(x) + (\theta\theta)\overline{\theta}\left[\overline{\lambda}(x) - \frac{i}{2}\overline{\sigma}^\mu\partial_\mu\phi(x)\right]$$
$$+ (\overline{\theta}\,\overline{\theta})\theta\left[\lambda(x) - \frac{i}{2}\sigma^\mu\partial_\mu\overline{\phi}(x)\right] + (\theta\theta)(\overline{\theta}\,\overline{\theta})\left[D(x) - \frac{1}{4}\Box C(x)\right].$$

$$(7.34)$$

This field again satisfies the condition (7.28) as can be verified using Eq. (1.134). The significance of the choice of components in Eq. (7.34) will become clear below. The following transformation of vector superfields (with V given by Eq. (7.34)) is the supersymmetric generalization of a *gauge transformation*:

$$V(x,\theta,\overline{\theta}) \longrightarrow V'(x,\theta,\overline{\theta})$$
$$= V(x,\theta,\overline{\theta}) + \Phi(x,\theta,\overline{\theta}) + \Phi^\dagger(x,\theta,\overline{\theta})$$
$$\equiv V(x,\theta,\overline{\theta}) + i\left[\Lambda(x,\theta,\overline{\theta}) - \Lambda^\dagger(x,\theta,\overline{\theta})\right]. \quad (7.35)$$

In this transformation $\Phi \equiv i\Lambda$ is any chiral superfield. The component expansion of the transformed vector superfield $V'(x,\theta,\overline{\theta})$ is seen to be (using Eq. (7.33))

$$V'(x,\theta,\overline{\theta}) = C(x) + A(x) + A^*(x)$$
$$+ \theta[\phi(x) + \sqrt{2}\psi(x)]$$
$$+ \overline{\theta}[\overline{\phi}(x) + \sqrt{2}\overline{\psi}(x)]$$
$$+ (\theta\theta)\left[M(x) + F(x)\right]$$
$$+ (\overline{\theta}\,\overline{\theta})\left[M^*(x) + F^*(x)\right]$$
$$+ (\theta\sigma^\mu\overline{\theta})\left[V_\mu(x) + i\partial_\mu\{A(x) - A^*(x)\}\right]$$
$$+ (\theta\theta)\overline{\theta}\left[\overline{\lambda}(x) - \frac{i}{2}\overline{\sigma}^\mu\partial_\mu\{\phi(x) + \sqrt{2}\psi(x)\}\right]$$
$$+ (\overline{\theta}\,\overline{\theta})\theta\left[\lambda(x) - \frac{i}{2}\sigma^\mu\partial_\mu\{\overline{\phi}(x) + \sqrt{2}\overline{\psi}(x)\}\right]$$
$$+ (\theta\theta)(\overline{\theta}\,\overline{\theta})\left[D(x) - \frac{1}{4}\Box\{C(x) + A(x) + A^*(x)\}\right]. \quad (7.36)$$

Hence the transformation (7.35) leads to the following transformation of the component fields:

$$
\begin{aligned}
C(x) &\longrightarrow C'(x) = C(x) + A(x) + A^*(x), \\
\phi(x) &\longrightarrow \phi'(x) + \sqrt{2}\psi(x), \\
M(x) &\longrightarrow M'(x) = M(x) + F(x), \\
V_\mu(x) &\longrightarrow V'_\mu(x) = V_\mu(x) + i\partial_\mu\{A(x) - A^*(x)\}, \\
\lambda(x) &\longrightarrow \lambda'(x) = \lambda(x), \\
D(x) &\longrightarrow D'(x) = D(x).
\end{aligned}
\tag{7.37}
$$

From Eq. (7.37) we see that the special choice of $V(x,\theta,\bar{\theta})$ given by Eq. (7.34) implies that the λ and D component fields are invariant under the transformation (7.35). We also observe that the field $V_\mu(x)$ transforms as

$$
V_\mu(x) \longrightarrow V'_\mu(x) = V_\mu(x) + i\partial_\mu\{A(x) - A^*(x)\},
\tag{7.38}
$$

which corresponds to an Abelian gauge transformation. We also see that

$$
F_{\mu\nu} = \partial_\mu V_\nu - \partial_\nu V_\mu
$$

is *super-gauge invariant*. Following J. Wess and B. Zumino [124], [125], one calls the transformation (7.35) the *supersymmetric extension of a gauge transformation*. Since, as stated above, the $(\theta\theta)(\bar{\theta}\,\bar{\theta})$-component, *i.e.* $D(x)$, is a good candidate for a supersymmetric Lagrangian (it transforms into a spacetime derivative under supersymmetric transformations), we see that the invariance of $D(x)$ under the transformation (7.35) as demonstrated by the relations (7.37) implies the invariance of this Lagrangian under supersymmetric gauge transformations.

From Eqs. (7.37) we see that one can choose a particular field Φ, *i.e.* choose a particular gauge, such that in the gauge transformed vector field $V'(x,\theta,\bar{\theta})$ the component fields C', ϕ' and M' vanish. This gauge is called the *Wess–Zumino gauge*. Taking

$$
\begin{aligned}
\sqrt{2}\psi(x) &= -\phi(x), \\
F(x) &= -M(x), \\
2\,\mathrm{Re}\,A(x) &= A(x) + A^*(x) = -C(x),
\end{aligned}
\tag{7.39}
$$

in Eq. (7.33), the transformed vector field $V_{\mathrm{WZ}}(x,\theta,\bar{\theta})$ assumes the form[2]

$$
V_{\mathrm{WZ}}(x,\theta,\bar{\theta}) = V(x,\theta,\bar{\theta}) + \Phi(x,\theta,\bar{\theta}) + \Phi^\dagger(x,\theta,\bar{\theta})
$$

---

[2]The index WZ indicating the *Wess–Zumino gauge*.

$$
\begin{aligned}
= \ & C(x) - C(x) + \theta[\phi(x) - \phi(x)] + \overline{\theta}[\overline{\phi}(x) - \overline{\phi}(x)] \\
& + (\theta\theta)[M(x) - M(x)] + (\overline{\theta}\,\overline{\theta})[M^*(x) - M^*(x)] \\
& + (\theta\sigma^\mu\overline{\theta})\Big[V_\mu(x) + i\partial_\mu\{A(x) - A^*(x)\}\Big] \\
& + (\overline{\theta}\,\overline{\theta})\theta\Big[\lambda(x) - \frac{i}{2}\sigma^\mu\partial_\mu\{\overline{\phi}(x) - \overline{\phi}(x)\}\Big] \\
& + (\theta\theta)\overline{\theta}\Big[\overline{\lambda}(x) - \frac{i}{2}\overline{\sigma}^\mu\partial_\mu\{\phi(x) - \phi(x)\}\Big] \\
& + (\theta\theta)(\overline{\theta}\,\overline{\theta})\Big[D(x) - \frac{1}{4}\Box\{C(x) - C(x)\}\Big].
\end{aligned}
$$

Hence

$$
\begin{aligned}
V_{\mathrm{WZ}}(x,\theta,\overline{\theta}) = \ & (\theta\sigma^\mu\overline{\theta})\Big[V_\mu(x) + i\partial_\mu\{A(x) - A^*(x)\}\Big] \\
& + (\theta\theta)\overline{\theta}\,\overline{\lambda}(x) + (\overline{\theta}\,\overline{\theta})\theta\lambda(x) + (\theta\theta)(\overline{\theta}\,\overline{\theta})D(x). \quad (7.40)
\end{aligned}
$$

Here $V_\mu(x)$ is the gauge field and λ is its supersymmetric partner. $D(x)$ is the so-called *auxiliary field*, the significance of which will become clear later.

It is important to observe that in Eq. (7.39) we have not fixed the imaginary part of the component scalar field $A(x)$ which causes the shift of the vector field $V_\mu(x)$. Hence the Wess–Zumino gauge does not fix the gauge freedom completely; one still has the gauge degree of freedom of conventional gauge theories. However, the Wess–Zumino gauge breaks supersymmetry in the sense that the supersymmetry variation of $\phi_A(x)$ and $M(x)$ violates the gauge condition

$$
C(x) = \phi_A(x) = M(x) = 0.
$$

For example, comparing Eq. (6.76) with Eq. (7.29) and using Eqs. (6.94) and (7.39), we have

$$
\delta'_s\phi_A(x) = 2\alpha_A M(x) + (\sigma^\mu\overline{\alpha})_A\{i\partial_\mu C(x) + V_\mu(x)\},
$$

and

$$
\delta'_s M(x) = \overline{\alpha}\overline{\lambda}(x) - \frac{i}{2}\partial_\mu\phi(x)\sigma^\mu\overline{\alpha}.
$$

Hence in the Wess–Zumino gauge ($\phi_A = C = M = 0$)

$$
\begin{aligned}
(\delta'_s\phi_A(x))_{WZ} &= \sigma^\mu_{A\dot{A}}\overline{\alpha}^{\dot{A}}V_\mu(x), \\
(\delta'_s M(x))_{WZ} &= \overline{\alpha}\overline{\lambda}(x).
\end{aligned}
$$

Thus the supersymmetry variations of $\phi_A(x)$ and $M(x)$ in this gauge do not vanish. From this we deduce that the Wess–Zumino gauge is noncovariant.

It is easy to compute powers of V_{WZ} by taking into account the anticommuting character of the Grassmann variables θ. With Im $A = 0$, we have (using Eq. (1.137) and $\theta^3 = 0$)

$$
\begin{aligned}
V_{WZ}(x, \theta, \bar{\theta}) &= (\theta \sigma^\mu \bar{\theta}) V_\mu + (\theta\theta) \bar{\theta}\,\bar{\lambda}(x) + (\bar{\theta}\,\bar{\theta})\theta\lambda(x) + (\theta\theta)(\bar{\theta}\,\bar{\theta}) D(x), \\
V_{WZ}^2(x, \theta, \bar{\theta}) &= \frac{1}{2}(\theta\theta)(\bar{\theta}\,\bar{\theta}) V_\mu(x) V^\mu(x), \\
V_{WZ}^3(x, \theta, \bar{\theta}) &= 0.
\end{aligned}
\tag{7.41}
$$

It is the last of these properties which makes the Wess–Zumino gauge a particularly convenient gauge to work in. Then we have, for instance,

$$
\begin{aligned}
\exp\{V\} &= 1 + V + \frac{1}{2}V^2 \\
&= 1 + (\theta\sigma^\mu\bar{\theta}) V_\mu(x) + (\theta\theta)\bar{\theta}\,\bar{\lambda}(x) + (\bar{\theta}\,\bar{\theta})\theta\lambda(x) \\
&\quad + (\theta\theta)(\bar{\theta}\,\bar{\theta})\left\{ D(x) + \frac{1}{4}V_\mu(x) V^\mu(x)\right\}.
\end{aligned}
\tag{7.42}
$$

This exponential of V will be used later in the construction of supersymmetric gauge theories (see Sec. 10.1).

Counting the number of bosonic and fermionic degrees of freedom of the vector super-multiplet before and after Wess–Zumino gauge fixing we have:

(a) In the case of the general vector superfield:

$C(x)$	:	real scalar field :	1	bosonic degree,
$\phi(x)$	:	complex two − spinor field :	4	fermionic degrees,
$M(x)$	:	complex scalar field :	2	bosonic degrees,
$V_\mu(x)$	:	real vector field :	4	bosonic degrees,
$\lambda(x)$	:	complex two − spinor field :	4	fermionic degrees,
$D(x)$	:	real scalar field :	1	bosonic degree.

Thus altogether the multiplet has sixteen degrees of freedom, eight bosonic degrees and eight fermionic degrees of freedom.

(b) In the case of the vector superfield in the Wess–Zumino gauge:[3]

$V_\mu(x)$	:	real vector field :	3	bosonic degrees,
$\lambda(x)$	:	complex two − spinor field :	4	fermionic degrees,
$D(x)$	:	real scalar field :	1	bosonic degree.

Thus altogether the multiplet has eight degrees of freedom, four bosonic degrees of freedom and four fermionic degrees of freedom.

---

[3]Three bosonic degrees since we are still free to choose the gauge of a conventional Abelian gauge theory, *e.g.* $V_0 = 0$.

We observe that the number of fermionic degrees of freedom is the same as the number of bosonic degrees of freedom in either case, as expected from the general considerations following Eq. (4.35).

7.3   The Supersymmetric Field Strength

The general supersymmetric field strength of an arbitrary vector superfield $V(x, \theta, \bar{\theta})$ — not in the special Wess–Zumino gauge — is defined by the components

$$W_A := -\frac{1}{4}(\overline{D}\,\overline{D})D_A V(x, \theta, \bar{\theta}), \tag{7.43}$$

$$\overline{W}_{\dot{A}} := -\frac{1}{4}(DD)\overline{D}_{\dot{A}} V(x, \theta, \bar{\theta}). \tag{7.44}$$

The quantities W_A and $\overline{W}_{\dot{A}}$ are examples of spinor superfields. We first show that W_A and $\overline{W}_{\dot{A}}$ are, in fact, chiral superfields and so represent irreducible representations of the Super–Poincaré algebra. The fact that superfields with spinor indices $A, \dot{A}$ can be chiral in the sense that they satisfy one of the chirality constraints (6.95), (6.96) like the superfields $\Phi, \Phi^\dagger$ without spinor indices is a reflection of the property of these constraints to select irreducible representations of the Super–Poincaré algebra. It may be recalled that in Sec. 4.2 we discussed the corresponding cases of the lowest-dimensional representations for bosonic and fermionic Clifford vacuum states.

Proposition 7.5:

> The field W_A is a left-handed chiral superfield, and $\overline{W}_{\dot{A}}$ is a right-handed chiral superfield. Moreover, both superfields are invariant under the supersymmetric gauge transformation (7.35).

Proof: (i) We first establish the chirality of $W_A, \overline{W}_{\dot{A}}$. This is readily shown since

$$\overline{D}_{\dot{A}}W_A = -\frac{1}{4}\overline{D}_{\dot{A}}(\overline{D}\,\overline{D})D_A V = 0, \tag{7.45a}$$

since $\overline{D}^3 = 0$ (see Eq. (6.54)). Similarly

$$D_A\overline{W}_{\dot{A}} = -\frac{1}{4}D_A(DD)\overline{D}_{\dot{A}} V = 0, \tag{7.45b}$$

since $D^3 = 0$.

(ii) In order to establish the gauge invariance of W_A we apply the transformation (7.35). Then

$$W_A \longrightarrow W_A' = -\frac{1}{4}(\overline{D}\,\overline{D})D_A V' = -\frac{1}{4}(\overline{D}\,\overline{D})D_A(V + \Phi + \Phi^\dagger).$$

Since D and $\overline{D}$ are linear operators we have

$$
\begin{aligned}
W'_A &= -\frac{1}{4}(\overline{D}\,\overline{D})D_A V - \frac{1}{4}(\overline{D}\,\overline{D})D_A \Phi - \frac{1}{4}(\overline{D}\,\overline{D})D_A \Phi^\dagger \\
&\overset{(7.43)}{\underset{(7.15)}{=}} W_A - \frac{1}{4}\overline{D}\,\overline{D}D_A \Phi \\
&\overset{(7.1)}{=} W_A - \frac{1}{4}\overline{D}_{\dot{A}}\overline{D}^{\dot{A}}D_A \Phi - \frac{1}{4}\overline{D}_{\dot{A}}D_A\overline{D}^{\dot{A}}\Phi \quad (\text{since } \overline{D}_{\dot{A}}\Phi = 0) \\
&= W_A - \frac{1}{4}\overline{D}_{\dot{A}}\{\overline{D}^{\dot{A}}, D_A\}\Phi.
\end{aligned}
$$

From Eq. (6.54) we know that $\{D, \overline{D}\}$ closes into P_μ. The latter, however, commutes with $\overline{D}_{\dot{A}}$ so that the last contribution vanishes on account of $\overline{D}_{\dot{A}}\Phi = 0$. Hence

$$
W'_A = W_A.
$$

In a similar way one demonstrates the invariance of $\overline{W}_{\dot{A}}$. Thus, with the help of Eq. (7.35),

$$
\begin{aligned}
\overline{W}_{\dot{A}} \longrightarrow \overline{W}'_{\dot{A}} &= -\frac{1}{4}DD\overline{D}_{\dot{A}}V' = -\frac{1}{4}DD\overline{D}_{\dot{A}}(V + \Phi + \Phi^\dagger) \\
&= -\frac{1}{4}DD\overline{D}_{\dot{A}}V - \frac{1}{4}DD\overline{D}_{\dot{A}}\Phi - \frac{1}{4}DD\overline{D}_{\dot{A}}\Phi^\dagger \\
&\overset{(7.1)}{=} \overline{W}_{\dot{A}} - \frac{1}{4}DD\overline{D}_{\dot{A}}\Phi^\dagger \\
&\overset{(7.9)}{=} \overline{W}_{\dot{A}} - \frac{1}{4}D^A D_A \overline{D}_{\dot{A}}\Phi^\dagger - \frac{1}{4}D^A \overline{D}_{\dot{A}}D_A \Phi^\dagger \\
&= \overline{W}_{\dot{A}} - \frac{1}{4}D^A \{D_A, \overline{D}_{\dot{A}}\}\Phi^\dagger.
\end{aligned}
$$

Using Eq. (6.54) and again Eq. (7.9) this is

$$
\overline{W}'_{\dot{A}} = \overline{W}_{\dot{A}} + \frac{i}{2}D^A \sigma^\mu_{A\dot{A}}\partial_\mu \Phi^\dagger = \overline{W}_{\dot{A}} + \frac{i}{2}\sigma^\mu_{A\dot{A}}\partial_\mu D^A \Phi^\dagger = \overline{W}_{\dot{A}}.
$$

This completes the proof of Proposition 7.5.

Our next task is the calculation of the component expansion of W_A in the Wess–Zumino gauge. This calculation is, of course, simplified if we use the variable

$$
y^\mu := x^\mu + i\theta\sigma^\mu\overline{\theta}
$$

(*cf.* Eq. (7.2)) and correspondingly

$$
z^\mu = x^\mu - i\theta\sigma^\mu\overline{\theta}
$$

(*cf.* Eq. (7.10)) for the calculation of the component expansion of $\overline{W}_{\dot{A}}$. Now,

$$V_{WZ}(x,\theta,\overline{\theta}) = (\theta\sigma^{\mu}\overline{\theta})V_{\mu}(x) + (\theta\theta)\overline{\theta}\,\overline{\lambda}(x) + (\overline{\theta}\,\overline{\theta})\theta\lambda(x) + (\theta\theta)(\overline{\theta}\,\overline{\theta})D(x).$$

Setting

$$x^{\mu} = y^{\mu} - i\theta\sigma^{\mu}\overline{\theta} = z^{\mu} + i\theta\sigma^{\mu}\overline{\theta},$$

we obtain the vector superfields in the coordinates y and z respectively, *i.e.*

$$
\begin{aligned}
V_{WZ}(x,\theta,\overline{\theta}) &= V_{WZ}(y - i\theta\sigma\overline{\theta},\theta,\overline{\theta}) \\
&= (\theta\sigma^{\mu}\overline{\theta})V_{\mu}(y - i\theta\sigma\overline{\theta}) + (\theta\theta)\overline{\theta}\,\overline{\lambda}(y - i\theta\sigma\overline{\theta}) \\
&\quad + (\overline{\theta}\,\overline{\theta})\theta\lambda(y - i\theta\sigma\overline{\theta}) + (\theta\theta)(\overline{\theta}\,\overline{\theta})D(y - i\theta\sigma\overline{\theta}) \\
&= (\theta\sigma^{\mu}\overline{\theta})V_{\mu}(y) - i(\theta\sigma^{\mu}\overline{\theta})(\theta\sigma^{\rho}\overline{\theta})\partial_{\rho}V_{\mu}(y) + (\theta\theta)(\overline{\theta}\,\overline{\lambda}(y)) \\
&\quad + (\overline{\theta}\,\overline{\theta})(\theta\lambda(y)) + (\theta\theta)(\overline{\theta}\,\overline{\theta})D(y) \\
&= (\theta\sigma^{\mu}\overline{\theta})V_{\mu}(y) + (\theta\theta)(\overline{\theta}\,\overline{\lambda}(y)) + (\overline{\theta}\,\overline{\theta})(\theta\lambda(y)) \\
&\quad + (\theta\theta)(\overline{\theta}\,\overline{\theta})\left[D(y) - \frac{i}{2}\partial_{\mu}V^{\mu}(y)\right] \\
&\equiv V_{WZ}^{(1)}(y,\theta,\overline{\theta}), \quad\quad\quad\quad\quad\quad\quad (7.46)
\end{aligned}
$$

where we used Eq. (1.137), and similarly

$$
\begin{aligned}
V_{WZ}(x,\theta,\overline{\theta}) &= V_{WZ}(z + i\theta\sigma\overline{\theta},\theta,\overline{\theta}) \\
&= (\theta\sigma^{\mu}\overline{\theta})V_{\mu}(z + i\theta\sigma\overline{\theta}) + (\theta\theta)\overline{\theta}\,\overline{\lambda}(z + i\theta\sigma\overline{\theta}) \\
&\quad + (\overline{\theta}\,\overline{\theta})\theta\lambda(z + i\theta\sigma\overline{\theta}) + (\theta\theta)(\overline{\theta}\,\overline{\theta})D(z + i\theta\sigma\overline{\theta}) \\
&= (\theta\sigma^{\mu}\overline{\theta})V_{\mu}(z) + i(\theta\sigma^{\mu}\overline{\theta})(\theta\sigma^{\nu}\overline{\theta})\partial_{\nu}V_{\mu}(z) + (\theta\theta)(\overline{\theta}\,\overline{\lambda}(z)) \\
&\quad + (\overline{\theta}\,\overline{\theta})(\theta\lambda(z)) + (\theta\theta)(\overline{\theta}\,\overline{\theta})D(z) \\
&= (\theta\sigma^{\mu}\overline{\theta})V_{\mu}(z) + (\theta\theta)(\overline{\theta}\,\overline{\lambda}(z)) + (\overline{\theta}\,\overline{\theta})\theta\lambda(z) \\
&\quad + (\theta\theta)(\overline{\theta}\,\overline{\theta})\left[D(z) + \frac{i}{2}\partial_{\mu}V^{\mu}(z)\right] \equiv V_{WZ}^{(2)}(z,\theta,\overline{\theta}), \quad (7.47)
\end{aligned}
$$

again using Eq. (1.137). Thus we have

$$V_{WZ}(x,\theta,\overline{\theta}) = V_{WZ}^{(1)}(y,\theta,\overline{\theta}) = V^{(2)}(z,\theta,\overline{\theta}).$$

A suitable set of coordinates to calculate the component expansion of W_A in Eq. (7.45a) is given by $(y,\theta,\overline{\theta})$. Then

$$W_A = -\frac{1}{4}\overline{D}\,\overline{D}D_A V_{WZ}(x,\theta,\overline{\theta}) = -\frac{1}{4}\overline{D}^{(1)}\overline{D}^{(1)}D_A^{(1)}V_{WZ}^{(1)}(y,\theta,\overline{\theta}),$$

since in terms of y the covariant derivatives D and $\overline{D}$ are given by $D^{(1)}$ and $\overline{D}^{(1)}$ of Eqs. (7.4a) and (7.4b). Considering the first derivatives we have:

$$
D_A^{(1)} V_{WZ}^{(1)}(y,\theta,\overline{\theta}) \overset{(6.48)}{=} (\partial_A + 2i\sigma_{A\dot{B}}^{\mu}\overline{\theta}^{\dot{B}}\partial_\mu) V_{WZ}^{(1)}(y,\theta,\overline{\theta})
$$

$$
\overset{(7.46)}{=} \partial_A \Big\{ (\theta\sigma^\mu\overline{\theta})V_\mu(y) + (\theta\theta)\overline{\theta}\,\overline{\lambda}(y) + (\overline{\theta}\,\overline{\theta})\theta\lambda(y)
$$

$$
+ (\theta\theta)(\overline{\theta}\,\overline{\theta})\Big[D(y) - \frac{i}{2}\partial_\mu V^\mu(y)\Big]\Big\}
$$

$$
+ 2i\sigma_{A\dot{B}}^{\rho}\overline{\theta}^{\dot{B}}\partial_\rho \Big\{ (\theta\sigma^\mu\overline{\theta})V_\mu(y) + (\theta\theta)\overline{\theta}\,\overline{\lambda}(y)
$$

$$
+ (\overline{\theta}\,\overline{\theta})\theta\lambda(y) + (\theta\theta)(\overline{\theta}\,\overline{\theta})\Big[D(y) - \frac{i}{2}\partial_\mu V^\mu(y)\Big]\Big\},
$$

and hence

$$
D_A^{(1)} V_{WZ}^{(1)}(y,\theta,\overline{\theta}) = (\partial_A\theta^B)\sigma_{B\dot{C}}^{\mu}\overline{\theta}^{\dot{C}}V_\mu(y) + \partial_A(\theta\theta)\overline{\theta}\,\overline{\lambda}(y)
$$

$$
+ (\overline{\theta}\,\overline{\theta})\partial_A\theta^B\lambda_B(y) + \partial_A(\theta\theta)(\overline{\theta}\,\overline{\theta})\Big[D(y) - \frac{i}{2}\partial_\mu V^\mu(y)\Big]
$$

$$
+ 2i\sigma_{A\dot{B}}^{\rho}\overline{\theta}^{\dot{B}}\theta^C\sigma_{C\dot{D}}^{\mu}\overline{\theta}^{\dot{D}}V_\mu(y)
$$

$$
+ 2i\sigma_{A\dot{B}}^{\rho}\overline{\theta}^{\dot{B}}(\theta\theta)\overline{\theta}\partial_\rho\overline{\lambda}(y) \quad \text{(since } (\overline{\theta}\,\overline{\theta})\overline{\theta} = 0)
$$

$$
= \sigma_{A\dot{C}}^{\mu}\overline{\theta}^{\dot{C}}V_\mu(y) + 2\theta_A\overline{\theta}\,\overline{\lambda}(y) + \overline{\theta}\,\overline{\theta}\lambda_A(y)
$$

$$
+ 2\theta_A(\overline{\theta}\,\overline{\theta})\Big[D(y) - \frac{i}{2}\partial_\mu V^\mu(y)\Big]
$$

$$
+ \underbrace{2i\sigma_{A\dot{B}}^{\rho}\overline{\theta}^{\dot{B}}\theta^C\sigma_{C\dot{D}}^{\mu}\overline{\theta}^{\dot{D}}\partial_\rho V_\mu(y)}_{T_1}
$$

$$
+ \underbrace{2i\sigma_{A\dot{B}}^{\rho}\overline{\theta}^{\dot{B}}(\theta\theta)\overline{\theta}_{\dot{C}}\partial_\rho\overline{\lambda}^{\dot{C}}(y)}_{T_2}. \tag{7.48}
$$

The underbraced term T_1 can be rewritten as follows (with $\mu \longleftrightarrow \rho, C \longleftrightarrow D$):

$$
T_1 = -2i\sigma_{A\dot{B}}^{\mu}\overline{\theta}^{\dot{B}}\overline{\theta}^{\dot{D}}\theta^C\sigma_{C\dot{D}}^{\rho}\partial_\mu V_\rho(y) \overset{(1.100c)}{=} -i\sigma_{A\dot{B}}(\overline{\theta}\,\overline{\theta})\epsilon^{\dot{B}\dot{D}}\theta^C\sigma_{C\dot{D}}^{\rho}\partial_\mu V_\rho(y)
$$

$$
= -i\sigma_{A\dot{B}}^{\mu}(\overline{\theta}\,\overline{\theta})\epsilon^{\dot{B}\dot{D}}\epsilon^{CD}\sigma_{C\dot{D}}^{\rho}\partial_\mu V_\rho(y)\theta_D \quad \text{(using Eq. (1.72))}
$$

$$
= i\sigma_{A\dot{B}}^{\mu}(\overline{\theta}\,\overline{\theta})\epsilon^{DC}\epsilon^{\dot{B}\dot{D}}\sigma_{C\dot{D}}^{\rho}\partial_\mu V_\rho(y)\theta_D \overset{(1.106a)}{=} i(\overline{\theta}\,\overline{\theta})\sigma_{A\dot{B}}^{\mu}\overline{\sigma}^{\rho\dot{B}D}\partial_\mu V_\rho(y)\theta_D
$$

$$
= i(\overline{\theta}\,\overline{\theta})(\sigma^\mu\overline{\sigma}^\rho)_A^{\ D}\partial_\mu V_\rho(y)\theta_D.
$$

The last term of Eq. (7.48), T_2, can be written (using $\overline{\theta}_{\dot{C}}\overline{\lambda}^{\dot{C}} = -\overline{\theta}^{\dot{C}}\overline{\lambda}_{\dot{C}}$):

$$
\begin{aligned}
T_2 &= 2i\sigma^\mu_{A\dot{B}}\overline{\theta}^{\dot{B}}(\theta\theta)\overline{\theta}_{\dot{C}}\partial_\mu\overline{\lambda}^{\dot{C}}(y) = -2i\sigma^\mu_{A\dot{B}}\overline{\theta}^{\dot{B}}\overline{\theta}^{\dot{C}}(\theta\theta)\partial_\mu\overline{\lambda}_{\dot{C}}(y) \\
&\overset{(1.100c)}{=} -i\sigma^\mu_{A\dot{B}}(\overline{\theta}\,\overline{\theta})\epsilon^{\dot{B}\dot{C}}(\theta\theta)\partial_\mu\overline{\lambda}_{\dot{C}}(y) = -i(\theta\theta)(\overline{\theta}\,\overline{\theta})\sigma^\mu_{A\dot{B}}\partial_\mu\overline{\lambda}^{\dot{B}}(y).
\end{aligned}
$$

Hence

$$
\begin{aligned}
D^{(1)}_A V^{(1)}_{WZ}(y,\theta,\overline{\theta}) &= \sigma^\mu_{A\dot{B}}\overline{\theta}^{\dot{B}}V_\mu(y) + 2\theta_A\overline{\theta}\,\overline{\lambda}(y) + (\overline{\theta}\,\overline{\theta})\lambda_A(y) + (\overline{\theta}\,\overline{\theta})\big[2\delta_A{}^B D(y) \\
&\quad + i(\sigma^\mu\overline{\sigma}^\nu)_A{}^B\partial_\mu V_\nu(y) - i\delta_A{}^B\eta^{\mu\nu}\partial_\mu V_\nu(y)\big]\theta_B \\
&\quad - i(\theta\theta)(\overline{\theta}\,\overline{\theta})(\sigma^\mu\partial_\mu\overline{\lambda}(y))_A.
\end{aligned} \tag{7.49}
$$

In rewriting this expression we make use of the following result.

Proposition 7.6:

If $\sigma^{\mu\nu}$ is given by Eq. (1.138a), it can be expressed in the following way:

$$
2\sigma^{\mu\nu}{}_A{}^B = i\big[-\delta_A{}^B\eta^{\mu\nu} + (\sigma^\mu\overline{\sigma}^\nu)_A{}^B\big]. \tag{7.50}
$$

Proof: From Eq. (1.141a) we know that

$$
(\sigma^\mu\overline{\sigma}^\nu)_A{}^B + (\sigma^\nu\overline{\sigma}^\mu)_A{}^B = 2\eta^{\mu\nu}\delta_A{}^B.
$$

Subtracting from both sides $2(\sigma^\mu\overline{\sigma}^\nu)_A{}^B$, we obtain

$$
-(\sigma^\mu\overline{\sigma}^\nu)_A{}^B + (\sigma^\nu\overline{\sigma}^\mu)_A{}^B = 2(\eta^{\mu\nu}\delta_A{}^B - \sigma^\mu\overline{\sigma}^\nu_A{}^B).
$$

But (see Eq. (1.138a))

$$
\frac{i}{4}(\sigma^\nu\overline{\sigma}^\mu - \sigma^\mu\overline{\sigma}^\nu)_A{}^B = (\sigma^{\nu\mu})_A{}^B.
$$

Hence

$$
(-\sigma^\mu\overline{\sigma}^\nu + \sigma^\nu\overline{\sigma}^\mu)_A{}^B = -2[-\delta_A{}^B\eta^{\mu\nu} + (\sigma^\mu\overline{\sigma}^\nu)_A{}^B],
$$

and so

$$
-4i(\sigma^{\nu\mu})_A{}^B = -2[-\delta_A{}^B\eta^{\mu\nu} + (\sigma^\mu\overline{\sigma}^\nu)_A{}^B],
$$

i.e.

$$
2(\sigma^{\mu\nu})_A{}^B = i[-\delta_A{}^B\eta^{\mu\nu} + (\sigma^\mu\overline{\sigma}^\nu)_A{}^B],
$$

which had to be shown.

Using Eq. (7.50) we can rewrite the coefficient of the term in $(\overline{\theta}\,\overline{\theta})$ of Eq. (7.49), *i.e.*

$$
\begin{aligned}
2\delta_A{}^B D(y) &+ i(\sigma^\mu\overline{\sigma}^\nu)_A{}^B \partial_\mu V_\nu(y) - i\delta_A{}^B \eta^{\mu\nu}\partial_\mu V_\nu(y)\\
&= 2\delta_A{}^B D(y) + i\bigl[-\delta_A{}^B \eta^{\mu\nu} + (\sigma^\mu\overline{\sigma}^\nu)_A{}^B\bigr]\partial_\mu V_\nu(y)\\
&= 2\delta_A{}^B D(y) + 2\sigma^{\mu\nu}{}_A{}^B \partial_\mu V_\nu(y)\\
&= 2\delta_A{}^B D(y) + \sigma^{\mu\nu}{}_A{}^B \partial_\mu V_\nu(y) + \sigma^{\nu\mu}{}_A{}^B \partial_\nu V_\mu(y)\\
&= 2\delta_A{}^B D(y) + \sigma^{\mu\nu}{}_A{}^B \bigl[\partial_\mu V_\nu(y) - \partial_\nu V_\mu(y)\bigr]\\
&= 2\delta_A{}^B D(y) + \sigma^{\mu\nu}{}_A{}^B F_{\mu\nu}(y),
\end{aligned}
$$

where we made use of the antisymmetry of $\sigma^{\mu\nu}$ in μ and ν, and we set

$$
F_{\mu\nu}(y) := \partial_\mu V_\nu(y) - \partial_\nu V_\mu(y),
$$

as the usual expression for the field strength tensor. Hence we obtain:

$$
\begin{aligned}
D_A^{(1)} V_{WZ}^{(1)}(y,\theta,\overline{\theta}) &= \sigma^\mu{}_{A\dot{B}}\overline{\theta}^{\dot{B}} V_\mu(y) + 2\theta_A\overline{\theta}\,\overline{\lambda}(y) + (\overline{\theta}\,\overline{\theta})\lambda_A(y)\\
&\quad + (\overline{\theta}\,\overline{\theta})\bigl[2\delta_A{}^B D(y) + (\sigma^{\mu\nu})_A{}^B F_{\mu\nu}(y)\bigr]\theta_B\\
&\quad - i(\theta\theta)(\overline{\theta}\,\overline{\theta})\sigma^\mu{}_{A\dot{B}}\partial_\mu\overline{\lambda}^{\dot{B}}(y). \tag{7.51}
\end{aligned}
$$

Then

$$
\begin{aligned}
W_A &= -\frac{1}{4}\overline{D}^{(1)}\overline{D}^{(1)} D_A^{(1)} V_{WZ}^{(1)}(y,\theta,\overline{\theta}) \overset{(6.49)}{=} -\frac{1}{4}\overline{\partial}_{\dot{A}}\overline{\partial}^{\dot{A}}\bigl(D_A^{(1)} W_{WZ}^{(1)}(y,\theta,\overline{\theta})\bigr)\\
&= -\frac{1}{4}\epsilon_{\dot{A}\dot{B}}\overline{\partial}^{\dot{B}}\overline{\partial}^{\dot{A}}\bigl(D_A^{(1)} V_{WZ}^{(1)}(y,\theta,\overline{\theta})\bigr) = \frac{1}{4}\epsilon_{\dot{A}\dot{B}}\overline{\partial}^{\dot{A}}\overline{\partial}^{\dot{B}}\bigl(D_A^{(1)} V_{WZ}^{(1)}(y,\theta,\overline{\theta})\bigr)\\
&= \frac{1}{4}\epsilon_{\dot{A}\dot{B}}\overline{\partial}^{\dot{A}}\overline{\partial}^{\dot{B}}\Bigl\{\sigma^\mu{}_{A\dot{C}}\overline{\theta}^{\dot{C}} V_\mu(y) + 2\theta_A\overline{\theta}\,\overline{\lambda}(y) + (\overline{\theta}\,\overline{\theta})\lambda_A(y)\\
&\quad + (\overline{\theta}\,\overline{\theta})\bigl[2\delta_A{}^B D(y) + \sigma^{\mu\nu}{}_A{}^B F_{\mu\nu}(y)\bigr]\theta_B - i(\theta\theta)(\overline{\theta}\,\overline{\theta})\sigma^\mu{}_{A\dot{B}}\partial_\mu\overline{\lambda}^{\dot{B}}(y)\Bigr\}\\
&= \lambda_A(y) + 2\theta_A D(y) + \sigma^{\mu\nu}{}_A{}^B \theta_B F_{\mu\nu} - i(\theta\theta)\sigma^\mu{}_{A\dot{B}}\partial_\mu\overline{\lambda}^{\dot{B}}(y).
\end{aligned}
$$

Hence the component expansion of W_A is given by

$$
W_A = \lambda_A(y) + 2D(y)\theta_A + (\sigma^{\mu\nu}\theta)_A F_{\mu\nu}(y) - i(\theta\theta)\sigma^\mu{}_{A\dot{B}}\partial_\mu\overline{\lambda}^{\dot{B}}(y). \tag{7.52}
$$

In a similar way we can find that the component expansion of $\overline{W}_{\dot{A}}$ is given by

$$
\overline{W}_{\dot{A}} = \overline{\lambda}_{\dot{A}}(z) + 2D(z)\overline{\theta}_{\dot{A}} - \epsilon_{\dot{A}\dot{B}}(\overline{\sigma}^{\mu\nu}\overline{\theta})^{\dot{B}} F_{\mu\nu}(z) + i(\overline{\theta}\,\overline{\theta})(\partial_\mu\lambda(z)\sigma^\mu)_{\dot{A}}. \tag{7.53}
$$

In order to verify this result we recall that $(z^\mu, \theta, \overline{\theta})$ is a convenient set of coordinates in which we can compute the component expansion of $\overline{W}_{\dot{A}}$. Thus

$$\overline{W}_{\dot{A}} = -\frac{1}{4}DD\overline{D}_{\dot{A}}V_{WZ}(x, \theta, \overline{\theta}) = -\frac{1}{4}D^{(2)}D^{(2)}\overline{D}_{\dot{A}}^{(2)}V_{WZ}^{(2)}(z, \theta, \overline{\theta}).$$

We first calculate the expansion of $\overline{D}_{\dot{A}}^{(2)}V_{WZ}^{(2)}$, *i.e.*

$$\overline{D}_{\dot{A}}^{(2)}V_{WZ}^{(2)}(z, \theta, \overline{\theta}) \overset{(6.51)}{=} \{-\overline{\partial}_{\dot{A}} - 2i\theta^A\sigma_{A\dot{A}}^\mu\partial_\mu\}V_{WZ}^{(2)}(z, \theta, \overline{\theta})$$

$$= -\overline{\partial}_{\dot{A}}V_{WZ}^{(2)}(z, \theta, \overline{\theta}) - 2i(\theta\sigma^\mu)_{\dot{A}}\partial_\mu V_{WZ}^{(2)}(z, \theta, \overline{\theta})$$

$$\overset{(7.47)}{=} -\overline{\partial}_{\dot{A}}\left\{\theta\sigma^\mu\overline{\theta}V_\mu(z) + (\theta\theta)\overline{\theta}\,\overline{\lambda}(z) + (\overline{\theta}\,\overline{\theta})\theta\lambda(z)\right.$$

$$+ (\theta\theta)(\overline{\theta}\,\overline{\theta})\left[D(z) + \frac{i}{2}\partial_\mu V^\mu(z)\right]\Big\}$$

$$- 2i(\theta\sigma^\mu)_{\dot{A}}\partial_\mu\left\{\theta\sigma^\rho\overline{\theta}V_\rho(z) + (\theta\theta)\overline{\theta}\,\overline{\lambda}(z) + (\overline{\theta}\,\overline{\theta})\theta\lambda(z)\right.$$

$$+ (\theta\theta)(\overline{\theta}\,\overline{\theta})\left[D(z) + \frac{i}{2}\partial_\mu V^\mu(z)\right]\Big\}$$

and so

$$\overline{D}_{\dot{A}}^{(2)}V_{WZ}^{(2)}(z, \theta, \overline{\theta}) = \theta^A\sigma_{A\dot{B}}^\mu\overline{\partial}_{\dot{A}}\overline{\theta}^{\dot{B}}V_\mu(z) - (\theta\theta)\overline{\partial}_{\dot{A}}\overline{\theta}_{\dot{B}}\overline{\lambda}^{\dot{B}}(z)$$

$$- \overline{\partial}_{\dot{A}}(\overline{\theta}\,\overline{\theta})\theta\lambda(z) - (\theta\theta)\overline{\partial}_{\dot{A}}(\overline{\theta}\,\overline{\theta})\left[D(z) + \frac{i}{2}\partial_\mu V^\mu(z)\right]$$

$$\underbrace{- 2i(\theta\sigma^\mu)_{\dot{A}}(\theta\sigma^\rho\overline{\theta})\partial_\mu V_\rho(z)}_{T_1}$$

$$\underbrace{- 2i(\theta\sigma^\mu)_{\dot{A}}(\overline{\theta}\,\overline{\theta})\theta\partial_\mu\lambda(z)}_{T_2} + O(\theta^3)$$

$$= (\theta\sigma^\mu)_{\dot{A}}V_\mu(z) + (\theta\theta)\overline{\lambda}_{\dot{A}}(z) + 2\overline{\theta}_{\dot{A}}\theta\lambda(z)$$

$$+ 2(\theta\theta)\overline{\theta}_{\dot{A}}\left[D(z) + \frac{i}{2}\partial_\mu V^\mu(z)\right]$$

$$- i(\theta\theta)\epsilon_{\dot{A}\dot{B}}(\overline{\sigma}^\mu\sigma^\nu)^{\dot{B}}{}_{\dot{C}}\partial_\mu V_\nu(z)\overline{\theta}^{\dot{C}} + i(\theta\theta)(\overline{\theta}\,\overline{\theta})(\partial_\mu\lambda(z)\sigma^\mu)_{\dot{A}},$$

where the terms of order θ^3 vanish and we used the following relations:

(i) $\overline{\partial}_{\dot{A}}\overline{\theta}_{\dot{B}} = -\epsilon_{\dot{A}\dot{B}}$,   (see Eq. (6.42)),

(ii) $\overline{\partial}_{\dot{A}}(\overline{\theta}\,\overline{\theta}) = -2\overline{\theta}_{\dot{A}}$,   (as in Eq. (6.5p)).

(iii) Considering the expression underbraced T_1, we have

$$
\begin{aligned}
T_1 &= -2i\theta^B\theta^C\sigma^\mu_{B\dot{A}}\sigma^\nu_{C\dot{D}}\overline{\theta}^{\dot{D}}\partial_\mu V_\nu(z)\\
&\overset{(1.100a)}{=} i(\theta\theta)\epsilon^{BC}\sigma^\mu_{B\dot{A}}\sigma^\nu_{C\dot{D}}\overline{\theta}^{\dot{D}}\partial_\mu V_\nu(z)\\
&= -i(\theta\theta)\epsilon^{BC}\sigma^\mu_{B\dot{B}}\delta^{\dot{B}}_{\dot{A}}\sigma^\nu_{C\dot{D}}\overline{\theta}^{\dot{D}}\partial_\mu V_\nu(z)\\
&\overset{(1.78b)}{=} i(\theta\theta)\epsilon^{BC}\sigma^\mu_{B\dot{B}}\epsilon^{\dot{B}\dot{C}}\epsilon_{\dot{C}\dot{A}}\sigma^\nu_{C\dot{D}}\overline{\theta}^{\dot{D}}\partial_\mu V_\nu(z)\\
&= -i(\theta\theta)\epsilon_{\dot{A}\dot{C}}\epsilon^{\dot{C}\dot{B}}\epsilon^{CB}\sigma^\mu_{B\dot{B}}\sigma^\nu_{C\dot{D}}\overline{\theta}^{\dot{D}}\partial_\mu V_\nu(z)\\
&\overset{(1.106a)}{=} -i(\theta\theta)\epsilon_{\dot{A}\dot{C}}(\overline{\sigma}^\mu)^{\dot{C}C}(\sigma^\nu)_{C\dot{D}}\overline{\theta}^{\dot{D}}\partial_\mu V_\nu(z)\\
&= -i(\theta\theta)\epsilon_{\dot{A}\dot{C}}(\overline{\sigma}^\mu\sigma^\nu)^{\dot{C}}{}_{\dot{D}}\overline{\theta}^{\dot{D}}\partial_\mu V_\nu(z).
\end{aligned}
$$

(iv) Considering the expression underbraced T_2:

$$
\begin{aligned}
T_2 &= -2i\theta^B\sigma^\mu_{B\dot{A}}(\overline{\theta}\,\overline{\theta})\theta^C\partial_\mu\lambda_C(z) = -2i\theta^B\theta^C\sigma^\mu_{B\dot{A}}(\overline{\theta}\,\overline{\theta})\partial_\mu\lambda_C(z)\\
&\overset{(1.100a)}{=} -i(\theta\theta)(\overline{\theta}\,\overline{\theta})\epsilon^{BC}\sigma^\mu_{B\dot{A}}\partial_\mu\lambda_C(z) = -i(\theta\theta)(\overline{\theta}\,\overline{\theta})\partial_\mu\lambda_C(z)\epsilon^{CB}\sigma^\mu_{B\dot{A}}\\
&= i(\theta\theta)(\overline{\theta}\,\overline{\theta})\epsilon^{BC}\partial_\mu\lambda_C(z)\sigma^\mu_{B\dot{A}} = i(\theta\theta)(\overline{\theta}\,\overline{\theta})(\partial_\mu\lambda(z)\sigma^\mu)_{\dot{A}}.
\end{aligned}
$$

Hence

$$
\begin{aligned}
\overline{D}^{(2)}_{\dot{A}}V^{(2)}_{WZ}(z,\theta,\overline{\theta}) &= (\theta\sigma^\mu)_{\dot{A}}V_\mu(z) + (\theta\theta)\overline{\lambda}_{\dot{A}}(z) + 2\overline{\theta}_{\dot{A}}(\theta\lambda(z))\\
&\quad +(\theta\theta)\{2\epsilon_{\dot{A}\dot{C}}D(z) + i\epsilon_{\dot{A}\dot{C}}\eta^{\mu\nu}\partial_\mu V_\nu(z)\\
&\quad -i\epsilon_{\dot{A}\dot{B}}(\overline{\sigma}^\mu\sigma^\nu)^{\dot{B}}{}_{\dot{C}}\partial_\mu V_\nu(z)\}\overline{\theta}^{\dot{C}} + i(\theta\theta(\overline{\theta}\,\overline{\theta})(\partial_\mu\lambda(z)\sigma^\mu)_{\dot{A}}\\
&= (\theta\sigma^\mu)_{\dot{A}}V_\mu(z) + (\theta\theta)\overline{\lambda}_{\dot{A}}(z) + 2\overline{\theta}_{\dot{A}}(\theta\lambda(z))\\
&\quad +(\theta\theta)\Big\{2\epsilon_{\dot{A}\dot{C}}D(z) - i\epsilon_{\dot{A}\dot{B}}\big[-\delta^{\dot{B}}_{\dot{C}}\eta^{\mu\nu}\\
&\quad +(\overline{\sigma}^\mu\sigma^\nu)^{\dot{B}}{}_{\dot{C}}\big]\partial_\mu V_\nu(z)\Big\}\overline{\theta}^{\dot{C}} + i(\theta\theta)(\overline{\theta}\,\overline{\theta})(\partial_\mu\lambda(z)\sigma^\mu)_{\dot{A}}.
\end{aligned}
\tag{7.54}
$$

Next we need the following result.

Proposition 7.7:

The following relation holds:

$$
2\overline{\sigma}^{\mu\nu\,\dot{B}}{}_{\dot{C}} = i\big[-\delta^{\dot{B}}_{\dot{C}}\eta^{\mu\nu} + (\overline{\sigma}^\mu\sigma^\nu)^{\dot{B}}{}_{\dot{C}}\big].
\tag{7.55}
$$

Proof: From Eq. (1.141b) we have

$$(\overline{\sigma}^\mu \sigma^\nu + \overline{\sigma}^\nu \sigma^\mu)^{\dot{B}}{}_{\dot{C}} = 2\eta^{\mu\nu} \delta^{\dot{B}}{}_{\dot{C}}.$$

Subtracting from both sides $2(\overline{\sigma}^\mu \sigma^\nu)^{\dot{B}}{}_{\dot{C}}$, we obtain

$$-(\overline{\sigma}^\mu \sigma^\nu - \overline{\sigma}^\nu \sigma^\mu)^{\dot{B}}{}_{\dot{C}} = 2\eta^{\mu\nu} \delta^{\dot{B}}{}_{\dot{C}} - 2(\overline{\sigma}^\mu \sigma^\nu)^{\dot{B}}{}_{\dot{C}}.$$

Using Eq. (1.138b),

$$4i(\overline{\sigma}^{\mu\nu})^{\dot{B}}{}_{\dot{C}} = 2\eta^{\mu\nu} \delta^{\dot{B}}{}_{\dot{C}} - 2(\overline{\sigma}^\mu \sigma^\nu)^{\dot{B}}{}_{\dot{C}},$$

i.e.

$$2\overline{\sigma}^{\mu\nu\,\dot{B}}{}_{\dot{C}} = i\big[-\delta^{\dot{B}}{}_{\dot{C}} \eta^{\mu\nu} + (\overline{\sigma}^\mu \sigma^\nu)^{\dot{B}}{}_{\dot{C}}\big],$$

which is Eq. (7.55).

Using Eq. (7.55) we can rewrite Eq. (7.54):

$$
\begin{aligned}
\overline{D}^{(2)}_{\dot{A}} V^{(2)}_{WZ}(z,\theta,\overline{\theta}) &= (\theta\sigma^\mu)_{\dot{A}} V_\mu(z) + (\theta\theta)\overline{\lambda}_{\dot{A}}(z) + 2\overline{\theta}_{\dot{A}}(\theta\lambda(z))\\
&\quad +(\theta\theta)\{2\epsilon_{\dot{A}\dot{C}}D(z) - 2\epsilon_{\dot{A}\dot{B}}(\overline{\sigma}^{\mu\nu})^{\dot{B}}{}_{\dot{C}}\partial_\mu V_\nu(z)\}\overline{\theta}^{\dot{C}}\\
&\quad +i(\theta\theta)(\overline{\theta}\,\overline{\theta})(\partial_\mu\lambda(z)\sigma^\mu)_{\dot{A}}.
\end{aligned}
$$

Using the antisymmetry of $\overline{\sigma}^{\mu\nu}$ in μ and ν, we obtain

$$
\begin{aligned}
\overline{\sigma}^{\mu\nu}\partial_\mu V_\nu(z) &= \frac{1}{2}\overline{\sigma}^{\mu\nu}\partial_\mu V_\nu(z) + \frac{1}{2}\overline{\sigma}^{\nu\mu}\partial_\nu V_\mu(z)\\
&= \frac{1}{2}\overline{\sigma}^{\mu\nu}[\partial_\mu V_\nu(z) - \partial_\nu V_\mu(z)] = \frac{1}{2}\overline{\sigma}^{\mu\nu} F_{\mu\nu}(z),
\end{aligned}
$$

where as before

$$F_{\mu\nu} = \partial_\mu V_\nu(z) - \partial_\nu V_\mu(z).$$

Hence

$$
\begin{aligned}
\overline{D}^{(2)}_{\dot{A}} V^{(2)}_{WZ}(z,\theta,\overline{\theta}) &= (\theta\sigma^\mu)_{\dot{A}} V_\mu(z) + (\theta\theta)\overline{\lambda}_{\dot{A}}(z) + 2\overline{\theta}_{\dot{A}}\theta\lambda(z)\\
&\quad +(\theta\theta)\{2\epsilon_{\dot{A}\dot{C}}D(z) - \epsilon_{\dot{A}\dot{B}}(\overline{\sigma}^{\mu\nu})^{\dot{B}}{}_{\dot{C}}F_{\mu\nu}(z)\}\overline{\theta}^{\dot{C}}\\
&\quad +i(\theta\theta)(\overline{\theta}\,\overline{\theta})(\partial_\mu\lambda(z)\sigma^\mu)_{\dot{A}}. \qquad (7.56)
\end{aligned}
$$

Now we can easily obtain $\overline{W}_{\dot{A}}$, *i.e.*

$$
\begin{aligned}
\overline{W}_{\dot{A}} &= -\frac{1}{4} D^{(2)} D^{(2)} \overline{D}^{(2)}_{\dot{A}} V^{(2)}_{WZ}(z,\theta,\overline{\theta})\\
&\overset{(6.50)}{=} -\frac{1}{4}\partial^A\partial_A \overline{D}^{(2)}_{\dot{A}} V^{(2)}_{WZ}(z,\theta,\overline{\theta}) = \frac{1}{4}\epsilon^{AB}\partial_A\partial_B\{\overline{D}^{(2)}_{\dot{A}} V^{(2)}_{WZ}(z,\theta,\overline{\theta})\}\\
&\overset{(6.5q)}{=} \overline{\lambda}_{\dot{A}}(z) + 2\epsilon_{\dot{A}\dot{C}}D(z)\overline{\theta}^{\dot{C}} - \epsilon_{\dot{A}\dot{B}}(\overline{\sigma}^{\mu\nu})^{\dot{B}}{}_{\dot{C}}F_{\mu\nu}(z)\overline{\theta}^{\dot{C}} + i(\overline{\theta}\,\overline{\theta})(\partial_\mu\lambda(z)\sigma^\mu)_{\dot{A}}.
\end{aligned}
$$

We have thus established Eq. (7.53).

We see from Eqs. (7.52) and (7.53) that the superfields W_A and $\overline{W}_{\dot{A}}$ contain only the gauge-invariant fields D, λ and $F_{\mu\nu} = \partial_\mu V_\nu - \partial_\nu V_\mu$. Furthermore, as shown with Eqs. (7.45a) and (7.45b), these fields are chiral. We can also prove the following result.

Proposition 7.8:

The fields $W_A, \overline{W}^{\dot{A}}$ obey the following relation:

$$\overline{D}_{\dot{A}}\overline{W}^{\dot{A}} = D^A W_A. \tag{7.57}$$

Proof: We have:

$$
\begin{aligned}
\overline{D}_{\dot{A}}\overline{W}^{\dot{A}} &= \epsilon^{\dot{A}\dot{B}}\overline{D}_{\dot{A}}\overline{W}_{\dot{B}} \\
&\overset{(7.44)}{=} \epsilon^{\dot{A}\dot{B}}\left\{ -\frac{1}{4}\overline{D}_{\dot{A}}(DD)\overline{D}_{\dot{B}}V(x,\theta,\overline{\theta}) \right\} \\
&= -\frac{1}{4}\overline{D}_{\dot{A}}(DD)\overline{D}^{\dot{A}}V(x,\theta,\overline{\theta}) \\
&\overset{(6.71)}{=} -\frac{1}{4}D^A(\overline{D}\,\overline{D})D_A V(x,\theta,\overline{\theta}) \overset{(7.43)}{=} D^A W_A.
\end{aligned}
$$

Chapter 8

Supersymmetric Lagrangians

8.1 Grassmann Integration

As in the case of differentiation with respect to a Grassmann variable in Sec. 6.2, we begin by defining the relevant symbol, *i.e.* the integral[1]

$$\int da f(a) = I[f],$$

where "a" denotes a single Grassmann number. Since Grassmann variables are discrete objects, the integral does not represent the area under a curve $f(a)$, nor is any meaning attached to upper and lower limits of the integral. Rather we define a functional, which associates a c-number $I[f]$ with every element $f(a) \in G$ (G being the Grassmann algebra[2] with one element). Furthermore we demand
(i)

$$\int da f(a + b) = \int da f(a), \tag{8.1}$$

implying translation invariance of the integral,
(ii)

$$\int da \{\alpha f(a) + \beta g(a)\} = \alpha \int da f(a) + \beta \int da g(a), \tag{8.2}$$

implying complex linearity for every $\alpha, \beta \in \mathbb{C}$. If "$a$" denotes a single Grassmann number, any function $f(a)$ can be written

$$f(a) = f(0) + f^{(1)} a,$$

---

[1] See *e.g.* F.A. Berezin [10] or B. DeWitt [26].
[2] For a discussion of Grassmann algebras see also R. Penrose [87], Chap. 11.6.

since $a^2 = 0$. We now define the quantity $I[f]$ to be equal to $f^{(1)}$, *i.e.*

$$I[f] = \int da\, f(a) = \int da\,[f(0) + f^{(1)}a] = f(0) \int da\, \mathbb{1} + f^{(1)} \int da\, a := f^{(1)},$$

(8.3)

where we used Eq. (8.2). Thus our *definition* implies

$$\int da\, \mathbb{1} := 0, \quad \int da\, a := 1. \tag{8.4}$$

It should be observed that Eq. (8.4) satisfies Eqs. (8.1) and (8.2). We also observe that there is no difference between differentiation and integration with respect to a Grassmann variable, *i.e.*

$$\frac{\partial}{\partial a} f(a) = f^{(1)} = \int da\, f(a).$$

Next we consider the Grassmann algebra G_2 generated by two elements θ_1 and θ_2. The algebra G_2 has four independent elements, *i.e.*

$$\mathbb{1}_{G_2}, \ \theta_1, \ \theta_2, \ \theta_1\theta_2.$$

We wish to define the integral

$$\int d\theta_1 d\theta_2 f(\theta_1, \theta_2).$$

First we demand that the $d\theta_A$'s also *anticommute*, *i.e.*

$$\{d\theta_A, d\theta_B\} = \{d\theta_A, \theta_B\} = 0. \tag{8.5}$$

In order to preserve consistency with Eq. (8.4) we *define*:

$$\int d\theta_1 \int d\theta_2 \mathbb{1} = \int d\theta_1 \left[\int d\theta_2 \mathbb{1} \right] = 0, \tag{8.6a}$$

$$\int d\theta_1 \int d\theta_2 \theta_1 \overset{(8.5)}{=} -\int d\theta_2 \left[\int d\theta_1 \theta_1 \right] \overset{(8.4)}{=} -\int d\theta_2 \mathbb{1} \overset{(8.4)}{=} 0, \tag{8.6b}$$

$$\int d\theta_1 \int d\theta_2 \theta_2 = \int d\theta_1 \mathbb{1} \overset{(8.4)}{=} 0, \tag{8.6c}$$

$$\int d\theta_1 \int d\theta_2 \theta_1\theta_2 = -\int d\theta_1 \left[\int d\theta_2 \theta_2 \right] \theta_1 = -\int d\theta_1 \theta_1 = -1, \tag{8.6d}$$

again using Eq. (8.4). The integral of an arbitrary function $f(\theta_1, \theta_2) \in G_2$ is then obtained by linearity, *i.e.*

$$
\begin{aligned}
\int d\theta_1 \int d\theta_2 f(\theta_1, \theta_2) &= \int d\theta_1 \int d\theta_2 \left[f^{(0)} + \theta_1 f^{(1)} + \theta_2 f^{(2)} + \theta_1 \theta_2 f^{(3)} \right] \\
&\overset{(8.2)}{=} f^{(0)} \int d\theta_1 \int d\theta_2 + f^{(1)} \int d\theta_1 \int d\theta_2 \theta_1 \\
&\quad + f^{(2)} \int d\theta_1 \int d\theta_2 \theta_2 + f^{(3)} \int d\theta_1 \int d\theta_2 \theta_1 \theta_2 \\
&= -f^{(3)} \quad \text{(using Eqs. (8.6a) to (8.6d)).} \qquad (8.7)
\end{aligned}
$$

We see from this result that an integral

$$
\int d\theta_1 \int d\theta_2 f(\theta_1, \theta_2)
$$

corresponds to a projection such that the highest order component of the expansion of $f(\theta_1, \theta_2)$ is projected out.

Delta functions are defined by the relation

$$
\int da f(a) \delta(a) = f(0), \qquad f(a) \in G_1. \qquad (8.8)
$$

Since

$$
f(a) = f(0) + f^{(1)} a,
$$

and

$$
\begin{aligned}
\int da f(a) a &= \int da [f(0) + f^{(1)} a] a = \int da [f(0) a + f^{(1)} a^2] \\
&= f(0) \int da a \overset{(8.4)}{=} f(0),
\end{aligned}
$$

we conclude that

$$
\delta(a) = a. \qquad (8.9)
$$

We define *volume elements of superspace* by the relations:

$$
d^2\theta := -\frac{1}{4} d\theta^A d\theta^B \epsilon_{AB}, \qquad (8.10)
$$

$$
d^2\bar{\theta} := -\frac{1}{4} d\bar{\theta}_{\dot{A}} d\bar{\theta}_{\dot{B}} \epsilon^{\dot{A}\dot{B}}, \qquad (8.11)
$$

$$
d^4\theta := d^2\theta d^2\bar{\theta}. \qquad (8.12)
$$

Proposition 8.1:

With the definitions (8.10) to (8.12) we have

$$\int d^2\theta(\theta\theta) = 1,$$ (8.13a)

$$\int d^2\bar{\theta}(\bar{\theta}\,\bar{\theta}) = 1,$$ (8.13b)

and, of course,

$$\left. \begin{array}{ccccc} \int d^2\theta & = & 0 & = & \int d^2\bar{\theta} \\[2mm] \int d^2\theta\,\theta_A & = & 0 & = & \int d^2\bar{\theta}\,\bar{\theta}_{\dot{A}}. \end{array} \right\}$$ (8.13c)

Proof: (i) Consider, using the definition (8.10),

$$
\begin{aligned}
\int d^2\theta(\theta\theta) &= -\frac{1}{4}\int d\theta^A d\theta^B \epsilon_{AB}(\theta\theta) \overset{(1.99a)}{=} \frac{1}{2}\int d\theta^A d\theta^B \epsilon_{AB}\theta^1\theta^2 \\
&= \frac{1}{2}\int (\epsilon_{12}d\theta^1 d\theta^2 + \epsilon_{21}d\theta^2 d\theta^1)\theta^1\theta^2 \\
&\overset{(8.5)}{=} \frac{1}{2}\int(\epsilon_{12}-\epsilon_{21})d\theta^1 d\theta^2\theta^1\theta^2 \overset{(1.66)}{=} -\int d\theta^1 d\theta^2\theta^1\theta^2 \overset{(8.6d)}{=} 1.
\end{aligned}
$$

(ii) Using the definition (8.11) we have:

$$
\begin{aligned}
\int d^2\bar{\theta}(\bar{\theta}\,\bar{\theta}) &= -\frac{1}{4}\int d\bar{\theta}_{\dot{A}} d\bar{\theta}_{\dot{B}} \epsilon^{\dot{A}\dot{B}}(\bar{\theta}\,\bar{\theta}) \overset{(1.99b)}{=} -\frac{1}{2}\int d\bar{\theta}_{\dot{A}} d\bar{\theta}_{\dot{B}} \epsilon^{\dot{A}\dot{B}}\bar{\theta}_{\dot{1}}\bar{\theta}_{\dot{2}} \\
&= -\frac{1}{2}\int [d\bar{\theta}_{\dot{1}} d\bar{\theta}_{\dot{2}} - d\bar{\theta}_{\dot{2}} d\bar{\theta}_{\dot{1}}]\bar{\theta}_{\dot{1}}\bar{\theta}_{\dot{2}} \\
&\overset{(8.5)}{=} -\frac{1}{2}\int 2d\bar{\theta}_{\dot{1}} d\bar{\theta}_{\dot{2}}\bar{\theta}_{\dot{1}}\bar{\theta}_{\dot{2}} = \int d\bar{\theta}_{\dot{1}}\left[\int d\bar{\theta}_{\dot{2}}\bar{\theta}_{\dot{2}}\right]\bar{\theta}_{\dot{1}} = \int d\bar{\theta}_{\dot{1}}\bar{\theta}_{\dot{1}} = 1.
\end{aligned}
$$

Proposition 8.2:

The delta function on G_2, defined by

$$\int d^2\theta f(\theta)\delta^2(\theta) = f(0),$$ (8.14)

is given by

$$\delta^2(\theta) = (\theta\theta).$$ (8.15)

Proof: Let $f(\theta) \in G_2$. Then $f(\theta)$ has the expansion

$$f(\theta) = f(0) + \theta^A f_A^{(1)} + (\theta\theta)f^{(2)},$$

and

$$
\begin{aligned}
\int d^2\theta f(\theta) &= \int d^2\theta \Big[f(0) + \theta f^{(1)} + (\theta\theta)f^{(2)} \Big] \\
&\overset{(1.97)}{=} f(0) \int d^2\theta + f^{(1)A} \int d^2\theta \theta_A + f^{(2)} \int d^2\theta(\theta\theta) \\
&\overset{(8.13a)}{=} f^{(2)}.
\end{aligned}
$$

Since $(\theta\theta)\theta^A = 0$, we also have

$$
\begin{aligned}
\int d^2\theta f(\theta)\theta^A &= \int d^2\theta \big[f(0) + f^{(1)}\theta + (\theta\theta)f^{(2)} \big]\theta^A \\
&= \int d^2\theta \big[f(0)\theta^A + f^{(1)B}\theta_B\theta^A \big] \\
&\overset{(1.97)}{=} \int d^2\theta \big[f(0)\theta^A - f_B^{(1)}\theta^B\theta^A \big] \\
&\overset{(1.100a)}{=} \int d^2\theta \Big[f(0)\theta^A + \frac{1}{2}f_B^{(1)}\epsilon^{BA}(\theta\theta) \Big] \\
&= f(0) \int d^2\theta \theta^A + \frac{1}{2}f_B^{(1)}\epsilon^{BA} \int d^2\theta(\theta\theta) \\
&\overset{(8.13a)}{=} -\frac{1}{2}\epsilon^{AB}f_B^{(1)} = -\frac{1}{2}f^{(1)A},
\end{aligned}
$$

and

$$
\begin{aligned}
\int d^2\theta f(\theta)(\theta\theta) &= \int d^2\theta [f(0) + \theta f^{(1)} + (\theta\theta)f^{(2)}](\theta\theta) \\
&= f(0) \int d^2\theta(\theta\theta) = f(0).
\end{aligned}
$$

Comparing this result with that of Eq. (8.14) we find

$$(\theta\theta) = \delta^2(\theta),$$

which had to be shown. Similarly we obtain the following result.

Proposition 8.3:

The delta function $\delta^2(\bar{\theta})$ is given by

$$\delta^2(\overline{\theta}) = \overline{\theta}\,\overline{\theta}, \tag{8.16}$$

and then

$$\int d^2\overline{\theta}(\overline{\theta}\,\overline{\theta}) = \int d^2\overline{\theta}\delta^2(\overline{\theta}) = 1.$$

We then have, using Eq. (8.12),

$$
\begin{aligned}
\int d^4\theta f(\theta)\delta^2(\overline{\theta}) &= \int d^2\theta \int d^2\overline{\theta} f(\theta)\delta^2(\overline{\theta}) \\
&= \int d^2\theta \int d^2\overline{\theta}[f^{(0)} + \theta f^{(1)} + (\theta\theta)f^{(2)}]\delta^2(\overline{\theta}) \\
&\overset{(8.16)}{=} \int d^2\theta [f^{(0)} + \theta f^{(1)} + (\theta\theta)f^{(2)}] \int d^2\overline{\theta}(\overline{\theta}\,\overline{\theta}) \\
&\overset{(8.13b)}{=} \int d^2\theta [f^{(0)} + \theta f^{(1)} + (\theta\theta)f^{(2)}] \\
&\overset{(1.97)}{=} f^{(0)} \int d^2\theta + f^{(1)A} \int d^2\theta \theta_A + f^{(2)} \int d^2\theta(\theta\theta) \\
&\overset{(8.13a)}{=} f^{(2)}.
\end{aligned}
$$

Hence

$$\int d^4\theta f(\theta)\delta^2(\overline{\theta}) = f^{(2)}. \tag{8.17}$$

We now consider a general superfield with component expansion given by Eq. (6.76), and we evaluate the following integral:

$$
\begin{aligned}
\int d^4\theta \Phi(x,\theta,\overline{\theta}) &= \int d^4\theta \Big\{ f(x) + \theta\phi(x) + \overline{\theta}\overline{\chi}(x) + (\theta\theta)m(x) \\
&\quad + (\overline{\theta}\,\overline{\theta})n(x) + \theta\sigma^\mu\overline{\theta}V_\mu(x) + (\theta\theta)(\overline{\theta}\,\overline{\lambda}(x)) \\
&\quad + (\overline{\theta}\,\overline{\theta})\theta\psi(x) + (\theta\theta)(\overline{\theta}\,\overline{\theta})d(x) \Big\} \\
&= f(x) \int d^4\theta\,\mathbb{1} + \phi^A(x) \int d^4\theta\theta_A \\
&\quad + \overline{\chi}_{\dot{A}}(x) \int d^2\theta d^2\overline{\theta}\,\overline{\theta}^{\dot{A}} + m(x) \int d^2\theta d^2\overline{\theta}(\theta\theta) \\
&\quad + n(x) \int d^2\theta d^2\overline{\theta}(\overline{\theta}\,\overline{\theta}) + V_\mu(x) \int d^2\theta d^2\overline{\theta}(\theta\sigma^\mu\overline{\theta}) \\
&\quad - \overline{\lambda}^{\dot{A}}(x) \int d^2\theta d^2\overline{\theta}(\theta\theta)\overline{\theta}_{\dot{A}} \\
&\quad - \psi_A(x) \int d^2\theta d^2\overline{\theta}(\overline{\theta}\,\overline{\theta})\theta^A \\
&\quad + d(x) \int d^2\theta d^2\overline{\theta}(\overline{\theta}\,\overline{\theta})(\theta\theta) = d(x), \tag{8.18}
\end{aligned}
$$

using previous formulas and Eq. (1.97). Thus integrating any superfield with respect to the Grassmannn supercoordinates always projects out the highest order component field.

8.2 Lagrangians and Actions

We have already mentioned before, that the highest order component of a superfield transforms under supersymmetry transformations into a total derivative. A spacetime integral of such a quantity is therefore invariant under supersymmetry transformations. This fundamental property provides the criterion for constructing supersymmetric Lagrangian densities. In order to construct a supersymmetric coupling of several superfields, one considers simply the resulting superfield obtained by multiplication and selects the highest order component field. But this highest order component field is obtained, as explained in Chapter 7, by integration with respect to the Grassmann numbers θ and $\bar{\theta}$. Hence a supersymmetric action integral can be described by

$$\mathcal{A} := \int d^4x \int d^4\theta \mathcal{L} = \int d^4x \int d^2\theta d^2\bar{\theta}\mathcal{L} \qquad (8.19)$$

(using Eq. (8.12)) where $\mathcal{L}$ is a supersymmetric Lagrangian density (*i.e.* supersymmetric up to a total divergence).

8.2.1 Construction of Lagrangians from Scalar Superfields

The most general supersymmetric and renormalizable Lagrangian is given by the following expression:

$$\begin{aligned}
\mathcal{L} = \ & \Phi_i^\dagger\Phi_i + \left(g_i\Phi_i + \frac{1}{2}m_{ij}\Phi_i\Phi_j + \frac{1}{3}\lambda_{ijk}\Phi_i\Phi_j\Phi_k\right)\delta^2(\bar{\theta}) \\
& + \left(g_i^*\Phi_i^\dagger + \frac{1}{2}m_{ij}^*\Phi_i^\dagger\Phi_j^\dagger + \frac{1}{3}\lambda_{ijk}^*\Phi_i^\dagger\Phi_j^\dagger\Phi_k^\dagger\right)\delta^2(\theta) \qquad (8.20)
\end{aligned}$$

(summation over i, j, k understood). Here $i, j = 1, 2, \ldots, N$, where N is the number of scalar superfields. It can be shown[3] that the condition of renormalizability forbids powers of Φ in Eq. (8.20) higher than the third; however, linear terms are permissible. To find a supersymmetric Lagrangian such as that of Eq. (8.20) for chiral superfields, one can use arguments from dimensional analysis. From previous considerations we know that the Grassmann

---

[3]See, for instance, J. Wess and J. Bagger [123].

number θ has dimension $+1/2$ in length. According to Eq. (7.5) the component fields of a chiral superfield are a scalar, a spinor and another scalar field. The spinor field has, as usual, the dimension $-3/2$, and the two complex scalar fields differ by one unit of dimension in view of the extra factor $(\theta\theta)$ in front of one of them. We therefore assign the chiral superfield Φ the dimension -1. From

$$\int d\theta\theta = 1$$

we deduce that $\int d\theta$ has dimension $-1/2$; this implies, in particular, that $\int d^4\theta$ has dimension -2 in length. This therefore leads to a unique choice for a free (quadratic) massless action without dimensional parameters, *i.e.* to the expression

$$\mathcal{A}_{\text{kin}} = \int d^4x d^4\theta \Phi_i \Phi_i^\dagger,$$

where the dimension of d^4x is 4, the dimension of $d^4\theta$ is -2, and the dimension of $\Phi_i\Phi_i^\dagger$ is -2, such that $\mathcal{A}_{\text{kin}}$ is dimensionless. These arguments also justify the mass and interaction terms of Eq. (8.20), *i.e.*

$$\mathcal{A}_{\text{mass}} = \frac{1}{2}\int d^4x d^2\theta m_{ij}\Phi_i\Phi_j,$$

where the dimension of m_{ij} is -1, the dimension of $\Phi_i\Phi_j$ is -2, the dimension of $d^2\theta$ is -1, and the dimension of d^4x is 4, such that $\mathcal{A}_{\text{mass}}$ is dimensionless, and

$$\mathcal{A}_{\text{int}} = \frac{1}{3}\int d^4x d^2\theta \lambda_{ijk}\Phi_i\Phi_j\Phi_k,$$

where the dimension of $\Phi_i\Phi_j\Phi_k$ is -3, the dimension of $d^2\theta$ is -1, the dimension of d^4x is 4, and the coupling λ_{ijk} is dimensionless.

The couplings m_{ij} and λ_{ijk} are symmetric in their indices.

Now from Eq. (7.16) we know that the product $\Phi_i\Phi_k$ has the expansion:

$$\begin{aligned}\Phi_i\Phi_k &= A_i(y)A_k(y) + \sqrt{2}\theta\{A_i(y)\psi_k(y) + \psi_i(y)A_k(y)\}\\ &+(\theta\theta)\{A_i(y)F_k(y) + F_i(y)A_k(y) - \psi_i(y)\psi_k(y)\},\end{aligned}$$

so that

$$\begin{aligned}\Phi_i^\dagger\Phi_k^\dagger &= A_i^*(z)A_k^*(z) + \sqrt{2}\bar\theta\{A_i^*(z)\bar\psi_k(z) + \bar\psi_i(z)A_k^*(z)\}\\ &+(\bar\theta\bar\theta)\{A_i^*(z)F_k^*(z) + F_i^*(z)A_k^*(z) - \bar\psi_i(z)\bar\psi_k(z)\}.\end{aligned}$$

The component field expansion of the product of three fields Φ is given by Eq. (7.17); the expansion of $\Phi_i^\dagger\Phi_l^\dagger\Phi_k^\dagger$ is then obtained by complex conjugation.

The expansion of $\Phi_i^\dagger \Phi_i^\dagger$ is given by Eq. (7.27a). We emphasize that according to Eq. (7.31) the expression $\Phi_i^\dagger \Phi_i$ is a vector superfield, the term of highest order in this multiplet being of the form

$$(\theta\theta)(\overline{\theta}\,\overline{\theta})D(x).$$

Inserting Eq. (8.20) into the action intergral of Eq. (8.19) we obtain

$$
\begin{aligned}
\mathcal{A} \;=\; & \int d^4x \int d^4\theta \Big\{ \Phi_i^\dagger \Phi_i + g_i \Phi_i \delta^2(\overline{\theta}) + \frac{1}{2} m_{ij} \Phi_i \Phi_j \delta^2(\overline{\theta}) \\
& + \frac{1}{3}\lambda_{ijk}\Phi_i\Phi_j\Phi_k\delta^2(\overline{\theta}) + g_i^* \Phi_i^\dagger \delta^2(\theta) \\
& + \frac{1}{2}m_{ij}^* \Phi_i^\dagger \Phi_j^\dagger \delta^2(\theta) + \frac{1}{3}\lambda_{ijk}^* \Phi_i^\dagger \Phi_j^\dagger \Phi_k^\dagger \delta^2(\theta) \Big\}.
\end{aligned}
\tag{8.21}
$$

Inserting the component field expansions for the various terms and using the rules of integration developed in Sec. 8.1, we obtain from the coefficient of $(\theta\theta)(\overline{\theta}\,\overline{\theta})$ in Eq. (7.27a) the contribution

$$
\begin{aligned}
\int d^4x \int d^4\theta \Phi_i^\dagger \Phi_i \;=\; & \int d^4x \Big\{ \frac{1}{2}\partial_\mu A_i^*(x)\partial^\mu A_i(x) - \frac{1}{4}A_i^*(x)\Box A_i(x) \\
& - \frac{1}{4}A_i(x)\Box A_i^*(x) + \frac{i}{2}\partial_\mu \psi_i(x)\sigma^\mu \overline{\psi}_i(x) \\
& - \frac{i}{2}\psi_i(x)\sigma^\mu \partial_\mu \overline{\psi}_i(x) + F_i^*(x)F_i(x) \Big\}.
\end{aligned}
\tag{8.22}
$$

From the coefficient of $(\theta\theta)$ in Eq. (7.16) we obtain:

$$
\begin{aligned}
\frac{m_{ij}}{2}\int d^4x d^4\theta \Phi_i \Phi_j \delta^2(\overline{\theta}) &= \frac{m_{ij}}{2}\int d^4x d^2\theta \Phi_i \Phi_j \\
&= \frac{m_{ij}}{2}\int d^4x [A_i(x)F_j(x) + F_j(x)A_i(x) - \psi_i(x)\psi_j(x)] \\
&= \int d^4x\, m_{ij}\Big[A_i(x)F_j(x) - \frac{1}{2}\psi_i(x)\psi_j(x) \Big],
\end{aligned}
\tag{8.23}
$$

since $m_{ij} = m_{ji}$. Similarly by complex conjugation of Eq. (8.23) we obtain

$$
\begin{aligned}
\frac{1}{2}m_{ij}^* \int d^4x d^4\theta \Phi_i^\dagger \Phi_j^\dagger \delta^2(\theta) &= \frac{1}{2}m_{ij}^* \int d^4x d^2\overline{\theta}\, \Phi_i^\dagger \Phi_j^\dagger \\
&= \int d^4x\, m_{ij}^* \Big[A_i^*(x)F_j^*(x) - \frac{1}{2}\overline{\psi}_i(x)\overline{\psi}_j(x) \Big].
\end{aligned}
\tag{8.24}
$$

Using Eq. (7.17) we obtain in a similar way:

$$\frac{1}{3}\lambda_{ijk}\int d^4x \int d^4\theta \Phi_i \Phi_j \Phi_k \delta^2(\overline{\theta}) = \frac{1}{3}\lambda_{ijk}\int d^4x \int d^2\theta \Phi_i \Phi_j \Phi_k$$

$$= \int d^4x \Big\{ \frac{1}{3}\lambda_{ijk}\Big[A_i(x)A_j(x)F_k(x) + A_i(x)F_j(x)A_k(x)$$

$$+ F_i(x)A_j(x)A_k(x) - \psi_i(x)\psi_j(x)A_k(x)$$

$$- \psi_i(x)\psi_k(x)A_j(x) - A_i(x)\psi_j(x)\psi_k(x)\Big] \Big\}$$

$$= \int d^4x \lambda_{ijk}\Big[A_i(x)A_j(x)F_k(x) - \psi_i(x)\psi_j(x)A_k(x)\Big], \quad (8.25)$$

since λ_{ijk} is symmetric in all three indices. Analogously we have

$$\frac{1}{3}\int d^4x \int d^4\theta\, \lambda^*_{ijk}\Phi^\dagger_i \Phi^\dagger_j \Phi^\dagger_k \delta^2(\theta) = \int d^4x \lambda^*_{ijk}[A*_i(x)A^*_j(x)F^*_k(x)$$

$$quad - \overline{\psi}_i(x)\overline{\psi}_j(x)A^*_k(x)]. \quad (8.26)$$

In addition

$$\int d^4x \int d^4\theta\, g_i \Phi_i \delta^2(\overline{\theta}) = \int d^4x \int d^4\theta\, g_i (A_i + \sqrt{2}\theta\psi_i + \theta\theta F_i)\delta^2(\overline{\theta})$$

$$= \int d^4x\, g_i F_i(x),$$

and

$$\int d^4x \int d^4\theta\, g^*_i \Phi^\dagger_i \delta^2(\overline{\theta}) = \int d^4x\, g^*_i F^*_i(x).$$

We now observe that we are free to shift the coordinates from y and z to x. Such a change of variables does not alter the Lagrangian density because the highest order component of any scalar superfield has, for example, the form $\theta\theta D(y)$ and

$$(\theta\theta)D(y) = (\theta\theta)D(x + i\theta\sigma\overline{\theta})$$
$$= (\theta\theta)[D(x) + i(\theta\sigma^\mu\overline{\theta})\partial_\mu D(x) - (\theta\sigma^\mu\overline{\theta})(\theta\sigma^\rho\overline{\theta})\partial_\mu\partial_\rho D(x)]$$
$$= (\theta\theta)D(x),$$

since $(\theta\theta)\theta = 0$. Hence, changing the variable from y and z to x does not change the Lagrangian. Then the action (8.21) becomes, on adding Eq. (8.22) to Eq. (8.26):

$$
\begin{aligned}
\mathcal{A} = \int d^4x \Bigg\{ &\frac{1}{2}\partial_\mu A_i^*(x)\partial^\mu A_i(x) - \frac{1}{4}A_i^*(x)\Box A_i(x) - \frac{1}{4}A_i(x)\Box A_i^*(x) \\
&+\frac{i}{2}\partial_\mu\psi_i(x)\sigma^\mu\overline{\psi}_i(x) - \frac{i}{2}\psi_i(x)\sigma^\mu\partial_\mu\overline{\psi}_i(x) \\
&+F_i^*(x(F_i(x) + \left[m_{ij}\left(A_i(x)F_j(x) - \frac{1}{2}\psi_i(x)\psi_j(x)\right)\right. \\
&+\lambda_{ijk}\left(A_i(x)A_j(x)F_k(x) - \psi_i(x)\psi_j(x)A_k(x)\right) \\
&\left. +g_iF_i(x) + \text{Hermitian conjugate}\right] \Bigg\}.
\end{aligned}
\tag{8.27}
$$

We observe that $\mathcal{A}$ does not contain a kinetic term for the field $F_i(x)$. Such fields are called "*auxiliary fields*". These fields can be eliminated from the theory with the help of their respective equations of motion; these fields thus provide potentials in terms of dynamical fields. The reason for the nonoccurrance of derivatives of the fields F_i is that these derivatives would be associated with θ^3 which is zero. We now rewrite a few terms of Eq. (8.27).

Proposition 8.4:

We can show that

$$
\frac{1}{2}\partial_\mu A_i^*(x)\partial^\mu A_i(x) - \frac{1}{4}A_i^*(x)\Box A_i(x) - \frac{1}{4}A_i(x)\Box A_i^*(x)
$$
$$
= -A_i^*(x)\Box A_i(x) + \text{ total derivatives.}
\tag{8.28}
$$

Verification: Consider:

$$
\frac{1}{2}\partial_\mu A_i^*(x)\partial^\mu A_i(x) = \frac{1}{2}\partial_\mu[A_i^*(x)\partial^\mu A_i(x)] - \frac{1}{2}A_i^*(x)\Box A_i(x),
$$

and

$$
\begin{aligned}
-\frac{1}{4}A_i(x)\Box A_i^*(x) &= -\frac{1}{4}A_i(x)\partial_\mu\partial^\mu A_i^*(x) \\
&= -\frac{1}{4}\partial_\mu[A_i(x)\partial^\mu A_i^*(x)] + \frac{1}{4}\partial_\mu A_i(x)\partial^\mu A_i^*(x) \\
&= -\frac{1}{4}\partial_\mu[A_i(x)\partial^\mu A_i^*(x)] + \frac{1}{4}\partial_\mu[A_i(x)\partial^\mu A_i^*(x)] \\
&\quad -\frac{1}{4}A_i^*(x)\Box A_i(x).
\end{aligned}
$$

Hence

$$\frac{1}{2}\partial_\mu A_i^*(x)\partial^\mu A_i(x) - \frac{1}{4}A_i^*(x)\Box A_i(x) = -A_i^*(x)\Box A_i(x) + \text{ total derivatives.}$$

Using Eq. (1.134) we have

$$\frac{i}{2}(\partial_\mu\psi_i(x))\sigma^\mu\overline{\psi}_i(x) - \frac{i}{2}\psi_i(x)\sigma^\mu(\partial_\mu\overline{\psi}_i(x))$$

$$= -\frac{i}{2}\overline{\psi}_i(x)\overline{\sigma}^\mu(\partial_\mu\psi_i(x)) + \frac{i}{2}(\partial_\mu\overline{\psi}_i(x))\overline{\sigma}^\mu\psi_i(x)$$

$$= -\frac{i}{2}\partial_\mu(\overline{\psi}_i(x)\overline{\sigma}^\mu\psi_i(x)) + i(\partial_\mu\overline{\psi}_i(x))\overline{\psi}^\mu\psi_i(x)$$

$$= i(\partial_\mu\overline{\psi}_i(x))\overline{\sigma}^\mu\psi_i(x) + \text{ total derivatives.} \qquad (8.29)$$

Inserting Eqs. (8.28) and (8.29) into Eq. (8.27) we obtain the final expression for the action (8.20) in terms of component fields, *i.e.*

$$\mathcal{A} = \int d^4x \bigg\{ F_i^*(x)F_i(x) + i(\partial_\mu\overline{\psi}_i(x))\overline{\sigma}^\mu\psi_i(x) - A_i^*(x)\Box A_i(x)$$

$$+ \bigg[m_{ij}\bigg(A_i(x)F_j(x) - \frac{1}{2}\psi_i(x)\psi_j(x) \bigg)$$

$$+ \lambda_{ijk}\bigg(A_i(x)A_j(x)F_k(x) - \psi_i(x)\psi_j(x)A_k(x) \bigg)$$

$$+ g_i F_i(x) + \text{ Hermitian conjugate} \bigg] \bigg\},$$

or

$$\mathcal{A} = \int d^4x \bigg\{ i(\partial_\mu\overline{\psi}_i(x))\overline{\sigma}^\mu\psi_i(x) - A_i^*(x)\Box A_i(x)$$

$$+ \big(m_{ik}A_i(x) + \lambda_{ijk}A_i(x)A_j(x) \big) F_k(x)$$

$$+ \big(m_{ik}^*A_i^*(x) + \lambda_{ijk}^*A_i^*(x)A_j^*(x) \big) F_k^*(x)$$

$$- \frac{1}{2}m_{ij}\psi_i(x)\psi_j(x) - \lambda_{ijk}\psi_i(x)\psi_j(x)A_k(x) - \frac{1}{2}m_{ij}^*\overline{\psi}_i(x)\overline{\psi}_j(x)$$

$$- \lambda_{ijk}^*\overline{\psi}_i(x)\overline{\psi}_j(x)A_k^*(x) + F_i^*(x)F_j^*(x)$$

$$+ gF_i(x) + g_i^*F_i^*(x) \bigg\}. \qquad (8.30)$$

This is the so-called *"off-shell"* form of the action which involves the auxiliary fields $F_i(x)$. The auxiliary fields $F_i(x)$ can be eliminated with the help of the equations of motion, thus giving the *"on-shell"* form of the action integral. Consider the Euler–Lagrange equations

$$\frac{\partial\mathcal{L}}{\partial F_i(x)} - \partial_\mu\frac{\partial\mathcal{L}}{\partial(\partial_\mu F_i(x))} = 0$$

for the fields $F_i(x)$. Since the Lagrangian density does not contain derivatives of the fields $F_i(x)$, we have

$$\frac{\partial \mathcal{L}}{\partial F_i(x)} = 0$$

Hence

$$
\begin{aligned}
\frac{\partial \mathcal{L}}{\partial F_i(x)} &= F_i^*(x) + g_i + m_{ij} A_j(x) + \lambda_{ijk} A_j(x) A_k(x) = 0, \\
\frac{\partial \mathcal{L}}{\partial F_i^*(x)} &= F_i(x) + g_i^* + m_{ij}^* A_j^*(x) + \lambda_{ijk}^* A_j^*(x) A_k^*(x) = 0. \quad (8.31)
\end{aligned}
$$

From these equations which determine the auxiliary fields $F_i(x)$ we see that if the scalar fields $A_i(x)$ all have dimension $(\text{lenght})^{-1}$, $m \sim (\text{length})^{-1}$ and λ is dimensionless, then $F_i(x)$ must have dimension $(\text{length})^{-2}$. This is consistent with the arguments given earlier.

Replacing $F_i(x), F_i^*(x)$ in Eq. (8.30) by their expressions in terms of $A_i(x), A_i^*(x)$, we obtain

$$
\begin{aligned}
\mathcal{A} = \int d^4x \Big\{ & i\partial_\mu \overline{\psi}_i(x) \overline{\sigma}^\mu \psi_i(x) - A_i^*(x) \Box A_i(x) - \frac{1}{2} m_{ij} \psi_i(x) \psi_j(x) \\
& - \frac{1}{2} m_{ij}^* \overline{\psi}_i(x) \overline{\psi}_j(x) - \lambda_{ijk} \psi_i(x) \psi_j(x) A_k(x) - \lambda_{ijk}^* \overline{\psi}_i(x) \overline{\psi}_j(x) A_k^*(x) \\
& + \Big[g_k + m_{ki} A_i(x) + \lambda_{kij} A_i(x) A_j(x) \Big] F_k(x) \\
& + \Big[g_k^* + m_{ki}^* A_i^*(x) + \lambda_{kij}^* A_i^*(x) A_j^*(x) \Big] F_k^*(x) + F_k^*(x) F_k(x) \Big\}.
\end{aligned}
$$

Replacing $F_i(x), F_i^*(x)$ in this expression by their equivalents in terms of $A_j(x), A_j^*(x)$ obtained from Eqs. (8.31), we obtain the *on-shell* form of the action integral, *i.e.*

$$
\begin{aligned}
\mathcal{A} = \int d^4x \Big\{ & i\partial_\mu \overline{\psi}_i(x) \overline{\sigma}^\mu \psi_i(x) - A_i^*(x) \Box A_i(x) - \frac{1}{2} m_{ij} \psi_i(x) \psi_j(x) \\
& - \frac{1}{2} m_{ij}^* \overline{\psi}_i(x) \overline{\psi}_j(x) - \lambda_{ijk} \psi_i(x) \psi_j(x) A_k(x) \\
& - \lambda_{ijk}^* \overline{\psi}_i(x) \overline{\psi}_j(x) A_k^*(x) - V(A_i, A_j^*) \Big\}, \quad (8.32)
\end{aligned}
$$

where

$$V(A_i, A_j^*) = F_k^*(x) F_k(x)$$

is the potential of this supersymmetric field theory, and $F_k(x)$ and $F_k^*(x)$ are functions of $A_i(x), A_i^*(x)$ given by Eq. (8.31). The potential $V = |F_k(x)|^2$

is always greater than or equal to zero (this being a consequence of super-symmetry). Configurations for which $F_k = 0$ are absolute minima of the potential. To extract the particle content of the field theory described by the action (8.32), we choose the simplest case of only one scalar superfield Φ, so that all indices can be dropped in Eq. (8.32) and our starting point is the Lagrangian density

$$
\begin{aligned}
\mathcal{L} &= i\partial_\mu \overline{\psi}(x)\overline{\sigma}^\mu \psi(x) - A^*(x)\Box A(x) - \frac{1}{2}m\psi^2(x) - \frac{1}{2}m^*\overline{\psi}^2(x) \\
&\quad - \lambda\psi^2(x)A(x) - \lambda^*\overline{\psi}^2(x)A^*(x) - V(A, A^*),
\end{aligned}
$$

where

$$
\begin{aligned}
V(A, A^*) &= |F(x)|^2 = F(x)F^*(x) \\
&= \big(g + mA(x) + \lambda A^2(x)\big)\big(g^* + m^*A^*(x) + \lambda^*A^{*2}(x)\big) \\
&= |g|^2 + gm^*A^*(x) + g\lambda^*A^{*2}(x) + mg^*A(x) + |m|^2|A(x)|^2 \\
&\quad + m\lambda^*A^*(x)|A(x)|^2 + \lambda g^*A^2(x) + m^*\lambda|A(x)|^2A(x) \\
&\quad + |\lambda|^2(|A(x)|^2)^2.
\end{aligned}
$$

Furthermore, assuming that g, m and λ are real, we obtain (using Eq. (1.134))

$$
\begin{aligned}
\mathcal{L} &= -i\psi(x)\sigma^\mu \partial_\mu \overline{\psi}(x) - A^*(x)\Box A(x) - g^2 - \frac{1}{2}m\big(\psi^2(x) + \overline{\psi}^2(x)\big) \\
&\quad - mg\big(A(x) + A^*(x)\big) - \lambda^2(|A(x)|^2)^2 - \lambda\big(\overline{\psi}^2(x)A^*(x) + \psi^2(x)A(x)\big) \\
&\quad - g\lambda\big(A^{*2}(x) + A^2(x)\big) - m\lambda|A(x)|^2\big(A(x) + A^*(x)\big). \qquad (8.33)
\end{aligned}
$$

Then the *Euler–Lagrange equations* for the various fields are:

(i) For $\psi(x)$:

$$
\frac{\partial \mathcal{L}}{\partial \psi(x)} - \partial_\mu \frac{\partial \mathcal{L}}{\partial(\partial_\mu \psi(x))} = 0,
$$

where

$$
\frac{\partial \mathcal{L}}{\partial \psi(x)} = -i\sigma^\mu \partial_\mu \overline{\psi}(x) - m\psi(x) - 2\lambda\psi(x)A(x), \qquad \frac{\partial \mathcal{L}}{\partial(\partial_\mu \psi)} = 0,
$$

so that

$$
-i\sigma^\mu \partial_\mu \overline{\psi}(x) = m\psi(x) + 2\lambda\psi(x)A(x); \qquad (8.34)
$$

(ii) For $\overline{\psi}(x)$:

$$
\frac{\partial \mathcal{L}}{\partial \overline{\psi}(x)} - \partial_\mu \frac{\partial \mathcal{L}}{\partial(\partial_\mu \overline{\psi}(x))} = 0,
$$

where

$$\frac{\partial \mathcal{L}}{\partial \overline{\psi}(x)} = -m\overline{\psi}(x) - 2\lambda \overline{\psi}(x)A^*(x), \qquad \frac{\partial \mathcal{L}}{\partial(\partial_\mu \overline{\psi}(x))} = -i\psi(x)\sigma^\mu,$$

so that

$$i\partial_\mu \psi(x)\sigma^\mu = m\overline{\psi}(x) + 2\lambda \overline{\psi}(x)A^*(x); \qquad (8.35)$$

(iii) For $A^*(x)$:

$$\frac{\partial \mathcal{L}}{\partial A^*(x)} - \partial_\mu \frac{\partial \mathcal{L}}{\partial(\partial_\mu A^*(x))} = 0,$$

where

$$\frac{\partial \mathcal{L}}{\partial A^*(x)} = -\Box A(x) - \lambda\overline{\psi}^2(x) - m^2 A(x) - 2m\lambda|A(x)|^2 - m\lambda A^2(x)$$
$$- 2\lambda^2 |A(x)|^2 A(x) - mg - 2g\lambda A^*(x),$$

$$\frac{\partial \mathcal{L}}{\partial(\partial_\mu A^*(x))} = 0,$$

so that

$$(\Box + m^2)A(x) = -\lambda\{\overline{\psi}^2(x) + 2m|A(x)|^2 + mA^2(x) + 2\lambda|A(x)|^2 A(x)$$
$$+ 2gA^*(x)\} + mg. \qquad (8.36)$$

If we set $M := m$ and $\lambda = 0 = g$ (free theory), we obtain from Eqs. (8.34) and (8.35)

$$i\sigma^\mu \partial_\mu \overline{\psi}(x) + M\psi(x) = 0, \qquad i\partial_\mu \psi(x)\sigma^\mu - M\overline{\psi}(x) = 0,$$

which are equivalent to

$$(i\gamma_W^\mu \partial_\mu - M)\psi_M(x) = 0,$$

where γ_W^μ are the γ-matrices in the Weyl representation and ψ_M is a four-component Majorana spinor with left- and right-handed components ψ and $\overline{\psi}$ respectively. Furthermore, Eq. (8.36) becomes (also with $g = 0$)

$$(\Box + M^2)A(x) = 0,$$

where $A(x)$ is a complex scalar field. Hence the Lagrangian (8.33) describes the field theory (in the case $g, \lambda = 0$) of a spinor field and a complex scalar field, each with the same mass M. This is, of course, simply the action of the *free Wess–Zumino model* (discussed in Chapter 5), formulated here in terms of two-component Weyl spinors.

In order to see that the real and imaginary parts of the complex "scalar" field $A(x)$ do, in fact, define the scalar and pseudoscalar components of the Wess–Zumino supermultiplet, we recall that the Weyl formulation employed here does not preserve a well defined parity property. Previously we saw that (in our present notation) for a chiral superfield repesenting the lowest supersymmetric multiplet (*cf.* Eq. (7.5)) the expansion of the corresponding superfield, *i.e.*

$$\Phi(y, \theta) = A(y) + \sqrt{2}\theta\psi(y) + \theta\theta F(y),$$

implies the following transformation of the component fields (*cf.* Eq. (7.19))

$$
\begin{aligned}
\delta'_s A(x) &= \sqrt{2}\alpha\psi(x), \\
\delta'_s(\sqrt{2}\psi_A(x)) &= 2\alpha_A F(x) + 2i(\sigma^\mu\overline{\alpha})_A \partial_\mu A(x), \\
\delta'_s F(x) &= -i\partial_\mu(\sqrt{2}\psi(x))\sigma^\mu\overline{\alpha},
\end{aligned}
$$

where $F^*(x) = -mA(x)$ from Eq. (8.31). Comparing these relations with Eq. (5.12), we see that indeed

$$
\begin{aligned}
\sqrt{2}\psi(x)|_{\text{here}} &= \psi(x)|_{\text{Eq. (5.12)}}, \\
A(x)|_{\text{here}} &= f^*(x)|_{\text{Eq. (5.12)}}, \\
\alpha|_{\text{here}} &= \epsilon|_{\text{Eq. (5.12)}}.
\end{aligned}
$$

Thus the real and imaginary parts of the complex field $A(x)$ of Eq. (8.36) are the real scalar and pseudoscalar fields of the Wess–Zumino model, the combination $(A+B)/2$ arising as a result of the decomposition of 4×4 Dirac quantities of well defined parity into 2×2 Weyl matrices.

8.2.2   Construction of Lagrangians from Vector Superfields

In the previous section we discussed Lagrangians constructed entirely from scalar superfields which describe spin zero and spin one-half particles. Of course, one also wants to construct model theories which describe spin-1 particles, the ultimate aim being the construction of supersymmetric Yang–Mills gauge theories. In this section we restrict ourselves to the construction of the Lagrangian of a supersymmetric Abelian gauge theory. As discussed in Sec. 7.3, the supersymmetric generalization of the electromagnetic field strength is given by the chiral superfields W_A and $\overline{W}_{\dot{A}}$ which have been constructed from vector superfields. Since W_A is chiral, as demonstrated by Eq. (7.45a), the expression $W^A W_A$ is a scalar field (*i.e.* a Lorentz scalar as in Eq. (1.92a)) and therefore the $(\theta\theta)$-component of $W^A W_A$ is of interest in the construction of supersymmetrically invariant Lagrangians. (We have seen that the $(\theta\theta)$-component of a scalar superfield Φ always transforms into a spacetime derivative). Hence our first task is to calculate the component expansion of the scalar superfield $W^A W_A$.

Proposition 8.5:

The component field expansion of the scalar superfield $W^A W_A$ is given by:

$$
\begin{aligned}
W^A W_A &= \lambda^2(y) + 4D(y)\lambda(y)\theta + \lambda(y)\sigma^{\mu\nu}\theta F_{\mu\nu}(y) \\
&\quad + \theta\theta\left\{ 4D^2(y) - 2i\lambda(y)\sigma^\mu\partial_\mu\overline{\lambda}(y) - \frac{1}{2}F_{\mu\nu}(y)F^{\mu\nu}(y) \right. \\
&\quad \left. - \frac{i}{2}F^*_{\mu\nu}(y)F^{\mu\nu}(y) \right\},
\end{aligned}
\tag{8.37a}
$$

where $F^*_{\mu\nu}(y)$ is the dual field strength of $F_{\mu\nu}(y)$.

Proof: Using Eq. (7.52) we have

$$
\begin{aligned}
W^A W_A &= \left\{ \lambda^A(y) + 2D(y)\theta^A + \epsilon^{AB}(\sigma^{\mu\nu})_B{}^C\theta_C F_{\mu\nu}(y) \right. \\
&\quad \left. - i(\theta\theta)\epsilon^{AB}(\sigma^\mu)_{B\dot{B}}\partial_\mu\overline{\lambda}^{\dot{B}}(y) \right\} \cdot \left\{ \lambda_A(y) + 2D(y)\theta_A \right. \\
&\quad + (\sigma^{\rho\sigma})_A{}^B\theta_B F_{\rho\sigma}(y) \\
&\quad \left. - i(\theta\theta)(\sigma^\mu)_{A\dot{C}}\partial_\mu\overline{\lambda}^{\dot{C}}(y) \right\}.
\end{aligned}
$$

Using various results obtained earlier, properties such as Eq. (1.97), the Grassmann property $(\theta\theta)\theta_A = 0$, *etc.* this becomes

$$
\begin{aligned}
W^A W_A &= \lambda^2(y) + 2D(y)(\lambda(y)\theta) + \lambda(y)\sigma^{\rho\sigma}\theta F_{\rho\sigma}(y) \\
&\quad - i(\theta\theta)\lambda(y)\sigma^\mu\partial_\mu\overline{\lambda}(y) + 2D(y)\theta\lambda(y) + 4D^2(y)\theta\theta \\
&\quad + 2D(y)(\theta\sigma^{\rho\sigma}\theta)F_{\rho\sigma}(y) + \underbrace{\epsilon^{AB}(\sigma^{\mu\nu})_B{}^C\theta_C\lambda_A(y)F_{\mu\nu}(y)}_{=\,T_1} \\
&\quad + \underbrace{\epsilon^{AB}(\sigma^{\mu\nu})_B{}^C\theta_C(\sigma^{\rho\sigma})_A{}^D\theta_D F_{\mu\nu}(y)F_{\rho\sigma}(y)}_{=\,T_2} \\
&\quad \underbrace{- i(\theta\theta)\epsilon^{AB}(\sigma^\mu)_{B\dot{B}}\partial_\mu\overline{\lambda}^{\dot{B}}(y)\lambda_A(y)}_{=\,T_3} \\
&\quad + \underbrace{2\epsilon^{AB}(\sigma^{\mu\nu})_B{}^C\theta_C\theta_A D(y)F_{\mu\nu}(y)}_{=\,T_4}.
\end{aligned}
$$

Now:

$$
\begin{aligned}
T_1 &= \epsilon^{AB}(\sigma^{\mu\nu})_B{}^C\theta_C\lambda_A(y)F_{\mu\nu}(y) = -\lambda_A(y)\epsilon^{AB}(\sigma^{\mu\nu})_B{}^C\theta_C F_{\mu\nu}(y) \\
&= \epsilon^{BA}\lambda_A(y)(\sigma^{\mu\nu})_B{}^C\theta_C F_{\mu\nu}(y) = (\lambda(y)\sigma^{\mu\nu}\theta)F_{\mu\nu}(y),
\end{aligned}
$$

and

$$
\begin{aligned}
T_3 &= -i(\theta\theta)\epsilon^{AB}\sigma^\mu{}_{B\dot{B}}\partial_\mu\bar{\lambda}^{\dot{B}}(y)\lambda_A(y) = i(\theta\theta)\lambda_A(y)\epsilon^{AB}(\sigma^\mu\partial_\mu\bar{\lambda}(y))_B \\
&= -i(\theta\theta)\lambda(y)\sigma^\mu\partial_\mu\bar{\lambda}(y),
\end{aligned}
$$

and

$$
T_4 = 2\epsilon^{AB}(\sigma^{\mu\nu})_B{}^C\theta_C\theta_A D(y)F_{\mu\nu}(y) = 2\theta\sigma^{\mu\nu}\theta D(y)F_{\mu\nu}(y) = 0,
$$

since (*cf.* Eq. (1.151))

$$
\theta\sigma^{\mu\nu}\theta = 0.
$$

Finally

$$
\begin{aligned}
T_2 &:= \epsilon^{AB}(\sigma^{\mu\nu})_B{}^C\theta_C(\sigma^{\rho\sigma})_A{}^D\theta_D F_{\mu\nu}(y)F_{\rho\sigma}(y) \\
&= \epsilon^{AB}(\sigma^{\mu\nu})_B{}^C\theta_C\theta_D(\sigma^{\rho\sigma})_A{}^D F_{\mu\nu}(y)F_{\rho\sigma}(y) \\
&\overset{(1.100b)}{=} \frac{1}{2}(\theta\theta)\epsilon^{AB}(\sigma^{\mu\nu})_B{}^C\epsilon_{CD}(\sigma^{\rho\sigma})_A{}^D F_{\mu\nu}(y)F_{\rho\sigma}(y) \\
&\overset{(1.149)}{=} \frac{1}{2}(\theta\theta)\epsilon^{AB}(\sigma^{\mu\nu})_B{}^C\epsilon_{AD}(\sigma^{\rho\sigma})_C{}^D F_{\mu\nu}(y)F_{\rho\sigma}(y) \\
&= \frac{1}{2}(\theta\theta)\epsilon^{AB}\epsilon_{AD}(\sigma^{\mu\nu}\sigma^{\rho\sigma})_B{}^D F_{\mu\nu}(y)F_{\rho\sigma}(y).
\end{aligned}
$$

Contracting the ϵ's this becomes

$$
\begin{aligned}
T_2 &= -\frac{1}{2}(\theta\theta)\delta^B{}_D(\sigma^{\mu\nu}\sigma^{\rho\sigma})_B{}^D F_{\mu\nu}(y)F_{\rho\sigma}(y) \\
&= -\frac{1}{2}(\theta\theta)\mathrm{Tr}[\sigma^{\mu\nu}\sigma^{\rho\sigma}]F_{\mu\nu}(y)F_{\rho\sigma}(y) \\
&\overset{(1.144)}{=} -\frac{1}{2}(\theta\theta)\left[\frac{1}{2}\{\eta^{\mu\rho}\eta^{\nu\sigma} - \eta^{\mu\sigma}\eta^{\nu\rho}\} + \frac{i}{2}\epsilon^{\mu\nu\rho\sigma}\right]F_{\mu\nu}(y)F_{\rho\sigma}(y) \\
&= (\theta\theta)\left[-\frac{1}{4}F_{\mu\nu}(y)F^{\mu\nu}(y) + \frac{1}{4}F_{\mu\nu}(y)F^{\nu\mu}(y) - \frac{i}{4}\epsilon^{\mu\nu\rho\sigma}F_{\mu\nu}(y)F_{\rho\sigma}(y)\right] \\
&= (\theta\theta)\left[-\frac{1}{2}F_{\mu\nu}(y)F^{\mu\nu}(y) - \frac{i}{4}F_{\mu\nu}(y)\epsilon^{\mu\nu\rho\sigma}F_{\rho\sigma}(y)\right] \\
&= (\theta\theta)\left[-\frac{1}{2}F_{\mu\nu}(y)F^{\mu\nu}(y) - \frac{i}{2}F_{\mu\nu}(y)F^{*\mu\nu}(y)\right],
\end{aligned}
$$

where $F^{*\mu\nu}$ is the *dual field strength* defined by

$$F^{\mu\nu*}(y) = \frac{1}{2}\epsilon^{\mu\nu\rho\sigma}F_{\rho\sigma}(y). \qquad (8.37b)$$

Collecting all terms we obtain

$$\begin{aligned}
W^A W_A &= \lambda^2(y) + 4D(y)\lambda(y)\theta + 2\lambda(y)\sigma^{\mu\nu}\theta F_{\mu\nu}(y) \\
&\quad + (\theta\theta)\Big[4D^2(y) - 2i\lambda(y)\sigma^\mu\partial_\mu\overline{\lambda}(y) - \frac{1}{2}F_{\mu\nu}(y)F^{\mu\nu}(y) \\
&\quad - \frac{i}{2}F_{\mu\nu}(y)F^{\mu\nu*}(y)\Big].
\end{aligned} \qquad (8.37c)$$

We have thus obtained expression (8.37a). We observe that the superfield $W^A W_A$ depends only on y and θ; this dependence is characteristic of a left-handed chiral superfield. We now derive the corresponding expansion for $\overline{W}_{\dot{A}}\overline{W}^{\dot{A}}$.

Proposition 8.6:

The component field expansion of the scalar superfield $\overline{W}_{\dot{A}}\overline{W}^{\dot{A}}$ is given by

$$\begin{aligned}
\overline{W}_{\dot{A}}\overline{W}^{\dot{A}} &= \overline{\lambda}^2(z) + 4D(z)\overline{\lambda}(z)\overline{\theta} - 2\overline{\lambda}(z)\overline{\sigma}^{\mu\nu}\overline{\theta}F_{\mu\nu}(z) \\
&\quad + (\overline{\theta}\,\overline{\theta})\Big[4D^2(z) + 2i(\partial_\mu\lambda(z))\sigma^\mu\overline{\lambda}(z) - \frac{1}{2}F_{\mu\nu}(z)F^{\mu\nu}(z) \\
&\quad + \frac{i}{2}F^*_{\mu\nu}(z)F^{\mu\nu}(z)\Big].
\end{aligned} \qquad (8.38)$$

Proof: Using Eq. (7.53) we have

$$\begin{aligned}
\overline{W}_{\dot{A}}\overline{W}^{\dot{A}} &= [\overline{\lambda}_{\dot{A}}(z) + 2D(z)\overline{\theta}_{\dot{A}} - \epsilon_{\dot{A}\dot{B}}(\overline{\sigma}^{\mu\nu}\overline{\theta})^{\dot{B}}F_{\mu\nu}(z) + i(\overline{\theta}\,\overline{\theta})(\partial_\mu\lambda(z)\sigma^\mu)_{\dot{A}}] \\
&\quad [\overline{\lambda}^{\dot{A}}(z) + 2D(z)\overline{\theta}^{\dot{A}} - (\overline{\sigma}^{\rho\sigma})^{\dot{A}}{}_{\dot{B}}\overline{\theta}^{\dot{B}}F_{\rho\sigma}(z) - i(\overline{\theta}\,\overline{\theta})(\partial_\mu\lambda(z)\sigma^\mu)_{\dot{B}}\epsilon^{\dot{B}\dot{A}}] \\
&= \overline{\lambda}^2(z) + 2D(z)\overline{\lambda}(z)\overline{\theta} - (\overline{\lambda}(z)\overline{\sigma}^{\mu\nu}\overline{\theta})F_{\mu\nu}(z) \\
&\quad \underbrace{-i(\overline{\theta}\,\overline{\theta})\overline{\lambda}_{\dot{A}}(z)(\partial_\mu\lambda(z)\sigma^\mu)_{\dot{B}}\epsilon^{\dot{B}\dot{A}}}_{= T_1} + 2D(z)\overline{\theta}\,\overline{\lambda}(z) + 4(\overline{\theta}\,\overline{\theta})D^2(z) \\
&\quad - 2D(z)(\overline{\theta}\overline{\sigma}^{\rho\sigma}\overline{\theta})F_{\rho\sigma}(z) \underbrace{-\epsilon_{\dot{A}\dot{B}}(\overline{\sigma}^{\mu\nu})^{\dot{B}}{}_{\dot{C}}\overline{\theta}^{\dot{C}}\overline{\lambda}^{\dot{A}}F_{\mu\nu}(z)}_{= T_2}
\end{aligned}$$

$$- 2\epsilon_{\dot{A}\dot{B}}(\overline{\sigma}^{\mu\nu})^{\dot{B}}{}_{\dot{C}}\overline{\theta}^{\dot{C}}\overline{\theta}^{\dot{A}}D(z)F_{\mu\nu}(z)$$

$$+ \underbrace{\epsilon_{\dot{A}\dot{B}}(\overline{\sigma}^{\mu\nu})^{\dot{B}}{}_{\dot{C}}\overline{\theta}^{\dot{C}}(\overline{\sigma}^{\rho\sigma})^{\dot{A}}{}_{\dot{D}}\overline{\theta}^{\dot{D}}F_{\mu\nu}(z)F_{\rho\sigma}(z)}_{= \, T_3} + i(\overline{\theta}\,\overline{\theta})\partial_\mu\lambda(z)\sigma^\mu\overline{\lambda}(z).$$

Now,

$$T_1 = -i(\overline{\theta}\,\overline{\theta})\overline{\lambda}_{\dot{A}}(z)(\partial_\mu\lambda(z)\sigma^\mu)_{\dot{B}}\epsilon^{\dot{B}\dot{A}} = i(\overline{\theta}\,\overline{\theta})(\partial_\mu\lambda(z)\sigma^\mu)_{\dot{B}}\epsilon^{\dot{B}\dot{A}}\overline{\lambda}_{\dot{A}}(z)$$

$$= i(\overline{\theta}\,\overline{\theta})(\partial_\mu\lambda(z)\sigma^\mu\overline{\lambda}(z)),$$

and

$$\overline{\theta}\overline{\sigma}^{\mu\nu}\overline{\theta} \quad = \quad \overline{\theta}_{\dot{A}}(\overline{\sigma}^{\mu\nu})^{\dot{A}}{}_{\dot{B}}\overline{\theta}^{\dot{B}} = \overline{\theta}_{\dot{A}}(\overline{\sigma}^{\mu\nu})^{\dot{A}}{}_{\dot{B}}\epsilon^{\dot{B}\dot{C}}\overline{\theta}_{\dot{C}} = \overline{\theta}_{\dot{A}}\overline{\theta}_{\dot{C}}(\overline{\sigma}^{\mu\nu})^{\dot{A}}{}_{\dot{B}}\epsilon^{\dot{B}\dot{C}}$$

$$\overset{(1.100d)}{=} \quad -\frac{1}{2}(\overline{\theta}\,\overline{\theta})\epsilon_{\dot{A}\dot{C}}(\overline{\sigma}^{\mu\nu})^{\dot{A}}{}_{\dot{B}}\epsilon^{\dot{B}\dot{C}} = \frac{1}{2}(\overline{\theta}\,\overline{\theta})(\overline{\sigma}^{\mu\nu})^{\dot{A}}{}_{\dot{B}}\epsilon^{\dot{B}\dot{C}}\epsilon_{\dot{C}\dot{A}}$$

$$= \quad \frac{1}{2}(\overline{\theta}\,\overline{\theta})(\overline{\sigma}^{\mu\nu})^{\dot{A}}{}_{\dot{A}} = 0,$$

since

$$\mathrm{Tr}[\overline{\sigma}^{\mu\nu}] = 0.$$

Also:

$$T_2 = -\epsilon_{\dot{A}\dot{B}}(\overline{\sigma}^{\mu\nu})^{\dot{B}}{}_{\dot{C}}\overline{\theta}^{\dot{C}}\overline{\lambda}^{\dot{A}}(z)F_{\mu\nu}(z) = \overline{\lambda}^{\dot{A}}(z)\epsilon_{\dot{A}\dot{B}}(\overline{\sigma}^{\mu\nu})^{\dot{B}}{}_{\dot{C}}\overline{\theta}^{\dot{C}}F_{\mu\nu}(z)$$

$$= -\epsilon_{\dot{B}\dot{A}}\overline{\lambda}^{\dot{A}}(z)(\overline{\sigma}^{\mu\nu})^{\dot{B}}{}_{\dot{C}}\overline{\theta}^{\dot{C}}F_{\mu\nu}(z) = -\overline{\lambda}(z)\overline{\sigma}^{\mu\nu}\overline{\theta}F_{\mu\nu}(z),$$

and

$$T_3 \quad = \quad \epsilon_{\dot{A}\dot{B}}(\overline{\sigma}^{\mu\nu})^{\dot{B}}{}_{\dot{C}}\overline{\theta}^{\dot{C}}(\overline{\sigma}^{\rho\sigma})^{\dot{A}}{}_{\dot{D}}\overline{\theta}^{\dot{D}}F_{\mu\nu}(z)F_{\rho\sigma}(z)$$

$$\overset{(1.100c)}{=} \quad \frac{1}{2}(\overline{\theta}\,\overline{\theta})\epsilon_{\dot{A}\dot{B}}(\overline{\sigma}^{\mu\nu})^{\dot{B}}{}_{\dot{C}}\epsilon^{\dot{C}\dot{D}}(\overline{\sigma}^{\rho\sigma})^{\dot{A}}{}_{\dot{D}}F_{\mu\nu}(z)F_{\rho\sigma}(z)$$

$$\overset{(1.149)}{=} \quad \frac{1}{2}(\overline{\theta}\,\overline{\theta})\epsilon_{\dot{C}\dot{B}}(\overline{\sigma}^{\mu\nu})^{\dot{B}}{}_{\dot{A}}(\overline{\sigma}^{\rho\sigma})^{\dot{A}}{}_{\dot{D}}\epsilon^{\dot{C}\dot{D}}F_{\mu\nu}(z)F_{\rho\sigma}(z)$$

$$= \quad -\frac{1}{2}(\overline{\theta}\,\overline{\theta})(\overline{\sigma}^{\mu\nu}\overline{\sigma}^{\rho\sigma})^{\dot{B}}{}_{\dot{D}}\epsilon^{\dot{D}\dot{C}}\epsilon_{\dot{C}\dot{B}}F_{\mu\nu}(z)F_{\rho\sigma}(z)$$

$$= \quad -\frac{1}{2}(\overline{\theta}\,\overline{\theta})\mathrm{Tr}[\overline{\sigma}^{\mu\nu}\overline{\sigma}^{\rho\sigma}]F_{\mu\nu}(z)F_{\rho\sigma}(z),$$

where we used Eq. (1.144). Now, using Eq. (8.39) which will be derived below, this becomes

$$T_3 \quad = \quad -\frac{1}{4}(\overline{\theta}\,\overline{\theta})\Big[F_{\mu\nu}(z)F^{\mu\nu}(z) - F_{\mu\nu}(z)F^{\nu\mu}(z) - i\epsilon^{\mu\nu\rho\sigma}F_{\mu\nu}(z)F_{\rho\sigma}(z)\Big]$$

$$\overset{(8.37a)}{=} \quad -\frac{1}{2}(\overline{\theta}\,\overline{\theta})\Big[F_{\mu\nu}(z)F^{\mu\nu}(z) - iF_{\mu\nu}(z)F^{*\mu\nu}(z)\Big].$$

As mentioned, in deriving the last expression we made use of the result of the following Proposition 8.7

Proposition 8.7:

The following relation holds:

$$\mathrm{Tr}[\overline{\sigma}^{\mu\nu}\overline{\sigma}^{\rho\sigma}] = \frac{1}{2}[\eta^{\mu\rho}\eta^{\nu\sigma} - \eta^{\mu\sigma}\eta^{\nu\rho}] - \frac{i}{2}\epsilon^{\mu\nu\rho\sigma}. \tag{8.39}$$

Proof: This proof is similar to that of formula (1.144). Consider, using Eq. (1.138b),

$$
\begin{aligned}
\mathrm{Tr}[\overline{\sigma}^{\mu\nu}\overline{\sigma}^{\rho\sigma}] &= -\frac{1}{16}\mathrm{Tr}[(\overline{\sigma}^{\mu}\sigma^{\nu} - \overline{\sigma}^{\nu}\sigma^{\mu})(\overline{\sigma}^{\rho}\sigma^{\sigma} - \overline{\sigma}^{\sigma}\sigma^{\rho})] \\
&= -\frac{1}{16}\mathrm{Tr}[\overline{\sigma}^{\mu}\sigma^{\nu}\overline{\sigma}^{\rho}\sigma^{\sigma}] + \frac{1}{16}\mathrm{Tr}[\overline{\sigma}^{\mu}\sigma^{\nu}\overline{\sigma}^{\sigma}\sigma^{\rho}] \\
&\quad +\frac{1}{16}\mathrm{Tr}[\overline{\sigma}^{\nu}\sigma^{\mu}\overline{\sigma}^{\rho}\sigma^{\sigma}] - \frac{1}{16}\mathrm{Tr}[\overline{\sigma}^{\nu}\sigma^{\mu}\overline{\sigma}^{\sigma}\sigma^{\rho}] \\
&= -\frac{1}{16}\mathrm{Tr}[\sigma^{\sigma}\overline{\sigma}^{\mu}\sigma^{\nu}\overline{\sigma}^{\rho}] + \frac{1}{16}\mathrm{Tr}[\sigma^{\rho}\overline{\sigma}^{\mu}\sigma^{\nu}\overline{\sigma}^{\sigma}] \\
&\quad +\frac{1}{16}\mathrm{Tr}[\sigma^{\sigma}\overline{\sigma}^{\nu}\sigma^{\mu}\overline{\sigma}^{\rho}] - \frac{1}{16}\mathrm{Tr}[\sigma^{\rho}\overline{\sigma}^{\nu}\sigma^{\mu}\overline{\sigma}^{\sigma}] \\
&\overset{(1.146)}{=} -\frac{1}{8}\{\eta^{\sigma\mu}\eta^{\nu\rho} + \eta^{\mu\nu}\eta^{\sigma\rho} - \eta^{\sigma\nu}\eta^{\mu\rho} - i\epsilon^{\sigma\mu\nu\rho}\} \\
&\quad +\frac{1}{8}\{\eta^{\rho\mu}\eta^{\nu\sigma} + \eta^{\mu\nu}\eta^{\rho\sigma} - \eta^{\rho\nu}\eta^{\mu\sigma} - i\epsilon^{\rho\mu\nu\sigma}\} \\
&\quad +\frac{1}{8}\{\eta^{\sigma\nu}\eta^{\mu\rho} + \eta^{\mu\nu}\eta^{\sigma\rho} - \eta^{\sigma\mu}\eta^{\nu\rho} - i\epsilon^{\sigma\nu\mu\rho}\} \\
&\quad -\frac{1}{8}\{\eta^{\rho\nu}\eta^{\mu\sigma} + \eta^{\nu\mu}\eta^{\sigma\rho} - \eta^{\mu\rho}\eta^{\nu\sigma} - i\epsilon^{\rho\nu\mu\sigma}\} \\
&= \frac{1}{2}\eta^{\mu\rho}\eta^{\nu\sigma} - \frac{1}{2}\eta^{\mu\sigma}\eta^{\nu\rho} - \frac{i}{2}\epsilon^{\mu\nu\rho\sigma},
\end{aligned}
$$

as claimed by Eq. (8.39).

With the rewritten versions of the above expressions we then obtain:

$$
\begin{aligned}
\overline{W}_{\dot{A}}\overline{W}^{\dot{A}} &= \overline{\lambda}^2(z) + 4D(z)\overline{\lambda}(z)\overline{\theta} - 2(\overline{\lambda}(z)\overline{\sigma}^{\mu\nu}\overline{\theta})F_{\mu\nu}(z) \\
&\quad +2i(\overline{\theta}\,\overline{\theta})(\partial_\mu\lambda(z)\sigma^\mu\overline{\lambda}(z)) + 4(\overline{\theta}\,\overline{\theta})D^2(z) \\
&\quad -\frac{1}{2}(\overline{\theta}\,\overline{\theta})\Big[F_{\mu\nu}(z)F^{\mu\nu}(z) - iF_{\mu\nu}(z)F^{*\mu\nu}(z)\Big]
\end{aligned}
$$

$$= \quad \overline{\lambda}^2(z) + 4D(z)\overline{\lambda}(z)\overline{\theta} - 2(\overline{\lambda}(z)\overline{\sigma}^{\mu\nu}\overline{\theta})F_{\mu\nu}(z)$$

$$+(\overline{\theta}\,\overline{\theta})\Big\{4D^2(z) + 2i\partial_\mu\lambda(z)\sigma^\mu\overline{\lambda}(z)$$

$$-\frac{1}{2}F_{\mu\nu}(z)F^{\mu\nu}(z) + \frac{i}{2}F_{\mu\nu}(z)F^{*\mu\nu}(z)\Big\}.$$

This is the expression claimed by Eq. (8.38).

For the construction of Lagrangians it is important to observe that the $(\theta\theta)$-component is invariant under the shift of coordinates

$$y^\mu = x^\mu + i\theta\sigma^\mu\overline{\theta} \longrightarrow x^\mu.$$

Thus

$$(\theta\theta)W^AW_A\Big|_{\theta\theta} \quad = \quad (\theta\theta)\Big[4D^2(y) - 2i\lambda(y)\sigma^\mu\partial_\mu\overline{\lambda}(y)$$

$$-\frac{1}{2}F_{\mu\nu}(y)F^{\mu\nu}(y) - \frac{i}{2}F_{\mu\nu}(y)F^{*\mu\nu}(y)\Big]$$

$$= \quad (\theta\theta)\Big[4D^2(x + i\theta\sigma\overline{\theta}) - 2i\lambda(x + i\theta\sigma\overline{\theta})\sigma^\mu\partial_\mu\overline{\lambda}(x + i\theta\sigma\overline{\theta})$$

$$-\frac{1}{2}F_{\mu\nu}(x + i\theta\sigma\overline{\theta})F^{\mu\nu}(x + i\theta\sigma\overline{\theta})$$

$$-\frac{i}{2}F_{\mu\nu}(x + i\theta\sigma\overline{\theta})F^{*\mu\nu}(x + i\theta\sigma\overline{\theta})\Big]$$

$$= \quad (\theta\theta)\Big[4D^2(x) - 2i\lambda(x)\sigma^\mu\partial_\mu\overline{\lambda}(x)$$

$$-\frac{1}{2}F_{\mu\nu}(x)F^{\mu\nu}(x) - \frac{i}{2}F_{\mu\nu}(x)F^{*\mu\nu}(x)\Big],$$

since a typical term in the expression can be expanded using Eq. (1.137) like

$$4(\theta\theta)D^2(x + i\theta\sigma\overline{\theta}) \quad = \quad 4(\theta\theta)\Big[D(x)D(x) + D(x)i(\theta\sigma\overline{\theta})\partial_\mu D(x)$$

$$-\frac{1}{4}D(x)(\theta\theta)(\overline{\theta}\,\overline{\theta})\Box D(x) + i(\theta\sigma^\mu\overline{\theta})(\partial_\mu D(x))D(x)\Big]$$

$$= \quad 4(\theta\theta)\Big[D^2(x) + 2iD(x)(\theta\sigma^\mu\overline{\theta})\partial_\mu D(x)$$

$$-\frac{1}{2}(\theta\theta)(\overline{\theta}\,\overline{\theta})(\Box D(x))D(x)\Big]$$

$$= \quad 4(\theta\theta)D^2(x),$$

since all powers of θ higher than the second vanish. The same invariance applies to the $(\overline{\theta}\,\overline{\theta})$-component of $\overline{W}_{\dot{A}}\overline{W}^{\dot{A}}$, where we can shift the coordinates according to

$$z^{\mu} = x^{\mu} - i(\theta\sigma^{\mu}\overline{\theta}) \longrightarrow x^{\mu}$$

without altering the $(\overline{\theta}\,\overline{\theta})$-component, *i.e.*

$$
\begin{aligned}
(\overline{\theta}\,\overline{\theta})\overline{W}_{\dot{A}}\overline{W}^{\dot{A}}\Big|_{\overline{\theta}\,\overline{\theta}} &= (\overline{\theta}\,\overline{\theta})\Big[4D^2(z) + 2i\partial_{\mu}\lambda(z)\sigma^{\mu}\overline{\lambda}(z) \\
&\qquad - \frac{1}{2}F_{\mu\nu}(z)F^{\mu\nu}(z) + \frac{i}{2}F_{\mu\nu}(z)F^{*\mu\nu}(z)\Big] \\
&= (\overline{\theta}\,\overline{\theta})\Big[4D^2(x) + 2i\partial_{\mu}\lambda(x)\sigma^{\mu}\overline{\lambda}(x) \\
&\qquad - \frac{1}{2}F_{\mu\nu}(x)F^{\mu\nu}(x) + \frac{i}{2}F_{\mu\nu}(x)F^{*\mu\nu}(x)\Big].
\end{aligned}
$$

We can now cosntruct the supersymmetric and gauge-invariant generalization of the action of a pure Abelian gauge theory. As demonstrated at the end of Sec. 8.1, integration with respect to θ or $\overline{\theta}$ corresponds to projecting out the highest order component of the corresponding superfield. As stated above, the highest order components of $W^A W_A$ and $\overline{W}_{\dot{A}}\overline{W}^{\dot{A}}$ are independent of any particular representation (*i.e.* are invariant under shifts of coordinates). We may therefore shift both scalar superfields to the x-representation and add the highest order components. This procedure gives the desired action

$$\mathcal{A} = \int d^4x \int d^4\theta \left\{ W^A W_A \delta^2(\overline{\theta}) + \overline{W}_{\dot{A}}\overline{W}^{\dot{A}}\delta^2(\theta) \right\}. \tag{8.40}$$

In terms of its component fields this integral becomes

$$\mathcal{A} = \int d^4x \Big[8D^2(x) - F_{\mu\nu}(x)F^{\mu\nu}(x) - 2i\lambda(x)\sigma^{\mu}\partial_{\mu}\overline{\lambda}(x) + 2i(\partial_{\mu}\lambda(x))\sigma^{\mu}\overline{\lambda}(x)\Big].$$

Integrating the last term by parts we find

$$
\begin{aligned}
2i\int d^4x(\partial_{\mu}\lambda(x))\sigma^{\mu}\overline{\lambda}(x) &= 2i\int d^4x[\partial_{\mu}(\lambda(x)\sigma^{\mu}\overline{\lambda}(x)) - \lambda(x)\sigma^{\mu}(\partial_{\mu}\overline{\lambda}(x))] \\
&= 2i\int d^3S_{\mu}\lambda(x)\sigma^{\mu}\overline{\lambda}(x) - 2i\int d^4x\lambda(x)\sigma^{\mu}\partial_{\mu}\overline{\lambda}(x).
\end{aligned}
$$

Assuming the fields $\lambda, \overline{\lambda}$ fall off sufficiently fast at infinity so that the surface integral vanishes, we obtain

$$\mathcal{A} = \int d^4x [8D^2(x) - F_{\mu\nu}(x)F^{\mu\nu}(x) - 4i\lambda(x)\sigma^\mu\partial_\mu\overline{\lambda}(x)]. \qquad (8.41)$$

This is the action for a pure supersymmetric Abelian gauge theory with the following properties:

(a) The field $D(x)$ is an auxiliary field which can be eliminated with the help of the equation of motion.

(b) Supersymmetry requires a massless fermionic partner $\lambda(x)$ of the massless gauge boson $V_\mu(x)$ which is called the "*gauge fermion*" or "*gaugino*" or "*photino*". Thus the fermion field $\lambda(x)$ (in nonsupersymmetric contexts a matter field) now becomes part of the gauge field.

(c) The invariance of Eq. (8.41) under supersymmetric transformations is manifest, since the integrand is a $(\theta\theta)$- or correspondingly $(\overline{\theta}\,\overline{\theta})$-component of a scalar superfield, which transforms under supersymmetry transformations into a total spacetime derivative.

(d) The invariance of Eq. (8.41) under gauge transformations (7.35) is also manifest since the action is a functional of the fields $D(x), F_{\mu\nu}(x) = \partial_\mu V_\nu(x) - \partial_\nu V_\mu(x)$, and $\lambda(x)$ which are gauge invariant (*cf.* Eq. (7.37)).

8.2.3   Remarks

Knowing the general procedure for constructing supersymmetric Lagrangians it is straight-forward to construct other supersymmetric theories like *e.g.* supersymmetric quantum mechanics or supersymmetric nonlinear σ-models, the latter being of considerable interest in connection with superstrings. The Lagrangian density of a nonsupersymmetric nonlinear σ-model is given by

$$\mathcal{L} = -\frac{1}{2}g_{ij}(\phi)\partial_\mu\phi^i(x)\partial^\mu\phi^j(x),$$

where $\phi^i(x), i = 1, 2, \ldots, N$, denotes a set of N scalar fields and g_{ij} is the metric of a manifold. To obtain the supersymmetric version one replaces the scalar fields $\phi^i(x)$ by scalar superfields $\Phi^i(x, \theta)$. In order to obtain the component field expansion of the resulting superfield Lagrangian, one performs the θ-integration to project out the highest component. Alternatively, one can use the projection technique of Sec. 9.2. The resulting Lagrangian density is[4]

$$\mathcal{L} = h_{ij*}(A, \overline{A}) \partial_\mu A^i(x) \partial^\mu \overline{A}^j(x) - \frac{i}{2} h_{ij*}(A, \overline{A}) \psi^I(x) \sigma^\mu \overset{\leftrightarrow}{D_\mu} \overline{\psi}^j(x)$$

$$+ \frac{1}{4} R_{j*ilk*}(A, \overline{A}) (\overline{\psi}^j(x) \overline{\psi}^k(x))(\psi^i(x) \psi^l(x)).$$

Here the scalar fields A^i of the supermultiplets $\Phi^i(x, \theta)$ and their conjugates $\overline{A}^i$ parametrize a complex manifold, h_{ij*} are the components of the Hermitian metric, and R_{j*ilk*} are the components of the curvature tensor. The quantity D_μ denotes a covariant derivative given by

$$D_\mu \psi^i = \partial_\mu \psi^i + \Gamma^i_{mk} \partial_\mu A^m \psi^k(x), \qquad (8.42)$$

where Γ^k_{mk} are the *Christoffel symbols*. As was shown by Zumino [134], in order to be able to describe a supersymmetric nonlinear σ-model, the complex manifold which is parametrized by the scalar fields A^i, must have an additional structure, *i.e.* the Hermitian metric must obey the relations:

$$\frac{\partial h_{ij*}(A, \overline{A})}{\partial A^k} = \frac{\partial h_{kj*}(A, \overline{A})}{\partial A^i},$$

$$\frac{\partial h_{ij*}(A, \overline{A})}{\partial \overline{A}^k} = \frac{\partial h_{ij*}(A, \overline{A})}{\partial \overline{A}^j}.$$

Such a complex manifold is called a *Kähler manifold*. It was shown by Zumino [134] that there is a one-to-one correspondence between supersymmetric non-linear σ-models and Kähler manifolds. These properties all depend on the dimension of the underlying Minkowski space (here $3 + 1$). If one is interested in a $(1 + 1)$-dimensional supersymmetric field theory, one obtains a similar result (see [128]); in this case the manifold is real and Riemannian. For a readable introduction into complex manifolds and Kähler geometry we refer to work of L. Alvarez–Gaumé and D.Z. Freedman [4], and to related literature.[5]

---

[4]See *e.g.* J.A. Bagger [6] and J.A. Bagger and E. Witten [7].
[5]See *e.g.* the books by S. Goldberg [48], R.O. Wells [119] and K. Yano [133]. See also the review article by J.W. van Holten [113].

Chapter 9

Spontaneous Breaking of Supersymmetry

9.1 The Superpotential

We now discuss some aspects of the so-called *superpotential* in relation to the spontaneous breaking of supersymmetry. Our motivation for investigating properties of the superpotential first is the usefulness of the latter in the study of specific models of spontaneous breaking of supersymmetry such as the models of L. O'Raifeartaigh [84] and P. Fayet and J. Iliopoulos [34].

We remarked in connection with Eq. (8.20) that the most general supersymmetric and renormalizable Lagrangian density which involves only scalar superfields is given by

$$\mathcal{L} = \mathcal{L}_{\text{kin}} + \mathcal{L}_{\text{pot}},$$

where the kinetic term is

$$\mathcal{L}_{\text{kin}} = \Phi_i^\dagger \Phi_i,$$

and

$$\mathcal{L}_{\text{pot}} = \left(g_i \Phi_i + \frac{1}{2} m_{ij} \Phi_i \Phi_j + \frac{1}{2} \lambda_{ijk} \Phi_i \Phi_j \Phi_k\right) \delta^2(\bar{\theta}) + \text{h.c}$$

is the potential term. We consider first the case of a theory with a single scalar field as in the Wess–Zumino model. Then

$$\mathcal{L} = \Phi^\dagger \Phi \big|_{\theta\theta\bar{\theta}\bar{\theta}} + \left[\left(g\Phi + \frac{1}{2} m\Phi^2 + \frac{1}{3} \lambda\Phi^3\right)\big|_{\theta\theta} + \text{h.c.}\right], \qquad (9.1)$$

where the superscripts indicate that the integration with respect to θ and $\bar{\theta}$ removes the delta function leaving only the coefficient of the appropriate term of the power series expansion. Denoting $W[\Phi]$ by

$$W[\Phi] := g\Phi + \frac{1}{2} m\Phi^2 + \frac{1}{3} \lambda\Phi^3, \qquad (9.2)$$

we can write the action in the form

$$A = \int d^4x \int d^4\theta \left\{ \Phi^\dagger \Phi + \left[W[\Phi]\delta^2(\bar{\theta}) + \text{h.c.} \right] \right\}. \tag{9.3}$$

The functional $W[\Phi]$, which is a polynomial of the superfield Φ, is called the *superpotential*. The superpotential contains mass terms and interactions. The integral

$$\int d^4\theta \, W[\Phi]\delta^2(\bar{\theta}) = \int d^2\theta \, W[\Phi] \tag{9.4}$$

appearing in the action (9.3) projects out the highest order component of the superpotential and is manifestly supersymmetric due to the fact that this component always transforms into a spacetime derivative. We now show that the integral in Eq. (9.4) is given by:

$$\int d^2\theta \, W[\Phi] = \frac{\partial W}{\partial A} F - \frac{1}{2} \frac{\partial^2 W}{\partial A^2} \psi\psi, \tag{9.5}$$

where

$$\frac{\partial W}{\partial A} := \frac{\partial W[A]}{\partial A(x)}.$$

Here $W[A]$ denotes the superpotential (9.2) with the superfield Φ replaced by the first component $A(x)$. The quantities $A(x)$, $\psi(x)$ and $F(x)$ being the component fields of the scalar superfield Φ, *i.e.*

$$\Phi(y,\theta) = A(y) + \sqrt{2}\theta\psi(y) + \theta\theta \, F(y). \tag{9.6}$$

The second term on the right hand side of Eq. (9.5) is a mass term of the fermionic part of the underlying model. A systematic derivation of Eq. (9.5) will be given in Sec. 9.2. We first demonstrate Eq. (9.5) explicitly for the particular case in which the superpotential is given by Eq. (9.2), *i.e.*

$$W[\Phi] = g\Phi + \frac{1}{2}m\Phi^2 + \frac{1}{3}\lambda\Phi^3.$$

Then

$$\int d^2\theta \, W[\Phi] = g\Phi \Big|_{\theta\theta} + \frac{1}{2}m\Phi^2 \Big|_{\theta\theta} + \frac{1}{3}\lambda\Phi^3 \Big|_{\theta\theta}$$

$$\overset{(*)}{=} gF(y) + \frac{1}{2}m\left[2F(y)A(y) - \psi(y)\psi(y) \right]$$

$$+ \frac{1}{3}\lambda\left[3A^2(y)F(y) - 3A(y)\psi(y)\psi(y) \right]$$

$$= gF(y) + m\left[F(y)A(y) - \frac{1}{2}\psi^2(y) \right]$$

$$+ \lambda\left[A^2(y)F(y) - A(y)\psi^2(y) \right].$$

In step (*) we made use of Eqs. (7.5), (7.16) and (7.17). On the other hand

$$W[A] = gA(y) + \frac{1}{2}mA^2(y) + \frac{1}{3}\lambda A^3(y), \tag{9.7}$$

and

$$\frac{\partial W[A]}{\partial A} = g + mA(y) + \lambda A^2(y), \tag{9.8}$$

$$\frac{\partial^2 W[A]}{\partial A^2} = m + 2\lambda A(y). \tag{9.9}$$

Then

$$\frac{\partial W[A]}{\partial A}F(y) - \frac{1}{2}\frac{\partial^2 W[A]}{\partial A^2}\psi^2(y) = gF(y) + mA(y)F(y) + \lambda A^2(y)F(y)$$

$$- \frac{1}{2}m\psi^2(y) - \lambda\psi^2(y)A(y)$$

$$= gF(y) + m\Big[F(y)A(y) - \frac{1}{2}\psi^2(y)\Big]$$

$$+ \lambda\Big[A^2(y)F(y) - \psi^2(y)A(y)\Big]$$

$$= \int d^2\theta\, W[\Phi].$$

In general, if we have a set of n scalar superfields Φ_i, $i = 1, 2, \ldots, n$, and the superpotential is a functional of these n superfields, *i.e.*

$$W[\Phi] = W[\Phi_1, \Phi_2, \ldots, \Phi_n].$$

Then

$$\int d^4\theta W[\Phi_1, \Phi_2, \ldots, \Phi_n]\delta^2(\bar{\theta}) = \int d^2\theta W[\Phi_1, \Phi_2, \ldots, \Phi_n]$$

$$= \sum_i W_i F_i - \frac{1}{2}\sum_{i,j} W_{ij}\psi_i\psi_j, \tag{9.10}$$

where

$$W_i := \frac{\partial W[A_1, A_2, \ldots, A_n]}{\partial A_i}, \tag{9.11}$$

$$W_{ij} := \frac{\partial^2 W[A_1, A_2, \ldots, A_n]}{\partial A_i \partial A_j}. \tag{9.12}$$

The matrix with elements W_{ij} is called the *fermionic mass matrix*. Setting

$$W[\Phi_1, \ldots, \Phi_n] := \sum_{i=1}^n g_i\Phi_i + \frac{1}{2}\sum_{i,j=1}^n m_{ij}\Phi_i\Phi_j + \frac{1}{3}\sum_{i,j,k=1}^n \lambda_{ijk}\Phi_i\Phi_j\Phi_k,$$

and

$$\overline{W}[\Phi_1^\dagger,\ldots,\Phi_n^\dagger] := \sum_{i=1}^n g_i^* \Phi_i^\dagger + \frac{1}{2}\sum_{i,j=1}^n m_{ij}^* \Phi_i^\dagger\Phi_j^\dagger + \frac{1}{3}\sum_{i,j,k=1}^n \lambda_{ijk}^* \Phi_i^\dagger\Phi_j^\dagger\Phi_k^\dagger,$$

we can write the action integral (8.21) in terms of superpotentials as

$$\mathcal{A} = \int d^4x \left\{ \int d^4\theta \Phi_i^\dagger\Phi_i + \int d^2\theta W[\Phi_1,\ldots,\Phi_n] + \int d^2\overline{\theta}\,\overline{W}[\Phi_1^\dagger,\ldots,\Phi_n^\dagger] \right\}$$

$$\overset{(9.10)}{=} \int d^4x \left\{ \int d^4\theta \Phi_i^\dagger\Phi_i + W_iF_i - \frac{1}{2}W_{ij}\psi_i\psi_j + \overline{W}_iF_i^* - \frac{1}{2}\overline{W}_{ij}\overline{\psi}_i\overline{\psi}_j \right\}$$

$$\overset{(7.27b)}{=} \int d^4x \left\{ i\partial_\mu\overline{\psi}_i(x)\overline{\sigma}^\mu\psi_i(x) - A_i^*(x)\Box A_i(x) + |F_i(x)|^2 \right.$$

$$+ W_iF_i(x) + \overline{W}_iF_i^*(x) - \frac{1}{2}W_{ij}\psi_i(x)\psi_j(x) - \frac{1}{2}\overline{W}_{ij}\overline{\psi}_i(x)\overline{\psi}_j(x)$$

$$\left. + \text{ total derivatives} \right\}. \tag{9.13}$$

The auxiliary fields $F_i(x)$ and $F_i^*(x)$ can be eliminated with the help of their respective Euler-Lagrange equations. Thus, since the action $\mathcal{A}$ does not contain derivatives of $F_i(x)$ or $F_i^*(x)$, we have

$$\left.\begin{aligned}\frac{\partial\mathcal{L}}{\partial F_i^*(x)} &= F_i(x) + W_i^* = 0,\\[2mm]\frac{\partial\mathcal{L}}{\partial F_i(x)} &= F_i^*(x) + W_i = 0,\end{aligned}\right\} \tag{9.14}$$

and we therefore obtain for the action (9.13):

$$\mathcal{A} = \int d^4x \left\{ i\partial_\mu\overline{\psi}_i(x)\overline{\sigma}^\mu\psi_i(x) - A_i^*(x)\Box A_i(x) \right.$$

$$\left. - \frac{1}{2}W_{ij}\psi_i(x)\psi_j(x) - \frac{1}{2}\overline{W}_{ij}\overline{\psi}_i(x)\overline{\psi}_j(x) - |W_i|^2 \right\}.$$

Here we use:

$$|W_i|^2 = \sum_{i=1}^n W_i\overline{W}_i \overset{(9.11)}{=} \sum_{i=1}^n \left|\frac{\partial W}{\partial A_i}\right|^2 \overset{(9.14)}{=} \sum_{i=1}^n |F_i(x)|^2 \overset{(8.32)}{=} V(A,A^*).$$

Hence the scalar potential, i.e. that part of the Lagrangian density which does not contain any derivative of the fermionic field, is related to the superpotential W by:

$$V(A,A^*) = \sum_{i=1}^n \left|\frac{\partial W}{\partial A_i}\right|^2. \tag{9.15}$$

In the particular case of a single scalar field Φ — *i.e.* in the case of the Wess–Zumino model — the scalar potential $V(A, A^*)$ is obtained by inserting Eq. (9.8) into Eq. (9.15), *i.e.*

$$V(A, A^*) = \left| g + mA(y) + \lambda A^2(y) \right|^2.$$

We observe that the scalar potential is always greater than or equal to zero.

Now the invariance of any theory — *i.e.* the action — with respect to supersymmetry transformations implies that the Hamiltonian H commutes with the generators of the supersymmetry transformations, *i.e.*

$$\left[H, Q_A \right] = 0 = \left[H, \overline{Q}_{\dot{A}} \right].$$

By definition the *ground state* $|0\rangle$ satisfies[1] $H|0\rangle = 0$. Thus, besides $H|0\rangle = 0$, a supersymmetric theory has the property that

$$Q_A|0\rangle = \overline{Q}_{\dot{A}}|0\rangle = 0,$$

i.e. the ground state is invariant under supersymmetry transformations. The energy of the ground state, *i.e.* the eigenvalue of $H|0\rangle$, is zero, *i.e.* minimized, if the scalar potential (9.15), which is positive definite, is zero. Thus in a theory constructed from chiral scalar superfields Φ_i, supersymmetry is unbroken if and only if the scalar potential (9.15) is zero and hence

$$\frac{\partial W}{\partial A_i} = 0, \qquad (9.16)$$

for all $i = 1, 2, \ldots, n$ at the minimum. Hence, a nonvanishing derivative $\partial W / \partial A_i$ for some index i at the absolute minimum, signals the breaking of supersymmetry.

9.2 Projection Technique

Before we discuss supersymmetry breaking in more detail, we develop a projection technique which is a useful tool for evaluating integrals over Grassmann variables. This technique is based on the fact that Grassmann integration is equivalent to Grassmann differentiation. We made already explicit use of this method in evaluating the Grassmann integral of the superpotential (see Eq. (9.5)).

---

[1] We see from Eq. (5.113) that this condition is trivially satisfied in the Wess–Zumino model due to the normal ordering.

We start by considering a chiral superfield

$$\Phi(y, \theta) = A(y) + \sqrt{2}\theta\psi(y) + (\theta\theta)F(y), \tag{9.17a}$$

which obeys the constraint

$$\overline{D}_{\dot{A}}^{(1)}(y)\Phi(y, \theta) = 0. \tag{9.17b}$$

We show that one can obtain the component fields of this supermultiplet by the following projection operations:

$$A(x) = \Phi(y, \theta)\big|_{\theta=\bar{\theta}=0} = \Phi(x, \theta, \bar{\theta})\big|_{\theta=\bar{\theta}=0}, \tag{9.18a}$$

$$\psi_A(x) = \frac{1}{\sqrt{2}}D_A^{(1)}\Phi(y, \theta)\big|_{\theta=\bar{\theta}=0} = \frac{1}{\sqrt{2}}D_A^{(1)}\Phi(x, \theta, \bar{\theta})\big|_{\theta=\bar{\theta}=0}, \tag{9.18b}$$

$$F(x) = -\frac{1}{4}D^{(1)2}\Phi(y, \theta)\big|_{\theta=\bar{\theta}=0} = -\frac{1}{4}D^2\Phi(x, \theta, \bar{\theta})\big|_{\theta=\bar{\theta}=0}, \tag{9.18c}$$

where (*cf.* Eq. (7.2))

$$y_\mu = x_\mu + i\theta\sigma_\mu\bar{\theta}.$$

The projections given by the relations (9.18a) to (9.18c) are seen to be independent of the particular field representation $\Phi(y, \theta)$ or $\Phi(x, \theta, \bar{\theta})$ since the change of representation is caused by this shift of variables.

We now verify the relations (9.18a) to (9.18c).

(i) Setting $\theta = \bar{\theta} = 0$ in Eq. (9.17a) yields the scalar field $A(x)$ trivially.

(ii) Using Eq. (6.48) we find:

$$\begin{aligned}
\frac{1}{\sqrt{2}}D_A^{(1)}\Phi(y, \theta) &= \frac{1}{\sqrt{2}}\left(\partial_A + 2i(\sigma^\mu\bar{\theta})_A\partial_\mu\right)\Phi(y, \theta)\big|_{\theta=\bar{\theta}=0} \\
&= \frac{1}{\sqrt{2}}\partial_A\left[A(y) + \sqrt{2}\theta^B\psi_B(y) + \theta\theta F(y)\right]\Big|_{\theta=\bar{\theta}=0} \\
&\quad + i\sqrt{2}(\sigma^\mu\bar{\theta})_A\left[\partial_\mu A(y) + \sqrt{2}\theta^B\partial_\mu\psi_B(y)\right. \\
&\quad \left. + \theta\theta\partial_\mu F(y)\right]\Big|_{\theta=\bar{\theta}=0} \\
&= \left(\psi_A(y) + \sqrt{2}\theta_A F(y)\right)\Big|_{\theta=\bar{\theta}=0} \\
&= \psi_A(x),
\end{aligned}$$

where in the next to last step we made use of Eqs. (6.5f) and (6.5p).

(iii) From Eq. (6.48) we have:

$$D^{(1)A} = \epsilon^{AB} D^{(1)}_{B} \overset{(6.5g)}{=} -\partial^A + 2i\left(\epsilon\sigma^\mu\overline{\theta}\right)^A \partial_\mu.$$

Then

$$D^{(1)2} = D^{(1)A} D^{(1)}_{A} = \left(-\partial^A + 2i\left(\epsilon\sigma^\mu\overline{\theta}\right)^A \partial_\mu\right)\left(\partial_A + 2i\left(\sigma^\nu\overline{\theta}\right)_A \partial_\nu\right)$$

$$= -\partial^A \partial_A + 2i\left(\sigma^\nu\overline{\theta}\right)_A \partial_\nu \partial^A$$

$$+ 2i\left(\epsilon\sigma^\mu\overline{\theta}\right)^A \partial_\mu \partial_A - 4\left(\epsilon\sigma^\mu\overline{\theta}\right)^A \left(\sigma^\nu\overline{\theta}\right)_A \partial_\mu \partial_\nu.$$

Now

$$\left(\epsilon\sigma^\mu\overline{\theta}\right)^A \left(\sigma^\nu\overline{\theta}\right)_A \partial_\mu \partial_\nu = \epsilon^{AB}(\sigma^\mu)_{B\dot{B}}\overline{\theta}^{\dot{B}}(\sigma^\nu)_{A\dot{C}}\overline{\theta}^{\dot{C}} \partial_\mu \partial_\nu$$

$$\overset{(1.100c)}{=} \frac{1}{2}\left(\overline{\theta}\,\overline{\theta}\right)\epsilon^{\dot{B}\dot{C}}\epsilon^{AB}(\sigma^\mu)_{B\dot{B}}(\sigma^\nu)_{A\dot{C}}\partial_\mu \partial_\nu$$

$$\overset{(1.106a)}{=} -\frac{1}{2}\left(\overline{\theta}\,\overline{\theta}\right)\,\mathrm{Tr}\left[\overline{\sigma}^\mu\sigma^\nu\right]\partial_\mu \partial_\nu \overset{(1.109a)}{=} -\left(\overline{\theta}\,\overline{\theta}\right)\square.$$

Furthermore

$$2i\left(\sigma^\nu\overline{\theta}\right)_A \partial_\nu \partial^A + 2i\left(\epsilon\sigma^\mu\overline{\theta}\right)^A \partial_\mu \partial_A = 2i\sigma^\nu_{\ A\dot{B}}\overline{\theta}^{\dot{B}} \partial_\nu \partial^A + 2i\left(\epsilon\sigma^\mu\overline{\theta}\right)^A \partial_\mu \partial_A$$

$$= 2i\delta_A^{\ B}\sigma^\nu_{\ B\dot{B}}\overline{\theta}^{\dot{B}} \partial_\nu \partial^A + 2i\left(\epsilon\sigma^\mu\overline{\theta}\right)^A \partial_\mu \partial_A$$

$$= 2i\epsilon_{AC}\epsilon^{CB}\sigma^\nu_{\ B\dot{B}}\overline{\theta}^{\dot{B}} \partial_\nu \partial^A + 2i\left(\epsilon\sigma^\mu\overline{\theta}\right)^A \partial_\mu \partial_A$$

$$= 2i\epsilon^{CB}\sigma^\nu_{\ B\dot{B}}\overline{\theta}^{\dot{B}} \partial_\nu \partial_C + 2i\left(\epsilon\sigma^\mu\overline{\theta}\right)^A \partial_\mu \partial_A$$

$$\overset{(6.5g)}{=} 2i\left(\epsilon\sigma^\mu\overline{\theta}\right)^A \partial_\mu \partial_A + 2i\left(\epsilon\sigma^\mu\overline{\theta}\right)^A \partial_\mu \partial_A$$

$$= 4i\left(\epsilon\sigma^\nu\overline{\theta}\right)^A \partial_\nu \partial_A.$$

Hence:

$$D^{(1)2} = -\partial^A \partial_A + 4i\left(\epsilon\sigma^\mu\overline{\theta}\right)^A \partial_\mu \partial_A + 4\left(\overline{\theta}\,\overline{\theta}\right)\square. \qquad (9.19)$$

Then the last component of the supermultiplet is obtained with:

$$D^{(1)2}\Phi(y,\theta)\Big|_{\theta=\overline{\theta}=0} \overset{(9.17a)}{\underset{(9.19)}{=}} -\partial^A \partial_A\left[A(y) + \sqrt{2}\theta\psi(y) + \theta\theta F(y)\right]\Big|_{\theta=\overline{\theta}=0}$$

$$+ 4i\left(\epsilon\sigma^\mu\overline{\theta}\right)^A \partial_\mu \partial_A \Phi(y,\theta)\Big|_{\theta=\overline{\theta}=0}$$

$$+ 4\left(\overline{\theta}\,\overline{\theta}\right)\square\,\Phi(y,\theta)\Big|_{\theta=\overline{\theta}=0}$$

$$\overset{(6.5g)}{=} \epsilon^{AB}\partial_B \partial_A \theta\theta F(y)\Big|_{\theta=0}$$

$$\overset{(6.5i)}{=} -\epsilon^{AB}\partial_A \partial_B \theta\theta F(y)\Big|_{\theta=0} \overset{(6.5q)}{=} -4F(x).$$

Hence:

$$F(x) = -\frac{1}{4}D^{(1)2}\Phi(y,\theta)\Big|_{\theta=\bar{\theta}=0},$$

which had to be shown.

From Grassmann integration we know that the integral $\int d^2\theta$ projects out the $\theta\theta$-component of any function integrated over; in particular for the chiral superfield $\Phi(y,\theta)$ we have

$$\int d^2\theta\, \Phi(y,\theta) = F(x). \tag{9.20}$$

Thus, comparing Eq. (9.18c) and Eq. (9.20) we have formally

$$\int d^2\theta = -\frac{1}{4}D^{(1)2}\Big|_{\theta=\bar{\theta}=0} = -\frac{1}{4}D^2\Big|_{\theta=\bar{\theta}=0}. \tag{9.21}$$

We now derive the expression given in Eq. (9.5) for the superpotential by using this projection technique. Thus

$$\int d^2\theta\, W[\Phi] = -\frac{1}{4}D^{(1)2}W[\Phi]\Big|_{\theta=\bar{\theta}=0} = -\frac{1}{4}D^{(1)A}D^{(1)}{}_A W[\Phi]\Big|_{\theta=\bar{\theta}=0}$$

$$= -\frac{1}{4}D^{(1)A}\left[\frac{\partial W[\Phi]}{\partial\Phi}D^{(1)}{}_A\Phi\right]\Big|_{\theta=\bar{\theta}=0}$$

$$= -\frac{1}{4}\left[\frac{\partial^2 W[\Phi]}{\partial\Phi\partial\Phi}D^{(1)A}\Phi D^{(1)}{}_A\Phi + \frac{\partial W[\Phi]}{\partial\Phi}D^{(1)2}\Phi\right]\Big|_{\theta=\bar{\theta}=0}$$

$$= -\frac{1}{4}\left[2\frac{\partial^2 W[A]}{\partial A^2}\psi^A\psi_A - 4\frac{\partial W[A]}{\partial A}F\right]$$

$$= \frac{\partial W[A]}{\partial A}F(x) - \frac{1}{2}\frac{\partial^2 W[A]}{\partial A^2}\psi(x)\psi(x). \tag{9.22}$$

It is now an easy matter to extend these calculations to the case of n chiral superfields (*cf.* Eq. (9.10)), *i.e.*

$$\int d^2\theta\, W[\Phi_1,\ldots,\Phi_n] = -\frac{1}{4}D^{(1)2}W[\Phi_1,\ldots,\Phi_n]\Big|_{\theta=\bar{\theta}=0}$$

$$= -\frac{1}{4}D^{(1)A}\left[\sum_{i=1}^{n}\frac{\partial W[\Phi_1,\ldots,\Phi_n]}{\partial\Phi_i}D^{(1)}{}_A\Phi_i\right]\Big|_{\theta=\bar{\theta}=0}$$

$$= -\frac{1}{4}\Big[\sum_{i,j=1}^{n} \frac{\partial^2 W[\Phi_1,\dots,\Phi_n]}{\partial\Phi_i\partial\Phi_j}\Big(D^{(1)A}\Phi_i\Big)\Big(D^{(1)}{}_A\Phi_j\Big)$$

$$+ \sum_{i=1}^{n} \frac{\partial W[\Phi_1,\dots,\Phi_n]}{\partial\Phi_i} D^{(1)2}\Phi_i\Big]\Big|_{\theta=\bar\theta=0}$$

$$= \sum_{i=1}^{n} \frac{\partial W[A]}{\partial A_i} F_i(x) - \frac{1}{2}\sum_{i,j=1}^{n} \frac{\partial^2 W[A]}{\partial A_i\partial A_j}\psi_i(x)\psi_j(x). \qquad (9.23)$$

For an antichiral (*i.e.* right-handed) superfield we have the expansion

$$\Phi^\dagger(z,\bar\theta) = A^*(z) + \sqrt{2}\,\bar\theta\,\bar\psi(z) + \bar\theta\,\bar\theta F^*(z), \qquad (9.24a)$$

obeying the defining constraint:

$$D_A\,\Phi^\dagger = 0. \qquad (9.24b)$$

In much the same way as in Eqs. (9.18a) to (9.18c) we obtain the component fields of the antichiral supermultiplet by the projections:

$$A^*(x) = \Phi^\dagger(z,\bar\theta)\Big|_{\theta=\bar\theta=0} = \Phi^\dagger(x,\theta,\bar\theta)\Big|_{\theta=\bar\theta=0}, \qquad (9.25a)$$

$$\bar\psi_{\dot A}(x) = \frac{1}{\sqrt{2}}\overline{D}^{(2)}_{\dot A}\,\Phi^\dagger(z,\bar\theta)\Big|_{\theta=\bar\theta=0} = \frac{1}{\sqrt{2}}\overline{D}_{\dot A}\,\Phi^\dagger(x,\theta,\bar\theta)\Big|_{\theta=\bar\theta=0}, \qquad (9.25b)$$

$$F^*(x) = -\frac{1}{4}\big(\overline{D}^{(2)}\big)^2\Phi^\dagger(z,\bar\theta)\Big|_{\theta=\bar\theta=0} = -\frac{1}{4}\overline{D}^2\Phi^\dagger(x,\theta,\bar\theta)\Big|_{\theta=\bar\theta=0}. \qquad (9.25c)$$

Here the variable z is given by (*cf.* Eq. (7.10)) $z^\mu = x^\mu - i\theta\sigma^\mu\bar\theta$.

We now verify relations (9.25a) to (9.25c).

(i) Obviously, setting $\theta = \bar\theta = 0$ in the component field expansion of the antichiral superfield $\Phi^\dagger(z,\bar\theta)$ gives:

$$A^*(x) = \Phi^\dagger(z,\bar\theta)\Big|_{\theta=\bar\theta=0}.$$

(ii) Using Eq. (6.51) we find:

$$\frac{1}{\sqrt{2}}\overline{D}^{(2)}_{\dot A}\,\Phi^\dagger(z,\bar\theta)\Big|_{\theta=\bar\theta=0} = \frac{1}{\sqrt{2}}\Big[-\bar\partial_{\dot A} - 2i\big(\theta\sigma^\mu\big)_{\dot A}\partial_\mu\Big]\Phi^\dagger(z,\bar\theta)\Big|_{\theta=\bar\theta=0}$$

$$= -\frac{1}{\sqrt{2}}\bar\partial_{\dot A}\Big[A^*(z) + \sqrt{2}\,\bar\theta\,\bar\psi(z) + \bar\theta\,\bar\theta F^*(z)\Big]\Big|_{\theta=\bar\theta=0}$$

$$- i\sqrt{2}\big(\theta\sigma^\mu\big)_{\dot A}\partial_\mu\Phi^\dagger(z,\bar\theta)\Big|_{\theta=\bar\theta=0}$$

$$= \left[-\overline{\partial}_{\dot{A}} \overline{\theta}_{\dot{B}} \overline{\psi}^{\dot{B}}(z) - \frac{1}{\sqrt{2}} \overline{\partial}_{\dot{A}}(\overline{\theta}\,\overline{\theta}) F^*(z) \right] \Big|_{\theta=\overline{\theta}=0}$$

$$\overset{(6.5\text{k})}{=} \left[\epsilon_{\dot{A}\dot{B}} \overline{\psi}^{\dot{B}}(z) + \sqrt{2}\,\overline{\theta}_{\dot{A}} F^*(z) \right] \Big|_{\theta=\overline{\theta}=0}$$

$$= \overline{\psi}_{\dot{A}}(x).$$

(iii) Before we demonstrate the last relation, it is useful to derive the following expression:

$$\overline{D}^{(2)2} = \overline{D}^{(2)}_{\dot{A}} \overline{D}^{(2)\dot{A}} = -\overline{\partial}_{\dot{A}} \overline{\partial}^{\dot{A}} - 4i(\theta\sigma^\mu\overline{\partial})\partial_\mu + 4\theta\theta\,\Box. \tag{9.26}$$

In order to verify Eq. (9.26) we recall that according to Eq. (6.42) rewritten with the help of Eq. (1.106b) we have

$$\overline{D}^{(2)\dot{A}} = \overline{\partial}^{\dot{A}} + 2i(\theta\sigma^\mu\overline{\epsilon})^{\dot{A}}\partial_\mu.$$

Then:

$$\overline{D}^{(2)}_{\dot{A}} \overline{D}^{(2)\dot{A}} \overset{(6.51)}{=} \left(-\overline{\partial}_{\dot{A}} - 2i(\theta\sigma^\mu)_{\dot{A}}\partial_\mu \right) \left(\overline{\partial}^{\dot{A}} + 2i(\theta\sigma^\nu\overline{\epsilon})^{\dot{A}}\partial_\nu \right)$$

$$= -\overline{\partial}\,\overline{\partial} - 2i(\theta\sigma^\mu\overline{\partial})\partial_\mu - 2i\overline{\partial}_{\dot{A}}(\theta\sigma^\mu\overline{\epsilon})^{\dot{A}}\partial_\mu$$

$$+ 4(\theta\sigma^\mu)_{\dot{A}}(\theta\sigma^\nu\overline{\epsilon})^{\dot{A}}\partial_\mu\partial_\nu$$

$$= -\overline{\partial}\,\overline{\partial} - 2i(\theta\sigma^\mu\overline{\partial})\partial_\mu + 2i(\theta\sigma^\mu)_{\dot{B}}\epsilon^{\dot{B}\dot{A}}\overline{\partial}_{\dot{A}}\partial_\mu$$

$$+ 4\theta^A\theta^B\sigma^\mu_{A\dot{A}}\sigma^\nu_{B\dot{B}}\epsilon^{\dot{B}\dot{A}}\partial_\mu\partial_\nu$$

$$\overset{(1.100\text{a})}{\underset{(6.5\text{j})}{=}} -\overline{\partial}\,\overline{\partial} - 2i(\theta\sigma^\mu\overline{\partial})\partial_\mu - 2i(\theta\sigma^\mu\overline{\partial})\partial_\mu$$

$$+ 2(\theta\theta)\epsilon^{AB}\epsilon^{\dot{A}\dot{B}}\sigma^\mu_{A\dot{A}}\sigma^\nu_{B\dot{B}}\partial_\mu\partial_\nu$$

$$\overset{(1.106\text{a})}{=} -\overline{\partial}\,\overline{\partial} - 4i(\theta\sigma^\mu\overline{\partial})\partial_\mu + 2(\theta\theta)\,\text{Tr}\left[\sigma^\mu\overline{\sigma}^\nu\right]\partial_\mu\partial_\nu$$

$$\overset{(1.109\text{a})}{=} -\overline{\partial}\,\overline{\partial} - 4i(\theta\sigma^\mu\overline{\partial})\partial_\mu + 4(\theta\theta)\,\Box.$$

Hence, with Eq. (9.26) we find:

$$\overline{D}^{(2)2}\Phi^\dagger(z,\overline{\theta})\Big|_{\theta=\overline{\theta}=0} = \left(-\overline{\partial}\,\overline{\partial} - 4i(\theta\sigma^\mu\overline{\partial})\partial_\mu + 4(\theta\theta)\,\Box \right)\Phi^\dagger(z,\theta)\Big|_{\theta=\overline{\theta}=0}$$

$$= -\overline{\partial}\,\overline{\partial}\Phi^\dagger(z,\theta)\Big|_{\theta=\overline{\theta}=0}$$

$$= -\overline{\partial}\,\overline{\partial}\left[A^*(z) + \sqrt{2}\,\overline{\theta}\,\overline{\psi}(x) + (\overline{\theta}\,\overline{\theta}) F^*(z) \right]\Big|_{\theta=\overline{\theta}=0}$$

$$= -\overline{\partial}\,\overline{\partial}(\overline{\theta}\,\overline{\theta}) F^*(z)\Big|_{\theta=\overline{\theta}=0}$$

$$\overset{(6.5j)}{=} \epsilon_{\dot{A}\dot{B}} \overline{\partial}^{\dot{B}} \overline{\partial}^{\dot{A}} (\theta\,\theta) F^*(z)\Big|_{\theta=\overline{\theta}=0}$$

$$= -\epsilon_{\dot{A}\dot{B}} \overline{\partial}^{\dot{A}} \overline{\partial}^{\dot{B}} (\overline{\theta}\,\overline{\theta}) F^*(z)\Big|_{\theta=\overline{\theta}=0}$$

$$\overset{(6.5q)}{=} -4F^*(x).$$

As in the case of chiral superfields we can project out the component field of the highest order term in two ways:

- Either by evaluation of the Grassmann integral

$$\int d^2\overline{\theta}\,\Phi^\dagger(z,\overline{\theta}) = F^*(x), \tag{9.27a}$$

- or by differentiation as above, *i.e.*

$$-\frac{1}{4}\overline{D}^{(2)2}\,\Phi^\dagger(z,\overline{\theta}) = F^*(x).$$

From these equivalent procedures we deduce by comparison of Eq. (9.25c) and Eq. (9.27a) the formal relation

$$\int d^2\overline{\theta} = -\frac{1}{4}\overline{D}^{(2)2}\Big|_{\theta=\overline{\theta}=0} = -\frac{1}{4}\overline{D}^2\Big|_{\theta=\overline{\theta}=0}. \tag{9.27b}$$

In the following we write simply $\Big|_0$ for $\Big|_{\theta=\overline{\theta}=0}$.

The projection technique is not restricted to the evaluation of the Grassmann integral of the superpotential but can also be applied to other cases and is particularly useful for finding the component expansion of supersymmetric actions formulated in terms of chiral and antichiral superfields. Consider an arbitrary superfield with the power series expansion of Eq. (6.7). From the general formalism of Grassmann integration we know that $\int d^4\theta$ projects out the $(\theta\theta)(\overline{\theta}\,\overline{\theta})$-component of this superfield, *i.e.* (*cf.* Eq. (6.7) and Eq. (8.18))

$$\int d^4\theta\,\Phi(x,\theta,\overline{\theta}) = d(x). \tag{9.28}$$

Proposition 9.1:

The *d*-component of a chiral superfield can be obtained by applying the operator $1/16\overline{D}_{\dot{A}} D^2 \overline{D}^{\dot{A}}$, *i.e.*

$$\frac{1}{16}\overline{D}_{\dot{A}}D^2\overline{D}^{\dot{A}}\,\Phi(x,\theta,\overline{\theta})\Big|_0 = d(x). \tag{9.29}$$

Proof: In order to verify Eq. (9.29) we consider:

$$
\begin{aligned}
\frac{1}{16}\overline{D}_{\dot{A}}D^2\overline{D}^{\dot{A}}\,\Phi(x,\theta,\overline{\theta})\Big|_0
&= \frac{1}{16}\overline{D}_{\dot{A}}D^A D_A\overline{D}^{\dot{A}}\,\Phi(x,\theta,\overline{\theta})\Big|_0 \\[4pt]
&\overset{\substack{(6.41)\\(6.42)}}{=} \frac{1}{16}(-\overline{\partial}_{\dot{A}})(-\partial^A)(\partial_A)(\overline{\partial}^{\dot{A}})\,\Phi(x,\theta,\overline{\theta})\Big|_0 \\[4pt]
&\quad + \text{terms which vanish at } \theta=\overline{\theta}=0 \\[4pt]
&= \frac{1}{16}\partial^A\partial_A\overline{\partial}_{\dot{A}}\overline{\partial}^{\dot{A}}\,\Phi(x,\theta,\overline{\theta})\Big|_0 \\[4pt]
&\overset{\substack{(6.5g)\\(6.5j)}}{=} \frac{1}{16}\epsilon^{AB}\partial_B\partial_A\epsilon_{\dot{A}\dot{B}}\overline{\partial}^{\dot{B}}\overline{\partial}^{\dot{A}}\,\Phi(x,\theta,\overline{\theta})\Big|_0 \\[4pt]
&\overset{(6.7)}{=} \frac{1}{16}\epsilon^{AB}\partial_B\partial_A\epsilon_{\dot{A}\dot{B}}\overline{\partial}^{\dot{B}}\overline{\partial}^{\dot{A}}\,(\theta\theta)(\overline{\theta}\,\overline{\theta})d(x)\Big|_0 \\[4pt]
&\overset{(6.5q)}{=} d(x).
\end{aligned}
$$

This completes the proof of Proposition 9.1.

Hence, comparing Eqs. (9.28) and (9.29) we see that formally we have

$$\int d^4\theta = \frac{1}{16}\overline{D}_{\dot{A}}D^2\overline{D}^{\dot{A}}\Big|_0. \tag{9.30}$$

With this correspondence we can rederive the kinetic part of the supersymmetric action (8.30) in a more elegant way. Consider

$$
\begin{aligned}
\int d^4\theta\,\Phi^\dagger\Phi
&\overset{(9.30)}{=} \frac{1}{16}\overline{D}_{\dot{A}}D^2\overline{D}^{\dot{A}}\,\Phi^\dagger\Phi\Big|_0 \\[4pt]
&= \frac{1}{16}\overline{D}_{\dot{A}}D^2\big\{(\overline{D}^{\dot{A}}\Phi^\dagger)\Phi + \Phi^\dagger(\overline{D}^{\dot{A}}\Phi)\big\}\Big|_0 \\[4pt]
&\overset{(9.17b)}{=} \frac{1}{16}\overline{D}_{\dot{A}}D^A D_A\big\{(\overline{D}^{\dot{A}}\Phi^\dagger)\Phi\big\}\Big|_0 \\[4pt]
&= \frac{1}{16}\overline{D}_{\dot{A}}D^A\big\{(D_A\overline{D}^{\dot{A}}\Phi^\dagger)\Phi - (\overline{D}^{\dot{A}}\Phi^\dagger)(D_A\Phi)\big\}\Big|_0 \\[4pt]
&= \frac{1}{16}\overline{D}_{\dot{A}}\big\{(D^2\overline{D}^{\dot{A}}\Phi^\dagger)\Phi + (D_A\overline{D}^{\dot{A}}\Phi^\dagger)(D^A\Phi) \\[4pt]
&\qquad - (D^A\overline{D}^{\dot{A}}\Phi^\dagger)(D_A\Phi) + (\overline{D}^{\dot{A}}\Phi^\dagger)(D^2\Phi)\big\}\Big|_0
\end{aligned}
$$

$$
= \frac{1}{16} \Big\{ \big(\overline{D}_{\dot{A}} D^2 \overline{D}^{\dot{A}} \Phi^\dagger \big) \Phi - \big(D^2 \overline{D}^{\dot{A}} \Phi^\dagger \big) \big(\overline{D}_{\dot{A}} \Phi \big)
$$
$$
+ \big(\overline{D}_{\dot{A}} D_A \overline{D}^{\dot{A}} \Phi^\dagger \big) \big(D^A \Phi \big) + \big(D_A \overline{D}^{\dot{A}} \Phi^\dagger \big) \big(\overline{D}_{\dot{A}} D^A \Phi \big)
$$
$$
- \big(\overline{D}_{\dot{A}} D^A \overline{D}^{\dot{A}} \Phi^\dagger \big) \big(D_A \Phi \big) - \big(D^A \overline{D}^{\dot{A}} \Phi^\dagger \big) \big(\overline{D}_{\dot{A}} D_A \Phi \big)
$$
$$
+ \big(\overline{D}^2 \Phi^\dagger \big) \big(D^2 \Phi \big) - \big(\overline{D}^{\dot{A}} \Phi^\dagger \big) \big(\overline{D}_{\dot{A}} D^2 \Phi \big) \Big\} \Big|_0
$$

$$
= \frac{1}{16} \Big\{ \big(D^A \overline{D}^2 D_A \Phi^\dagger \big) \Phi + \overline{D}_{\dot{A}} \Big(-\overline{D}^{\dot{A}} D_A + 2i \big(\sigma^\mu \bar{\epsilon} \big)_A{}^{\dot{A}} \partial_\mu \Big) \Phi^\dagger \big(D^A \Phi \big)
$$
$$
+ \Big(-\overline{D}^{\dot{A}} D_A + 2i \big(\sigma^\mu \bar{\epsilon} \big)_A{}^{\dot{A}} \partial_\mu \Big) \Phi^\dagger \Big(-D^A \overline{D}_{\dot{A}} - 2i \big(\epsilon \sigma^\nu \big)^A{}_{\dot{A}} \partial_\nu \Big) \Phi
$$
$$
- \Big[\overline{D}_{\dot{A}} \Big(-\overline{D}^{\dot{A}} D^A - 2i \big(\bar{\sigma}^\mu \big)^{\dot{A} A} \partial_\mu \Big) \Phi^\dagger \Big] \big(D_A \Phi \big)
$$
$$
- \Big[\Big(-\overline{D}^{\dot{A}} D^A - 2i \big(\bar{\sigma}^\mu \big)^{\dot{A} A} \partial_\mu \Big) \Phi^\dagger \Big]
$$
$$
\times \Big[\Big(-D_A \overline{D}_{\dot{A}} - 2i \big(\sigma^\nu \big)_{A \dot{A}} \partial_\nu \Big) \Phi \Big]
$$
$$
+ \big(\overline{D}^2 \Phi^\dagger \big) \big(D^2 \Phi \big) - \big(\overline{D}^{\dot{A}} \Phi^\dagger \big) \big(D^2 \overline{D}_{\dot{A}} + 4i D^A \big(\sigma^\rho \big)_{A \dot{A}} \partial_\rho \big) \Phi \Big\} \Big|_0 .
$$

In deriving this intermediate result we made use of Eqs. (9.17b), (6.66), (7.1), (6.54) and (6.59). Thus

$$
\int d^4 \theta \, \Phi^\dagger \Phi = \frac{1}{16} \Big\{ 2i \big(\sigma^\mu \bar{\epsilon} \big)_A{}^{\dot{A}} \partial_\mu \big(\overline{D}_{\dot{A}} \Phi^\dagger \big) \big(D^A \Phi \big)
$$
$$
+ 4 \big(\sigma^\mu \bar{\epsilon} \big)_A{}^{\dot{A}} \big(\partial_\mu \Phi^\dagger \big) \big(\epsilon \sigma^\nu \big)^A{}_{\dot{A}} \big(\partial_\nu \Phi \big)
$$
$$
+ 2i \overline{D}_{\dot{A}} \big(\bar{\sigma}^\mu \big)^{\dot{A} A} \big(\partial_\mu \Phi^\dagger \big) \big(D_A \Phi \big) + 4 \big(\bar{\sigma}^\mu \big)^{\dot{A} A} \big(\partial_\mu \Phi^\dagger \big) \big(\sigma^\nu \big)_{A \dot{A}} \big(\partial_\nu \Phi \big)
$$
$$
+ \big(\overline{D}^2 \Phi^\dagger \big) \big(D^2 \Phi \big) - 4i \big(\overline{D}^{\dot{A}} \Phi^\dagger \big) \big(\sigma^\mu \big)_{A \dot{A}} \partial_\mu \big(D^A \Phi \big) \Big\} .
$$

Here we used Eqs. (9.17b) and (9.24b). Hence inserting the projections (9.18a) to (9.18c) and (9.25a) to (9.25c) we get

$$
\int d^4 \theta \, \Phi^\dagger \Phi = \frac{1}{16} \Big\{ 4i \big(\sigma^\mu \big)_{A \dot{B}} \epsilon^{\dot{B} \dot{A}} \partial_\mu \overline{\psi}_{\dot{A}}(x) \psi^A(x)
$$
$$
+ 4 \big(\sigma^\mu \big)_{A \dot{B}} \epsilon^{\dot{B} \dot{A}} \epsilon^{AC} \big(\sigma^\nu \big)_{C \dot{A}} \partial_\mu A^*(x) \partial_\nu A(x)
$$
$$
+ 4i \big(\bar{\sigma}^\mu \big)^{\dot{A} A} \partial_\mu \overline{\psi}_{\dot{A}}(x) \psi_A(x)
$$
$$
+ 4 \, \mathrm{Tr} \big[\bar{\sigma}^\mu \sigma^\nu \big] \partial_\mu A^*(x) \partial_\nu A(x)
$$
$$
+ 16 F^*(x) F(x) - 8i \overline{\psi}^{\dot{A}}(x) \big(\sigma^\mu \big)_{A \dot{A}} \partial_\mu \psi^A(x) \Big\}
$$

$$\overset{(1.106a)}{=} \frac{1}{16} \Big\{ 16 F^*(x)F(x) - 4i\psi(x)\sigma^\mu \big(\partial_\mu \overline{\psi}(x)\big)$$

$$+ 4 \, \mathrm{Tr} \, \big[\overline{\sigma}^\mu \sigma^\nu\big] \partial_\mu A^*(x) \partial_\nu A(x)$$

$$+ 4i \big(\partial_\mu \overline{\psi}(x)\big)\overline{\sigma}^\mu \psi(x)$$

$$+ 4 \, \mathrm{Tr} \, \big[\overline{\sigma}^\mu \sigma^\nu\big] \partial_\mu A^*(x) \partial_\nu A(x)$$

$$+ 8i \big(\partial_\mu \psi(x)\big)\sigma^\mu \overline{\psi}(x) \Big\}$$

$$\overset{(1.134)}{\underset{(1.109a)}{=}} F^*(x)F(x) + \partial_\mu A^*(x)\partial^\mu A(x)$$

$$+ \frac{i}{2}\big(\partial_\mu \overline{\psi}(x)\big)\overline{\sigma}^\mu \psi(x) - \frac{i}{2}\overline{\psi}(x)\overline{\sigma}^\mu \big(\partial_\mu \psi(x)\big)$$

$$= F^*(x)F(x) - A^*(x)\,\square\, A(x) + i\big(\partial_\mu \overline{\psi}(x)\big)\overline{\sigma}^\mu \psi(x)$$

$$+ \text{total derivatives},$$

where we used Eqs. (8.28) and (8.29) in the last step. We see that the expression obtained agrees with the previous result Eq. (7.27b).

The projection technique to find component field expansions of supersymmetric actions can also be used to obtain the component field expansions of spinor superfields such as W_A, which are used in the construction of action integrals of supersymmetric gauge theories. Since the spinor superfield carries an external Lorentz index, *i.e.* W_A transforms according to the $(1/2, 0)$-representation of $SL(2, \mathbb{C})$, the separation into components by projection requires reduction with respect to the group $SL(2, \mathbb{C})$. Consider the component field expansion of the spinor superfield W_A, *i.e.* (*cf.* Eq. (7.52))

$$W_A(y, \theta) = \lambda_A(y) + 2D(y)\theta_A + \big(\sigma^{\mu\nu}\theta\big)_A F_{\mu\nu}(y) - i\big(\theta\theta\big)\big(\sigma^\mu\big)_{A\dot{B}}\partial_\mu \overline{\lambda}^{\dot{B}}(y). \quad (9.31)$$

Proposition 9.2:

As for scalar superfields one can obtain the component fields of a spinor superfield, *i.e.* the fields $\lambda_A(x), D(x)$, and $F_{\mu\nu}$, by projection:

$$\lambda_A(x) = W_A(y, \theta)\big|_0, \quad\quad\quad\quad\quad\quad (9.32a)$$

$$D(x) = -\frac{1}{4}D^{(1)A}W_A(y, \theta)\big|_0 = -\frac{1}{4}D^A W_A(x, \theta, \overline{\theta})\big|_0, \quad (9.32b)$$

$$\big(\sigma_{\mu\nu}\epsilon^\top\big)_{AB} F^{\mu\nu}(x) = -\frac{1}{2}\big(D_A^{(1)}W_B + D_B^{(1)}W_A\big)\big|_0 \equiv -\frac{1}{2}D_{(A}^{(1)}W_{B)}\big|_0. \quad (9.32c)$$

Remark: Projecting out the symmetric combination[2] $D_{(A}W_{B)}$ and the con-

---

[2]Equation (9.32c) defines this standard notation.

traction $D^A W_A$ corresponds to reduction with respect to the Lorentz group in the following sense. The spinor superfield W_A describes an irreducible supermultiplet (of mass zero) with the three component fields $\lambda_A(x)$, $D(x)$, and $F_{\mu\nu}(x)$. The Clifford vacuum of this multiplet corresponds to the field

$$W_A\big|_0 = \lambda_A(x),$$

i.e. the Clifford vacuum carries a spinor index.[3] According to the general procedure outlined in Chapter 4, the other components of this multiplet are obtained by application of the supercharge $\overline{Q}_{\dot{A}}$ to the Clifford vacuum. In the superfield formalism this operation is represented by the projection $D_A W_B\big|_0$. With respect to the Lorentz group the projection operation $D_A W_B\big|_0$ corresponds to the Kronecker product of two $(1/2, 0)$ representations with sum decomposition given by Eq. (1.152b), *i.e.*

$$(1/2, 0) \otimes (1/2, 0) = (0, 0) \oplus (1, 0),$$

where the $D(x)$-field transforms according to $(0, 0)$, and the $F_{\mu\nu}$-field transforms according to the $(1, 0)$-representation.

Proof of Proposition 9.2:

(i) Equation (9.32a) is obvious.

(ii) With Eq. (6.48) and then Eqs. (6.5g) and (9.31) we get:

$$
\begin{aligned}
D^{(1)A} W_A(y, \theta)\big|_0 &= \epsilon^{AB}\left(\partial_B + 2i(\sigma^\mu)_{B\dot{B}}\overline{\theta}^{\dot{B}}\partial_\mu\right)W_A\big|_0 \\
&= -\partial^A\Big[\lambda_A(y) + 2D(y)\theta_A + (\sigma^{\mu\nu}\theta)_A F_{\mu\nu}(y) \\
&\qquad - i(\theta\theta)(\sigma^\mu)_{A\dot{B}}\partial_\mu\overline{\lambda}^{\dot{B}}(y)\Big]\Big|_0 \\
&\qquad + 2i\epsilon^{AB}(\sigma^\mu)_{B\dot{B}}\overline{\theta}^{\dot{B}}\partial_\mu W_A(y, \theta)\big|_0 \\
&= \Big[-2D(y)\partial^A\theta_A - (\sigma^{\mu\nu})_A{}^B\partial^A\theta_B F_{\mu\nu}(y)\Big]\Big|_0 \\
&= \Big[-2D(y)\delta^A{}_A - (\sigma^{\mu\nu})_A{}^B\delta^A{}_B F_{\mu\nu}(y)\Big]\Big|_0 \\
&= \Big[-4D(y) - \mathrm{Tr}\,[\sigma^{\mu\nu}] F_{\mu\nu}(y)\Big]\Big|_0 \\
&= -4D(x).
\end{aligned}
$$

---

[3]See also the discussion following Eq. (4.35).

In the last step we used Eq. (1.138a) as well as Eq. (1.109a), leading to:

$$\text{Tr}\,[\sigma^{\mu\nu}] = \frac{i}{4}\Big[\text{Tr}\,[\sigma^{\mu}\bar{\sigma}^{\nu}] - \text{Tr}\,[\sigma^{\nu}\bar{\sigma}^{\mu}]\Big] = \frac{i}{2}\Big(\eta^{\mu\nu} - \eta^{\nu\mu}\Big) = 0.$$

(iii) Consider, using Eq. (6.48)

$$
\begin{aligned}
\frac{1}{2}D^{(1)}_{(A}W_{B)}\Big|_0 \;=\;& \frac{1}{2}\Big(D^{(1)}_A W_B + D^{(1)}_B W_A\Big)\\[4pt]
=\;& \frac{1}{2}\Big\{\Big(\partial_A + 2i(\sigma^{\mu})_{A\dot{B}}\bar{\theta}^{\dot{B}}\partial_{\mu}\Big)W_B\\
&+ \Big(\partial_B + 2i(\sigma^{\mu})_{B\dot{B}}\bar{\theta}^{\dot{B}}\partial_{\mu}\Big)W_A\Big\}\Big|_0\\[4pt]
=\;& \frac{1}{2}\Big(\partial_A W_B + \partial_B W_A\Big)\Big|_0\\
& + \text{terms which vanish at } \theta = \bar{\theta} = 0\\[4pt]
\overset{(9.31)}{=}\;& \frac{1}{2}\Big\{2D(y)\partial_A\theta_B + (\sigma^{\mu\nu})_B{}^C\partial_A\theta_C F_{\mu\nu}(y)\\
& + 2D(y)\partial_B\theta_A + (\sigma^{\mu\nu})_A{}^C\partial_B\theta_C F_{\mu\nu}(y)\Big\}\Big|_0\\[4pt]
\overset{(6.5f)}{=}\;& \Big\{-D(y)\epsilon_{AB} - \frac{1}{2}(\sigma^{\mu\nu})_B{}^C\epsilon_{AC}F_{\mu\nu}(y)\\
& -D(y)\epsilon_{BA} - \frac{1}{2}(\sigma^{\mu\nu})_A{}^C\epsilon_{BC}F_{\mu\nu}(y)\Big\}\Big|_0\\[4pt]
=\;& \frac{1}{2}\Big\{(\sigma^{\mu\nu})_B{}^C\epsilon_{CA} + (\sigma^{\mu\nu})_A{}^C\epsilon_{CB}\Big\}F_{\mu\nu}(y)\Big|_0\\[4pt]
=\;& -\frac{1}{2}\Big\{(\sigma^{\mu\nu}\epsilon^{\top})_{BA} + (\sigma^{\mu\nu}\epsilon^{\top})_{AB}\Big\}F_{\mu\nu}(x)\\[4pt]
\overset{(1.149)}{=}\;& -(\sigma^{\mu\nu}\epsilon^{\top})_{AB}F_{\mu\nu}(x).
\end{aligned}
$$

Hence

$$-\frac{1}{2}D^{(1)}_{(A}W_{B)}\Big|_0 = (\sigma^{\mu\nu}\epsilon^{\top})_{AB}F_{\mu\nu}(x).$$

This completes the proof of Proposition 9.2.

It is convenient to calculate another expression involving spinor superfields, which will be needed later. Thus consider

$$
\begin{aligned}
D^{(1)2}W_A\Big|_0 \;=\;& D^{(1)B}D^{(1)}_B W_A\Big|_0\\[4pt]
\overset{(6.48)}{=}\;& \Big(-\partial^B + 2i\epsilon^{BC}(\sigma^{\mu})_{C\dot{B}}\bar{\theta}^{\dot{B}}\partial_{\mu}\Big)\Big(\partial_B + 2i(\sigma^{\nu})_{B\dot{B}}\bar{\theta}^{\dot{B}}\partial_{\nu}\Big)W_A\Big|_0\\[4pt]
=\;& -\partial^B\partial_B W_A\Big|_0
\end{aligned}
$$

$$\stackrel{(9.31)}{=} -\partial^B \partial_B \Big[\lambda_A(y) + 2D(y)\theta_A + (\sigma^{\mu\nu})_A{}^C \theta_C F_{\mu\nu}(y)$$

$$- i(\theta\theta)(\sigma^\mu)_{A\dot{B}} \partial_\mu \overline{\lambda}^{\dot{B}}(y)\Big]\Big|_0$$

$$= i\partial^B \partial_B (\theta\theta)(\sigma^\mu)_{A\dot{B}} \partial_\mu \overline{\lambda}^{\dot{B}}(y)\Big|_0$$

$$\stackrel{(6.5q)}{=} 4i(\sigma^\mu)_{A\dot{B}} \partial_\mu \overline{\lambda}^{\dot{B}}(x).\tag{9.33}$$

Raising the spinor index with ϵ^{AB} yields:

$$D^{(1)2} W^A\Big|_0 = 4i\epsilon^{AB}(\sigma^\mu)_{B\dot{B}} \partial_\mu \overline{\lambda}^{\dot{B}}(x).\tag{9.34}$$

The same type of analysis can be applied to the antichiral spinor superfield $\overline{W}_{\dot{A}}$, which transforms according to the $(0, 1/2)$ representation of $SL(2,\mathbb{C})$. This superfield has the the component field expansion (*cf.* Eq. (7.53))

$$\overline{W}_{\dot{A}}(z,\overline{\theta}) = \overline{\lambda}_{\dot{A}}(z) + 2D(z)\overline{\theta}_{\dot{A}} - \epsilon_{\dot{A}\dot{B}}(\overline{\sigma}^{\mu\nu})^{\dot{B}}{}_{\dot{C}}\overline{\theta}^{\dot{C}} F_{\mu\nu}(z) + i(\overline{\theta}\,\overline{\theta})(\partial_\mu \lambda(z)\sigma^\mu)_{\dot{A}}.\tag{9.35}$$

The component fields of this antichiral spinor superfield are given by the projections

$$\overline{\lambda}_{\dot{A}}(x) = \overline{W}_{\dot{A}}(z,\overline{\theta})\Big|_0,\tag{9.36a}$$

$$D(x) = -\frac{1}{4}\overline{D}^{(2)}_{\dot{A}} \overline{W}^{\dot{A}}(z,\overline{\theta})\Big|_0, \qquad (cf.\ \text{Eq. (7.57)})\tag{9.36b}$$

$$(\overline{\epsilon}\,\overline{\sigma}^{\mu\nu})_{\dot{A}\dot{B}} F_{\mu\nu}(x) = \frac{1}{2}\Big(\overline{D}^{(2)}_{\dot{A}} \overline{W}_{\dot{B}}(z,\overline{\theta}) + \overline{D}^{(2)}_{\dot{B}} \overline{W}_{\dot{A}}(z,\overline{\theta})\Big)\Big|_0$$

$$\equiv \frac{1}{2}\overline{D}^{(2)}_{(\dot{A}} \overline{W}_{\dot{B})}(z,\overline{\theta})\Big|_0.\tag{9.36c}$$

Furthermore

$$\overline{D}^{(2)2} \overline{W}_{\dot{A}}(z,\overline{\theta})\Big|_0 = \overline{D}^2 \overline{W}_{\dot{A}}(x,\theta,\overline{\theta})\Big|_0 = -4i(\partial_\mu \lambda(x)\sigma^\mu)_{\dot{A}}.\tag{9.36d}$$

Here we verify the projection given by Eq. (9.36c). Using Eq. (6.51) we get

$$\frac{1}{2}\overline{D}^{(2)}_{(\dot{A}} \overline{W}_{\dot{B})}(z,\overline{\theta})\Big|_0 = \frac{1}{2}\overline{\partial}_{(\dot{A}} \overline{W}_{\dot{B})}(z,\overline{\theta})\Big|_0$$

$$+ \text{ terms which vanish for } \theta = \overline{\theta} = 0$$

$$\stackrel{(9.35)}{=} -\frac{1}{2}\Big[2D(z)\overline{\partial}_{(\dot{A}}\overline{\theta}_{\dot{B})} - \epsilon_{(\dot{B}\dot{D}}(\overline{\sigma}^{\mu\nu})^{\dot{D}}{}_{\dot{C}}\overline{\partial}_{\dot{A})}\overline{\theta}^{\dot{C}} F_{\mu\nu}(z)\Big]\Big|_0$$

$$\stackrel{(6.5f)}{\underset{(6.51)}{=}} -\frac{1}{2}\Big[2D(x)(\epsilon_{\dot{B}\dot{A}} + \epsilon_{\dot{A}\dot{B}}) - \epsilon_{\dot{B}\dot{D}}(\overline{\sigma}^{\mu\nu})^{\dot{D}}{}_{\dot{C}}\delta_{\dot{A}}{}^{\dot{C}} F_{\mu\nu}(x)$$

$$- \epsilon_{\dot{A}\dot{D}}(\overline{\sigma}^{\mu\nu})^{\dot{D}}{}_{\dot{C}}\delta_{\dot{B}}{}^{\dot{C}} F_{\mu\nu}(x)\Big]$$

$$= \frac{1}{2}\left[\epsilon_{\dot{B}\dot{D}}(\overline{\sigma}^{\mu\nu})^{\dot{D}}{}_{\dot{A}} + \epsilon_{\dot{A}\dot{D}}(\overline{\sigma}^{\mu\nu})^{\dot{D}}{}_{\dot{B}}\right]F_{\mu\nu}(x)$$

$$\overset{(1.150)}{=} \epsilon_{\dot{A}\dot{D}}(\overline{\sigma}^{\mu\nu})^{\dot{D}}{}_{\dot{B}}F_{\mu\nu}(x)$$

$$= (\overline{\epsilon}\,\overline{\sigma}^{\mu\nu})_{\dot{A}\dot{B}}F_{\mu\nu}(x).$$

This verifies Eq. (9.36c).

With the above projections it is now possible to compute the Grassmann integrated action of a pure supersymmetric Abelian gauge theory. From Eq. (8.40) we know that the supersymmetric action of this theory is — expressed in terms of the spinor superfields W_A and $\overline{W}_{\dot{A}}$:

$$\mathcal{A} = \int d^4x\,\widetilde{\mathcal{A}},$$

where

$$
\begin{aligned}
\widetilde{\mathcal{A}} &= \int d^2\theta\, W^A W_A + \int d^2\overline{\theta}\,\overline{W}_{\dot{A}}\overline{W}^{\dot{A}} \\
&\overset{(9.21)}{\underset{(9.27b)}{=}} -\frac{1}{4}D^2(W^A W_A)\Big|_0 - \frac{1}{4}\overline{D}^2(\overline{W}_{\dot{A}}\overline{W}^{\dot{A}})\Big|_0 \\
&= -\frac{1}{4}D^B\left[(D_B W^A)W_A - W^A(D_B W_A)\right]\Big|_0 \\
&\quad -\frac{1}{4}\overline{D}_{\dot{B}}\left[(\overline{D}^{\dot{B}}\overline{W}_{\dot{A}})\overline{W}^{\dot{A}} - \overline{W}_{\dot{A}}(\overline{D}^{\dot{B}}\overline{W}^{\dot{A}})\right]\Big|_0 \\
&= -\frac{1}{4}\Big[(D^2 W^A)W_A + (D_B W^A)(D^B W_A) \\
&\qquad -(D^B W^A)(D_B W_A) + W^A(D^2 W_A)\Big]\Big|_0 \\
&\quad -\frac{1}{4}\Big[(\overline{D}^2\overline{W}_{\dot{A}})\overline{W}^{\dot{A}} + (\overline{D}^{\dot{B}}\overline{W}_{\dot{A}})(\overline{D}_{\dot{B}}\overline{W}^{\dot{A}}) \\
&\qquad -(\overline{D}_{\dot{B}}\overline{W}_{\dot{A}})(\overline{D}^{\dot{B}}\overline{W}^{\dot{A}}) + \overline{W}_{\dot{A}}(\overline{D}^2\overline{W}^{\dot{A}})\Big]\Big|_0 \\
&= -\frac{1}{4}\Big[(D^2 W^A)W_A + W^A(D^2 W_A)\Big]\Big|_0 \\
&\quad +\frac{1}{2}(D^B W^A)(D_B W_A)\Big|_0 \\
&\quad -\frac{1}{4}\Big[(\overline{D}^2\overline{W}_{\dot{A}})\overline{W}^{\dot{A}} + \overline{W}_{\dot{A}}(\overline{D}^2\overline{W}^{\dot{A}})\Big]\Big|_0 \\
&\quad +\frac{1}{2}(\overline{D}_{\dot{B}}\overline{W}_{\dot{A}})(\overline{D}^{\dot{B}}\overline{W}^{\dot{A}})\Big|_0.
\end{aligned}
\tag{9.37}
$$

Using the projections of the spinor superfields W_A and $\overline{W}_{\dot{A}}$ calculated above, we obtain:

$$\left(D^2 W^A\right) W_A\Big|_0 \overset{(9.32a)}{\underset{(9.34)}{=}} 4i\epsilon^{AB}\left(\sigma^\mu\right)_{B\dot{B}}\partial_\mu \overline{\lambda}^{\dot{B}}(x)\lambda_A(x)$$

$$= -4i\lambda_A(x)\epsilon^{AB}\left(\sigma^\mu\right)_{B\dot{B}}\partial_\mu \overline{\lambda}^{\dot{B}}(x)$$

$$= 4i\epsilon^{BA}\lambda_A(x)\left(\sigma^\mu\right)_{B\dot{B}}\partial_\mu \overline{\lambda}^{\dot{B}}(x)$$

$$= 4i\lambda(x)\sigma^\mu\partial_\mu \overline{\lambda}(x).$$

In a similar way we get for the second term of Eq. (9.37):

$$W^A\left(D^2 W_A\right)\Big|_0 \overset{(9.33)}{=} \lambda^A(x)4i\left(\sigma^\mu\right)_{A\dot{B}}\partial_\mu \overline{\lambda}^{\dot{B}}(x) = 4i\lambda(x)\sigma^\mu\partial_\mu \overline{\lambda}(x).$$

Hence

$$-\frac{1}{4}\left[\left(D^2 W^A\right)W_A + W^A\left(D^2 W_A\right)\right]\Big|_0 = -2i\lambda(x)\sigma^\mu\partial_\mu \overline{\lambda}(x). \qquad (9.38)$$

The next contribution to be evaluated is

$$\frac{1}{2}\left(D^B W^A\right)\left(D_B W_A\right) = \frac{1}{4}\left(D^B W^A + D^A W^B\right)\left(D_B W_A\right)$$

$$+\frac{1}{4}\left(D^B W^A - D^A W^B\right)\left(D_B W_A\right)$$

$$= \frac{1}{8}\left(D^B W^A + D^A W^B\right)\left(D_B W_A + D_A W_B\right)$$

$$+\frac{1}{8}\left(D^B W^A + D^A W^B\right)\left(D_B W_A - D_A W_B\right)$$

$$+\frac{1}{8}\left(D^B W^A - D^A W^B\right)\left(D_B W_A + D_A W_B\right)$$

$$+\frac{1}{8}\left(D^B W^A - D^A W^B\right)\left(D_B W_A - D_A W_B\right)$$

$$= \frac{1}{8}D^{(B}W^{A)}D_{(B}W_{A)} + \frac{1}{8}D^{(B}W^{A)}D_{[B}W_{A]}$$

$$+\frac{1}{8}D^{[B}W^{A]}D_{(B}W_{A)} + \frac{1}{8}D^{[B}W^{A]}D_{[B}W_{A]},$$

where

$$D_{(B}W_{A)} := D_B W_A + D_A W_B, \qquad D_{[B}W_{A]} := D_B W_A - D_A W_B.$$

Now, contracting a symmetric second rank spinor with an antisymmetric one always gives zero. Furthermore, using Eqs. (1.152c) and (1.152f) we have

$$
\begin{aligned}
\frac{1}{8} D^{[B} W^{A]} D_{[B} W_{A]}\Big|_0 &= -\frac{1}{8}(DW)(DW)\epsilon^{BA}\epsilon_{BA}\Big|_0 \\
&= \frac{1}{8}(DW)(DW)\epsilon^{BA}\epsilon_{AB}\Big|_0 \\
&= \frac{1}{8}(DW)(DW)\delta^A{}_A\Big|_0 = \frac{1}{4}(DW)(DW)\Big|_0 = 4D^2(x),
\end{aligned}
$$

$$(9.39)$$

where in the last step we use Eq. (9.32b). Finally the product of two symmetric expressions gives

$$
\begin{aligned}
\frac{1}{8} D^{(B} W^{A)} D_{(B} W_{A)}\Big|_0 &= \frac{1}{8}\epsilon^{BC}\epsilon^{AD} D_{(C} W_{D)} D_{(B} W_{A)}\Big|_0 \\
&\overset{(9.32c)}{=} \frac{1}{8}\epsilon^{BC}\epsilon^{AD}\left(-2\sigma_{\mu\nu}\epsilon^{\top}\right)_{CD} F^{\mu\nu}(x)\left(-2\sigma_{\rho\sigma}\epsilon^{\top}\right)_{BA} F^{\rho\sigma}(x) \\
&\overset{(1.149)}{=} \frac{1}{2}\epsilon^{BC}\left(\sigma_{\mu\nu}\epsilon^{\top}\right)_{CD}\left(-\epsilon^{DA}\right)\left(\sigma_{\rho\sigma}\epsilon^{\top}\right)_{AB} F^{\mu\nu}(x) F^{\rho\sigma}(x) \\
&= -\frac{1}{2}\epsilon^{BC}\left(\sigma_{\mu\nu}\right)_C{}^E \epsilon^{\top}_{ED}\epsilon^{DA}\left(\sigma_{\rho\sigma}\right)_A{}^L \epsilon^{\top}_{LB} F^{\mu\nu}(x) F^{\rho\sigma}(x) \\
&= -\frac{1}{2}\left(\epsilon^{\top}_{LB}\epsilon^{BC}\right)\left(\epsilon^{\top}_{ED}\epsilon^{DA}\right)\left(\sigma_{\mu\nu}\right)_C{}^E\left(\sigma_{\rho\sigma}\right)_A{}^L F^{\mu\nu}(x) F^{\rho\sigma}(x) \\
&= -\frac{1}{2}\left(\epsilon^{CB}\epsilon^{\top}_{BL}\right)\left(\epsilon^{AD}\epsilon^{\top}_{DE}\right)\left(\sigma_{\mu\nu}\right)_C{}^E\left(\sigma_{\rho\sigma}\right)_A{}^L F^{\mu\nu}(x) F^{\rho\sigma}(x) \\
&= -\frac{1}{2}\delta^C{}_L\delta^A{}_E\left(\sigma_{\mu\nu}\right)_C{}^E\left(\sigma_{\rho\sigma}\right)_A{}^L F^{\mu\nu}(x) F^{\rho\sigma}(x) \\
&= -\frac{1}{2}\left(\sigma_{\mu\nu}\right)_L{}^E\left(\sigma_{\rho\sigma}\right)_E{}^L F^{\mu\nu}(x) F^{\rho\sigma}(x) \\
&= -\frac{1}{2}\left(\sigma_{\mu\nu}\sigma_{\rho\sigma}\right)_L{}^L F^{\mu\nu}(x) F^{\rho\sigma}(x) \\
&= -\frac{1}{2}\mathrm{Tr}\left[\sigma_{\mu\nu}\sigma_{\rho\sigma}\right] F^{\mu\nu}(x) F^{\rho\sigma}(x) \\
&\overset{(1.144)}{=} -\left\{\frac{1}{4}\left(\eta^{\mu\rho}\eta^{\nu\sigma}-\eta^{\mu\sigma}\eta^{\nu\rho}\right)+\frac{i}{4}\epsilon^{\mu\nu\rho\sigma}\right\} F_{\mu\nu}(x) F_{\rho\sigma}(x) \\
&\overset{(8.37a)}{=} -\frac{1}{2} F_{\mu\nu}(x) F^{\mu\nu}(x) - \frac{i}{2} F_{\mu\nu}(x) F^{*\mu\nu}(x).
\end{aligned}
$$

$$(9.40)$$

Similarly, from the $\overline{W}_{\dot{A}}\overline{W}^{\dot{A}}$-term in the action integral (9.37) one obtains the following contributions to the component field expansion: First

$$
-\frac{1}{4}\left(\overline{D}^2\overline{W}_{\dot{A}}\right)\overline{W}^{\dot{A}}\Big|_0 \overset{(9.36a)}{\underset{(9.36d)}{=}} i\left(\partial_\mu\lambda(x)\sigma^\mu\right)_{\dot{A}}\overline{\lambda}^{\dot{A}}(x) = i\left(\partial_\mu\lambda(x)\right)\sigma^\mu\overline{\lambda}(x). \quad (9.41)
$$

Furthermore

$$
\begin{aligned}
\tfrac{1}{2}\left(\overline{D}_{\dot{B}}\overline{W}_{\dot{A}}\right)\left(\overline{D}^{\dot{B}}\overline{W}^{\dot{A}}\right)\Big|_{0} \;=\; & \tfrac{1}{8}\left(\overline{D}_{(\dot{B}}\overline{W}_{\dot{A})}\right)\left(\overline{D}^{(\dot{B}}\overline{W}^{\dot{A})}\right)\Big|_{0} \\
& +\tfrac{1}{8}\left(\overline{D}_{[\dot{B}}\overline{W}_{\dot{A}]}\right)\left(\overline{D}^{[\dot{B}}\overline{W}^{\dot{A}]}\right)\Big|_{0} \\
& +\tfrac{1}{8}\left(\overline{D}_{(\dot{B}}\overline{W}_{\dot{A})}\right)\epsilon^{\dot{B}\dot{C}}\epsilon^{\dot{A}\dot{D}}\left(\overline{D}_{(\dot{C}}\overline{W}_{\dot{D})}\right)\Big|_{0} \\
& +\tfrac{1}{8}\left(\overline{D}_{[\dot{B}}\overline{W}_{\dot{A}]}\right)\left(\overline{D}^{[\dot{B}}\overline{W}^{\dot{A}]}\right)\Big|_{0} \\
\overset{(9.36\mathrm{b})}{\underset{(9.36\mathrm{c})}{=}} \; & \tfrac{1}{2}\left(\overline{\epsilon}\,\overline{\sigma}^{\mu\nu}\right)_{\dot{B}\dot{A}}\epsilon^{\dot{B}\dot{C}}\epsilon^{\dot{A}\dot{D}}\left(\overline{\epsilon}\,\overline{\sigma}^{\rho\sigma}\right)_{\dot{C}\dot{D}}F_{\mu\nu}(x)F_{\rho\sigma}(x) \\
& -2D^{2}(x)\epsilon_{\dot{B}\dot{A}}\epsilon^{\dot{B}\dot{A}} \\
\overset{(1.150)}{=} \; & \tfrac{1}{2}\left(\overline{\epsilon}\,\overline{\sigma}^{\mu\nu}\right)_{\dot{B}\dot{A}}\epsilon^{\dot{B}\dot{C}}\epsilon^{\dot{A}\dot{D}}\left(\overline{\epsilon}\,\overline{\sigma}^{\rho\sigma}\right)_{\dot{D}\dot{C}}F_{\mu\nu}(x)F_{\rho\sigma}(x) \\
& +2D^{2}(x)\delta_{\dot{A}}{}^{\dot{A}} \\
= \; & \tfrac{1}{2}\epsilon_{\dot{B}\dot{E}}\left(\overline{\sigma}^{\mu\nu}\right)^{\dot{E}}{}_{\dot{A}}\epsilon^{\dot{B}\dot{C}}\epsilon^{\dot{A}\dot{D}}\epsilon_{\dot{D}\dot{F}}\left(\overline{\sigma}^{\rho\sigma}\right)^{\dot{F}}{}_{\dot{C}}F_{\mu\nu}(x)F_{\rho\sigma}(x) \\
& +4D^{2}(x) \\
= \; & -\tfrac{1}{2}\epsilon_{\dot{E}\dot{B}}\epsilon^{\dot{B}\dot{C}}\delta^{\dot{A}}{}_{\dot{F}}\left(\overline{\sigma}^{\mu\nu}\right)^{\dot{E}}{}_{\dot{A}}\left(\overline{\sigma}^{\rho\sigma}\right)^{\dot{F}}{}_{\dot{C}}F_{\mu\nu}(x)F_{\rho\sigma}(x) \\
& +4D^{2}(x) \\
= \; & -\tfrac{1}{2}\delta_{\dot{E}}{}^{\dot{C}}\left(\overline{\sigma}^{\mu\nu}\overline{\sigma}^{\rho\sigma}\right)^{\dot{E}}{}_{\dot{C}}F_{\mu\nu}(x)F_{\rho\sigma}(x)+4D^{2}(x) \\
= \; & -\tfrac{1}{2}\mathrm{Tr}\left[\overline{\sigma}^{\mu\nu}\overline{\sigma}^{\rho\sigma}\right]F_{\mu\nu}(x)F_{\rho\sigma}(x)+4D^{2}(x) \\
\overset{(1.145)}{=} \; & -\left[\tfrac{1}{4}\left(\eta^{\mu\rho}\eta^{\nu\sigma}-\eta^{\mu\sigma}\eta^{\nu\rho}\right)-\tfrac{i}{4}\epsilon^{\mu\nu\rho\sigma}\right]F_{\mu\nu}(x)F_{\rho\sigma}(x) \\
& +4D^{2}(x) \\
\overset{(8.37\mathrm{a})}{=} \; & -\tfrac{1}{2}F_{\mu\nu}(x)F^{\mu\nu}(x)+\tfrac{i}{2}F_{\mu\nu}(x)F^{*\mu\nu}(x)+4D^{2}(x).
\end{aligned}
$$

$$(9.42)$$

Hence, inserting Eqs. (9.38) to (9.42) into the action (9.37) we finally obtain:

$$
\begin{aligned}
\int d^{2}\theta\, W^{A}W_{A} + \int d^{2}\overline{\theta}\,\overline{W}_{\dot{A}}\overline{W}^{\dot{A}} \\
= 2i\lambda(x)\sigma^{\mu}\partial_{\mu}\overline{\lambda}(x) + 4D^{2}(x) - \tfrac{1}{2}F_{\mu\nu}(x)F^{\mu\nu}(x) \\
- \tfrac{i}{2}F_{\mu\nu}(x)F^{*\mu\nu}(x) + 2i\left(\partial_{\mu}\lambda(x)\right)\sigma^{\mu}\overline{\lambda}(x)
\end{aligned}
$$

$$-\frac{1}{2}F_{\mu\nu}(x)F^{\mu\nu}(x) + \frac{i}{2}F_{\mu\nu}(x)F^{*\mu\nu}(x) + 4D^2(x)$$
$$= -F_{\mu\nu}(x)F^{\mu\nu}(x) + 8D^2(x) - 4i\lambda(x)\sigma^\mu\partial_\mu\overline{\lambda}(x)$$
$$+ \text{ total derivatives} . \tag{9.43}$$

This expression — obtained by applying the projection technique — is seen to agree with Eq. (8.41) which was obtained by Grassmann integration.

9.3   Spontaneous Symmetry Breaking

If supersymmetric gauge theories are to find realistic applications in high energy physics, both supersymmetry and the gauge symmetry must be broken spontaneously. Supersymmetry must be broken, because in experiments one does not observe degenerate Bose-Fermi multiplets.

The spontaneous breaking of gauge symmetry is well understood [93], [78], [94], [60], but supersymmetry imposes additional conditions which require further discussion. In this section we restrict ourselves to the case of supersymmetric theories constructed from scalar superfields, this is the so-called *O'Raifeartaigh mechanism of supersymmetry breaking*. The breaking of supersymmetry in supersymmetric gauge theory — *the Fayet–Iliopoulos mechanism of supersymmetry breaking* — will be considered later.[4]

The pecularity of the spontaneous breakdown of supersymmetry is due to the fact that the Hamiltonian of any supersymmetric theory is related to the supercharges Q_A and $\overline{Q}_{\dot{A}}$ by (see below)

$$H = \frac{1}{4}\left\{Q_1\overline{Q}_{\dot{1}} + \overline{Q}_{\dot{1}}Q_1 + Q_2\overline{Q}_{\dot{2}} + \overline{Q}_{\dot{2}}Q_2\right\}. \tag{9.44}$$

In order to verify this relation we recall that (*cf.* Eq. (6.2)):

$$\{Q_A, \overline{Q}_{\dot{B}}\} = 2(\sigma^\mu)_{A\dot{B}}P_\mu.$$

Multiplying from the right by $(\overline{\sigma}^\nu)^{\dot{B}A}$ and summing over the spinor indices A and $\dot{B}$, we obtain:

$$\{Q_A, \overline{Q}_{\dot{B}}\}(\overline{\sigma})^{\nu\dot{B}A} = 2(\sigma^\mu)_{A\dot{B}}(\overline{\sigma}^\nu)^{\dot{B}A}P_\mu = 2\text{Tr}\left[\sigma^\mu\overline{\sigma}^\nu\right]P_\mu$$
$$\overset{(1.109a)}{=} 4\eta^{\mu\nu}P_\mu = 4P^\nu.$$

---

[4]For further discussions of aspects of supersymmetry breaking such as dynamical supersymmetry breaking see the review article by Y. Shadmi [107] and references therein and the lectures of K. Intriligator and N. Seiberg [59].

Taking the 0-component, we find:

$$P^0 = H = \frac{1}{4}\{Q_A, \overline{Q}_{\dot{B}}\}(\overline{\sigma}^0)^{\dot{B}A} = \frac{1}{4}(Q_A\overline{Q}_{\dot{B}} + \overline{Q}_{\dot{B}}Q_A)(\overline{\sigma}^0)^{\dot{B}A}$$
$$= \frac{1}{4}\{Q_1\overline{Q}_{\dot{1}} + \overline{Q}_{\dot{1}}Q_1 + Q_2\overline{Q}_{\dot{2}} + \overline{Q}_{\dot{2}}Q_2\},$$

where

$$\overline{\sigma}^0 = \mathbb{1}_{2\times2}.$$

Equation (9.44) tells us that the spectrum of the Hamiltonian is semi-positive definite, *i.e.*

$$\langle\psi|H|\psi\rangle \geq 0 \qquad\qquad (9.45)$$

for every state $|\psi\rangle$. Since H is semipositive definite, those states with vanishing energy density are the lowest lying eigenstates and therefore the supersymmetric ground state of the theory. Such states — denoted by $|0\rangle$ — are supersymmetric, because as pointed out above

$$\langle0|H|0\rangle = 0$$

implies in view of Eq. (9.44)

$$Q_A|0\rangle = 0 \qquad \text{for } A = 1, 2, \qquad \text{and} \qquad \overline{Q}_{\dot{A}}|0\rangle = 0 \qquad \text{for } \dot{A} = 1, 2.$$

This fact, however, has the consequence that according to Eq. (6.23)

$$\delta_S|0\rangle = (\alpha^A Q_A + \overline{\alpha}_{\dot{A}}\overline{Q}^{\dot{A}})|0\rangle = 0. \qquad\qquad (9.46)$$

Hence a supersymmetric vacuum state $|0\rangle$ implies that the supercharges Q_A and $\overline{Q}_{\dot{A}}$ annihilate the vacuum, as expected on general grounds. Therefore the vacuum energy and consequently $V(A, A^*)$ are not only well defined, but are also bound to vanish, *i.e.*

$$E_{\text{vac}} = \langle0|H|0\rangle = 0. \qquad\qquad (9.47)$$

This situation is illustrated in Fig. 9.1 for a theory with scalar superfields.

Supersymmetric ground states are always at $E_{\text{vac}} = 0$ and may still be degenerate with other states which have $E_{\text{vac}} = 0$ as illustrated in Fig. 9.2, indicating the possible breakdown of some internal symmetry. However, in the case illustrated supersymmetry is unbroken because the ground state energy, the minimum of the potential, is zero.

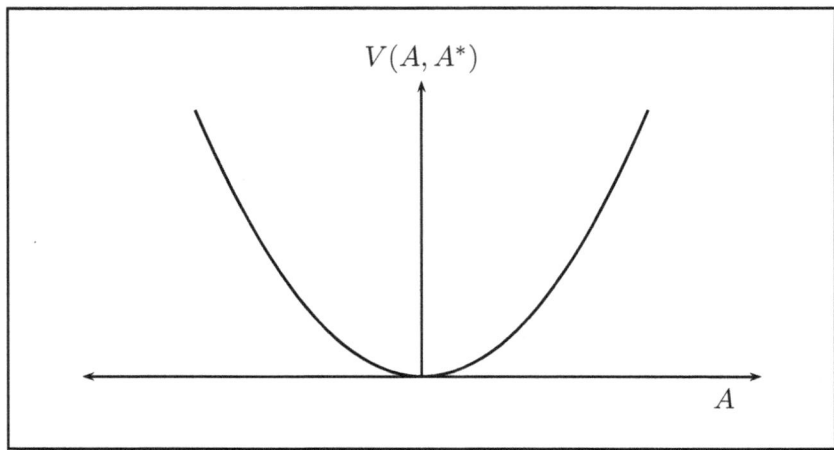

Figure 9.1: The vacuum or ground state has $V(A, A^*) = 0$. No breaking of supersymmetry and no breaking of an internal symmetry.

We have previously discussed an example which illustrates this general feature of supersymmetric theories. In Chapter 5 we demonstrated the normal ordering of the Hamiltonian of the Wess–Zumino model (*cf.* Eq. (5.113). This, of course, is equivalent[5] to fixing the energy of the vacuum to be zero.

We now consider the case in which the vacuum state is not annihilated by the supercharges, *i.e.*

$$Q_A|0\rangle \neq 0, \tag{9.48}$$

which implies in view of Eq. (9.44):

$$E_{\text{vac}} = \langle 0|H|0\rangle \neq 0. \tag{9.49}$$

We therefore arrive at the following conclusion: If supersymmetry is not spontaneously broken, *i.e.* if the vacuum is invariant under supersymmetry transformations (*cf.* Eq. (9.46)) the energy of the vacuum is zero. Conversely, if there exists a state for which the expectation value of the Hamiltonian is zero, supersymmetry is not spontaneously broken. Furthermore, if supersymmetry is spontaneously broken, the vacuum energy is positive (*cf.* Eq. (9.49)). The case of broken supersymmetry is illustrated in Fig. 9.3, where the expectation value of the scalar field is zero, but supersymmetry is spontaneously broken because the eneregy of the ground state is greater than zero.

---

[5]See *e.g.* J.D. Bjorken and S.D. Drell [18], Chap. 1.

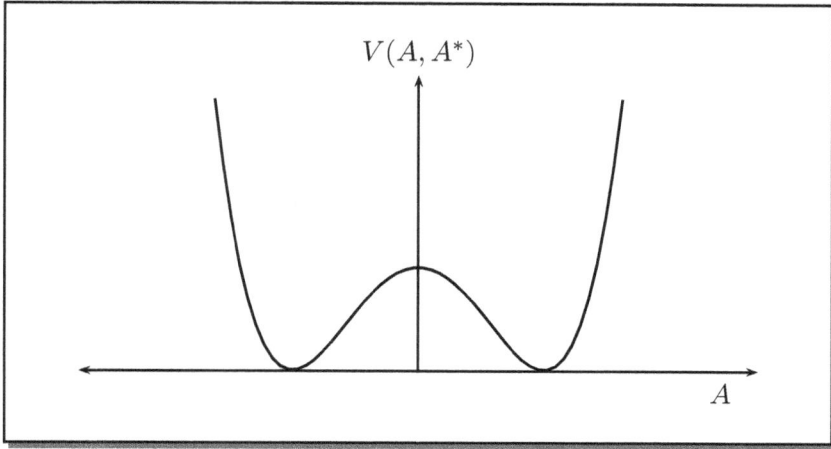

Figure 9.2: Degeneracy of the supersymmetric ground state due to the breakdown of some internal symmetry. Thus for the ground state no breaking of supersymmetry but breaking of the internal symmetry.

9.3.1 The Goldstone Theorem

The fact that a positive nonzero vacuum energy indicates supersymmetry breaking is a special case of the more general case that supersymmetry is spontaneously broken if any anticommutator $\{Q, X\}$, X denoting some operator, has a nonvanishing vacuum expectation value. The basic anticommutator of a supersymmetric theory (*cf.* Eq. (6.2)),

$$\{Q_A, \overline{Q}_{\dot{B}}\} = 2(\sigma^\mu)_{A\dot{B}} P_\mu,$$

can be considered as arising from integration of the local relation

$$\{\overline{Q}_{\dot{A}}, J_A^\mu(x)\} = 2(\sigma_\nu)_{A\dot{A}} T^{\mu\nu}(x) + \text{ S.T.} \tag{9.50}$$

where $J_A^\mu(x)$ is the conserved supercurrent introduced in Sec. 5.2. This supercurrent is related to the supercharges by (*cf.* Eq. (5.20))

$$Q_A = \int d^3x \, J_A^0(x), \quad \partial_\mu J_A^\mu(x) = 0. \tag{9.51}$$

In Eq. (9.50) $T^{\mu\nu}(x)$ is the energy-momentum tensor of the corresponding theory and S.T. denotes additional Schwinger terms which are of no interest in the present context since their vacuum expectation values vanish.[6] Using

---

[6]For a discussion of this topic see *e.g.* C. Itzykson and J.-B. Zuber [60], Chap. 11.3.1 or Cheng and Li [21], Chap. 5.1.

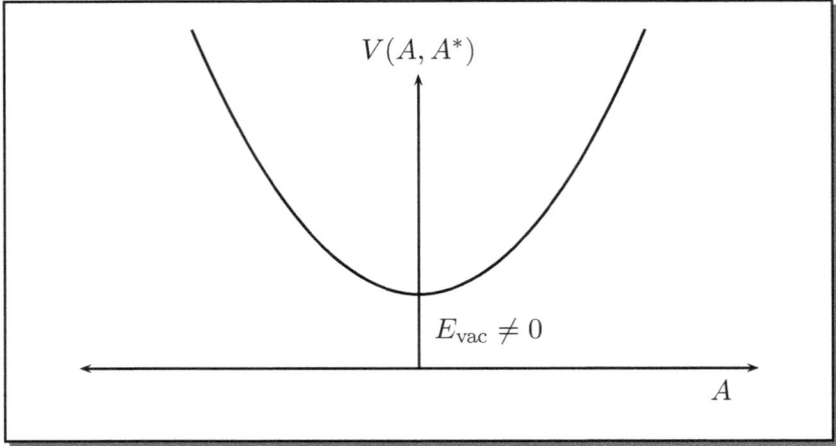

Figure 9.3: The case of broken supersymmetry but no breaking of internal symmetry.

Eq. (1.109b) we can reformulate Eq. (9.49) in the form:

$$T^{\mu\nu}(x) = \frac{1}{4}(\bar{\sigma}^\nu)^{\dot{A}A}\{\bar{Q}_{\dot{A}}, J^\mu_A(x)\} + \text{S.T.} \qquad (9.52)$$

If the supercharges Q_A annihilate the vacuum, *i.e.* if $|0\rangle$ is a supersymmetric ground state, Eq. (9.52) implies that the vacuum expectation value of the energy-momentum tensor vanishes, *i.e.*

$$\langle 0|T^{\mu\nu}(x)|0\rangle = 0. \qquad (9.53)$$

However, when supersymmetry is spontaneously broken, Eq. (9.53) ceases to be true, even in the so-called *tree approximation*, *i.e.* at the classical level. Instead of Eq. (9.53) we write in this case

$$\langle 0|T^{\mu\nu}(x)|0\rangle = E_{\text{vac}}\eta^{\mu\nu}. \qquad (9.54)$$

Inserting Eq. (9.52) into Eq. (9.54) we find

$$E_{\text{vac}}\eta^{\mu\nu} = \frac{1}{4}(\bar{\sigma}^\nu)^{\dot{A}A}\langle 0|\{\bar{Q}_{\dot{A}}, J^\mu_A(x)\}|0\rangle \qquad (9.55)$$

and a nonvanishing value of this expression means that supersymmetry is spontaneously broken. In other words, the vacuum state is not invariant,

$$\bar{Q}_{\dot{A}}|0\rangle \neq 0,$$

where $\overline{Q}_{\dot{A}}$ — or more generally simply Q — is the generator of the appropriate continuous symmetry transformation. Since for a scalar field $A(x)$ (*cf.* Eq. (5.118a))

$$\delta A(x) = -i\bar{\epsilon}_a [Q_a, A(x)],$$

we have

$$\delta \langle 0|A(x)|0\rangle = -i\bar{\epsilon}_a \langle 0|[Q_a, A(x)]|0\rangle,$$

and hence $Q|0\rangle \neq 0$ implies

$$\delta \langle 0|A(x)|0\rangle \neq 0.$$

We now show that this implies the existence of a massless fermion, which is called the *Goldstone fermion* or *goldstino*.[7]

If $j^\mu(x)$ is a conserved current, such as the supercurrent (5.17) which implies a conserved Majorana spinor current $k_a^\mu(x)$, *i.e.*

$$\partial_\mu j^\mu(x) = 0, \quad \partial_\mu k_a^\mu(x) = 0,$$

with a Majorana index $a = 1, 2, 3, 4$, or equivalently

$$\partial_\mu J^\mu{}_A(x) = 0, \quad \partial_\mu \overline{J}^\mu{}_{\dot{A}}(x) = 0,$$

in the Weyl formulation, then for any local operator $A(y)$

$$0 = \int_V d^3x \left[\partial_\mu j^\mu(x), A(y)\right]$$
$$= \frac{d}{dt} \int_V d^3x \left[j^0(\mathbf{x}, t), A(y)\right] + \int_S d\mathbf{s} \left[\mathbf{j}(\mathbf{x}, t), A(y)\right]$$
$$= \frac{d}{dt} \left[Q(t), A(y)\right],$$

if the surface S enclosing the volume V is made large enough so that on S:

$$\left[\mathbf{j}(\mathbf{x}, t), A(y)\right] \sim 0.$$

It follows that the expectation value satisfies

$$\frac{d}{dt} \langle 0|[Q(t), A(y)]|0\rangle = 0. \tag{9.56}$$

Inserting a complete set of four-momentum eigenstates $|p_n\rangle$ and using translation invariance, *i.e.*[8]

$$\langle 0|j_0(\mathbf{x}, t)|p_n\rangle = \langle 0|j_0(0)|p_n\rangle e^{-ip_n x}$$

---

[7]See *e.g.* C. Itzykson and J.-B. Zuber [60] or T.-P. Cheng and L.-F. Li [21], Chap. 5.3.

[8]This follows from

$$e^{icP} F(x) e^{-icP} = F(x + c)$$

we have

$$
\begin{aligned}
\langle 0|\left[Q(t), A(x')\right]|0\rangle &= \int d^3x \langle 0|\left[j_0(\mathbf{x}, t), A(x')\right]|0\rangle \\
&= \int d^3x \sum_n \Big[\langle 0|j_0(\mathbf{x}, t)|p_n\rangle\langle p_n|A(x')|0\rangle \\
&\quad - \langle 0|A(x')|p_n\rangle\langle p_n|j_0(\mathbf{x}, t)|0\rangle\Big] \\
&= \int d^3x \sum_n \Big[\langle 0|j_0(0)|p_n\rangle\langle p_n|A(0)|0\rangle e^{-ip_n(x-x')} \\
&\quad - \langle 0|A(0)|p_n\rangle\langle p_n|j_0(0)|0\rangle e^{ip_n(x-x')}\Big] \\
&= \sum_n (2\pi)^3\delta(\mathbf{p}_n)\Big[\langle 0|j_0(0)|p_n\rangle\langle p_n|A(0)|0\rangle e^{-iE_n(t-t')} \\
&\quad - \langle 0|A(0)|p_n\rangle\langle p_n|j_0(0)|0\rangle e^{iE_n(t-t')}\Big].
\end{aligned}
$$

Hence, Eq. (9.56) implies:

$$
\begin{aligned}
0 = \sum_n (2\pi)^3\delta(\mathbf{p}_n)\left(-iE_n\right) \\
\times \Big[\langle 0|j_0(0)|p_n\rangle\langle p_n|A(0)|0\rangle e^{-iE_n(t-t')} \\
- \langle 0|A(0)|p_n\rangle\langle p_n|j_0(0)|0\rangle e^{iE_n(t-t')}\Big].
\end{aligned}
$$

Thus, if a state $|p_n\rangle$ exists such that

$$
\langle 0|j_0(0)|p_n\rangle\langle p_n|A(0)|0\rangle \neq 0,
$$

it must have

$$
E_n\delta(\mathbf{p}_n) = 0, \quad \text{with} \quad E_n = \left(m_n^2 + \mathbf{p}_n^2\right)^{1/2},
$$

i.e. the mass m_n of the appropriate particle state must be zero. Since

$$
\langle 0|j_0(0)|p_n\rangle
$$

---

i.e.

$$
\partial_\mu F(x) = i\left[P_\mu, F(x)\right],
$$

and so

$$
\begin{aligned}
\langle a|\partial_\mu F(x)|b\rangle &= i\left(p_\mu^a - p_\mu^b\right)\langle a|F(x)|b\rangle \\
&= i\left(p_\mu^a - p_\mu^b\right)\langle a|F(0)|b\rangle e^{i(p_\mu^a - p_\mu^b)x^\mu}
\end{aligned}
$$

for an arbitrary function $F(x)$ in selfevident terminology.

must be invariant under Lorentz transformations U, *i.e.*

$$\langle 0|j_0(0)|p_n\rangle = \langle 0|U^\dagger (Uj_0(0)U^\dagger)U|p_n\rangle,$$

the operator $j_0(0)$ must transform with respect to the same representation of the Lorentz group as the state $|p_n\rangle$. Hence, if $j_0(0)$ is a spinor current, $|p_n\rangle$ must be a spinor state. This therefore proves the Goldstone theorem for either of the supersymmetric or nonsupersymmetric cases.

9.3.2 Remarks on the Wess–Zumino Model

We now investigate the Wess–Zumino model with respect to the possible breakdown of supersymmetry. To this end we have to evaluate the scalar potential V and then calculate the derivative $\partial W/\partial A_i$; supersymmetry is unbroken, if $\partial W/\partial A_i = 0$ as stated by Eq. (9.16). The action of the Wess–Zumino model in terms of superfields is given by (*cf.* Eq. (9.3))

$$\mathcal{A}_{\mathrm{WZ}} = \int d^4x \int d^4\theta\, \Phi^\dagger\Phi + \int d^4x \int d^2\theta W[\Phi] + \int d^4x \int d^2\bar\theta\, \overline{W}[\Phi],$$

where the superpotential $W[\Phi]$ takes the form

$$W[\Phi] = g\Phi + \frac{1}{2}m\Phi^2 + \frac{1}{3}\lambda\Phi^3.$$

Then we obtain

$$W[A] = gA(x) + \frac{1}{2}mA^2(x) + \frac{1}{3}\lambda A^3(x),$$

where $A(x)$ is the scalar field of the supermultiplet Φ, *i.e.*

$$\Phi = A(x) + \sqrt{2}\theta\psi(x) + (\theta\theta)F(x).$$

Differentiating $W[A]$ with respect to the scalar field $A(x)$ yields:

$$\frac{dW[A]}{dA(x)} = g + mA(x) + \lambda A^2(x).$$

Now, it is always possible to find solutions $A_\pm$ (*i.e.* degenerate bosonic ground states) to the equation

$$\frac{dW[A]}{dA(x)} = 0,$$

i.e.

$$A^2(x) + \frac{m}{\lambda}A(x) + \frac{g}{\lambda} = 0, \tag{9.57}$$

with

$$A_{\pm} = -\frac{m}{2\lambda} \pm \left[\frac{m^2}{4\lambda^2} - \frac{g}{\lambda}\right]^{1/2},$$

no matter how we adjust the three parameters m, g, and λ of the model — recall that $A(x)$ is a complex scalar field. Hence, according to the above discussions the vacuum energy is zero, if we choose $A(x)$ as solution of Eq. (9.57) and there is no supersymmetry breaking. This case is illustrated in Fig. 9.1 and Fig. 9.2 respectively, depending on the choice of the parameters, *i.e.* for $g = 0, m = 0$ Fig. 9.1 applies, whereas for $m = 0, g \neq 0, \lambda \neq 0$ Fig. 9.2 applies. In the latter case the "internal symmetry" of

$$V(A, A^*) = \left|\frac{dW[A]}{dA}\right|^2,$$

which is violated is the phase symmetry

$$A(x) \longrightarrow e^{i\alpha} A(x).$$

9.4   The O'Raifeartaigh Model

9.4.1   Spontaneous Breaking of Supersymmetry

The model of O'Raifeartaigh [84] is a supersymmetric field theory constructed from chiral superfields. In [84], O'Raifeartaigh demonstrated that at least three different scalar superfields are required to yield a model which exhibits the spontaneous breakdown of supersymmetry.

We denote the three chiral scalar superfields of the O'Raifeartaigh model by A, X, and Y with component field expansions

$$\left.\begin{array}{rcl} A & = & a + \sqrt{2}\theta\psi_A + (\theta\theta)F_A, \\ X & = & x + \sqrt{2}\theta\psi_X + (\theta\theta)F_X, \\ Y & = & y + \sqrt{2}\theta\psi_Y + (\theta\theta)F_Y. \end{array}\right\} \tag{9.58}$$

Corresponding expansions can be written down for the conjugate fields $A^{\dagger}$, $X^{\dagger}$, and $Y^{\dagger}$. The kinetic part of the Lagrangian of the O'Raifeartaigh model is given by

$$\mathcal{L}_{\text{kin}} = \left(A^{\dagger}A + X^{\dagger}X + Y^{\dagger}Y\right)\Big|_{\theta\theta\bar{\theta}\bar{\theta}-\text{component}}, \tag{9.59}$$

and the superpotential of the model is taken to be

$$W[A, X, Y] := \lambda AY + gX(A^2 - M^2), \tag{9.60}$$

where λ, g, and M are three real, nonvanishing parameters. The auxiliary fields of the component field expansions (9.58), *i.e.* the scalars F_A, F_X, and F_Y, are — according to Eqs. (9.11) and (9.14) — given by the partial derivatives of the superpotential, *i.e.*

$$F_A^* = -\frac{\partial W(a,x,y)}{\partial a}, \quad F_X^* = -\frac{\partial W(a,x,y)}{\partial x}, \quad F_Y^* = -\frac{\partial W(a,x,y)}{\partial y}.$$

Hence

$$\frac{\partial W(a,x,y)}{\partial a} = \lambda y + 2gax, \tag{9.61a}$$

$$\frac{\partial W(a,x,y)}{\partial x} = g(a^2 - M^2), \tag{9.61b}$$

$$\frac{\partial W(a,x,y)}{\partial y} = \lambda a. \tag{9.61c}$$

According to our general discussion (*cf.* Eq. (9.16)), supersymmetry is unbroken if and only if we can find a simultaneous solution to the set of equations

$$\frac{\partial W(a,x,y)}{\partial a} = \frac{\partial W(a,x,y)}{\partial x} = \frac{\partial W(a,x,y)}{\partial y} = 0. \tag{9.62}$$

However, the O'Raifeartaigh model has been constructed in such a way that such a simultaneous solution of Eqs. (9.62) does not exist. According to Eq. (9.61a) the equation

$$\frac{\partial W(a,x,y)}{\partial a} = 0 \tag{9.63}$$

is solved by

$$y = -\frac{2g}{\lambda}ax. \tag{9.64}$$

But the two equations

$$\frac{\partial W(a,x,y)}{\partial x} = 0 \quad \text{and} \quad \frac{\partial W(a,x,y)}{\partial y} = 0$$

are inconsistent with this solution, since $\partial W/\partial y = 0$ implies $a = 0$, since by assumption $\lambda \neq 0$. Inserting this solution into $\partial W/\partial x = 0$ leads to $gM^2 = 0$. However, this is not possible, since the parameters g and M are assumed to be nonzero. Hence, it is not possible to find a simultaneous solution of the three equations (9.62), therefore, supersymmetry must be broken spontaneously.

Since we cannot find a simultaneous solution to the set of equations (9.62), at least one of the auxiliary fields F_A, F_X, or F_Y acquires a nonvanishing vacuum expectation value. This is the general feature of spontaneously broken supersymmetry.

To obtain the ground state of the model, we have to minimize the scalar potential $V(a, x, y; a^*, x^*, y^*)$. According to Eq. (9.15) the scalar potential is related to the superpotential by

$$V(a, x, y; a^*, x^*, y^*) = \left| \frac{\partial W}{\partial a} \right|^2 + \left| \frac{\partial W}{\partial x} \right|^2 + \left| \frac{\partial W}{\partial y} \right|^2$$

$$= \left| \lambda y + 2gax \right|^2 + g^2 \left| a^2 - M^2 \right|^2 + \lambda^2 |a|^2, \qquad (9.65)$$

where we used Eqs. (9.61a) to (9.61c). Differentiating the scalar potential (9.65) with repsect to x and y, we obtain (setting these expressions equal to zero)

$$\frac{\partial V}{\partial x} = (\lambda y^* + 2gx^* a^*) 2ga \stackrel{!}{=} 0,$$

$$\frac{\partial V}{\partial y} = (\lambda y^* + 2gx^* a^*) \lambda \stackrel{!}{=} 0.$$

The second equation is satisfied if Eq. (9.64) holds. Then also the derivative $\partial V/\partial x$ vanishes and Eq. (9.64) gives the minimum value of the field y, but x remains undetermined. Such a kind of 'degeneracy' is a general feature of spontaneously broken supersymmetry. Hence, at the minimum we have

$$V(a, a^*) = g^2 \left| a^2 - M^2 \right|^2 + \lambda^2 |a|^2, \qquad (9.66)$$

and

$$\frac{\partial V(a, a^*)}{\partial a} = a^* \lambda^2 + 2g^2 (a^{*2} - M^2) a = 0. \qquad (9.67)$$

Equation (9.66) implies that the field a must be real, for setting

$$a = k + ih,$$

then Eq. (9.67) implies

$$\lambda^2 (k - ih) - 2g^2 M^2 (k + ih) + 2g^2 (k^2 + h^2)(k - ih) = 0.$$

Equating real and imaginary parts of the left hand side of this equation to zero, we obtain the two equations

$$k \left[\lambda^2 - 2g^2 M^2 + 2g^2 (k^2 + h^2) \right] = 0,$$

$$-h \left[\lambda^2 + 2g^2 M^2 + 2g^2 (k^2 + h^2) \right] = 0.$$

The latter of these equations is satisfied if and only if h is zero, since the expression in the bracket is a sum of positive terms which cannot vanish. Thus, Eq. (9.67) becomes

$$\frac{\partial V(a)}{\partial a} = a \left\{ \lambda^2 + 2g^2 (a^2 - M^2) \right\} = 0. \qquad (9.68)$$

One solution to this equation is $a = a_1 = 0$. Inserting this into the scalar potential Eq. (9.66) we obtain the value of the scalar potential V at the extremum $a = a_1 = 0$, $i.e.$

$$V(a_1 = 0) = g^2 M^4. \tag{9.69}$$

But $a = a_1 = 0$ is the position of a local maximum of V, since

$$\frac{\partial^2 V}{\partial a^2} = \lambda^2 + 6g^2 a^2 - 2g^2 M^2,$$

and

$$\left. \frac{\partial^2 V}{\partial a^2} \right|_{a=0} = 2g^2 \left(\frac{\lambda^2}{2g^2} - M^2 \right) < 0$$

for $M^2 > \lambda^2/2g^2$. The other solutions to Eq. (9.68) are obtained from

$$\lambda^2 + 2g^2 a^2 - 2g^2 M^2 = 0,$$

$i.e.$ the roots are

$$a_2, a_3 = \pm \left(M^2 - \frac{\lambda^2}{2g^2} \right)^{1/2},$$

provided $M^2 > \lambda^2/2g^2$, in which case a is real. The value of the scalar potential at the extrema a_2 and a_3 is:

$$
\begin{aligned}
V[a_2, a_3] &= g^2 \left| M^2 - \frac{\lambda^2}{2g^2} - M^2 \right|^2 + \lambda^2 \left| M^2 - \frac{\lambda^2}{2g^2} \right| \\
&= \frac{\lambda^4}{4g^2} + \lambda^2 \left(M^2 - \frac{\lambda^2}{2g^2} \right) = \lambda^2 \left(M^2 - \frac{\lambda^2}{4g^2} \right).
\end{aligned}
\tag{9.70}
$$

The second derivative at these extremas is

$$
\left. \frac{\partial^2 V}{\partial a^2} \right|_{a_2, a_3} = \lambda^2 + 6g^2 \left(M^2 - \frac{\lambda^2}{2g^2} \right) - 2g^2 M^2
$$

$$
= 4g^2 \left(M^2 - \frac{\lambda^2}{2g^2} \right) > 0 \quad \text{for } M^2 > \lambda^2/2g^2.
$$

Hence, a_2, a_3 are the locations of minima of the scalar potential (9.65), provided we adjust the parameters λ, M, and g such that

$$M^2 > \frac{\lambda^2}{2g^2}. \tag{9.71}$$

Furthermore, the difference between $V(a_1)$ and $V(a_2)$ or $V(a_3)$ is

$$V(a_1) - V(a_i)\Big|_{i=2,3} = g^2 M^4 - \lambda^2\left(M^2 - \frac{\lambda^2}{4g^2}\right) = g^2\left(M^2 - \frac{\lambda^2}{2g^2}\right)^2 > 0.$$

Hence with our choice of parameters, Eq. (9.71), the absolute minimum is degenerate and located at

$$a_2 = -\left(M^2 - \frac{\lambda^2}{2g^2}\right)^{1/2}, \qquad a_3 = +\left(M^2 - \frac{\lambda^2}{2g^2}\right)^{1/2},$$

and

$$V(a_2) = V(a_3) > 0.$$

Hence, the scalar potential of the O'Raifeartaigh model has the shape shown in Fig. 9.4.

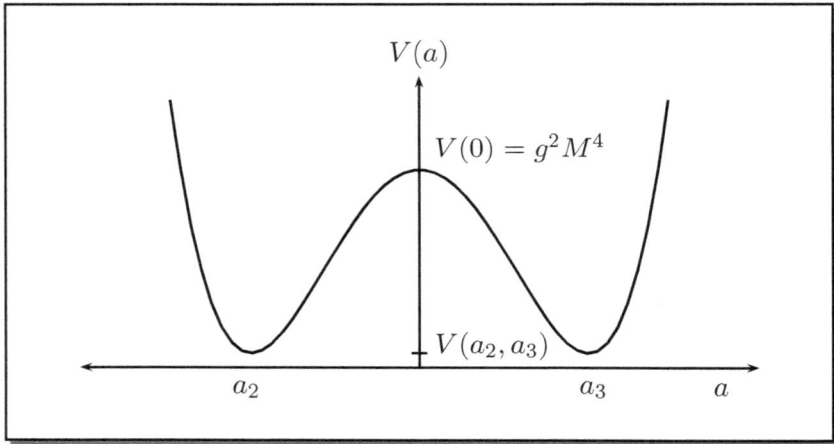

Figure 9.4: The scalar potential $V(a)$ of the O'Raifeartaigh model in the range $M^2 > \lambda^2/2g^2$. In the case shown both supersymmetry and an internal symmetry are broken.

The scalar potential given by Eq. (9.66) has the internal $U(1)$ phase symmetry

$$a \longrightarrow e^{i\alpha}a. \tag{9.72}$$

Now from Eq. (9.66) we have

$$\begin{aligned}
V(a) &= g^2(a^2 - M^2)^2 + \lambda^2 a^2 \\
&= g^2 M^4 + a^2(\lambda^2 - 2M^2 g^2) + g^2 a^4.
\end{aligned}$$

The spontaneous breaking of the symmetry (9.72) occurs if and only if the factor $\lambda^2 - 2M^2g^2$ is negative.[9] However, this case corresponds exactly to the choice (9.70) for the three parameters M, g, and λ.

In the case $M^2 < \lambda^2/2g^2$ or $\lambda^2 - 2M^2g^2 > 0$ no breaking of the internal symmetry appears[10] and the only solution to Eq. (9.68) is $a = 0$. In this case the second derivative of the scalar potential has the property

$$\left. \frac{d^2 V}{da^2} \right|_{a=0} > 0$$

and $a = 0$ is therefore the position of the absolute minimum of the scalar potential $V(a)$. It may be observed that for

$$M^2 < \frac{\lambda^2}{2g^2}$$

the solutions a_2 and a_3 are purely imaginary, thus violating the condition that the scalar field a be real. For

$$M^2 < \frac{\lambda^2}{2g^2}$$

the scalar potential $V(a)$ has the graphic representation shown in Fig. 9.5. According to Eq. (9.69) the potential energy of the minimum is

$$V(a = 0) = g^2 M^4 > 0$$

indicating the breakdown of supersymmetry.

9.4.2   The Mass Spectrum of the O'Raifeartaigh Model

In general if supersymmetry is broken, then according to Eq. (9.16)

$$\frac{\partial W}{\partial A_i} := W_i \neq 0$$

for some values of i. On the other hand at the minimum of the scalar potential we have

$$\frac{\partial V}{\partial A_i} = 0.$$

---

[9]This corresponds to the well-known case of a negative (mass)^2-term in the simple Higgs model. See *e.g.* L.H. Ryder [98], Chap. 8, for more details.

[10]See e.g. T.-P. Cheng and L.-F. Li, [21], Chap. 5.3.

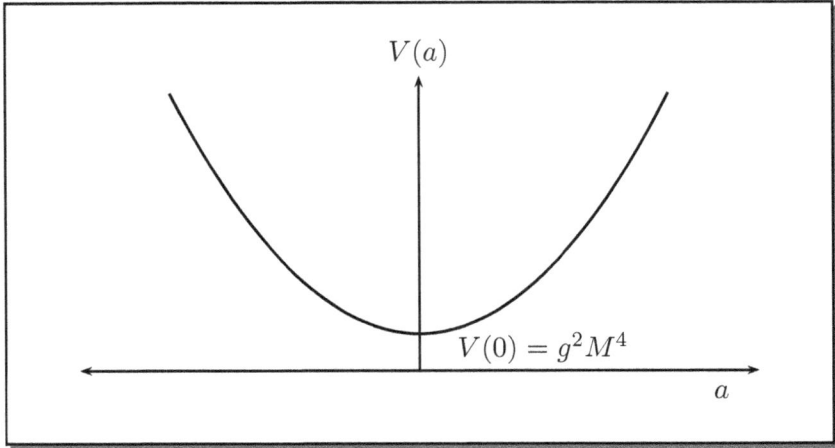

Figure 9.5: The scalar potential $V(a)$ of the O'Raifeartaigh model in the range $M^2 < \lambda^2/2g^2$. Supersymmetry is broken but no internal symmetry.

Using Eq. (9.15) this implies:

$$\frac{\partial V}{\partial A_i} = \frac{\partial}{\partial A_i}\sum_j\left|\frac{\partial W}{\partial A_j}\right|^2 = \frac{\partial}{\partial A_i}\sum_j W_j^*(A)W_j(A) = \sum_j\left[W_{ij}^*W_j + W_j^*W_{ij}\right] = 0,$$

where (*cf.* Eqs. (9.11) and (9.12))

$$W_{ij} := \frac{\partial^2 W}{\partial A_i \partial A_j}.$$

Hence, at the minimum of the scalar potential we have

$$W_{ij}W_j^* = 0. \tag{9.73}$$

Now, since W_{ij} is the fermionic mass matrix (see Eq. (9.10)), the diagonalized matrix and thus the eigenvalues of W_{ij} determine the masses of the fermions. If supersymmetry is spontaneously broken, then $W_j \neq 0$ for some j, and hence some of these eigenvalues must be zero according to Eq. (9.73). Thus, in order to determine the fermionic mass spectrum of the O'Raifeartaigh model we have to diagonalize the fermionic mass matrix W_{ij}. The detailed analysis given below leads to the following results for the case $M^2 < \lambda^2/2g^2$:

- One eigenvalue of the fermionic mass matrix vanishes.

- The second and third eigenvalues of the fermionic mass matrix are λ.

- The spinor ψ_X, *i.e.* the supersymmetric partner of the auxiliary field F_X, turns out to be the Goldstone fermion.

- There are two massive fermions of mass λ, which are linear combinations of the spinors ψ_A and ψ_Y.

- The associated bosonic mass spectrum is the following

  - The scalar field $x(x)$ is massless.
  - The scalar field $y(x)$ has mass λ.
  - A mass splitting occurs in the case of the scalars $a(x)$, *i.e.* setting as before
    $$a(x) = k(x) + ih(x),$$
    where $k(x)$ and $h(x)$ are real scalar fields, we find the following mass relations squares:
    $$m_k^2 = \lambda^2 - 4g^2 M^2,$$
    $$m_h^2 = \lambda^2 + 4g^2 M^2.$$

In order to obtain these results it is advantageous to write the Lagrangian of the O'Raifeartaigh model in terms of the component fields. The Lagrangian density is given by (*cf.* Eqs. (9.59) and (9.60))

$$
\begin{aligned}
\mathcal{L} &= \left(A^\dagger A + X^\dagger X + Y^\dagger Y \right)\Big|_{\theta\theta\,\bar\theta\bar\theta} + W[A, X, Y]\Big|_{\theta\theta} + \overline{W}[A^\dagger, X^\dagger, Y^\dagger]\Big|_{\bar\theta\bar\theta} \\
&= \left(A^\dagger A + X^\dagger X + Y^\dagger Y \right)\Big|_{\theta\theta\,\bar\theta\bar\theta} + \left(\lambda AY + gXA^2 - gM^2X \right)\Big|_{\theta\theta} \\
&\quad + \left(\lambda A^\dagger Y^\dagger + gX^\dagger A^{\dagger 2} - gM^2 X^\dagger \right)\Big|_{\bar\theta\bar\theta},
\end{aligned}
\tag{9.74}
$$

where $\big|_{...}$ indicates that the appropriate components are to be taken. Inserting the component field expansions (9.58) of the superfields A, X, and Y, and using Eqs. (7.27b), (8.27), and (8.29), we obtain for the kinetic part of the Lagrangian:

$$
\begin{aligned}
\left(A^\dagger A + X^\dagger X + Y^\dagger Y \right)\Big|_{\theta\theta\,\bar\theta\bar\theta} &= -\big[a^*(x)\Box a(x) + x^*(x)\Box x(x) + y^*(x)\Box y(x) \big] \\
&\quad + F_A^*(x)F_A(x) + F_X^*(x)F_X(x) + F_Y^*(x)F_Y(x) \\
&\quad + i\big[\partial_\mu \bar\psi_A(x)\bar\sigma^\mu \psi_A(x) + \partial_\mu \bar\psi_X(x)\bar\sigma^\mu \psi_X(x) \\
&\quad + \partial_\mu \bar\psi_Y(x)\bar\sigma^\mu \psi_Y(x) \big].
\end{aligned}
\tag{9.75}
$$

Now the $\theta\theta$-component of the product of two left-handed chiral superfields Φ_i and Φ_k is (*cf.* Eq. (7.16))

$$\Phi_i\Phi_k\Big|_{\theta\theta} = A_i(x)F_k(x) + A_k(x)F_i(x) - \psi_i(x)\psi_k(x),$$

which implies in our particular case

$$\lambda AY\Big|_{\theta\theta} = \lambda\Big(a(x)F_Y(x) + F_A(x)y(x) - \psi_A(x)\psi_Y(x)\Big). \tag{9.76}$$

The $\theta\theta$-term of a product of three scalar superfields has been obtained in Eq. (7.17) and is given by

$$\Phi_l\Phi_i\Phi_k\Big|_{\theta\theta} = A_l(x)A_i(x)F_k(x) + A_l(x)F_i(x)A_k(x) + F_l(x)A_i(x)A_k(x)$$
$$- \psi_l(x)\psi_k(x)A_i(x) - \psi_l(x)\psi_i(x)A_k(x) - A_l(x)\psi_i(x)\psi_k(x),$$

so that in the O'Raifeartaigh model

$$gXA^2\Big|_{\theta\theta} = g\Big[2x(x)a(x)F_A(x) + F_X(x)a^2(x)$$
$$- 2a(x)\psi_X(x)\psi_A(x) - x(x)\psi_A^2(x)\Big]. \tag{9.77}$$

Finally, using Eq. (9.58) we get

$$gM^2X\Big|_{\theta\theta} = gM^2F_X(x). \tag{9.78}$$

Similar expressions can be obtained for $\overline{W}[A^\dagger, X^\dagger, Y^\dagger]$.

Hence with Eqs. (9.75) to (9.78) the component field expansion of the Lagrangian density of the O'Raifeartaigh model is:[11]

$$\mathcal{L} = \big|\partial_\mu a(x)\big|^2 + \big|\partial_\mu x(x)\big|^2 + \big|\partial_\mu y(x)\big|^2$$
$$+ \big|F_A(x)\big|^2 + \big|F_X(x)\big|^2 + \big|F_Y(x)\big|^2$$
$$+ i\big\{\big(\partial_\mu\overline{\psi}_A(x)\big)\overline{\sigma}^\mu\psi_A(x) + \big(\partial_\mu\overline{\psi}_X(x)\big)\overline{\sigma}^\mu\psi_X(x) + \big(\partial_\mu\overline{\psi}_Y(x)\big)\overline{\sigma}^\mu\psi_Y(x)\big\}$$
$$+ \lambda a(x)F_Y(x) + \lambda F_A(x)y(x) - \lambda\psi_A(x)\psi_Y(x) + 2gx(x)a(x)F_A(x)$$
$$+ gF_X(x)a^2(x) - 2ga(x)\psi_X(x)\psi_A(x) - \frac{1}{2}x(x)\psi_A^2(x) - gM^2F_X(x)$$
$$+ \lambda a^*(x)F_Y^*(x) + \lambda F_A^*(x)y^*(x) - \lambda\overline{\psi}_A(x)\overline{\psi}_Y(x) + 2gx^*(x)a^*(x)F_A^*(x)$$
$$+ gF_X^*(x)a^{*2}(x) - 2ga^*(x)\overline{\psi}_X(x)\overline{\psi}_A(x) - gx^*(x)\overline{\psi}_A^2(x) - gM^2F_X^*(x). \tag{9.79}$$

---

[11]We rewrite terms of the form $a^*\Box a$ as $-(\partial_\mu a^*)(\partial^\mu a)$ as we can since

$$\int d^4x\, a^*\Box a = \int_{S_\infty} dS_\mu a^*\partial^\mu a - \int d^4x(\partial_\mu a^*)(\partial^\mu a),$$

where the first term vanishes when integrated over an infinitely large surface.

The three auxiliary fields $F_A(x)$, $F_X(x)$, and $F_Y(x)$ as well as their complex conjugates can be eliminated with the help of their respective Euler–Lagrange equations. We have

$$\frac{\partial \mathcal{L}}{\partial F_A(x)} = F_A^*(x) + 2gx(x)a(x) + \lambda y(x) = 0,$$

so that

$$F_A^*(x) = -\frac{\partial W}{\partial a} = -2gx(x)a(x) - \lambda y(x).$$

In the same manner we get

$$\frac{\partial \mathcal{L}}{\partial F_A^*(x)} = F_A(x) + 2gx^*(x)a^*(x) + \lambda y^*(x) = 0,$$

so that

$$F_A(x) = -\frac{\partial \overline{W}}{\partial a^*} = -2gx^*(x)a^*(x) - \lambda y^*(x).$$

The Euler-Lagrange equation for the F_X-field is

$$\frac{\partial \mathcal{L}}{\partial F_X(x)} = F_X^*(x) + g\left(a^2(x) - M^2\right) = 0,$$

implying:

$$F_X^*(x) = -\frac{\partial W}{\partial x(x)} = -g\left(a^2(x) - M^2\right),$$

and

$$F_X(x) = -\frac{\partial \overline{W}}{\partial x^*(x)} = -g\left(a^{*2}(x) - M^2\right).$$

Finally

$$\frac{\partial \mathcal{L}}{\partial F_Y(x)} = F_Y^*(x) + \lambda a(x) = 0,$$

so that

$$F_Y^*(x) = -\frac{\partial W}{\partial y(x)} = -\lambda a(x),$$

and

$$F_Y(x) = -\frac{\partial \overline{W}}{\partial y^*(x)} = -\lambda a^*(x).$$

We can therefore rewrite the scalar potential of the O'Raifeartaigh model in the form:

$$
\begin{aligned}
V &= \left|F_A(x)\right|^2 + \left|F_X(x)\right|^2 + \left|F_Y(x)\right|^2 \\
&= \big(2gx(x)a(x) + \lambda y(x)\big)\big(2gx^*(x)a^*(x) + \lambda y^*(x)\big) \\
&\quad + g^2\big(a^2(x) - M^2\big)\big(a^{*2}(x) - M^2\big) + \lambda^2 a(x)a^*(x) \\
&= \left|\lambda y(x) + 2gx(x)a(x)\right|^2 + g^2\left|a^2(x) - M^2\right|^2 + \lambda^2\left|a(x)\right|^2 \\
&= V(a, a^*, x, x^*, y, y^*),
\end{aligned}
\tag{9.80}
$$

where we made use of Eq. (9.65). Inserting this result into the Lagrangian (9.79) we obtain a Lagragian which depends only on the dynamical fields $a(x)$, $x(x)$, $y(x)$, $\psi_A(x)$, $\psi_X(x)$, and $\psi_Y(x)$, i.e.

$$
\begin{aligned}
\mathcal{L} &= \left|\partial_\mu a(x)\right|^2 + \left|\partial_\mu x(x)\right|^2 + \left|\partial_\mu y(x)\right|^2 \\
&\quad - \left|\lambda y(x) + 2gx(x)a(x)\right|^2 - g^2\left|a^2(x) - M^2\right|^2 - \lambda\left|a(x)\right|^2 \\
&\quad + i\Big[\big(\partial_\mu \overline{\psi}_A(x)\big)\overline{\sigma}^\mu \psi_A(x) + \big(\partial_\mu \overline{\psi}_X(x)\big)\overline{\sigma}^\mu \psi_X(x) + \big(\partial_\mu \overline{\psi}_Y(x)\big)\overline{\sigma}^\mu \psi_Y(x)\Big] \\
&\quad - \lambda\psi_A(x)\psi_Y(x) - 2ga(x)\psi_X(x)\psi_A(x) \\
&\quad - gx(x)\psi_A^2(x) - \lambda\overline{\psi}_A(x)\overline{\psi}_Y(x) \\
&\quad - 2ga^*(x)\overline{\psi}_X(x)\overline{\psi}_A(x) - gx^*(x)\overline{\psi}_A^2(x) \\
&\quad + g^2 M^2\big(a^2(x) + a^{*2}(x) - 2M^2\big).
\end{aligned}
\tag{9.81}
$$

Next we derive the Euler–Lagrange equations of the dynamical fields of the O'Raifeartaigh model.

(i) $a^*(x)$: Working out

$$
\frac{\partial \mathcal{L}}{\partial a^*(x)} - \partial_\mu \frac{\partial \mathcal{L}}{\partial\big(\partial_\mu a^*(x)\big)} = 0
$$

gives

$$
\begin{aligned}
-\Box a(x) - \lambda^2 a(x) &+ 4g^2 M^2 a^*(x) \\
&= 2g\overline{\psi}_X \overline{\psi}_A(x) + 2\lambda gy(x)x^*(x) \\
&\quad + 4g^2\left|x(x)\right|^2 a(x) + 2g^2\left|a(x)\right|^2 a(x),
\end{aligned}
\tag{9.82}
$$

i.e.

$$
\begin{aligned}
(\Box + \lambda^2)a(x) - 4g^2 M^2 a^*(x) &= -2g\overline{\psi}_X(x)\overline{\psi}_A(x) - 2\lambda gy(x)x^*(x) \\
&\quad - 4g^2\left|x(x)\right|^2 a(x) - 2g^2\left|a(x)\right|^2 a(x).
\end{aligned}
$$

(ii) $a(x)$: In a similar way we obtain:

$$(\Box + \lambda^2)a^*(x) - 4g^2M^2a(x) = -2g\psi_X(x)\psi_A(x) - 2\lambda gy^*(x)x(x)$$
$$- 4g^2|x(x)|^2a^*(x) - 2g^2|a(x)|^2a^*(x). \tag{9.83}$$

(iii) $x^*(x)$: The Euler–Lagrange equation

$$\frac{\partial\mathcal{L}}{\partial x^*(x)} - \partial_\mu \frac{\partial\mathcal{L}}{\partial(\partial_\mu x^*(x))} = 0$$

leads to

$$\Box x(x) = -2\lambda gy(x)a^*(x) - 4g^2|a(x)|^2x(x) - g\overline{\psi}_A^2(x). \tag{9.84}$$

(iv) $y^*(x)$: Here, the Euler–Lagrange equation

$$\frac{\partial\mathcal{L}}{\partial y^*(x)} - \partial_\mu \frac{\partial\mathcal{L}}{\partial(\partial_\mu y^*(x))} = 0$$

leads to

$$(\Box + \lambda^2)y(x) = -2\lambda ga(x)x(x). \tag{9.85}$$

(v) $\overline{\psi}_A(x)$: In this case the Euler–Lagrange equation

$$\frac{\partial\mathcal{L}}{\partial\overline{\psi}_A(x)} - \partial_\mu \frac{\partial\mathcal{L}}{\partial(\partial_\mu\overline{\psi}_A(x))} = 0$$

leads to

$$\frac{\partial\mathcal{L}}{\partial(\partial_\mu\overline{\psi}_A(x))} = i\overline{\sigma}^\mu\psi_A(x),$$

and

$$\frac{\partial\mathcal{L}}{\partial\overline{\psi}_A(x)} = -\lambda\overline{\psi}_Y(x) - 2ga^*(x)\overline{\psi}_X(x) - 2gx^*(x)\overline{\psi}_A(x).$$

Then the equation of motion for the field $\overline{\psi}_A(x)$ becomes

$$i\overline{\sigma}^\mu\partial_\mu\psi_A(x) = -\lambda\overline{\psi}_Y(x) - 2ga^*(x)\overline{\psi}_X(x) - 2gx^*(x)\overline{\psi}_A(x). \tag{9.86}$$

(vi) $\overline{\psi}_Y(x)$: The equation of motion for the field $\overline{\psi}_Y(x)$ is

$$i\overline{\sigma}^\mu\partial_\mu\psi_Y(x) = -\lambda\overline{\psi}_A(x). \tag{9.87}$$

(vii) $\overline{\psi}_X(x)$: The equation of motion for $\overline{\psi}_X(x)$ is

$$i\overline{\sigma}^\mu \partial_\mu \psi_X(x) = -2ga^*(x)\overline{\psi}_A(x). \tag{9.88}$$

In order to obtain the mass spectrum of the O'Raifeartaigh model, we have to consider small oscillations of the component fields around the absolute minimum of the potential $V(a, a^*, x, x^*, y, y^*)$ which we choose to be at

$$a \equiv \langle a \rangle = 0, \quad y \equiv \langle y \rangle = 0, \quad \text{and } x \text{ arbitrary.}$$

This choice corresponds to the case illustrated in Fig. 9.5, where

$$\min V = g^2 M^4, \quad \text{and } M^2 < \frac{\lambda^2}{2g^2},$$

and the internal $U(1)$ phase symmetry is not broken.

Then from Eq. (9.84) we conclude that the scalar field $x(x)$ is massless, and from Eq. (9.85) we find that the scalar $y(x)$ has mass λ. From Eqs. (9.86) and (9.87) we see[12] that there are two massive fermions of mass $-\lambda$ for λ negative, which are linear combinations of $\psi_A(x)$ and $\psi_Y(x)$. The equation of motion (9.88) demonstrates that $\psi_X(x)$ is a massless fermion, the *goldstino*. Decomposing the scalar field $a(x)$ into real and imaginary parts

$$\begin{aligned} a(x) &= k(x) + ih(x), \\ a^*(x) &= k(x) - ih(x), \end{aligned} \Bigg\} \tag{9.89}$$

where $k(x)$ and $h(x)$ are real scalar fields and inserting Eqs. (9.89) into Eq. (9.82) or Eq. (9.83), we arrive at the remarkable result that the mass-squared of the field $k(x)$ is

$$m_k^2 = \lambda^2 - 4g^2 M^2 = \lambda^2 - 4\frac{V(0)}{M^2}, \tag{9.90}$$

and that of the field $h(x)$ is

$$m_h^2 = \lambda^2 + 4g^2 M^2 = \lambda^2 + 4\frac{V(0)}{M^2}. \tag{9.91}$$

Here $V(0) = g^2 M^4$. We now see that the boson-fermion mass degeneracy — a characteristic property of supersymmetry — is destroyed. The quantity $V(0)$ is a measure of the magnitude of the breaking of supersymmetry.

We conclude these considerations with a number of remarks.

---

[12]By comparison with the Weyl equations (1.167b) for Weyl components of a Majorana spinor.

(i) At the tree graph level only the bosonic part of the mass spectrum is affected by broken supersymmetry. Here it is the scalar field $a(x)$ which shows a mass splitting. The reason for that is that $a(x)$ is the only scalar which couples to the Goldstone fermion field $\psi_X(x)$ *via* the term gXA^2 in the superpotential (9.60).

(ii) An interesting property of the mass spectrum at the classical, *i.e.* tree level, is the fact that

$$m_k^2 + m_h^2 = 2\lambda^2,$$

irrespective of the supersymmetry breaking. At the tree graph level the fermion masses remain unchanged, whereas the scalars are shifted by the same amount in opposite directions. This is a particular case of a more general result [39] which states that the supertrace (*cf.* Eq. (2.32)) of the squared mass matrix, *i.e.*

$$\text{STr } M^2 = \sum_J (-1)^{2J} (2J+1) M_J^2 = 0, \qquad (9.92)$$

remains zero in the presence of spontaneous symmetry breaking. Here, J denotes the spin of the particle. It can be shown, however, that this result does not hold if one includes radiative corrections [47].

(iii) In the above discussion we considered the case $M^2 < \lambda^2/2g^2$. This choice of the parameters leads to the correct sign of the mass term m_k^2 (*cf.* Eq. (9.90)). On the other hand, for $M^2 > \lambda^2/2g^2$ it is the scalar field $h(x) = \text{Im } a(x)$ which is responsible for the spontaneous breakdown of the internal $U(1)$ symmetry.

(iv) We state without further calculation that the same fermionic mass spectrum is obtained by finding the eigenvalues of the fermionic mass matrix W_{ij} (see Eqs. (9.12) and (9.73)).

Chapter 10

Supersymmetric Gauge Theories

10.1 Minimal Coupling

We now consider the supersymmetric generalization of various gauge theories and in addition an alternative to the O'Raifeartaigh model as a method to obtain spontaneous breaking of supersymmetry.[1] This alternative theory is known as the *Fayet–Iliopoulos model* [34].

In Sec. 7.2 we investigated vector superfields and the supersymmetric extension of Abelian gauge transformations. In addition, in Sec. 8.2.2 we calculated the action integral of a free Abelian gauge field, using the super-symmetric extension of the field strength

$$W_A = -\frac{1}{4}\overline{D}\,\overline{D}D_A V_{\mathrm{WZ}}(x, \theta, \overline{\theta}),\tag{10.1}$$

where $V_{\mathrm{WZ}}(x, \theta, \overline{\theta})$ is a vector superfield in the Wess–Zumino gauge (see Eq. (7.40)). The gauge invariant and supersymmetry invariant action is, as we have seen (*cf.* Eq. (8.40)),

$$\mathcal{A} = \int d^4x \int d^4\theta \left\{ W^A W_A \delta^2(\overline{\theta}) + \overline{W}_{\dot{A}} \overline{W}^{\dot{A}} \delta^2(\theta) \right\}.\tag{10.2}$$

In the following we write this action and similar expressions in the Grassmann

---

[1]The first formulation of a supersymmetric generalization of Abelian gauge theories in the component field formalism was performed by J. Wess and B. Zumino [126]. S. Ferrara and B. Zumino [41] and A. Salam and J. Strathdee [100] extended these constructions to supersymmetric non-Abelian gauge theories. See also Chapters 6 and 7 of P. West [127].

integrated form:

$$\mathcal{A} = \int d^4x \left\{ W^A W_A \Big|_{\theta\theta} + \overline{W}_{\dot{A}} \overline{W}^{\dot{A}} \Big|_{\overline{\theta}\,\overline{\theta}} \right\}.$$

The component field expansion yields the following action (*cf.* Eq. (8.41))

$$\mathcal{A} = \int d^4x \left\{ 8D^2(x) - F_{\mu\nu}(x)F^{\mu\nu}(x) - i\lambda(x)\sigma^\mu\partial_\mu\overline{\lambda}(x) \right\}, \qquad (10.3)$$

where

$$F_{\mu\nu}(x) = \partial_\mu V_\nu(x) - \partial_\nu V_\mu(x)$$

is the ususal field strength tensor, $D(x)$ is an auxiliary scalar field and $\lambda(x)$ is the supersymmetric partner (spin 1/2 field) of $V_\mu(x)$, now part of the gauge field.

The question now arises how one can include matter fields. This implies, of course, that we must search for gauge invariant interactions of scalar and vector superfields. We begin with the simplest case, the Abelian $U(1)$ group. We define the transformation of scalar superfields under global $U(1)$ gauge transformations by the relation

$$\Phi_i' = \exp\{-iq_i\lambda\}\Phi_i. \qquad (10.4)$$

Here the q_i are the $U(1)$ charges of the superfields Φ_i, and λ is the global $U(1)$ rotation angle. Both q_i and λ are real and constant. Since constants are particular cases of superfields satisfying the constraints

$$D_A\lambda = 0 \quad \text{and} \quad \overline{D}_{\dot{A}}\lambda = 0,$$

it follows that the transformed scalar superfield Φ_i' — being the product of such fields — is again a scalar superfield satisfying the constraint

$$\overline{D}_{\dot{A}}\Phi_i' = 0. \qquad (10.5)$$

It is now not difficult to construct a Lagrangian density in superspace which is manifestly invariant under the transformation (10.4) for constant λ. We set:

$$\mathcal{L} = \mathcal{L}_{\text{kin}} + W[\Phi]\Big|_{\theta\theta} + \overline{W}[\Phi^\dagger]\Big|_{\overline{\theta}\,\overline{\theta}}$$

$$= \Phi_i^\dagger\Phi_i\Big|_{\theta\theta\,\overline{\theta}\,\overline{\theta}} + \left[\frac{1}{2}m_{ij}\Phi_i\Phi_j + \frac{1}{3}\lambda_{ijk}\Phi_i\Phi_j\Phi_k + g_i\Phi_i\right]\Big|_{\theta\theta} + \text{h.c.} \qquad (10.6)$$

Here h.c. denotes the Hermitian conjugate of the term in square brackets. We then have (using Eq. (10.4)):

$$\mathcal{L}' = \Phi_i'^\dagger \Phi_i' \Big|_{\theta\theta\,\bar\theta\bar\theta} + \left[\frac{1}{2} m_{ij} \Phi_i' \Phi_j' + \frac{1}{3}\lambda_{ijk}\Phi_i'\Phi_j'\Phi_k' + g_i\Phi_i'\right]\Big|_{\theta\theta} + \text{ h.c.}$$

$$= \left(\Phi_i^\dagger e^{iq_i\lambda}\right)\left(e^{-iq_i\lambda}\Phi_i\right)\Big|_{\theta\theta\,\bar\theta\bar\theta}$$

$$+ \left[\frac{1}{2}m_{ij}e^{-i(q_i+q_j)\lambda}\Phi_i\Phi_j + g_i e^{-i\lambda q_i}\Phi_i\right.$$

$$\left. + \frac{1}{3}\lambda_{ijk}e^{-i(q_i+q_j+q_k)\lambda}\Phi_i\Phi_j\Phi_k\right]\Big|_{\theta\theta} + \text{ h.c.}$$

$$= \Phi_i^\dagger \Phi_i \Big|_{\theta\theta\,\bar\theta\bar\theta} + \left[\frac{1}{2}m_{ij}e^{-i(q_i+q_j)\lambda}\Phi_i\Phi_j + g_i e^{-i\lambda q_i}\Phi_i\right.$$

$$\left. + \frac{1}{3}\lambda_{ijk}e^{-i(q_i+q_j+q_k)\lambda}\Phi_i\Phi_j\Phi_k\right]\Big|_{\theta\theta} + \text{ h.c.} \qquad (10.7)$$

To obtain global $U(1)$ phase invariance of the superpotential we have to demand

$$\left.\begin{array}{rcll} g_i &= 0 & \text{if} & q_i \neq 0, \\ m_{ij} &= 0 & \text{if} & q_i + q_j \neq 0, \\ \lambda_{ijk} &= 0 & \text{if} & q_i + q_j + q_k \neq 0. \end{array}\right\} \qquad (10.8)$$

The kinetic term of the Lagrangian is automatically $U(1)$ invariant without any restrictions.

As demonstrated above, a global $U(1)$ phase transformation maps scalar superfields onto scalar superfields, this map depending crucially on the fact that the rotation angle λ may be looked at as a scalar superfield which satisfies the constraints $D\lambda = 0$ and $\overline{D}\lambda = 0$.

If we want to introduce *local* $U(1)$ phase transformations the above construction must be altered in the following way. Assume that in a local theory the parameter λ is a function of the spacetime variable x, *i.e.*

$$\lambda = \lambda(x),$$

and let

$$\Phi_i' = \exp\{-i\lambda(x)q_i\}\Phi_i.$$

Then the transformed quantity Φ_i' is no longer a scalar superfield, since

$$\overline{D}_{\dot A}\Phi_i' = \overline{D}_{\dot A}\left[\exp\{-iq_i\lambda(x)\}\Phi_i\right]$$

$$= \exp\{-iq_i\lambda(x)\}\overline{D}_{\dot A}\Phi_i - iq_i\overline{D}_{\dot A}(\lambda(x))\exp\{-iq_i\lambda(x)\}\Phi_i$$

$$= -iq_i\overline{D}_{\dot A}(\lambda(x))\exp\{-iq_i\lambda(x)\}\Phi_i.$$

In order to obtain a transformed scalar superfield Φ'_i, satisfying the constraint equation

$$\overline{D}_{\dot{A}}\Phi'_i = 0,$$

we must introduce a quantity

$$\Lambda = \Lambda(x,\theta,\overline{\theta}),$$

obeying the constraint equation, *i.e.*

$$\overline{D}_{\dot{A}}\Lambda(x,\theta,\overline{\theta}) = 0. \tag{10.9}$$

However, the quantity $\Lambda(x,\theta,\overline{\theta})$ satisfying the constraint (10.9) is nothing but a full scalar multiplet. Hence, in order to replace a global $U(1)$ phase transformation by a local phase transformation one is forced to introduce a new scalar superfield $\Lambda(x,\theta,\overline{\theta})$ such that transformed scalar superfields are again scalar superfields. Thus, in superspace, the local $U(1)$ phase transformation is given by

$$\left.\begin{aligned}
\Phi'_i &= \exp\{-iq_i\Lambda(x,\theta,\overline{\theta})\}\Phi_i, \\
\overline{D}_{\dot{A}}\Lambda(x,\theta,\overline{\theta}) &= 0, \\
\Phi'^\dagger_i &= \Phi^\dagger_i\exp\{iq_i\Lambda^\dagger(x,\theta,\overline{\theta})\}, \\
D_A\Lambda^\dagger(x,\theta,\overline{\theta}) &= 0.
\end{aligned}\right\} \tag{10.10}$$

As in the case of ordinary local phase transformations, the Lagrangian density (10.6) is not invariant under the transformations (10.10). The superpotential $W[\Phi]$ remains invariant with respect to the restrictions (10.8) but the kinetic part does not, since

$$\mathcal{L}'_{\text{kin}} = \Phi'^\dagger_i\Phi'_i\big|_{\theta\theta\,\overline{\theta}\,\overline{\theta}} = \exp\{iq_i(\Lambda^\dagger - \Lambda)\}\Phi^\dagger_i\Phi_i\big|_{\theta\theta\,\overline{\theta}\,\overline{\theta}} \neq \Phi^\dagger_i\Phi_i\big|_{\theta\theta\,\overline{\theta}\,\overline{\theta}}.$$

To obtain an invariant kinetic part of the Lagrangian we have to introduce a compensating field as in ordinary gauge theories. In the supersymmetric case we must choose a vector superfield

$$V = V(x,\theta,\overline{\theta}),$$

which has the transformation law under supersymmetric gauge transformations[2]

$$V'(x,\theta,\overline{\theta}) = V(x,\theta,\overline{\theta}) + i\big[\Lambda(x,\theta,\overline{\theta}) - \Lambda^\dagger(x,\theta,\overline{\theta})\big]. \tag{10.11}$$

---

[2]See Eq. (7.35) with $\Phi = i\Lambda$.

Then the invariant kinetic part of the superspace Lagrangian density is

$$\mathcal{L}_{\text{kin}} = \Phi_i^\dagger e^{q_i V} \Phi_i \Big|_{\theta\theta\,\bar\theta\bar\theta}. \tag{10.12}$$

In order to check the invariance of the kinetic term we consider:

$$
\begin{aligned}
\mathcal{L}'_{\text{kin}} &= \Phi_i'^\dagger e^{q_i V'} \Phi_i' \Big|_{\theta\theta\,\bar\theta\bar\theta} \\
&\overset{(10.10)}{\underset{(10.11)}{=}} \Phi_i^\dagger e^{iq_i\Lambda^\dagger} \exp\{q_i(V + i\Lambda - i\Lambda^\dagger)\} e^{-iq_i\Lambda} \Phi_i \Big|_{\theta\theta\,\bar\theta\bar\theta} \\
&= \Phi_i^\dagger e^{iq_i\Lambda^\dagger} e^{-iq_i\Lambda^\dagger} e^{q_i V} e^{iq_i\Lambda} e^{-iq_i\Lambda} \Phi_i \Big|_{\theta\theta\,\bar\theta\bar\theta} \\
&= \Phi_i^\dagger e^{q_i V} \Phi_i \Big|_{\theta\theta\,\bar\theta\bar\theta} = \mathcal{L}_{\text{kin}},
\end{aligned}
$$

as had to be shown. Hence, as we shall demonstrate below, Eq. (10.12) is the supersymmetric generalization of the principle of *minimal coupling* in ordinary gauge theories.

With the kinetic term Eq. (10.12) and the action (8.40) for pure Abelian supersymmetric gauge theories, the most general $U(1)$ gauge invariant supersymmetric action integral is

$$
\begin{aligned}
\mathcal{A} = \int d^4x \int d^4\theta \Big\{ & W^A W_A \delta^2(\bar\theta) + \overline{W}_{\dot A} \overline{W}^{\dot A} \delta^2(\theta) \\
& + \Phi_i^\dagger e^{iq_i V} \Phi_i + W[\Phi]\delta^2(\bar\theta) + \overline{W}[\Phi^\dagger]\delta^2(\theta) \Big\}. \tag{10.13}
\end{aligned}
$$

Here, the superpotential $W[\Phi]$ is a polynomial of the superfield Φ. Renormalizabilty allows $W[\Phi]$ to be at most of order three in Φ as we remarked in Sec. 8.2.1.

In order to demonstrate that the formal expression Eq. (10.12) is indeed the supersymmetric minimal coupling, we must re-express the kinetic term (10.12) in terms of component fields and show that the component field $V_\mu(x)$ of the vector superfield $V(x,\theta,\bar\theta)$ appears in Eq. (10.12) only in the covariant derivative. This is most easily done in the Wess–Zumino gauge Eq. (7.40), in which case we can use Eqs. (7.41) and (7.42). For simplicity we consider only the case of a single scalar superfield Φ with component field expansion

$$
\begin{aligned}
\Phi &= A(x) + \sqrt{2}\,\theta\psi(x) + \theta\theta F(x), \\
\Phi^\dagger &= A^*(x) + \sqrt{2}\,\bar\theta\,\bar\psi(x) + \bar\theta\,\bar\theta F^*(x).
\end{aligned}
$$

Then we find

$$
\Phi^\dagger e^{qV}\Phi\Big|_{\theta\theta\,\overline{\theta}\,\overline{\theta}} \overset{(7.42)}{=} \Phi^\dagger\Big\{1 + q\Big[\big(\theta\sigma^\mu\overline{\theta}\big)V_\mu(x) + (\theta\theta)\big(\overline{\theta}\,\overline{\lambda}(x)\big)
$$
$$
+ \big(\overline{\theta}\,\overline{\theta}\big)\big(\theta\lambda(x)\big) + (\theta\theta)\big(\overline{\theta}\,\overline{\theta}\big)D(x)
$$
$$
+ (\theta\theta)\big(\overline{\theta}\,\overline{\theta}\big)\frac{q}{4}V_\mu(x)V^\mu(x)\Big]\Big\}\Phi\Big|_{\theta\theta\,\overline{\theta}\,\overline{\theta}}
$$
$$
= \Phi^\dagger\Phi\Big|_{\theta\theta\,\overline{\theta}\,\overline{\theta}} + q\big(\theta\sigma^\mu\overline{\theta}\big)V_\mu(x)\Phi^\dagger\Phi\Big|_{\theta\theta\,\overline{\theta}\,\overline{\theta}}
$$
$$
+ q\big(\theta\theta\overline{\theta}\,\overline{\lambda}(x)\Phi^\dagger\Phi\big)\Big|_{\theta\theta\,\overline{\theta}\,\overline{\theta}} + q\big(\overline{\theta}\,\overline{\theta}\theta\lambda(x)\Phi^\dagger\Phi\big)\Big|_{\theta\theta\,\overline{\theta}\,\overline{\theta}}
$$
$$
+ q\Big(\theta\theta\overline{\theta}\,\overline{\theta}\big[D(x) + \frac{q}{4}V_\mu(x)V^\mu(x)\big]\Phi^\dagger\Phi\Big)\Big|_{\theta\theta\,\overline{\theta}\,\overline{\theta}}.
$$

The field $\Phi^\dagger$ can be shifted to the right because this action involves only an even number of anticommutations of Grassmann variables. Using Eqs. (7.27b), (8.28) and (8.29) for the product $\Phi^\dagger\Phi$ we obtain

$$
\Phi^\dagger e^{qV}\Phi\Big|_{\theta\theta\,\overline{\theta}\,\overline{\theta}} = \big|F(x)\big|^2 + i\partial_\mu\overline{\psi}(x)\overline{\sigma}^\mu\psi(x) - A^*(x)\Box A(x)
$$
$$
+ iq\big(\theta\sigma^\mu\overline{\theta}\big)\big(\theta\sigma^\nu\overline{\theta}\big)V_\mu(x)
$$
$$
\times \big\{\big(\partial_\nu A(x)\big)A^*(x) - \big(\partial_\nu A^*(x)\big)A(x)\big\}\Big|_{\theta\theta\,\overline{\theta}\,\overline{\theta}}
$$
$$
+ 2q\big(\theta\sigma^\mu\overline{\theta}\big)V_\mu(x)\big(\overline{\theta}\,\overline{\psi}(x)\big)\big(\theta\psi(x)\big)\Big|_{\theta\theta\,\overline{\theta}\,\overline{\theta}}
$$
$$
+ \sqrt{2}q(\theta\theta)\big(\overline{\theta}\,\overline{\lambda}(x)\big)\big(\overline{\theta}\,\overline{\psi}(x)\big)A(x)\Big|_{\theta\theta\,\overline{\theta}\,\overline{\theta}}
$$
$$
+ \sqrt{2}q\big(\overline{\theta}\,\overline{\theta}\big)\big(\theta\lambda(x)\big)\big(\theta\psi(x)\big)A^*(x)\Big|_{\theta\theta\,\overline{\theta}\,\overline{\theta}}
$$
$$
+ q\big[D(x) + \frac{q}{4}V_\mu(x)V^\mu(x)\big]\big|A(x)\big|^2
$$
$$
+ \text{ total derivatives.} \tag{10.14}
$$

The goal now is to rewrite various terms of Eq. (10.14) as terms of the form $(\theta\theta)(\overline{\theta}\,\overline{\theta})$. To this end we use the following expressions. With Eq. (1.137) we get

$$
\big(\theta\sigma^\mu\overline{\theta}\big)\big(\theta\sigma^\nu\overline{\theta}\big) = \frac{1}{2}\eta^{\mu\nu}(\theta\theta)\big(\overline{\theta}\,\overline{\theta}\big), \tag{10.15}
$$

and

$$
\big(\theta\sigma^\mu\overline{\theta}\big)\big(\overline{\theta}\,\overline{\psi}(x)\big)\big(\theta\psi(x)\big) = \theta^A(\sigma^\mu)_{A\dot{A}}\overline{\theta}^{\dot{A}}\overline{\theta}_{\dot{B}}\overline{\psi}^{\dot{B}}(x)\theta^B\psi_B(x)
$$
$$
= -\theta^A\theta^B(\sigma^\mu)_{A\dot{A}}\overline{\theta}^{\dot{A}}\epsilon_{\dot{B}\dot{C}}\overline{\theta}^{\dot{C}}\overline{\psi}^{\dot{B}}(x)\psi_B(x)
$$
$$
\overset{(1.100a)}{=} -\frac{1}{2}(\theta\theta)\epsilon^{AB}(\sigma^\mu)_{A\dot{A}}\overline{\theta}^{\dot{A}}\overline{\theta}^{\dot{C}}\epsilon_{\dot{B}\dot{C}}\overline{\psi}^{\dot{B}}(x)\psi_B(x)
$$

$$\stackrel{(1.100c)}{=} \frac{1}{4}(\theta\theta)(\overline{\theta}\,\overline{\theta})\epsilon^{AB}(\sigma^\mu)_{A\dot{A}}\epsilon^{\dot{A}\dot{C}}\epsilon_{\dot{B}\dot{C}}\overline{\psi}^{\dot{B}}(x)\psi_B(x)$$

$$= \frac{1}{4}(\theta\theta)(\overline{\theta}\,\overline{\theta})\epsilon^{BA}\epsilon^{\dot{C}\dot{A}}(\sigma^\mu)_{A\dot{A}}\epsilon_{\dot{B}\dot{C}}\overline{\psi}^{\dot{B}}(x)\psi_B(x)$$

$$\stackrel{(1.106a)}{=} -\frac{1}{4}(\theta\theta)(\overline{\theta}\,\overline{\theta})(\overline{\sigma}^\mu)^{\dot{C}B}\overline{\psi}_{\dot{C}}(x)\psi_B(x)$$

$$= -\frac{1}{4}(\theta\theta)(\overline{\theta}\,\overline{\theta})(\overline{\psi}(x)\overline{\sigma}^\mu\psi(x)). \tag{10.16}$$

Furthermore

$$(\theta\theta)(\overline{\theta}\,\overline{\lambda}(x))(\overline{\theta}\,\overline{\psi}(x)) = (\theta\theta)\overline{\theta}_{\dot{A}}\overline{\lambda}^{\dot{A}}(x)\overline{\theta}_{\dot{B}}\overline{\psi}^{\dot{B}}(x)$$

$$= -(\theta\theta)\overline{\theta}_{\dot{A}}\overline{\theta}_{\dot{B}}\overline{\lambda}^{\dot{A}}(x)\overline{\psi}^{\dot{B}}(x)$$

$$\stackrel{(1.100d)}{=} \frac{1}{2}(\theta\theta)(\overline{\theta}\,\overline{\theta})\epsilon_{\dot{A}\dot{B}}\overline{\lambda}^{\dot{A}}(x)\overline{\psi}^{\dot{B}}(x)$$

$$= -\frac{1}{2}(\theta\theta)(\overline{\theta}\,\overline{\theta})(\overline{\lambda}(x)\,\overline{\psi}(x)). \tag{10.17}$$

Similarly

$$(\overline{\theta}\,\overline{\theta})(\theta\lambda(x))(\theta\psi(x)) = -\frac{1}{2}(\theta\theta)(\overline{\theta}\,\overline{\theta})(\lambda(x)\psi(x)). \tag{10.18}$$

Inserting Eqs. (10.15) to (10.18) into Eq. (10.14) we obtain:

$$\Phi^\dagger e^{qV}\Phi\Big|_{\theta\theta\,\overline{\theta}\,\overline{\theta}} = |F(x)|^2 + i\partial_\mu\overline{\psi}(x)\overline{\sigma}^\mu\psi(x) - A^*(x)\Box A(x)$$

$$+ \frac{1}{2}iqV_\mu(x)\big\{(\partial_\mu A(x))A^*(x) - (\partial_\mu A^*(x))A(x)\big\}\Big|_{\theta\theta\,\overline{\theta}\,\overline{\theta}}$$

$$- \frac{1}{2}qV_\mu(x)(\overline{\psi}(x)\overline{\sigma}^\mu\psi(x))$$

$$- \frac{1}{\sqrt{2}}q\big[\overline{\lambda}(x)\overline{\psi}(x)A(x) + \lambda(x)\psi(x)A^*(x)\big]$$

$$+ q\Big[D(x) + \frac{q}{4}V_\mu(x)V^\mu(x)\Big]|A(x)|^2$$

$$+ \text{ total derivatives.} \tag{10.19}$$

Now we consider the following expression, which defines the covariant derivative D_μ of the scalar field $A(x)$:

$$|D_\mu A(x)|^2 \equiv \Big|\big(\partial_\mu - \frac{1}{2}iqV_\mu(x)\big)A(x)\Big|^2$$

$$= \Big\{\big(\partial_\mu - \frac{1}{2}iqV_\mu(x)\big)A(x)\Big\}\Big\{\big(\partial^\mu + \frac{1}{2}iqV^\mu(x)\big)A^*(x)\Big\}$$

$$= \Big\{\partial_\mu A(x) - \frac{1}{2}iqV_\mu(x)A(x)\Big\}\Big\{\partial^\mu A^*(x) + \frac{1}{2}iqV^\mu(x)A^*(x)\Big\}$$

$$= |\partial_\mu A(x)|^2 + \frac{1}{4}q^2 V_\mu(x)V^\mu(x)|A(x)|^2$$

$$+ \frac{1}{2} iq V^\mu(x) \left\{ \left(\partial_\mu A(x)\right) A^*(x) - A(x) \left(\partial_\mu A^*(x)\right) \right\}$$

$$= -A^*(x) \Box A(x) + \frac{1}{4} q^2 V_\mu(x) V^\mu(x) \big| A(x) \big|^2$$

$$+ \frac{1}{2} iq V^\mu(x) \left\{ \left(\partial_\mu A(x)\right) A^*(x) - A(x) \left(\partial_\mu A^*(x)\right) \right\}$$

$$+ \text{ total derivatives.} \tag{10.20}$$

We define

$$\boxed{D_\mu := \partial_\mu - iq\left(\frac{V_\mu(x)}{2}\right)} \tag{10.21}$$

as the gauge covariant derivate. The factor $1/2$, which at first seems unusual, has its origin in the choice of coefficients of the component field expansion (7.34) and can be removed by a redefinition of the vector field $V_\mu(x)$. With the gauge covariant derivative (10.21) we can rewrite the following spinor terms, appearing in Eq. (10.19):

$$i\left(\partial_\mu \overline{\psi}(x)\right)\overline{\sigma}^\mu \psi(x) - \frac{1}{2} q V_\mu(x)\overline{\psi}(x)\overline{\sigma}^\mu \psi(x) = \left[i\left(\partial_\mu + \frac{1}{2} iq V_\mu(x)\right)\overline{\psi}(x) \right]\overline{\sigma}^\mu \psi(x)$$

$$= i\left(D_\mu^* \overline{\psi}(x)\right)\overline{\sigma}^\mu \psi(x). \tag{10.22}$$

Hence, with Eqs. (10.20) and (10.22) we find for the gauge covariant kinetic term (10.19):

$$\Phi^\dagger e^{qV} \Phi \Big|_{\theta\theta\,\overline{\theta}\,\overline{\theta}} = \big| F(x) \big|^2 + i\left(D_\mu^* \overline{\psi}(x)\right)\overline{\sigma}^\mu \psi(x) + \big| D_\mu A(x) \big|^2$$

$$- \frac{1}{\sqrt{2}} q\left[\overline{\lambda}(x)\overline{\psi}(x) A(x) + \lambda(x)\psi(x) A^*(x) \right]$$

$$+ q D(x)\big| A(x) \big|^2. \tag{10.23}$$

This component field expansion of the modified kinetic term demonstrates that Eq. (10.12) is the correct supersymmetric extension of the principle of minimal coupling, *i.e.* in that term of the Lagrangian density $\mathcal{L}$, which couples the vector superfield $V(x,\theta,\overline{\theta})$ to the superfield Φ, the vector field $V_\mu(x)$ contained in the vector superfield $V(x,\theta,\overline{\theta})$ is completely absorbed in the gauge covariant derivative (10.21).

We mention in passing that the $(\theta\theta)(\overline{\theta}\,\overline{\theta})$-component of

$$\Phi^\dagger e^{qV} \Phi$$

contains no terms of dimension higher than the fourth which ensures that the Lagrangian (10.13) remains renormalizable. However, this depends crucially on the fact that we expanded the term $\exp\{qV\}$ in the Wess–Zumino gauge, in which $V_{WZ}^3 = 0$, (*cf.* Eq. (7.41)).

10.2 Super Quantum Electrodynamics

As an application of the ideas developed in Sec. 10.1 we extend ordinary electrodynamics — coupled to matter fields — to a supersymmetric theory.[3] We have to introduce two scalar superfields Φ_1 and Φ_2 which transform under local $U(1)$ gauge transformations according to Eq. (10.10), *i.e.*

$$\left. \begin{aligned} \Phi_1 \longrightarrow \Phi_1' &= \exp\{-iq_1\Lambda(x,\theta,\overline{\theta})\}\Phi_1, \\ \Phi_2 \longrightarrow \Phi_2' &= \exp\{-iq_2\Lambda(x,\theta,\overline{\theta})\}\Phi_2, \end{aligned} \right\} \qquad (10.24)$$

where $q_1 = +e$ and $q_2 = -e$. Then according to the restrictions, given in Eq. (10.8), the couplings and masses of the superpotential of the superaction (10.13) are subject to the following constraints:

$$g_1 = g_2, \qquad \text{since} \qquad q_1 = 0 \neq 0 \quad \text{and} \quad q_2 = -e \neq 0,$$
$$m_{12} =: m \neq 0, \qquad \text{since} \qquad q_1 + q_2 = e - e = 0,$$

and all couplings λ_{ijk} vanish since

$$\left. \begin{aligned} 2q_1 + q_2 \neq 0, \qquad 3q_1 \neq 0, \\ 2q_2 + q_1 \neq 0, \qquad 3q_2 \neq 0. \end{aligned} \right\} \qquad (10.25)$$

Inserting these restrictions into the superpotential $W[\Phi]$, we obtain a $U(1)$ gauge invariant action in superspace, *i.e.*

$$\begin{aligned} \mathcal{A}_{\text{SQED}} = \int d^4x \int d^4\theta \Big\{ &W^A W_A \delta^2(\overline{\theta}) + \overline{W}_{\dot{A}} \overline{W}^{\dot{A}} \delta^2(\theta) \\ &+ \Phi_1^\dagger e^{eV} \Phi + \Phi_2^\dagger e^{-eV} \Phi_2 \\ &- m\big(\Phi_1\Phi_2\delta^2(\overline{\theta}) + \Phi_1^\dagger\Phi_2^\dagger\delta^2(\theta)\big) \Big\}, \end{aligned} \qquad (10.26)$$

where we made use of Eqs. (10.6) and (10.12).

We now work out the component field representation of the superspace action $\mathcal{A}_{\text{SQED}}$. According to Eq. (8.41) we have

$$\begin{aligned} \int d^4x \int d^4\theta \Big\{ &W^A W_A \delta^2(\overline{\theta}) + \overline{W}_{\dot{A}} \overline{W}^{\dot{A}} \delta^2(\theta) \Big\} \\ &= \int d^4x \Big\{ 8D^2(x) - F_{\mu\nu}(x)F^{\mu\nu}(x) - 4i\lambda(x)\sigma^\mu\partial_\mu\overline{\lambda}(x) \Big\}. \end{aligned} \qquad (10.27)$$

---

[3]See J. Wess and B. Zumino [126].

Using Eq. (10.23) we obtain for the kinetic terms of the matter fields Φ_1 and Φ_2 the following component field expansion:

$$\int d^4x \int d^4\theta \, \Phi_1^\dagger e^{V(x,\theta,\bar\theta)} \Phi_1 = |F_1(x)|^2 + i(D_\mu^* \bar\psi_1(x))\bar\sigma^\mu \psi_1(x)$$
$$+ |D_\mu A_1(x)|^2$$
$$- \frac{e}{\sqrt{2}} \left[\bar\lambda(x)\bar\psi_1(x)A_1(x) + \lambda(x)\psi_1(x)A_1^*(x)\right]$$
$$+ eD(x)|A_1(x)|^2. \tag{10.28}$$

Here the covariant derivatives are according to Eq. (10.21):

$$D_\mu = \partial_\mu - \frac{1}{2}ieV_\mu(x), \qquad D_\mu^* = \partial_\mu + \frac{1}{2}ieV_\mu(x).$$

A similar expansion holds for the kinetic term of the Φ_2-field:

$$\int d^4x \int d^4\theta \, \Phi_2^\dagger e^{-eV(x,\theta,\bar\theta)} \Phi_2 = |F_2(x)|^2 + i(D_\mu \bar\psi_2(x))\bar\sigma^\mu \psi_2(x)$$
$$+ |D_\mu^* A_2(x)|^2$$
$$+ \frac{e}{\sqrt{2}} \left[\bar\lambda(x)\bar\psi_2(x)A_2(x) + \lambda(x)\psi_2(x)A_2^*(x)\right]$$
$$- eD(x)|A_2(x)|^2. \tag{10.29}$$

Finally the mass terms give, using Eq. (7.16):

$$\int d^4x \int d^4\theta \, m\Phi_1\Phi_2\delta^2(\bar\theta) = \int d^4x \, m\Phi_1\Phi_2\Big|_{\theta\theta}$$
$$= \int d^4x \, m\left[A_1(x)F_2(x) + A_2(x)F_1(x)\right.$$
$$\left. - \psi_1(x)\psi_2(x)\right], \tag{10.30}$$

and

$$\int d^4x \int d^4\theta \, m\Phi_1^\dagger\Phi_2^\dagger\delta^2(\theta) = \int d^4x \, m\Phi_1^\dagger\Phi_2^\dagger\Big|_{\bar\theta\bar\theta}$$
$$= \int d^4x \, m\left[A_1^*(x)F_2^*(x) + A_2^*(x)F_1^*(x)\right.$$
$$\left. - \bar\psi_1(x)\bar\psi_2(x)\right]. \tag{10.31}$$

Collecting all expansions, thus inserting Eqs. (10.27) to (10.31) into the superspace action (10.26), we obtain

$$
\mathcal{A}_{\mathrm{SQED}} = \int d^4x \Big\{ 8D^2(x) - F_{\mu\nu}(x)F^{\mu\nu}(x)
$$

$$
- 4i\lambda(x)\sigma^\mu \partial_\mu \overline{\lambda}(x) + \left|F_1(x)\right|^2 + \left|F_2(x)\right|^2
$$

$$
+ i\big(D_\mu^*\overline{\psi}_1(x)\big)\overline{\sigma}^\mu \psi_1(x) + i\big(D_\mu \overline{\psi}_2(x)\big)\overline{\sigma}^\mu \psi_2(x)
$$

$$
+ \left|D_\mu A_1(x)\right|^2 + \left|D_\mu^* A_2(x)\right|^2 + eD(x)\big(\left|A_1(x)\right|^2 - \left|A_2(x)\right|^2\big)
$$

$$
- \frac{e}{2^{1/2}} \Big[\overline{\lambda}(x)\big(\overline{\psi}_1(x)A_1(x) - \overline{\psi}_2(x)A_2(x)\big)
$$

$$
+ \lambda(x)\big(\psi_1(x)A_1^*(x) - \psi_2(x)A_2^*(x)\big)\Big]
$$

$$
- m\Big[A_1(x)F_2(x) + A_1^*(x)F_2^*(x) + A_2(x)F_1(x)
$$

$$
+ A_2^*(x)F_1^*(x) - \psi_1(x)\psi_2(x) - \overline{\psi}_1(x)\overline{\psi}_2(x)\Big]\Big\}. \tag{10.32}
$$

Our next task consists in eliminating the three auxiliary field $F_1(x)$, $F_2(x)$, and $D(x)$ using their respective Euler–Lagrange equations.

(i) $F_1(x)$: The Euler–Lagrange equation for the F_1-field is

$$
\frac{\partial \mathcal{L}}{\partial F_1(x)} = F_1^*(x) - mA_2(x) = 0,
$$

leading to
$$
F_1^*(x) = mA_2(x), \quad F_1(x) = mA_2^*(x).
$$

(ii) $F_2(x)$: The Euler–Lagrange equation for the F_2-field is

$$
\frac{\partial \mathcal{L}}{\partial F_2(x)} = F_2^*(x) - mA_1(x) = 0,
$$

so that
$$
F_2^*(x) = mA_1(x), \quad F_2(x) = mA_1^*(x).
$$

Hence
$$
\left|F_1(x)\right|^2 + \left|F_2(x)\right| = m^2\big(\left|A_1(x)\right|^2 + \left|A_2(x)\right|^2\big),
$$

and

$$
-m\big[A_1(x)F_2(x) + A_1^*(x)F_2^*(x) + A_2(x)F_1(x) + A_2^*(x)F_1^*(x)\big]
$$
$$
= -2m^2\big(\left|A_1(x)\right|^2 + \left|A_2(x)\right|^2\big).
$$

(iii) $D(x)$: We have

$$\frac{\partial \mathcal{L}}{\partial D(x)} = 16D(x) + e\big(|A_1(x)|^2 - |A_2(x)|^2\big) = 0,$$

so that

$$D(x) = -\frac{e}{16}\big(|A_1(x)|^2 - |A_2(x)|^2\big).$$

Then:

$$8D^2(x) + eD(x)\big(|A_1(x)|^2 - |A_2(x)|^2\big) = -\frac{e^2}{32}\Big(|A_1(x)|^2 - |A_2(x)|^2\Big).$$

Thus, in terms of the dynamical fields the supersymmetric and Abelian gauge invariant action (10.32) becomes:

$$
\begin{aligned}
\mathcal{A}_{\mathrm{SQED}} = \int d^4x \Big\{ &-F_{\mu\nu}(x)F^{\mu\nu}(x) - 4i\lambda(x)\sigma^\mu\partial_\mu\overline{\lambda}(x) \\
&+ i\big(D_\mu^*\overline{\psi}_1(x)\big)\overline{\sigma}^\mu\psi_1(x) + i\big(D_\mu\overline{\psi}_2(x)\big)\overline{\sigma}^\mu\psi_2(x) \\
&+ |D_\mu A_1(x)|^2 + |D_\mu^* A_2(x)|^2 \\
&- \frac{e}{2^{1/2}}\Big[\overline{\lambda}(x)\big(\overline{\psi}_1(x)A_1(x) - \overline{\psi}_2(x)A_2(x)\big) \\
&\qquad + \lambda(x)\big(\psi_1(x)A_1^*(x) - \psi_2(x)A_2^*(x)\big)\Big] \\
&+ m\big[\psi_1(x)\psi_2(x) + \overline{\psi}_1(x)\overline{\psi}_2(x)\big] \\
&- m^2\big(|A_1(x)|^2 + |A_2(x)|^2\big) \\
&- \frac{e^2}{32}\big(|A_1(x)|^2 - |A_2(x)|^2\big)^2 \Big\}.
\end{aligned}
\tag{10.33}
$$

In order to see that the action (10.33) describes the correct field content, *i.e.* in particular a massive Dirac field, we define the four-component Dirac spinor $\Psi(x)$ by (*cf.* Eq. (1.153))

$$\Psi(x) := \begin{pmatrix} \psi_{1A}(x) \\ \overline{\psi}_2^{\dot{A}}(x) \end{pmatrix}. \tag{10.34}$$

The Dirac adjoint of the spinor (10.34) is (*cf.* Eq. (1.201a)) with a suitable change of nomenclature:

$$\overline{\Psi}(x) = \big(\psi_2^B(x),\ \overline{\psi}_{1\dot{B}}(x)\big). \tag{10.35}$$

We now consider the following terms of the action (10.33):

$$i\Big(D_\mu^*\overline\psi_1(x)\Big)\overline\sigma^\mu\psi_1(x) + i\Big(D_\mu\overline\psi_2(x)\Big)\overline\sigma^\mu\psi_2(x)$$

$$= i\Big[\partial_\mu + \frac{ie}{2}V_\mu(x)\Big]\overline\psi_1(x)\overline\sigma^\mu\psi_1(x)$$

$$+ i\Big[\partial_\mu - \frac{ie}{2}V_\mu(x)\Big]\overline\psi_2(x)\overline\sigma^\mu\psi_2(x)$$

$$= i(\partial_\mu\overline\psi_1(x))\overline\sigma^\mu\psi_1(x) - \frac{e}{2}\overline\psi_1(x)\overline\sigma^\mu V_\mu(x)\psi_1(x)$$

$$+ i(\partial_\mu\overline\psi_2(x))\overline\sigma^\mu\psi_2(x) + \frac{e}{2}\overline\psi_2(x)\overline\sigma^\mu V_\mu(x)\psi_2(x)$$

$$= -i\overline\psi_1(x)\overline\sigma^\mu(\partial_\mu\psi_1(x)) - \frac{e}{2}\overline\psi_1(x)\overline\sigma^\mu V_\mu(x)\psi_1(x)$$

$$+ i(\partial_\mu\overline\psi_2(x))\overline\sigma^\mu\psi_2(x) + \frac{e}{2}\overline\psi_2(x)\overline\sigma^\mu V_\mu(x)\psi_2(x)$$

$$+ \text{ total derivatives.}$$

Now using Eq. (1.134) we find:

$$i\Big(D_\mu^*\overline\psi_1(x)\Big)\overline\sigma^\mu\psi_1(x) + i\Big(D_\mu\overline\psi_2(x)\Big)\overline\sigma^\mu\psi_2(x)$$

$$= -i\overline\psi_1(x)\overline\sigma^\mu\Big[\partial_\mu - \frac{1}{2}ieV_\mu(x)\Big]\psi_1(x)$$

$$- i\psi_2(x)\sigma^\mu(\partial_\mu\overline\psi_2(x)) - \frac{e}{2}\psi_2(x)\sigma^\mu V_\mu(x)\overline\psi_2(x)$$

$$+ \text{ total derivatives}$$

$$= -i\overline\psi_1(x)\overline\sigma^\mu\Big[\partial_\mu - \frac{1}{2}ieV_\mu(x)\Big]\psi_1(x)$$

$$- i\psi_2(x)\sigma^\mu\Big[\partial_\mu - \frac{1}{2}ieV_\mu(x)\Big]\overline\psi_2(x)$$

$$+ \text{ total derivatives}$$

$$= -i\overline\psi_1(x)\overline\sigma^\mu D_\mu\psi_1(x) - i\psi_2(x)\sigma^\mu D_\mu\overline\psi_2(x)$$

$$+ \text{ total derivatives}$$

$$= -i\left(\psi_2(x),\ \overline\psi_1(x)\right)\begin{pmatrix} 0 & \sigma^\mu D_\mu \\ \overline\sigma^\mu D_\mu & 0 \end{pmatrix}\begin{pmatrix}\psi_1(x) \\ \overline\psi_2(x)\end{pmatrix}$$

$$+ \text{ total derivatives}$$

$$= -i\overline\Psi(x)\gamma_W^\mu D_\mu\Psi(x) + \text{ total derivatives}$$

$$= -i\overline\Psi(x)\slashed{D}\Psi(x) + \text{ total derivatives.} \qquad (10.36)$$

Here we made use of Eqs. (1.159), (10.34), and (10.35). The gauge covariant derivative is given by

$$D_\mu = \partial_\mu - \frac{1}{2} i e V_\mu. \tag{10.37}$$

Hence, in the action (10.33) we can rewrite the terms quadratic in the spinor field ψ as

$$
\begin{aligned}
i\big(D_\mu^*\overline\psi_1(x)\big)\overline\sigma^\mu\psi_1(x) &+ i\big(D_\mu\overline\psi_2(x)\big)\overline\sigma^\mu\psi_2(x) \\
&+ m\big[\psi_1(x)\psi_2(x) + \overline\psi_1(x)\overline\psi_2(x)\big] \\
&= -i\overline\Psi(x)\gamma_W^\mu D_\mu\Psi(x) + m\,\big(\psi_2(x),\ \ \overline\psi_1(x)\big)\begin{pmatrix}\psi_1(x)\\ \overline\psi_2(x)\end{pmatrix} \\
&= -\Big\{i\overline\Psi(x)\gamma_W^\mu D_\mu\Psi(x) - m\overline\Psi(x)\Psi(x)\Big\} \\
&= -\overline\Psi(x)\Big(i\slashed{D} - m\Big)\Psi(x). \tag{10.38}
\end{aligned}
$$

This part of the action (10.33) is the Lagrangian density for a massive electron field (of mass m and charge $q = -e$) which is minimally coupled to a bosonic $U(1)$ gauge field $V_\mu(x)$.[4]

10.3   The Fayet–Iliopoulos Model

In this section we investigate an alternative mechanism of spontaneous breaking of supersymmetry which is called the *Fayet–Iliopoulos mechanism of spontaneous breaking of supersymmetry*.[5] This mechanism is an alternative to the O'Raifeartaigh model discussed in Chapter 9 as a mechanism to obtain spontaneous breaking of supersymmetry. In the Fayet–Iliopoulos model one considers the breaking of supersymmetry in gauge theories with Abelian gauge groups. The basic ingredients of this theory are the Lagrangian of supersymmetric quantum electrodynamics, *i.e.* the Lagrangian of the action (10.26), and the $(\theta\theta)(\overline\theta\,\overline\theta)$-component of a vector superfield. Thus, the starting point of our considerations is the Lagrangian density

$$
\begin{aligned}
\mathcal{L}_{\text{FI}} = W^A W_A\big|_{\theta\theta} &+ \overline W_{\dot A}\overline W^{\dot A}\big|_{\overline\theta\,\overline\theta} \\
&+ \Phi_1^\dagger e^{eV}\Phi_1\big|_{\theta\theta\overline\theta\,\overline\theta} + \Phi_2^\dagger e^{-eV}\Phi_2\big|_{\theta\theta\overline\theta\,\overline\theta} \\
&- m\Big(\Phi_1\Phi_2\big|_{\theta\theta} + \Phi_1^\dagger\Phi_2^\dagger\big|_{\overline\theta\,\overline\theta}\Big) + 2kV\big|_{\theta\theta\overline\theta\,\overline\theta}. \tag{10.39}
\end{aligned}
$$

---

[4]For a detailed discussion see *e.g.* T.P. Cheng and L.-F. Li [21], Chap. 5.3.
[5]See P. Fayet and J. Iliopoulos [34] or E. Witten [130].

Expanding this superspace Lagrangian in terms of its component fields, using Eqs. (7.29) and (10.32), we obtain:

$$
\begin{aligned}
\mathcal{L}_{\mathrm{FI}} = {}& -F_{\mu\nu}(x)F^{\mu\nu}(x) - 4i\lambda(x)\sigma^\mu\partial_\mu\overline{\lambda}(x) + iD_\mu^*\overline{\psi}_1(x)\overline{\sigma}^\mu\psi_1(x) \\
& + \left|D_\mu A_1(x)\right|^2 + iD_\mu\overline{\psi}_2(x)\overline{\sigma}^\mu\psi_2(x) + \left|D_\mu^*A_2(x)\right|^2 \\
& - \frac{e}{2^{1/2}}\Big[\overline{\lambda}(x)\big(\overline{\psi}_1(x)A_1(x) - \overline{\psi}_2(x)A_2(x)\big) \\
& + \lambda(x)\big(\psi_1(x)A_1^*(x) - \psi_2(x)A_2^*(x)\big)\Big] \\
& + eD(x)\Big(\left|A_1(x)\right|^2 - \left|A_2(x)\right|^2\Big) - m\Big[A_1(x)F_2(x) + A_1^*(x)F_2^*(x) \\
& + A_2(x)F_1(x) + A_2^*(x)F_1^*(x)\Big] \\
& + m\big(\psi_1(x)\psi_2(x) + \overline{\psi}_1(x)\overline{\psi}_2(x)\big) \\
& + 2kD(x) + \left|F_1(x)\right|^2 + \left|F_2(x)\right|^2 + 8D^2(x).
\end{aligned} \tag{10.40}
$$

For later convenience we write out explicitly the expressions of the covariant derivatives of the scalar fields $A_1(x)$ and $A_2(x)$ in Eq. (10.40) (all expressions are complete except for total derivative terms which are deleted):

$$
\begin{aligned}
\left|D_\mu A_1(x)\right|^2 = {}& -A_1^*(x)\square A_1(x) + \frac{1}{4}e^2\left|A_1(x)\right|^2 V_\mu(x)V^\mu(x) \\
& + \frac{1}{2}ieV^\mu(x)\big[(\partial_\mu A_1(x))A_1^*(x) - A_1(x)(\partial_\mu A_1^*(x))\big],
\end{aligned}
$$

where we made use of Eqs. (10.10) and (10.21). In the same way we find the explicit expression for the gauge covariant derivative of the A_2-field:

$$
\begin{aligned}
\left|D_\mu^*A_2(x)\right|^2 = {}& -A_2^*(x)\square A_2(x) + \frac{1}{4}e^2\left|A_2(x)\right|^2 V_\mu(x)V^\mu(x) \\
& - \frac{1}{2}ieV^\mu(x)\big[(\partial_\mu A_2(x))A_2^*(x) - A_2(x)(\partial_\mu A_2^*(x))\big].
\end{aligned}
$$

Then the Lagrangian density of the Fayet–Iliopoulos model becomes

$$
\begin{aligned}
\mathcal{L}_{\mathrm{FI}} = {}& -F_{\mu\nu}(x)F^{\mu\nu}(x) - 4i\lambda(x)\sigma^\mu\partial_\mu\overline{\lambda}(x) \\
& + i\big(D_\mu^*\overline{\psi}_1(x)\big)\overline{\sigma}^\mu\psi_1(x) + i\big(D_\mu\overline{\psi}_2(x)\big)\overline{\sigma}^\mu\psi_2(x) \\
& - A_1^*(x)\square A_1(x) + \frac{1}{4}e^2\left|A_1(x)\right|^2 V_\mu(x)V^\mu(x) \\
& + \frac{1}{2}ieV^\mu(x)\big[(\partial_\mu A_1(x))A_1^*(x) - A_1(x)(\partial_\mu A_1^*(x))\big]
\end{aligned}
$$

$$- A_2^*(x)\Box A_2(x) + \frac{1}{4}e^2 |A_2(x)|^2 V_\mu(x)V^\mu(x)$$

$$- \frac{1}{2}ieV^\mu(x)\big[(\partial_\mu A_2(x))A_2^*(x) - A_2(x)(\partial_\mu A_2^*(x))\big]$$

$$- \frac{e}{2^{1/2}}\big[\overline{\lambda}(x)(\overline{\psi}_1(x)A_1(x) - \overline{\psi}_2(x)A_2(x))$$

$$+ \lambda(x)(\psi_1(x)A_1^*(x) - \psi_2(x)A_2^*(x))\big]$$

$$+ m\big[\psi_1(x)\psi_2(x) + \overline{\psi}_1(x)\overline{\psi}_2(x)\big]$$

$$+ 8D^2(x) + |F_1(x)|^2 + |F_2(x)|^2$$

$$- m\Big[A_1(x)F_2(x) + A_1^*(x)F_2^*(x)$$

$$+ A_2(x)F_1(x) + A_2^*(x)F_1^*(x)\Big]$$

$$+ D(x)\Big(2k + e(|A_1(x)|^2 - |A_2(x)|^2)\Big). \tag{10.41}$$

In the next step we eliminate the three auxiliary fields $F_1(x)$, $F_2(x)$, and $D(x)$ via their equations of motion. Thus

$$\frac{\partial \mathcal{L}_{\mathrm{FI}}}{\partial F_1^*(x)} = F_1(x) - mA_2^*(x) = 0,$$

leading to

$$F_1(x) = mA_2^*(x), \qquad \text{and similarly} \quad F_1^*(x) = mA_2(x). \tag{10.42}$$

Next:

$$\frac{\partial \mathcal{L}_{\mathrm{FI}}}{\partial F_2^*(x)} = F_2(x) - mA_1^*(x) = 0,$$

so that

$$F_2(x) = mA_1^*(x), \qquad \text{and} \quad F_2^*(x) = mA_1(x). \tag{10.43}$$

Finally

$$\frac{\partial \mathcal{L}_{\mathrm{FI}}}{\partial D(x)} = 16D(x) + 2k + e(|A_1(x)|^2 - |A_2(x)|^2) = 0,$$

such that:

$$D(x) = -\frac{1}{16}\big[2k + e(|A_1(x)|^2 - |A_2(x)|^2)\big]. \tag{10.44}$$

Then we have (with Eq. (10.44)):

$$8D^2(x) + D(x)\big[2k + e(|A_1(x)|^2 - |A_2(x)|^2)\big] = 8D^2(x) - 16D^2(x) = -8D^2(x),$$

and using Eqs. (10.42) and (10.43):

$$|F_1(x)|^2 + |F_2(x)|^2$$
$$- m\Big\{A_1(x)F_2(x) + A_1^*(x)F_2^*(x) + A_2(x)F_1(x) + A_2^*(x)F_1^*(x)\Big\}$$
$$= |F_1(x)|^2 + |F_2(x)|^2 - 2|F_1(x)|^2 - 2|F_2(x)|^2$$
$$= -|F_1(x)|^2 - |F_2(x)|^2$$

With these substitutions the Lagrangian density (10.41) takes the form:

$$\mathcal{L}_{\mathrm{FI}} = -F_{\mu\nu}(x)F^{\mu\nu}(x) - 4i\lambda(x)\sigma^\mu\partial_\mu\overline{\lambda}(x)$$
$$+ i\big(D_\mu^*\overline{\psi}_1(x)\big)\overline{\sigma}^\mu\psi_1(x) + i\big(D_\mu\overline{\psi}_2(x)\big)\overline{\sigma}^\mu\psi_2(x)$$
$$- A_1^*(x)\Box A_1(x) + \frac{1}{4}e^2|A_1(x)|^2 V_\mu(x)V^\mu(x)$$
$$+ \frac{1}{2}ieV^\mu(x)\big[(\partial_\mu A_1(x))A_1^*(x) - A_1(x)(\partial_\mu A_1^*(x))\big]$$
$$- A_2^*(x)\Box A_2(x) + \frac{1}{4}e^2|A_2(x)|^2 V_\mu(x)V^\mu(x)$$
$$- \frac{1}{2}ieV^\mu(x)\big[(\partial_\mu A_2(x))A_2^*(x) - A_2(x)(\partial_\mu A_2^*(x))\big]$$
$$- \frac{e}{2^{1/2}}\Big[\overline{\lambda}(x)\big(\overline{\psi}_1(x)A_1(x) - \overline{\psi}_2(x)A_2(x)\big)$$
$$+ \lambda(x)\big(\psi_1(x)A_1^*(x) - \psi_2(x)A_2^*(x)\big)\Big]$$
$$+ m\big(\psi_1(x)\psi_2(x) + \overline{\psi}_1(x)\overline{\psi}_2(x)\big) - \mathcal{V}(A_1, A_2), \qquad (10.45)$$

where

$$\mathcal{V}(A_1, A_2) = 8D^2(x) + |F_1(x)|^2 + |F_2(x)|^2 \qquad (10.46)$$

is the scalar potential of the Fayet–Iliopoulos model, where the three auxiliary fields $D(x)$, $F_1(x)$, and $F_2(x)$ are solutions of Eqs. (10.42), (10.43), and (10.44). Using these Euler–Lagrange equations, the explicit form of the scalar potential is

$$\mathcal{V}(A_1, A_2) = \frac{1}{32}\big[2k + e\big(|A_1(x)|^2 - |A_2(x)|^2\big)\big]^2 + m^2|A_1(x)|^2 + m^2|A_2(x)|^2.$$

Hence

$$\mathcal{V}(A_1, A_2) = \frac{1}{8}k^2 + \Big(m^2 + \frac{ek}{8}\Big)|A_1(x)|^2 + \Big(m^2 - \frac{ek}{8}|A_2(x)|^2\Big)$$
$$+ \frac{1}{32}e^2\Big[|A_1(x)|^2 - |A_2(x)|^2\Big]^2. \qquad (10.47)$$

We observe that there is no solution to the Euler–Lagrange equations of the auxiliary fields, *i.e.* Eqs. (10.42) to (10.44), such that the scalar potential given by Eq. (10.46) or Eq. (10.47) vanishes. Thus, according to the general discussions of Sec. 9.3 supersymmetry must be broken spontaneously.

We now distinguish two cases depending on the choice of the parameters and discuss each case separately.

(i) The case $4m^2 > ek/2$:

The absolute minimum of the scalar potential is determined by the vanishing of the partial derivatives

$$\frac{\partial \mathcal{V}(A_1, A_2)}{\partial A_1^*(x)} = \left[\left(m^2 + \frac{ek}{8}\right) + \frac{e^2}{16}\left(|A_1(x)|^2 - |A_2(x)|^2\right)\right] A_1(x),$$
(10.48a)

$$\frac{\partial \mathcal{V}(A_1, A_2)}{\partial A_2^*(x)} = \left[\left(m^2 - \frac{ek}{8}\right) - \frac{e^2}{16}\left(|A_1(x)|^2 - |A_2(x)|^2\right)\right] A_2(x),$$
(10.48b)

i.e. by the equations

$$\frac{\partial \mathcal{V}(A_1, A_2)}{\partial A_1^*(x)} = 0, \quad \frac{\partial \mathcal{V}(A_1, A_2)}{\partial A_2^*(x)} = 0.$$
(10.49)

Considering solutions with $A_1^{\min}(x) = 0$ — the value of the scalar field $A_1(x)$ at the minimum — we see that $A_2^{\min}(x)$, determined by Eq. (10.48b), is given by

$$\left[\left(m^2 - \frac{ek}{8}\right) + \frac{e^2}{16}\left|A_2^{\min}(x)\right|^2\right] A_2^{\min}(x) = 0.$$

Thus, if $A_2^{\min}(x) \neq 0$, we must have $m^2 < ek/8$, and in the domain $m^2 > ek/8$ we can only have the solution $A_2^{\min}(x) = 0$. Furthermore, at the absolute minimum $A_1^{\min}(x) = 0$ and $A_2^{\min}(x) = 0$, the scalar potential $\mathcal{V}(A_1, A_2)$ has the value

$$\mathcal{V}(0,0) = \frac{k^2}{8} > 0,$$
(10.50)

which indicates that supersymmetry is spontaneously broken. Hence we conclude, in the case $m^2 > ek/8$, both scalar fields $A_1(x)$ and $A_2(x)$ have real masses and vanishing vacuum expectation values, indicating that the internal $U(1)$ phase symmetry is unbroken. Within this range of parameters the Fayet–Iliopoulos model with Lagrangian given by

Eq. (10.45), *i.e* $4m^2 > ek/s$, describes two complex scalar fields $A_1(x)$ and $A_2(x)$, one of mass m_1 with

$$m_1^2 = m^2 + \frac{ek}{8},$$

and the other of mass m_2 with

$$m_2^2 = m^2 - \frac{ek}{8},$$

as well as three spinor fields $\psi_1(x)$, $\psi_2(x)$, and $\lambda(x)$, and one real vector field $V_\mu(x)$. The masses of the spinor fields and the vector field remain unchanged by the breaking of supersymmetry. In particular — as was demonstrated in Sec. 10.2 — the two two-component spinors $\psi_1(x)$ and $\psi_2(x)$ combine to a massive Dirac spinor field with mass m, whereas the spinor field $\lambda(x)$ and the gauge vector field $V_\mu(x)$ remain massless. We observe again that, just as in case of the O'Raifeartaigh model (see remark (ii) at the end of Chapter 9) the bosonic mass spectrum gives

$$m_1^2 + m_2^2 = 2m^2, \tag{10.51}$$

and it is only the bosonic mass spectrum which is affected by the breaking of supersymmetry.

The vector field $V_\mu(x)$ plays the role of a gauge field for the unbroken internal $U(1)$ phase symmetry. The spinor field $\lambda(x)$ is the Goldstone fermion which arises from spontaneously broken supersymmetry. From the transformation law of the field $\lambda(x)$ under supersymmetry transformations (see Eq. (6.94) with the identifications $\psi_A(x) \to \lambda_A(x)$ and $d(x) \to D(x)$), *i.e.*

$$\delta_s \lambda_A(x) = 2\alpha_A D(x) - \frac{i}{2}\alpha_A \partial^\mu V_\mu(x) + \cdots,$$

we see that $\lambda(x)$ transforms inhomogeneously as soon as the auxiliary field $D(x)$ acquires a nonvanishing vacuum expectation value

$$D(x) = -k/8,$$

which follows from Eq. (10.44) with $A_1(x) = A_2(x) = 0$, *i.e.*

$$\delta_s \lambda_A(x) = -\alpha_A \frac{k}{4} + \cdots . \tag{10.52}$$

This is a characteristic feature of the Goldstone fermion. In general we can say: *Nonzero expectation values of auxiliary fields induce the breakdown of supersymmetry.*

(ii) The case $4m^2 < ek/2$.

If we choose the parameters m, e, and k to satisfy $4m^2 < ek/2$, then $A_1^{\min}(x) = A_2^{\min}(x) = 0$ does not represent the location of the absolute minimum of the potential (10.47). Instead, as we saw above, choosing $A_1^{\min}(x) = 0$, the value of $A_2^{\min}(x)$ is determined by the equation

$$4m^2 - \frac{ek}{2} + \frac{e^2}{4}\left|A_2^{\min}(x)\right|^2 = 0,$$

or

$$\vartheta^2 := \left|A_2^{\min}(x)\right|^2 = \frac{4}{e^2}\left(\frac{ek}{2} - 4m^2\right) = \frac{16}{e^2}\left(\frac{ek}{8} - m^2\right) > 0. \qquad (10.53)$$

Expanding the potential (10.47) around the minimum at $A_1^{\min}(x) = 0$ and $A_2^{\min}(x) = \vartheta$ breaks the internal $U(1)$ gauge symmetry spontaneously. We consider small oscillations around the absolute minimum and set

$$\begin{aligned} A_1(x) &= A(x), \\ A_2(x) &= \vartheta + \widetilde{A}(x), \end{aligned} \qquad (10.54)$$

where ϑ is real and constant. To evaluate the particle spectrum, that is described by the Fayet–Iliopoulos model with these parameters, we insert Eqs. (10.54) into the Lagrangian (10.45). We get

$$\begin{aligned} \mathcal{L}_{\mathrm{FI}} = &-F_{\mu\nu}(x)F^{\mu\nu}(x) - 4i\lambda(x)\sigma^\mu\partial_\mu\overline{\lambda}(x) \\ &+ i\big(D_\mu^*\overline{\psi}_1(x)\big)\overline{\sigma}^\mu\psi_1(x) + i\big(D_\mu\overline{\psi}_2(x)\big)\overline{\sigma}^\mu\psi_2(x) \\ &- A^*(x)\Box A(x) + \frac{e^2}{4}|A(x)|^2 V_\mu(x)V^\mu(x) \\ &+ \frac{ie}{2}V^\mu(x)\big[\big(\partial_\mu A(x)\big)A^*(x) - A(x)\big(\partial_\mu A^*(x)\big)\big] \\ &- \big(\vartheta + \widetilde{A}^*(x)\big)\Box\big(\vartheta + \widetilde{A}(x)\big) \\ &+ \frac{e^2}{4}\big|\vartheta + \widetilde{A}(x)\big|^2 V_\mu(x)V^\mu(x) \\ &- \frac{ie}{2}V^\mu(x)\Big\{\big[\partial_\mu\big(\vartheta + \widetilde{A}(x)\big)\big]\big(\vartheta + \widetilde{A}^*(x)\big) \\ &- \big(\vartheta + \widetilde{A}(x)\big)\partial_\mu\big(\vartheta + \widetilde{A}^*(x)\big)\Big\} \\ &- \frac{e}{2^{1/2}}\Big[\overline{\lambda}(x)\big(\overline{\psi}_1(x)A(x) - \overline{\psi}_2(x)\big(\vartheta + \widetilde{A}(x)\big)\big) \end{aligned}$$

$$+ \lambda(x)\big(\psi_1(x)A^*(x) - \psi_2(x)(\vartheta + \widetilde{A}^*(x))\big)\Big]$$
$$+ m\Big[\psi_1(x)\psi_2(x) + \overline{\psi}_1(x)\overline{\psi}_2(x)\Big]$$
$$- \frac{1}{8}k^2 - \Big(m^2 + \frac{ek}{8}\Big)|A(x)|^2$$
$$- \Big(m^2 - \frac{ek}{8}\Big)\big|\vartheta + \widetilde{A}(x)\big|^2$$
$$- \frac{e^2}{32}\Big\{|A(x)|^2 - |\vartheta + \widetilde{A}(x)|^2\Big\}^2.$$

Hence

$$\begin{aligned}
\mathcal{L}_{\text{FI}} = {}& -F_{\mu\nu}(x)F^{\mu\nu}(x) - 4i\lambda(x)\sigma^\mu\partial_\mu\overline{\lambda}(x) \\
& + i\big(D^*_\mu\overline{\psi}_1(x)\big)\overline{\sigma}^\mu\psi_1(x) + i\big(D_\mu\overline{\psi}_2(x)\big)\overline{\sigma}^\mu\psi_2(x) \\
& - A^*(x)\Box A(x) - \widetilde{A}^*(x)\Box\widetilde{A}(x) \\
& + \frac{ie}{2}V^\mu(x)\big[\big(\partial_\mu A(x)\big)A^*(x) - A(x)\big(\partial_\mu A^*(x)\big)\big] \\
& + \frac{e^2}{4}|A(x)|^2 V_\mu(x)V^\mu(x) - \vartheta\Box\widetilde{A}(x) \\
& - \frac{ie}{2}V^\mu(x)\Big\{\big(\partial_\mu\widetilde{A}(x)\big)\widetilde{A}^*(x) - \widetilde{A}(x)\big(\partial_\mu\widetilde{A}^*(x)\big) \\
& + \vartheta\,\partial_\mu\big(\widetilde{A}(x) - \widetilde{A}^*(x)\big)\Big\} \\
& + \frac{e^2}{4}|\widetilde{A}(x)|^2 V_\mu(x)V^\mu(x) \\
& + \frac{e^2}{4}\vartheta\big[\widetilde{A}(x) + \widetilde{A}^*(x)\big]V_\mu(x)V^\mu(x) \\
& - \frac{e}{2^{1/2}}\Big[\overline{\lambda}(x)\big(\overline{\psi}_1(x)A(x) - \overline{\psi}_2(x)\widetilde{A}(x)\big) \\
& + \lambda(x)\big(\psi_1(x)A^*(x) - \psi_2(x)\widetilde{A}^*(x)\big)\Big] \\
& + \frac{e}{2^{1/2}}\,\vartheta\,\overline{\lambda}(x)\overline{\psi}_2(x) + \frac{e}{2^{1/2}}\,\vartheta\,\lambda(x)\psi_2(x) \\
& + m\Big[\psi_1(x)\psi_2(x) + \overline{\psi}_1(x)\overline{\psi}_2(x)\Big] \\
& - \frac{1}{8}k^2 - \Big(m^2 + \frac{ek}{8}\Big)|A(x)|^2 - \Big(m^2 - \frac{ek}{8}\Big)|\widetilde{A}(x)|^2 \\
& - \Big(m^2 - \frac{ek}{8}\Big)\vartheta\,\big(\widetilde{A}(x) + \widetilde{A}^*(x)\big) \\
& - \Big(m^2 - \frac{ek}{8}\Big)\vartheta^2 + \frac{1}{4}e^2\vartheta^2 V_\mu(x)V^\mu(x)
\end{aligned}$$

$$-\frac{e^2}{32}\left\{|A(x)|^2 - |\tilde{A}(x)|^2\right\}^2$$
$$-\frac{e^2}{32}\vartheta^2\left(\tilde{A}(x) + \tilde{A}^*(x)\right)^2 - \frac{e^2}{32}\vartheta^4$$
$$+\frac{e^2}{16}\vartheta\left\{|A(x)|^2 - |\tilde{A}(x)|^2\right\}\left(\tilde{A}(x) + \tilde{A}^*(x)\right)$$
$$+\frac{e^2}{16}\vartheta^2\left\{|A(x)|^2 - |\tilde{A}(x)|^2\right\}$$
$$-\frac{e^2}{16}\vartheta^3\left\{\tilde{A}(x) + \tilde{A}^*(x)\right\}. \tag{10.55}$$

We can simplify the sums of several terms. Thus

$$\frac{1}{8}k^2+\left(m^2 - \frac{ek}{8}\right)\vartheta^2 + \frac{e^2}{32}\vartheta^4$$
$$\overset{(10.53)}{=}\frac{1}{8}k^2 + \frac{4}{e^2}\left(m^2 - \frac{ek}{8}\right)\left(\frac{ek}{2} - 4m^2\right) + \frac{1}{2e^2}\left(\frac{ek}{2} - 4m^2\right)^2$$
$$= \frac{2m^2}{e^2}\left(ek - 4m^2\right). \tag{10.56}$$

Furthermore

$$-\left(m^2 + \frac{ek}{8}\right)|A(x)|^2 + \frac{e^2}{16}\vartheta^2\,|A(x)|^2 = -2m^2|A(x)|^2, \tag{10.57}$$

where we made use of Eq. (10.53). In addition we have

$$-\left(m^2 - \frac{ek}{8}\right)|\tilde{A}(x)|^2 - \frac{e^2}{16}\vartheta^2\,|\tilde{A}(x)|^2 = 0, \tag{10.58}$$

again using Eq. (10.53). Hence, with Eqs. (10.56) to (10.58), the Fayet–Iliopoulos Lagrangian (10.55) becomes

$$\begin{aligned}
\mathcal{L}_{\mathrm{FI}} = &-F_{\mu\nu}(x)F^{\mu\nu}(x) - 4i\lambda(x)\sigma^\mu\partial_\mu\overline{\lambda}(x)\\
&+ i\big(D_\mu^*\overline{\psi}_1(x)\big)\overline{\sigma}^\mu\psi_1(x) + i\big(D_\mu\overline{\psi}_2(x)\big)\overline{\sigma}^\mu\psi_2(x)\\
&- \vartheta\,\Box\tilde{A}(x) - \tilde{A}^*(x)\Box\tilde{A}(x)\\
&- \frac{ie}{2}V^\mu(x)\big\{\big(\partial_\mu\tilde{A}(x)\big)\tilde{A}^*(x) - \tilde{A}(x)\big(\partial_\mu\tilde{A}^*(x)\big)\\
&+ \vartheta\,\partial_\mu\big(\tilde{A}(x) - \tilde{A}^*(x)\big)\big\}
\end{aligned}$$

$$+ \frac{e^2}{4} |\tilde{A}(x)|^2 V_\mu(x) V^\mu(x)$$

$$+ \frac{e^2}{4} \vartheta \left[\tilde{A}(x) + \tilde{A}^*(x) \right] V_\mu(x) V^\mu(x)$$

$$- \frac{e}{2^{1/2}} \Big[\overline{\lambda}(x) \big(\overline{\psi}_1(x) A(x) - \overline{\psi}_2(x) \tilde{A}(x) \big)$$

$$+ \lambda(x) \big(\psi_1(x) A^*(x) - \psi_2(x) \tilde{A}^*(x) \big) \Big]$$

$$+ \frac{e}{2^{1/2}} \vartheta \, \overline{\lambda}(x) \overline{\psi}_2(x) + \frac{e}{2^{1/2}} \vartheta \, \lambda(x) \psi_2(x)$$

$$+ m \Big[\psi_1(x) \psi_2(x) + \overline{\psi}_1(x) \overline{\psi}_2(x) \Big]$$

$$- \tilde{\mathcal{V}}, \tag{10.59}$$

where the potential $\tilde{\mathcal{V}}$ is given by

$$\tilde{\mathcal{V}} = \frac{2m^2}{e^2} (ek - 4m^2) + 2m^2 |A(x)|^2 - \frac{e^2}{4} \vartheta^2 \, V_\mu(x) V^\mu(x)$$

$$+ \frac{e^2}{32} \vartheta^2 \left[\tilde{A}(x) + \tilde{A}^*(x) \right]^2 - \frac{e^2}{16} \vartheta^3 \left[\tilde{A}(x) + \tilde{A}^*(x) \right]$$

$$+ \frac{e^2}{32} \big\{ |A(x)|^2 - |\tilde{A}(x)|^2 \big\}^2 + \frac{e^2}{16} \vartheta^3 \big\{ \tilde{A}(x) + \tilde{A}^*(x) \big\}$$

$$- \frac{e^2}{16} \vartheta \big\{ |A(x)|^2 - |\tilde{A}(x)|^2 \big\} \big(\tilde{A}(x) + \tilde{A}^*(x) \big)$$

$$+ \left(m^2 - \frac{ek}{8} \right) \vartheta \, \big(\tilde{A}(x) + \tilde{A}^*(x) \big). \tag{10.60}$$

We now discuss the implications of these results:

(i) Supersymmetry is spontaneously broken as in the case $4m^2 > ek/2$, since the constant $2m^2(ek - 4m^2)/e^2$, appearing in the scalar potential (10.60) is strictly positive for $4m^2 < ek/2$.

(ii) The internal $U(1)$ gauge symmetry is spontaneously broken because the complex scalar field acquires a nonvanishing vacuum expectation value (*cf.* Eq. (10.53)).

(iii) The usual Higgs–Kibble mechanism[6] takes place, *i.e.* the vector field $V_\mu(x)$ acquires a mass, as can be seen from the term

$$\frac{e^2}{4} \vartheta^2 \, V_\mu(x) V^\mu(x)$$

---

[6]See *e.g.* T.-P. Cheng and L.-F. Li [21], Chap. 8.3.

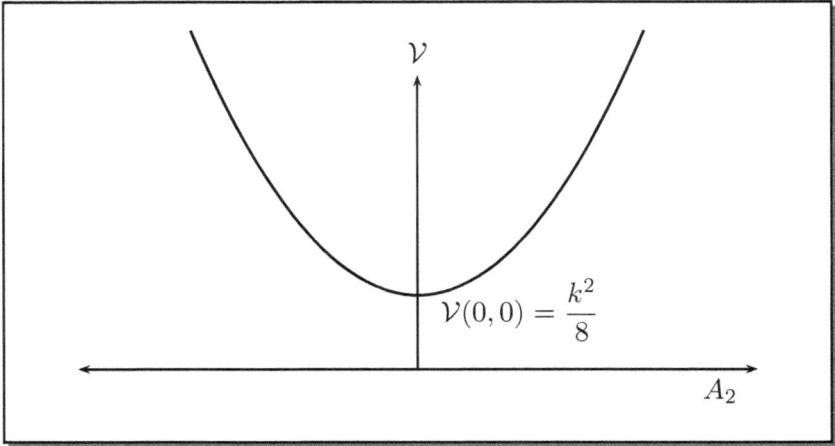

Figure 10.1: Supersymmetry breaking in the Fayet–Iliopoulos model for the range of parameters $8m^2 > ek$.

in the potential (10.60), by "eating up" the Goldstone boson field

$$\widetilde{A}(x) - \widetilde{A}^*(x)$$

which denotes the imaginary part of the complex scalar field $A_2(x)$. The real scalar field, described by

$$\widetilde{A}(x) + \widetilde{A}^*(x) = 2 \operatorname{Re} \widetilde{A}(x)$$

has a mass term $(e^2/8\vartheta^2)^{1/2}$.

(iv) In addition the theory contains a complex scalar field $A(x)$ with mass squared $8m^2$.

(v) We can completely eliminate the field

$$\widetilde{A}(x) - \widetilde{A}^*(x)$$

from the Lagrangian (10.59) by fixing the gauge (unitary gauge) and transform the field $\widetilde{A}(x) - \widetilde{A}^*(x)$ into a longitudinal degree of freedom of the vector field $V_\mu(x)$ (Higgs mechanism).[7]

(vi) The breaking of the internal $U(1)$ gauge symmetry also modifies the spinor mass terms, *i.e.*

$$m\Big[\psi_1(x)\psi_2(x) + \overline{\psi}_1(x)\overline{\psi}_2(x)\Big] + \frac{e}{2^{1/2}}\vartheta \Big[\overline{\lambda}(x)\overline{\psi}_2(x) + \lambda(x)\psi_2(x)\Big].$$

---

[7] See again T.-P. Cheng and L.-F. Li [21], Chap. 8.3, or C. Itzykson and J.-B. Zuber [60], Chap. 12.5.3.

Diagonalization of the mass terms leads to the following qualitative picture. The model describes one massless spinor field, which is a linear combination of $\lambda(x)$ and $\psi_1(x)$, and two massive spinor fields, i.e. $\psi(x) = \psi_2(x)$ and $\check{\psi}(x)$, the latter being a linear combination of $\psi_1(x)$ and $\lambda(x)$.

(vii) From the above discussion we see that nonvanishing vacuum expectation values of auxiliary fields induce supersymmetry breaking, whereas nonzero vacuum expectation values of dynamical scalar fields lead to the breakdown of gauge symmetry. These features are illustrated in Fig. 10.1 and Fig 10.2.

(viii) There is a fundamental difference between supersymmetry breaking in the O'Raifeartaigh model and that in the Fayet–Iliopoulos model. In the case of the former, supersymmetry breaking is the consequence of a nonvanishing vacuum expectation value of an auxiliary field, whereas in the case of the latter it is the consequence of the nonvanishing vacuum expectation value of the component field $D(x)$ of a vector superfield.

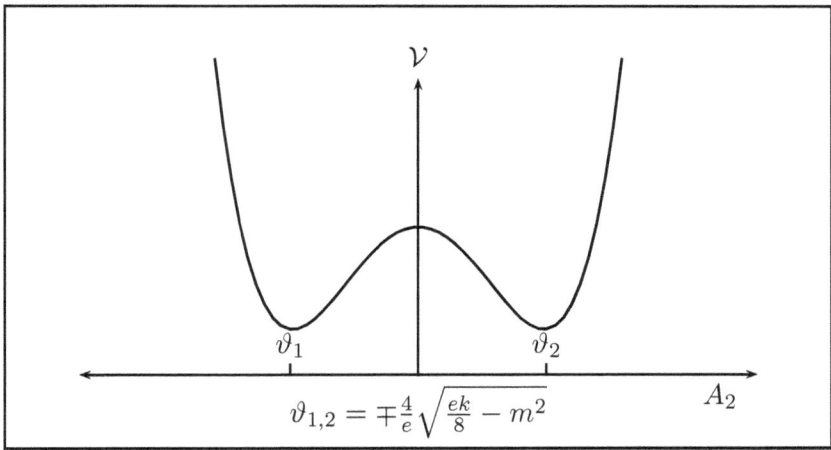

Figure 10.2: Breaking of supersymmetry and gauge symmetry in the Fayet–Iliopoulos model for parameters satisfying $8m^2 < ek$.

10.4   Supersymmetric Non-Abelian Gauge Theory

The subject of supersymmetric non-Abelian gauge theory[8] is too extensive
to permit a detailed treatment in the present context. We restrict ourselves
therefore here to the introduction of global and local non-Abelian gauge
transformations. The construction of appropriate gauge covariant derivatives
and Lagrangians then proceeds along lines similar to those outlined above in
the Abelian case.

It is straightforward to generalize the transformation laws (10.4) and
(10.10) to non-Abelian groups. Let G be the gauge group with Lie algebra $\widetilde{g}$.
For simplicity we shall assume $\widetilde{g}$ to be semi-simple.[9] The basis elements of
the Lie algebra are the Hermitian operators T_a, $a = 1, 2, \ldots, N$, N being the
dimension of the Lie algebra $\widetilde{g}$. The superfields Φ and $\Phi^\dagger$ transform tenso-
rially with respect to $U(g)$, where $U(g)$ is any nontrivial unitary continuous
representation of the internal symmetry group G. That is, for $g \in G$:

$$\Phi' = U(g)\Phi, \qquad \Phi'^\dagger = \Phi^\dagger U^\dagger(g), \tag{10.61}$$

or in components

$$\Phi'_i = U_{ij}(g)\Phi_j, \qquad \Phi'^\dagger_i = \Phi^\dagger_j U^\dagger_{ji}(g). \tag{10.62}$$

Here, the indices i, j run from 1 to the dimension of the representation. With
the help of the exponential map we represent any group element $g \in G$ in
the form

$$g = \exp\{-i\Lambda^a T_a\},$$

where T_a are the generators — *i.e.* the basis elements of the Lie algebra —
and Λ^a are N parameters. Hence, the exponential map, *i.e.*

$$\exp \; : \; \widetilde{g} \longrightarrow G$$

maps the Lie algebra $\widetilde{g}$ onto the group G. Then the representation matrix
$U(g)$ of Eq. (10.61) can be written as

$$\begin{aligned} U(g) &= U\big(\exp\{-i\Lambda^a T_a\}\big) \\ &\equiv \exp\big[-i\Lambda\big], \end{aligned} \tag{10.63}$$

where Λ is an $N \times N$-square matrix (N is the dimension of the representation)
given by:

$$\big(\Lambda_{ij}\big) = \big(\Lambda^a (T_a)_{ij}\big). \tag{10.64}$$

---

[8]See S. Ferrara and B. Zumino [41], A. Salam and J. Strathdee [100] or S. Weinberg [118],
Chap. 27.

[9]See *e.g.* A.S. Sciarrino and P. Sorba [104] or A.O. Barut and R. Rączka [8], Chap. 3.5.

The matrices $\left((T_a)_{ij}\right)$ are the generators T_a of the Lie algebra $\widetilde{g}$ in the particular representation.

As an example we consider the group $G = SU(2, \mathbb{C})$. If we choose the superfields Φ, $\Phi^\dagger$ to lie in the fundamental representation, the matrices $\left((T_a)_{ij}\right)$ of Eq. (10.64) are given by the three Pauli matrices. If we choose the adjoint representation, then the matrices $\left((T_a)_{ij}\right)$ are given by the structure constants ϵ_{ijk} of the group $SU(2, \mathbb{C})$.

Inserting Eq. (10.63) into Eq. (10.61) we obtain the generalization of the transformation law (10.4) to non-Abelian global phase transformations, *i.e.*

$$\Phi' = e^{-i\Lambda}\Phi, \qquad \Phi'^\dagger = \Phi^\dagger e^{i\Lambda^\dagger}, \tag{10.65a}$$

where (*cf.* Eq. (10.5))

$$\overline{D}_{\dot{A}}\Lambda = 0, \qquad D_A\Lambda^\dagger = 0, \tag{10.65b}$$

so that the chiral superfield remains a chiral superfield under the transformation. Now let X be any element of the Lie algebra $\widetilde{g}$ and $\widetilde{gl}$ the algebra associated with the general linear group $GL(n, \mathbb{C})$. Then we define the *adjoint representation* by the map [104]

$$\begin{aligned} \text{ad } X \; : \; & \widetilde{g} \longrightarrow \widetilde{gl}(\widetilde{g}), \\ & Y \longmapsto (\text{ad } X)(Y) := [X, Y]. \end{aligned} \tag{10.66}$$

Furthermore, the *Killing form B* is a bilinear map defined on the algebra $\widetilde{g}$, *i.e.* (see *e.g.* [104])

$$\begin{aligned} B \; : \; & \widetilde{g} \times \widetilde{g} \longrightarrow \mathbb{R}, \\ & (X, Y) \longmapsto B(X, Y) := \text{Tr}\left[\text{ad } X \text{ ad } Y\right]. \end{aligned} \tag{10.67}$$

To obtain a coordinate representation of the adjoint representation and the Killing form $B(X, Y)$, we calculate

$$\left(\text{ad } T_a\right)\left(T_b\right) \overset{(10.66)}{=} \left[T_a, T_b\right] = it_{ab}{}^c T_c, \tag{10.68}$$

where $t_{ab}{}^c$ are the *structure constants* of the internal symmetry group G. Hence, according to Eq. (10.68) the matrix corresponding to $\left(\text{ad } T_a\right) \in \widetilde{gl}(\widetilde{g})$ is given by the structure constants $\left(it_a\right)_b{}^c$ of the group G. Then the coordinate representation of the Killing form (10.67) becomes:

$$
\begin{aligned}
b_{ab} &= B(T_a, T_b) \\
&\overset{(10.67)}{=} \text{Tr} \left[\text{ad} \, T_a \, \text{ad} \, T_b \right] \\
&\overset{(10.68)}{=} \text{Tr} \left[i(t_a)_c{}^d \, i(t_b)_d{}^c \right] \\
&= -(t_a)_c{}^d (t_b)_d{}^c.
\end{aligned}
\tag{10.69}
$$

Since we assumed G to be semi-simple, the Killing form B is nondegenerate[10] and therefore we can always normalize the generators T_a such that

$$
B(T_a, T_b) = \text{Tr} \left[\text{ad} \, T_a \, \text{ad} \, T_b \right] = k \, \delta_{ab}, \quad k > 0.
\tag{10.70}
$$

In the literature,[11] the normalization is written

$$
\text{Tr} \left[T_a T_b \right] = k \delta_{ab},
\tag{10.71}
$$

where T_a, T_b are the generators in the adjoint representation.

Proposition 10.1:

> With the normalization (10.70) of the generators in the adjoint representation, the structure constants of the gauge group G, *i.e.* t_{abc}, are completely antisymmetric in their indices, since
>
> $$
> t_{bca} = -\frac{i}{k} \, \text{Tr} \left[T_a [T_b, T_c] \right].
> \tag{10.72}
> $$

Proof: Consider

$$
-\frac{i}{k} \, \text{Tr} \left[T_a [T_b, T_c] \right] = -\frac{i}{k} \, \text{Tr} \left[T_a i t_{bc}{}^d T_d \right] = \frac{1}{k} t_{bc}{}^d k \delta_{ad} = t_{bca}.
$$

In terms of the Killing form B, given by Eq. (10.67), this results reads:

$$
B(T_a, [T_b, T_c]) = ikt_{bca},
$$

and from the property $\text{Tr} \left[AB \right] = \text{Tr} \left[BA \right]$ one derives that t_{abc} is completely antisymmetric.

The Lagrangian of the supersymmetric action (10.13) is invariant under local non-Abelian gauge transformations provided we extend the transformation

---

[10]See *e.g.* T. Bröcker and T. tom Dieck [19], Proposition (5.13).
[11]See J. Wess and J. Bagger [123], Chap. VII, Eq. (7.13).

law of the vector superfield (10.11) in the following, representation independent way *i.e.*

$$e^{V'} = e^{-i\Lambda^\dagger} e^V e^{i\Lambda}, \quad \text{where } V_{ij} = V^a (T_a)_{ij}. \tag{10.73}$$

Proposition 10.2:

The expression

$$\text{Tr} \left[\Phi^\dagger e^V \Phi \right]$$

is invariant under local non-Abelian gauge transformations.

Proof: Using the transformations (10.65a) of the superfields and that of the vector superfield, given by Eq. (10.73), we find:

$$\text{Tr} \left[\Phi'^\dagger e^{V'} \Phi' \right] = \text{Tr} \left[\Phi^\dagger e^{i\Lambda^\dagger} e^{-i\Lambda^\dagger} e^V e^{i\Lambda} e^{-i\Lambda} \Phi \right] = \text{Tr} \left[\Phi^\dagger e^V \Phi \right]. \tag{10.74}$$

This completes the proof of Proposition 10.2.

Proposition 10.3:

The transformation of the vector superfield under non-Abelian gauge transformations, given by Eq. (10.73), is independent of the chosen representation.

Proof: In order to show that the transformation property (10.73) is independent of a particular representation, we use the Baker–Campbell–Hausdorff formula (*cf.* Eq. (6.10)) to evaluate the product of exponentials with matrix arguments. Thus

$$
\begin{aligned}
e^{-i\Lambda^\dagger} e^V e^{i\Lambda} &= \exp\left[-i\Lambda^\dagger\right] \exp\left\{ V + i\Lambda + \frac{i}{2}[V, \Lambda] \right. \\
&\quad \left. + \frac{1}{12}[\Lambda, [\Lambda, V]] + \frac{i}{12}[V, [V, \Lambda]] + \cdots \right\} \\
&= \exp\left\{ V + i\Lambda - i\Lambda^\dagger + \frac{i}{2}[V, \Lambda] + \frac{1}{12}[\Lambda, [\Lambda, V]] \right. \\
&\quad + \frac{i}{12}[V, [V, \Lambda]] - \frac{i}{2}\left[\Lambda^\dagger, V + i\Lambda + \frac{i}{2}[V, \Lambda]\right. \\
&\quad \left. + \frac{1}{12}[\Lambda, [\Lambda, V]] + \frac{i}{12}[V, [V, \Lambda]]\right] + \cdots \right\} \\
&= \exp\left\{ V + i(\Lambda - \Lambda^\dagger) + \frac{i}{2}[V, \Lambda] - \frac{i}{2}[\Lambda^\dagger, V] + \frac{1}{12}[\Lambda^\dagger, \Lambda] \right. \\
&\quad \left. + \frac{1}{4}[\Lambda^\dagger, [V, \Lambda]] + \frac{1}{12}[\Lambda, [\Lambda, V]] + \frac{i}{12}[V, [V, \Lambda]] + \cdots \right\}
\end{aligned}
$$

$$= \exp\Big\{ V^a T_a + i\Big(\Lambda^a - \Lambda^{\dagger a}\Big) T_a$$

$$+ \frac{i}{2} V^c \Lambda^b [T_c, T_b] - \frac{i}{2} \Lambda^{\dagger c} V^b [T_c, T_b]$$

$$+ \frac{1}{2} \Lambda^{\dagger c} \Lambda^b [T_c, T_b] + \frac{1}{4} \Lambda^{\dagger c} V^b \Lambda^d \Big[T_c, [T_b, T_d] \Big]$$

$$+ \frac{1}{12} \Lambda^c \Lambda^b V^d \Big[T_c, [T_b, T_d] \Big]$$

$$+ \frac{i}{12} V^c V^b \Lambda^d \Big[T_c, [T_b, T_d] \Big] + \cdots \Big\}$$

where we used the Hermiticity of the generators T_a. Making use of the basic commutation relation (10.68) we get

$$e^{-i\Lambda^\dagger} e^V e^{i\Lambda} = \exp\Big\{ \Big[V^a + i\Big(\Lambda^a - \Lambda^{\dagger a}\Big) \Big] T_a$$

$$- \frac{1}{2} V^c \Lambda^b t_{cb}{}^a T_a + \frac{1}{2} \Lambda^{\dagger c} V^b t_{cb}{}^a T_a$$

$$+ \frac{i}{2} \Lambda^{\dagger c} \Lambda^b t_{cb}{}^a T_a + \Big[-\frac{1}{4} \Lambda^{\dagger c} V^b \Lambda^d - \frac{1}{12} \Lambda^c \Lambda^b V^d$$

$$- \frac{i}{12} V^c V^b V^d \Big] t_{bd}{}^e t_{ce}{}^a T_a + \cdots \Big\}$$

$$= \exp\Big\{ \Big[V^a + i\big(\Lambda^a - \Lambda^{\dagger a}\big)$$

$$+ \frac{1}{2} t_{cb}{}^a \Big(\Lambda^{\dagger c} V^b + i\Lambda^{\dagger c} \Lambda^b - V^c \Lambda^b \Big)$$

$$+ \frac{1}{4} t_{bd}{}^e t_{ce}{}^a \Big(-\Lambda^{\dagger c} V^b \Lambda^d$$

$$- \frac{1}{3} \Lambda^c \Lambda^b V^d - \frac{i}{3} V^c V^b \Lambda^d \Big) + \cdots \Big] T_a \Big\}. \tag{10.75}$$

Hence we obtain only commutators of group generators in computing the product of exponentials of Eq. (10.73). This shows that we can write the transformed vector field V' in the form

$$V' = V'^a T_a. \tag{10.76}$$

The transformation (10.76) demonstrates that the transformation law (10.73) is independent of the particular representation of the group since the information about the chosen representation is contained in the generators T_a. This completes the proof of Proposition 10.3.

We now calculate the infinitesimal form of the transformation (10.73), *i.e.*

$$\delta V = V' - V,$$

or

$$\delta\left(e^V\right) = e^{V'} - e^V$$
$$= e^{-i\Lambda^\dagger} e^V e^{i\Lambda} - e^V$$
$$\approx -i\Lambda^\dagger e^V + i e^V \Lambda. \tag{10.77}$$

We begin our computation by defining the Lie derivative:

Definition:

If V and X are elements of the algebra $\tilde{g}$, the *Lie derivative* of X with respect to V is defined by

$$L_V X := [V, X]. \tag{10.78}$$

Proposition 10.4:

With the Lie derivative, defined by Eq. (10.78), we have

$$e^{L_V} X = e^V X e^{-V}. \tag{10.79}$$

Proof: We restrict ourselves here to a formal verification. A rigorous proof of Eq. (10.79) can be found in the literature.[12] The left hand side of Eq. (10.79) is defined by the power series expansion of the exponential, thus

$$e^{L_V} X = \left[\sum_{n=0}^{\infty} \frac{1}{n!} (L_V)^n\right] X$$
$$= X + [V, X] + \frac{1}{2}[V, [V, X]] + \cdots$$
$$\cdots + \frac{1}{k!}\left[V, \left[V, [\cdots [V, X] \cdots]\right]\right] + \cdots$$
$$= X + VX - XV + \frac{1}{2}V^2 X + \frac{1}{2}XV^2 - VXV + \cdots$$
$$= \left(1 + V + \frac{1}{2}V^2 + \cdots\right) X \left(1 - V + \frac{1}{2}V^2 \pm \cdots\right)$$
$$= e^V X e^{-V},$$

which had to be demonstrated.

Now obviously

$$[V, e^V] = 0, \tag{10.80}$$

---

[12]See *e.g.* W. Miller [72], Lemma 3.1.

and the variation of this expression yields

$$\delta\left[V, e^V\right] = 0,$$

i.e.

$$\delta\left(V e^V - e^V V\right) = 0.$$

This implies

$$(\delta V)e^V + V\delta\left(e^V\right) - \delta\left(e^V\right)V - e^V\delta V = 0,$$

or

$$(\delta V)e^V - e^V(\delta V) + \left[V, \delta\left(e^V\right)\right] = 0.$$

Multiplying this expression from the left and the right by $e^{-V/2}$, we obtain

$$e^{-V/2}(\delta V)e^{V/2} - e^{V/2}(\delta V)e^{-V/2} + e^{-V/2}\left[V, \delta\left(e^V\right)\right]e^{-V/2} = 0.$$

Hence

$$e^{-V/2}(\delta V)e^{V/2} - e^{V/2}(\delta V)e^{-V/2} = -e^{-V/2}\left(L_V\delta\,e^V\right)e^{-V/2},$$

where we used the definition of the Lie derivative Eq. (10.78). Using the property (10.79) of the Lie derivative, we can rewrite the left hand side as:

$$e^{-L_{V/2}}(\delta V) - e^{L_{V/2}}(\delta V) = -e^{-V/2}\left(L_V\delta\,e^V\right)e^{-V/2},$$

so that

$$
\begin{aligned}
2\sinh\left(L_{V/2}\right)(\delta V) \;\; &= \;\; -e^{-V/2}\left(L_V\delta\,e^V\right)e^{-V/2}\\[4pt]
&= \;\; -e^{-V/2}\left(L_V\left[e^{V'} - e^V\right]\right)e^{-V/2}\\[4pt]
&\overset{(10.77)}{=} -e^{-V/2}\left(L_V\left[-i\Lambda^\dagger e^V + ie^V\Lambda\right]\right)e^{-V/2}\\[4pt]
&\overset{\substack{(10.78)\\(10.80)}}{=} -iL_V\left[e^{-V/2}\Lambda^\dagger e^{V/2} - e^{V/2}\Lambda e^{-V/2}\right]\\[4pt]
&\overset{(10.79)}{=} -iL_V\left[e^{-L_{V/2}}\Lambda^\dagger - e^{L_{V/2}}\Lambda\right]
\end{aligned}
$$

$$\begin{aligned}
&= -\frac{i}{2}L_V\Big[e^{L_V/2}\Lambda^\dagger - e^{L_V/2}\Lambda \\
&\quad + e^{-L_V/2}\Lambda^\dagger - e^{-L_V/2}\Lambda - e^{L_V/2}\Lambda^\dagger \\
&\quad - e^{L_V/2}\Lambda + e^{-L_V/2}\Lambda^\dagger + e^{-L_V/2}\Lambda\Big] \\
&= -iL_V\Big\{\frac{1}{2}\Big(e^{L_V/2}+e^{-L_V/2}\Big)(\Lambda^\dagger-\Lambda) \\
&\quad -\frac{1}{2}\Big(e^{L_V/2}-e^{-L_V/2}\Big)(\Lambda^\dagger+\Lambda)\Big\} \\
&= -iL_V\big\{\cosh L_{V/2}(\Lambda^\dagger-\Lambda) - \sinh L_{V/2}(\Lambda^\dagger+\Lambda)\big\}.
\end{aligned}$$

Hence we obtain the result

$$2\sinh\big(L_{V/2}\big)(\delta V) = iL_V\big\{\sinh L_{V/2}(\Lambda^\dagger+\Lambda) - \cosh L_{V/2}(\Lambda^\dagger-\Lambda)\big\}.$$

Now with the help of Eq. (10.78) we can write

$$\frac{1}{2}L_V X = \frac{1}{2}[V,X] = [\tfrac{1}{2}V,X] = L_{V/2}X,$$

so that formally

$$\frac{1}{2}L_V = L_{V/2}.$$

Then

$$\delta V = \frac{i}{2}L_V\Big[\frac{1}{\sinh L_{V/2}}\sinh L_{V/2}(\Lambda^\dagger+\Lambda) \\
- \frac{1}{\sinh L_{V/2}}\cosh L_{V/2}(\Lambda^\dagger-\Lambda)\Big].$$

Hence

$$\delta V = V' - V = iL_{V/2}\Big[(\Lambda^\dagger+\Lambda) + \coth L_{V/2}(\Lambda-\Lambda^\dagger)\Big]. \tag{10.81}$$

From Eq. (10.81) we obtain the infinitesimal transformation by expanding coth in its power series, *i.e.*

$$\delta V = V' - V$$
$$= i(\Lambda-\Lambda^\dagger) + \frac{i}{2}[V,\Lambda^\dagger+\Lambda] + O(\Lambda^2). \tag{10.82}$$

This is the desired infinitesimal transformation.

The supersymmetric extension of the field strength is now defined in analogy to Eqs. (7.43) and (7.44) by

$$\begin{aligned}
W_A &:= -\tfrac{1}{4}\overline{D}\,\overline{D}e^{-V}D_A e^V, \\
\overline{W}_{\dot A} &:= -\tfrac{1}{4}D\,De^{-V}\overline{D}_{\dot A}e^V,
\end{aligned} \tag{10.83}$$

where according to Eq. (10.73) the vector superfields V are matrix quantities

$$V_{ij} = V^a \left(T_a\right)_{ij},$$

and the generators T_a are given in the adjoint representation of a gauge group G.

Proposition 10.5:

The field strength W_A transforms according to the rule

$$W_A \longrightarrow W'_A = e^{-i\Lambda(x,\theta,\bar{\theta})} W_A \, e^{i\Lambda(x,\theta,\bar{\theta})}, \tag{10.84}$$

and a similar relation holds for $\overline{W}_{\dot{A}}$.

Proof: We observe first that if e^V transforms according to Eq. (10.73), then e^{-V} transforms as

$$e^{-V'} = e^{-i\Lambda} e^{-V} e^{i\Lambda^\dagger}, \tag{10.85}$$

such that

$$e^{V'} e^{-V'} = e^{-i\Lambda^\dagger} e^{V} e^{i\Lambda} e^{-i\Lambda} e^{-V} e^{i\Lambda^\dagger} = 1 = e^{-V'} e^{V'}.$$

Then the transformed field strength W_A can be written as:

$$
\begin{aligned}
W'_A &= -\frac{1}{4} \overline{D}\,\overline{D} e^{-V'} D_A e^{V'} \\
&\overset{(10.73)}{\underset{(10.85)}{=}} -\frac{1}{4} \overline{D}\,\overline{D} \left(e^{-i\Lambda} e^{V} e^{i\Lambda^\dagger} \right) D_A \left(e^{-i\Lambda^\dagger} e^{V} e^{i\Lambda} \right) \\
&\overset{(10.65b)}{=} -\frac{1}{4} e^{-i\Lambda} \overline{D}\,\overline{D} \left(e^{-V} e^{i\Lambda^\dagger} \right) D_A \left(e^{-i\Lambda^\dagger} e^{V} e^{i\Lambda} \right) \\
&= -\frac{1}{4} e^{-i\Lambda} \overline{D}\,\overline{D} e^{-V} \left(D_A e^{V} \right) e^{i\Lambda} - \frac{1}{4} e^{-i\Lambda} \overline{D}\,\overline{D} D_A e^{i\Lambda} \\
&\overset{(10.83)}{=} e^{-i\Lambda} W_A e^{i\Lambda} - \frac{1}{4} e^{-i\Lambda} \overline{D}\,\overline{D} \left(D_A e^{i\Lambda} \right) \\
&\overset{(10.65b)}{=} e^{-i\Lambda} W_A e^{i\Lambda} - \frac{1}{4} e^{-i\Lambda} \overline{D}\,\overline{D} D_A e^{i\Lambda} - \frac{1}{4} e^{-i\Lambda} \overline{D} D_A \overline{D} e^{i\Lambda} \\
&= e^{-i\Lambda} W_A e^{i\Lambda} - \frac{1}{4} e^{-i\Lambda} \overline{D} \{ \overline{D}, D_A \} e^{i\Lambda} \\
&\overset{(6.54)}{=} e^{-i\Lambda} W_A e^{i\Lambda} - \frac{1}{2} e^{-i\Lambda} \overline{D}^{\dot{A}} \left(\sigma^\mu \right)_{A\dot{A}} P_\mu e^{i\Lambda} \\
&= e^{-i\Lambda} W_A e^{i\Lambda} + \frac{1}{2} e^{-i\Lambda} P_\mu \left(\sigma^\mu \right)_{A\dot{A}} \overline{D}^{\dot{A}} e^{i\Lambda} \\
&\overset{(10.65b)}{=} e^{-i\Lambda} W_A e^{i\Lambda}.
\end{aligned}
$$

Note that in the next to last step we made use of the commutator $\left[\overline{D}, P_\mu\right] = 0$.

The transformation of the second of relations (10.83) is demonstrated in a similar fashion.

For a treatment of further properties of supersymmetric non-Abelian gauge theories — in particular for the introduction of gauge covariant derivatives and Lagrangians — we refer to the literature.[13] The introduction presented in this section provides the basics tools with which the study of these theories can be pursued along the lines of the previous discussions of Abelian gauge theories.

---

[13]See J. Wess and J. Bagger [123], P. Fayet and S. Ferrara [35] or P. West [127], Chap. 15.2. See also S. Weinberg [118], Chap. 27 and D.S. Berman and E. Rabinovici [12].

Bibliography

[1] I.J.R. Aitchison, *Supersymmetry and the MSSM, An Elementary Introduction*, hep–ph0505105v1.

[2] I.J.R. Aitchison, *Supersymmetry in Particle Physics, An Elementary Introduction*, Cambridge University Press, Cambridge, 2007.

[3] L. Alvarez-Gaumé, *Supersymmetry and the Atiyah-Singer Index Theorem*, Commun. Math. Phys. **90**, 161 (1983).

[4] L. Alvarez-Gaumé and D.Z. Freedman, *A Simple Introduction to Complex Manifolds*, published in: *Unification of the Fundamental Particle Interactions*, Erice 1980, ed. S. Ferarra, J. Ellis and P. van Nieuwenhuizen, Plenum Press, New York, 1980.

[5] L. Alvarez-Gaumé and S.L. Hassan, *Introduction to S-Duality in $N = 2$ Supersymmetric Gauge Theories*, hep–th/9701069 (1997).

[6] J.A. Bagger, *Supersymmetric σ-models*, SLAC Report No. SLAC-PUB-3461, SLAC, Stanford, 1984.

[7] J.A. Bagger and E. Witten, *The Gauge Invariant Supersymmetric Nonlinear Sigma Model*, Phys. Lett. **118B**, 103 (1982).

[8] A.O. Barut and R. Rączka, *Theory of Group Representations and Applications*, Second revised edition, Polish Scientific Publishers, Warsaw, 1980.

[9] I.M. Benn and R.W. Tucker, *An Introduction to Spinors and Geometry with Applications in Physics*, Adam Hilger, Bristol, New York, 1987.

[10] F.A. Berezin, *The Method of Second Quantization*, Translation by M. Mugibayashi and A. Jeffrey, Academic Press, New York, 1966.

[11] F.A. Berezin and G.I. Kac, *Lie Groups with commuting and anticommuting parameters*, Mat. Sbornik **82**, 343 (1970), engl. translation in Math Sbornik **11**, 311 (1970).

[12] D.S. Berman and E. Rabinovici, *Les Houches lectures on supersymmetric gauge theories*, hep–th/0210044.

[13] M. Bianchi, S.Kovacs, and G. Rossi, *Instantons and Supersymmetry*, Lect. Notes Phys. **737**, 303 – 470 (2008).

[14] A. Bilal, *Duality in $N = 2$ SUSY SU(2) Yang–Mills Theory. A pedagogical introduction to the work of Seiberg and Witten*, hep–th/9601007 (1996).

[15] A. Bilal, *Introduction to Supersymmetry*, hep–th/0101055v1.

[16] P. Binetruy and K. Hentschel, *Supersymmetry: Theory, Experiment, and Cosmology*, Oxford University Press, Oxford, 2006.

[17] J.D. Bjorken and S.D. Drell, *Relativistic Quantum Mechanics*, McGraw Hill, 1964, Chapter 1.

[18] J.D. Bjorken and S.D. Drell, *Relativistic Quantum Fields*, McGraw Hill, 1964.

[19] T. Bröcker and T. tom Dieck, *Representations of Compact Lie Groups*, Springer-Verlag, New York, 1985, GTM 98, Chapter III.8.

[20] B.A. Campell and G. Fogleman, *Supersymmetry and Supergravity: A Short Review*, TRIUMF Report No. TRI-PP-126, TRIUMF, Vancouver, 1983.

[21] T.-P. Cheng and L.-F. Li, *Gauge Theory of Elementary Particle Physics*, The Clarendon Press, Oxford, 1984.

[22] F. Coester, M. Hammermesh and W.D. McGlinn, *Internal Symmetry and Lorentz Invariance*, Phys. Rev. **B135**, 451 (1964).

[23] S. Coleman and J. Mandula, *All Possible Symmetries of the S-Matrix*, Phys. Rev. **159**, 1251 (1967).

[24] F. Cooper and B. Freedman, *Aspects of Supersymmetric Quantum Mechanics*, Annals of Physics **146**, 262 (1983).

[25] M. de Roo, *Supersymmetry*, Lectures given at the University of Groningen, Internal Report No. 168, 1980/81, unpublished.

[26] B. DeWitt, *Supermanifolds*, Cambridge University Press, Cambridge, 1984.

[27] B. De Wit and D.Z. Freedman, *Phenomenology of Goldstone Neutrinos*, Phys. Rev. Lett. **35**, 827 (1975).

[28] M. Dine, *Supersymmetry Phenomenology*, hep-ph/9612389v1.

[29] M. Dine, *Supersymmetry and String Theory, Beyond the Standard Model*, Cambridge Universiry Press, Cambridge, 2007.

[30] R. Di Stefano, *Notes on the Conceptual Development of Supersymmetry*, in Ref. [63], pp. 169.

[31] N. Dragon, U. Ellwanger and M.G. Schmidt, *Supersymmetry and Supergravity*, Prog. Part. Nucl. Phys. **18**, 1 (1987).

[32] M. Drees, *An Introduction to Supersymmetry*, hep–ph/9611409v1.

[33] S. Duplij, W. Siegel, and J. Bagger (Eds.), *Concise Encyclopedia of Supersymmetry and Noncommutative Structures in Mathematics and Physics*, Springer, Netherlands, 2005.

[34] P. Fayet and J. Iliopoulos, *Spontaneously Broken Supergauge Symmetries and Goldstone Spinors*, Phys. Lett. **51B**, 461 (1974).

[35] P. Fayet and S. Ferrara, *Supersymmetry*, Physics Reports **32**, 249 (1977).

[36] G. Feldman and P.T. Matthews, *Super Particles, Superfields and Yang–Mills Lagrangians*, Cambridge University Report No. DAMPT 85-11, Cambridge, 1985.

[37] S. Ferrara, *Supersymmetry and Supergravity for Nonpractitioners*, Proceedings of International School of Physics *Enrico Fermi*, Course 81, Ed. G. Costa and R.R. Gatto, North Holland, Amsterdam, 1982, pp. 237.

[38] S. Ferrara (Ed.), *Supersymmetry*, Vols. 1,2, North Holland/World Scientific, Amsterdam/Singapore, 1987.

[39] S. Ferrara, L. Giradello and F. Palumbo, *General Mass Formula in Broken Supersymmetry*, Phys. Rev. **D20**, 403 (1979).

[40] S. Ferrara, J. Wess and B. Zumino, *Supergauge Multiplets and Superfields*, Phys. Lett. **51B**, 239, (1974).

[41] S. Ferrara and B. Zumino, *Supergauge Invariant Yang–Mills Theories*, Nucl. Phys. **B79**, 413 (1974).

[42] J.M. Figuera–O'Farrill, *BUSSTEPP Lectures on Supersymmetry*, hep–th/0109172v1.

[43] L. Frappat, P. Sorba, and A.S. Sciarrino, *Dictionary on Lie Superalgebras*, hep–th/9607161.

[44] P.G.O. Freund, *Introduction to Supersymmetry*, Cambridge University Press, Cambridge, 1986.

[45] D. Friedan and P. Windey, *Supersymmetric Derivation of the Atiyah–Singer Index and the Chiral Anomaly*, Nucl. Phys. **B235**, 395 (1984).

[46] S.J. Gates, Jr., M.T. Grisaru, M. Roček and W. Siegel, *Superspace or One Thousand and One Lessons in Supersymmetry*, Benjamin/Cummings, London, 1983.

[47] L. Giradello and J. Iliopoulos, *Quantum Corrections to a Mass Formula in Broken Supersymmetry*, Phys. Lett. **B88**, 85 (1975).

[48] S. Goldberg, *Curvature and Holonomy*, Dover, New York, 1982.

[49] H. Goldstein, *Classical Mechanics*, Sixth Edition, Addison Wesley, Reading, Mass., 1959.

[50] Yu.A. Gol'fand and E.P. Likhtman, *Extension of the Algebra of Poincaré Group Generators and Violation of P Invariance*, JETP Lett. **13**, 323, (1971).

[51] M.B. Green, J.H. Schwarz and E. Witten, *Superstring theory*, Vols. 1,2, Cambridge University Press, Cambridge, 1987.

[52] B. Greene, *The Elegant Universe. Superstrings, Hidden Dimensions and the Quest for the Ultimate Theory*, Norton, New York, 1999.

[53] B. Greene, *The Fabric of the Cosmos: Time, Space, and the Texture of Reality*, Knopf, New York, 2004.

[54] R. Haag, J.T. Łopuszański and M.F. Sohnius, *All Possible Generators of Supersymmetries of the S-Matrix*, Nucl. Phys. **B88**, 257 (1975).

[55] H.E. Haber and G.L. Kane, *Is Nature Supersymmetric?* Scientific American **254**, 42, (1986).

[56] H.E. Haber and G.L. Kane, *The Search for Supersymmetry: Probing Physics beyond the Standard Model*, Physics Reports **117**, 75 (1985).

[57] H.E. Haber, *Introducory Low–Energy Supersymmetry*, hep–ph/9306207v1.

[58] D. Hooper, *Nature's Blueprint, Supersymmetry and the Search for a Unified Theory of Matter and Force*, HarperCollins, New York, 2008.

[59] K. Intriligator and N. Seiberg, *Lectures on Supersymmetry Breaking*, Class. Quantum Grav. **24**, 741 (2007).

[60] C. Itzykson and J.-B. Zuber, *Quantum Field Theory*, McGraw-Hill, 1980.

[61] M. Kaku, *Introduction to Superstrings*, Springer, New York, 1988.

[62] G. Kane, *Supersymmetry, Unveiling the Ultimate Laws of Nature*, Basics Books, New York, 2000.

[63] G. Kane and M. Shifman (ed.), *The Supersymmetric World, The Beginnings of the Theory*, World Scientific, Singapore, 2000.

[64] S. Lang, *Linear Algebra*, Springer-Verlag, New York, 1987.

[65] S. Lang, *Algebra*, Third Edition, Addison Wesley, 1993.

[66] S. Lang, *Undergraduate Algebra*, Third Edition, UTM, Springer-Verlag, New York, 2005.

[67] F. Legovini, *Supersymmetry*, International School for Advanced Studies, Trieste, Report No. 24/83/E.P., 1983, unpublished.

[68] W. Lerche, *Introduction to Seiberg–Witten Theory and its Stringy Origin*, hep–th/9611190 (1996).

[69] U. Lindström, *Supersymmetry, a Biased Review*, hep–th/0204016.

[70] J.D. Lykken, *Introduction to Supersymmetry*, hep-th/9612114.

[71] S.P. Martin, *A Supersymmetry Primer*, hep-ph/9709356v4, June 2006.

[72] W. Miller, *Symmetry Groups and Their Applications*, Academic Press, New York, 1972.

[73] S.P. Misra, *Introduction to Supersymmetry and Supergravity*, Wiley Eastern Limited, New Delhi, 1992.

[74] H.J.W. Müller-Kirsten and A. Wiedemann, *Dirac Quantization of a Supersymmetric Field Theory*, Zeitschr. f. Physik C **35**, 471 (1987).

[75] H.J.W. Müller-Kirsten, *Classical Mechanics and Relativity*, World Scientific Press, Singapore, 2008.

[76] W. Nahm, *Supersymmetries and Their Representations*, Nucl. Phys. **B135** (1978), 149. This article is reprinted in [102].

[77] H.P. Nilles, *Supersymmetry, Supergravity and Particle Physics*, Physics Reports **110**, 1 (1984).

[78] L. Okun, *Leptons and Quarks*, North Holland Publ. Co., Amsterdam, 1982.

[79] D.I. Olive and P.C. West (Eds.), *Duality and Supersymmetric Theories*, Cambridge University Press, 1999.

[80] K.A. Olive, *Introduction to Supersymmetry: Astrophysical and Phenomenological Constraints*, hep-ph/9911307v1.

[81] L. O'Raifeartaigh, *Lorentz Invariance and Internal Symmetry*, Phys. Rev. B **139**, 1052 (1964).

[82] L. O'Raifeartaigh, *Internal Symmetry and Lorentz Invariance*, Phys. Rev. Lett. **14**, 332 (1965).

[83] L. O'Raifeartaigh, *Mass Differences and Lie Algebras of Finite Order*, Phys. Rev. Lett. **14**, 575 (1965).

[84] L. O'Raifeartaigh, *Spontaneous Symmetry Breaking for Chiral Scalar Superfields*, Nucl. Phys. **B96**, 331 (1975).

[85] A. Pais, *On Spinors in n Dimensions*, Jour. Math. Phys. **3**, 6, 1135 (1962).

[86] A. Pais, *Inward Bound, Of Matter and Forces in the Physical World*, Oxford University Press, 1986.

[87] R. Penrose, *The Road to Reality, A Complete Guide to the Laws of the Universe*, Vintage Books, London, 2004.

[88] J. Polchinski, *String Theory*, Vols, 1,2, Cambridge University Press, Cambridge, 2005.

[89] N. Polonsky, *Supersymmetry: Structure and Phenomena, Extension of the Standard Model*, Springer Verlag, Berlin, Heidelberg, 2001.

[90] Proceedings of the 28th Scottish Universities Summer School in Physics, *Superstrings and Supergravity*, Ed. A.T. Davies and D.G. Sutherland, SUSSP Publication, Edinburgh, 1986.

[91] Proceedings of the NATO Advanced Study Institute on Supersymmetry, Ed. K. Dietz, R. Flume, G.v. Gehlen and V. Rittenberg, Plenum Press, New York, 1985.

[92] Proceedings of the Thirteenth SLAC Summer Institute on Particle Physics, Supersymmetry, SLAC Report No. 296, Ed. E.C: Brennan, SLAC, Stanford, 1985.

[93] C. Quigg, *Gauge Theories of the Strong, Weak, and Electromagnetic Interactions*, Benjamin/Cummings, Reading, Massachussetts, 1983.

[94] P. Ramond, *Field Theory, A Modern Primer*, Benjamin/Cummings, Reading, Massachusetts, 1981.

[95] L. Randall, *Warped Passages. Unraveling the Mysteries of the Universe's Hidden Dimensions*, Harper Collins, New York, 2005.

[96] V. Rittenberg, *A Guide to Lie Superalgebras*, in: VI International Colloquium on Group Theoretical Methods in Physics, Tübingen, pp. 3, (1977). This article is reprinted in Ref. [102].

[97] P. Roman, *Theory of Elementary Particles*, North Holland Publ. Co., Amsterdam, 1960, p. 75.

[98] L.H. Ryder, *Quantum Field Theory*, Cambridge University Press, Cambridge, 1985.

[99] A. Salam and J. Strathdee, *Supergauge Transformations*, Nucl. Phys. **B76**, 477 (1974).

[100] A. Salam and J. Strathdee, *Supersymmetry and Nonabelian Gauges*, Phys. Lett. **51B**, 353 (1974).

[101] A. Salam and J. Strathdee, *Supersymmetry and Superfields*, Fortschr. Phys. **26**, 57 (1978).

[102] A. Salam and E. Sezgin (Eds.), *Supergravities in Diverse Dimensions*, Vols. 1,2, North–Holland/World Scientific, Amsterdam/Singapore, 1989.

[103] M.D. Scadron, *Advanced Quantum Theory*, Springer-Verlag, New York, 1979.

[104] A.S. Sciarrino and P. Sorba: *Group Theory in Particle Physics*, LAPP Report No. TH-79, to be published in: Methods and Formulae in High Energy Physics Data Analysis, LAPP, Annecy-le-Vieux, 1979.

[105] L.I. Schiff, *Quantum Mechanics*, McGraw-Hill Publ. Co., New York, 1955, Chapter 52.

[106] R.U. Sexl and H.K. Urbantke, *Relativität, Gruppen, Teilchen*, Springer-Verlag, New York, 1979. Engl. translation: *Relativity, Groups, Particles: Special Relativity and Relativistic Symmetry in Field and Particle Physics*, Springer, Wien, 2000.

[107] Y. Shadmi, *Supersymmetry Breaking*, hep–th/0601076v1.

[108] R. Slanski, *Symmetries of Theories in Higher Dimension*, Los Alamos Report No. LA-UR-85-2768, Los Alamos, 1985, to be published in Proceedings of Lectures at 1985 Les Houches Summer School.

[109] M.F. Sohnius, *Introducing Supersymmetry*, Physics Reports **128**, 29 (1985).

[110] P.P. Srivastava, *Supersymmetry, Superfields and Supergravity: An Introduction*, Adam Hilger, London, 1986.

[111] J. Terning, *TASI–2002 Lectures: Non-perturbative Supersymmetry*, hep–th/0306119v2 (2008).

[112] J. Terning, *Modern Supersymmetry: Dynamics and Duality*, Oxford University Press, Oxford, 2009.

[113] J.W. van Holten, *Kähler Manifolds and Supersymmetry*, hep–th/0309094v1, 2008.

[114] P. van Nieuwenhuizen, *An Introduction to Simple Supergravity and the Kaluza–Klein Program*, in Les Houches Lectures, Session XL, *Relativity, Groups and Topology II*, B.S. DeWitt and R. Stora (eds.), North Holland, Amsterdam, 1984.

[115] V.S. Varadarajan, *Lie Groups, Lie Algebras and Their Representations*, GTM 102, Springer-Verlag, New York, 1984.

[116] D.V. Volkov and V.P. Akulov, *Possible Universal Neutrino Interaction*, JETP Lett. **16** (1972), 438.

[117] D.V. Volkov and V.P. Akulov, *Is the Neutrino a Goldstone Particle?*, Phys. Lett. **46B**, 109 (1973).

[118] S. Weinberg, *The Quantum Theory of Fields, Vol. 3, Supersymmetry*, Cambridge University Press, Cambridge, 2000.

[119] R.O. Wells, *Differential Analysis on Complex Manifolds*, GTM 65, Springer Verlag, New York, 1980.

[120] J. Wess, *Supersymmetry*, Lectures given at XV. Int. Universitäts-wochen, Schladming, Austria, Acta Physica Austriaca, Suppl XV., 475 (1976).

[121] J. Wess, *Supersymmetry, Supergravitation*, Physikalische Blätter **43**, 1 (1987).

[122] J. Wess, *From Symmetry to Supersymmetry*, in: SUSY 2007, Proceed. of the 15th International Conference on Supersymmetry and the Unification of Fundamental Interactions, Karlsruhe, 2007.

[123] J. Wess and J. Bagger, *Supersymmetry and Supergravity*, Princeton University Press, Princeton, 1983.

[124] J. Wess and B. Zumino, *A Lagrangian Model Invariant under Super-gauge Transformations*, Phys. Lett. **49B**, 52 (1974).

[125] J. Wess and B. Zumino, *Supergauge Transformations in Four Dimensions*, Nucl. Phys. **B70**, 39 (1974).

[126] J. Wess and B. Zumino, *Supergauge Invariant Extension of Quantum Electrodynamics*, Nucl. Phys. **B78**, 1 (1974).

[127] P. West, *Introduction to Supersymmetry and Supergravity*, World Scientific Press, Singapore, 1986.

[128] A. Wiedemann and H.J.W. Müller-Kirsten, *Explicit Construction of Supercharges of Supersymmetric Nonlinear Sigma Models in 1+1 Spacetime Dimensions*, Fortschr. Phys. **41**, 447 (1993).

[129] A. Wipf, *Non-perturbative Methods in Supersymmetric Theories*, hep–th/0504180v2 (2005).

[130] E. Witten, *Dynamical Breaking of Supersymmetry*, Nucl. Phys. **B185**, 513 (1981).

[131] E. Witten, *Constraints on Supersymmetry Breaking*, Nucl. Phys. **B202**, 253, (1982).

[132] E. Witten, *Introduction to Supersymmetry*, in: The Unity of the Fundamental Interactions, Ed. A. Zichichi, Plenum Press, New York, 1983, pp. 305.

[133] K. Yano, *Differential Geometry on Complex and Almost Complex Spaces*, Pergamon Press, New York, 1965.

[134] B. Zumino, *Supersymmetry and Kähler Manifolds*, Phys. Lett. **87B**, 203, (1979).

Index

www.ingramcontent.com/pod-product-compliance
Lightning Source LLC
Chambersburg PA
CBHW081225220326
41598CB00037B/6883